MASS SPECTROMETRY OF HETEROCYCLIC COMPOUNDS

GENERAL HETEROCYCLIC CHEMISTRY SERIES

Arnold Weissberger and Edward C. Taylor, Editors

MASS SPECTROMETRY OF HETEROCYCLIC COMPOUNDS
by Q. N. Porter and J. Baldas

Mass Spectrometry of Heterocyclic Compounds

Q. N. PORTER
University of Melbourne

J. BALDAS
University of Melbourne

Wiley-Interscience, a Division of John Wiley & Sons, Inc.
New York · London · Sydney · Toronto

INTRODUCTION TO THE SERIES

General Heterocyclic Chemistry

The series, "The Chemistry of Heterocyclic Compounds," published since 1950 by Wiley-Interscience, is organized according to classes of compound. Each volume deals with syntheses, reactions, properties, structure, physical chemistry, etc., of compounds belonging to a specific class, such as pyridines, thiophenes, and pyrimidines, three-membered ring systems. This series has become the basic reference collection for information on heterocyclic compounds.

Many aspects of heterocyclic chemistry have been established as disciplines of *general* significance and application. Furthermore, many reactions, transformations, and uses of heterocyclic compounds have specific significance. We plan, therefore, to publish monographs that will treat such topics as nuclear magnetic resonance of heterocyclic compounds, mass spectra of heterocyclic compounds, photochemistry of heterocyclic compounds, X-Ray structure determination of heterocyclic compounds, UV and IR spectroscopy of heterocyclic compounds, and the utility of heterocyclic compounds in organic synthesis. These treatises should be of interest to *all* organic chemists as well as to those whose particular concern is heterocyclic chemistry. The new series, organized as described above, will survey under each title *the whole field of heterocyclic chemistry* and is entitled "General Heterocyclic Chemistry." The editors express their profound gratitude to Dr. D. J. Brown of Canberra for his invaluable help in establishing the new series.

Department of Chemistry
Princeton University
Princeton, New Jersey

Research Laboratories
Eastman Kodak Company
Rochester, New York

Edward C. Taylor

Arnold Weissberger

JUN 15 1972

PREFACE

There is no doubt that mass spectrometry is one of the major tools used by those interested in problems of structural organic chemistry. This volume is directed at aiding the practicing organic chemist in his interpretation of mass spectra of the major classes of heterocyclic compounds, as set out in the last paragraph of the Introduction. The literature has been covered up to 1967 inclusive, and some selected recent advances (up to early 1970) are described briefly at the end of each chapter.

We are greatly indebted to Mrs. Joan Porter for her valuable assistance at all stages of preparation of the manuscript, and to the publishers, editors, and authors of *Analytical Biochemistry, Analytical Chemistry, Arkiv för Kemi, Australian Journal of Chemistry, Biochemische Zeitschrift, Bulletin de la Société chimique de France, Canadian Journal of Chemistry, Chemical Communications, Helvetica Chimica Acta, Journal of the American Chemical Society, Journal of the Chemical Society, Journal of Organic Chemistry, Monatshefte für Chemie, Organic Mass Spectrometry, Recueil des Travaux chimiques des Pays-Bas, Tetrahedron,* and *Tetrahedron Letters* for permission to reproduce diagrams.

Q. N. PORTER
J. BALDAS

Parkville, Victoria
June 1970

CONTENTS

INTRODUCTION

During the last ten years there has been an almost explosive growth in the use of mass spectrometry by organic chemists, and indeed today there can be few who do not have access to the technique and do not use it more or less regularly to help solve structural problems of varying degrees of complexity.

This rapid growth of the technique has been accompanied by, and to a large extent engendered by, the development of relatively simple theories for the rationalization of the observed fragmentations (1–5). In this volume we have assumed the reader is acquainted with the basic ideas concerning ionization by electron impact, with the principle of charge localization immediately preceding fragmentation and triggering off specific fragmentations, and with the structural factors that influence the stabilities of ionic and neutral fragments.

We have adhered to the soundly based and generally accepted convention of representing an odd-electron molecular radical cation as $[M]^{+\cdot}$ and an even-electron carbonium (or oxonium or sulfonium) ion as $[R]^{+}$. Likewise we have used the convention of a "fishhook" ($\frown$) to represent (a) a homolytic bond fission or (b) delocalization involving single-electron shifts (6, 7). Examples of these processes are

(a)

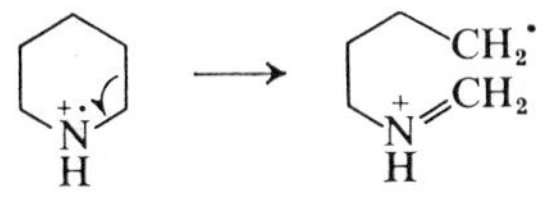

(b)

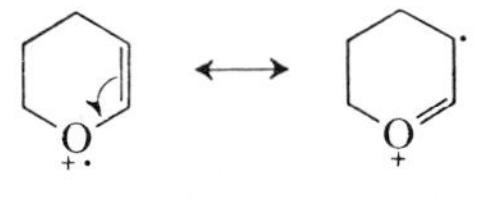

1

A normal arrow is used to show (a) heterolytic bond cleavage or (b) delocalization involving movement of an electron pair:

(a) $CH_3\!-\!\overset{+}{O}\!\!=\!\!CH_2 \longrightarrow CH_3^+ + O\!\!=\!\!CH_2$

(b)

The term "α-cleavage" has been used in the sense defined by Budzikiewicz, Djerassi, and Williams (6) as involving fission of a bond originating at an atom adjacent to the one assumed to carry the charge. β-Cleavage thus involves a bond one further removed from the atom carrying the charge, and so on:

α-Cleavage

β-Cleavage

γ-Cleavage

We have attempted to represent fragmentations as fully as possible by a series of reasonable ion structures. We do not pretend that the formulations for ions widely used in this book represent more than convenient rationalizations of fragmentations in "mechanistic" terms familiar to organic chemists. The structures used are clearly only approximate representations, and the problems involved in deciding upon the best structural representation for a particular ion will be touched on from time to time.

We have necessarily been rather selective in the material we have chosen to illustrate fragmentations. We have confined ourselves to the heteroatoms N, O, and S, and within the limits imposed by this choice we have concentrated on examples that demonstrate the fragmentations of basic heterocyclic systems. Thus we have limited our discussions of the fragmentations of various complex natural products containing heterocyclic rings, and in particular we have completely avoided discussion of the fragmentation of carbohydrates.

REFERENCES

1. J. H. Beynon, *Mass Spectrometry and Its Applications to Organic Chemistry*, Elsevier, Amsterdam, 1960.
2. K. Biemann, *Mass Spectrometry*, McGraw-Hill, New York, 1962.
3. H. Budzikiewicz, C. Djerassi, and D. H. Williams, *Interpretation of Mass Spectra of Organic Compounds*, Holden-Day, San Francisco, 1964.
4. H. Budzikiewicz, C. Djerassi, and D. H. Williams, *Structure Elucidation of Natural Products by Mass Spectrometry*, Vols. I and II, Holden-Day, San Francisco, 1964.
5. H. Budzikiewicz, C. Djerassi, and D. H. Williams, *Mass Spectrometry of Organic Compounds*, Holden-Day, San Francisco, 1967.
6. Ref. 5. pp. 1–4.
7. J. S. Shannon, *Proc. Roy. Aust. Chem. Inst.*, **1964**, 328.

1 OXIRANS (ETHYLENE OXIDES) AND OXETANS

I. OXIRANS (ETHYLENE OXIDES)

Ethylene oxides, or oxirans, are an important and versatile class of organic compounds. It is surprising that only recently have they been subjected to detailed mass-spectrometric study.

A. Aliphatic Epoxides

The earlier literature contains only scattered references to the mass spectra of aliphatic epoxides. The low-resolution spectra of ethylene oxide, 1,2-epoxypropane, epichlorhydrin and *cis*- and *trans*-2,3-epoxybutane are available in the API tables (1–5). High-resolution data are available for ethylene oxide and 1,2-epoxypropane (6); and appearance potentials for the principal ions of these two compounds (7) and four other simple epoxides (8) have been recorded as have the principal ions (without intensities) for neopentyl- and *t*-butyloxiran (9). The ionization potential of ethylene oxide has been determined by spectroscopic, photoionization, and electron impact methods, and the values obtained are 10.81 (10), 10.56 (11), and 10.65 ($\pm$0.1) (7) eV, respectively. The ionization potential of 1,2-epoxypropane, however, is considerably lower; 9.80 ($\pm$0.1) eV [also by electron impact (7)]. The mass spectrum of ethylene oxide, shown in Fig. 1.1, is relatively simple; there is a strong molecular ion and two major fragments at m/e 29 (CHO$^+$) and m/e 15 (CH$_3$$^+$). The initial step in the fragmentations may be rearrangement to acetaldehyde from which the methyl and formyl ions may be readily obtained.

$$\begin{array}{c} \text{CH}_2\text{---CH}_2 \longrightarrow \text{CH}_3\text{--CHO}^{+\cdot} \longrightarrow \text{CH}_3{}^+ + {}^{\cdot}\text{CHO} \\ \diagdown \diagup \qquad\qquad\qquad\qquad m/e\ 15 \\ \text{O} \\ {}^{+\cdot} \qquad\qquad\qquad \Big\downarrow \\ \text{M}^{+\cdot},\ m/e\ 44 \\ \qquad\qquad \text{H---C}{\equiv}\text{O}^+ + \text{CH}_3{}^{\cdot} \\ \qquad\qquad\qquad m/e\ 29 \end{array}$$

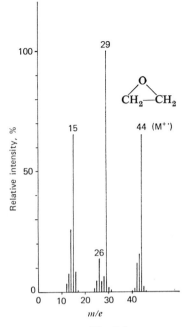

Fig. 1.1

This behavior has parallels in the pyrolysis and photolysis (both directly or using Hg sensitization) of ethylene oxide, which result in formation of methyl and formyl radicals (12–14).

The mass spectrum of perfluoroethylene oxide resembles that of ethylene oxide, as both CF_3^+ and CFO^+ are observed. The base peak here, however, is $CF_2^{+\cdot}$, presumably formed by transannular cleavage, indicating the much

$$\begin{array}{c} CF_2\!\!-\!\!CF_2 \\ \diagdown\!\!/ \\ O \\ M^{+\cdot},\ m/e\ 116 \end{array} \bigg]^{+\cdot} \longrightarrow \quad \underset{m/e\ 50\ (100\%)}{CF_2^{+\cdot}} \ + \ \underset{m/e\ 66\ (2.8\%)}{CF_2\!=\!O^{+\cdot}}$$

greater stability of the difluorocarbene radical ion compared with that of carbene itself (15).

Recently there has appeared a detailed study of epoxides of a series of terminal and nonterminal alkenes with high-resolution and deuterium-labeling data (16). Note that stereochemistry of the epoxide ring does not seem to be of importance, for *cis* and *trans* pairs give virtually identical spectra. The molecular ion peak from ethylene oxide is rather intense (65%), but the intensity is reduced to about 5% in 2,3-epoxybutane and becomes very low in more highly substituted epoxides and those with a long alkyl side chain. The major fragmentation processes of aliphatic epoxides may be classified into three groups, simple bond fissions, rearrangement reactions, and transannular cleavages, and are discussed accordingly.

1.   Simple Fissions

a.   α-CLEAVAGE.   α-Cleavage is a common and mechanistically simple process in mass spectrometry and is shown here for 2,3-epoxyheptane (**1**) together with the percent total ionization values (Σ_{40}). The larger *n*-butyl

radical is lost much more readily than the methyl radical (some of the M-CH$_3$· ion of **1** is due to δ-cleavage), and in general the loss of the largest alkyl substituent is the preferred α-cleavage process. α-Cleavage may also lead to retention of charge on the hydrocarbon fragment (fragmentation iii for **1**). α-Cleavage appears to be a much less favored process in epoxides than in acyclic ethers; it has been suggested that this difference may be due to the increase in strain in the formation of oxirenium ions of type **1a** or **1b**. It is also possible that the molecular ion may be better represented by **2** rather than **3**, and structure **2** would not favor α-cleavage. The heats of

formation, however, of the two α-cleavage ions of 1,2-epoxypropane, (M^{+}·-H·) and (M^{+}·-CH$_3$·), favor (at least in this case) a cyclic structure. α-Cleavage has been observed in some complex epoxy compounds such as melianone (**4**), which is reported to give peaks at M-71 and *m/e* 71 (17, 18), and turraeanthin lactone (**5**), which gives an intense ion at *m/e* 71 (19).

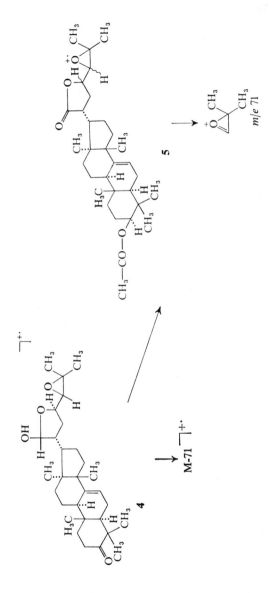

b.   β-CLEAVAGE.   This type of cleavage is of the same order of importance as α-cleavage in these epoxides, a marked contrast to acyclic ethers
in which α-cleavage is the predominant fission reaction and β-cleavage is not
important (20). In a typical case such as **1**, β-cleavage may be visualized as
involving the molecular ion **1c**.

CH$_3$—CH=$\overset{+}{O}$—$\overset{\cdot}{C}$H—CH$_2$—CH$_2$—CH$_2$—CH$_3$

1c

CH$_3$—CH=$\overset{+}{O}$—CH=CH$_2$
m/e 71, Σ$_{40}$ 7.0%

C$_3$H$_7^+$
m/e 43,
Σ$_{40}$ 5.1%

1

c.   γ-CLEAVAGE.   γ-Cleavage is in general far more important than either
α- or β-cleavage, especially in terminal epoxides, and may give rise to the base
peak of the spectrum as is the case for 1,2-epoxyhexane (**6**). Its spectrum
is shown in Fig. 1.2. Mass measurement has confirmed the composition of
the ion *m/e* 71 in **6**, and deuterium labeling has clearly shown that the γ-bond
is broken. This unexpected cleavage has been explained in three ways (16).

6a

−C$_2$H$_5^{\cdot}$

6c, *m/e* 71, Σ$_{40}$ 19.05%

6

6b

−C$_2$H$_5^{\cdot}$

6d, *m/e* 71

If the molecular ion is represented as **6b**, the stable cyclic oxonium ion **6d**
would result, and formulation **6a** would result in **6c**; such an ionic structure
is not unreasonable considering that 30% of the M-1 ion of tetrahydrofuran
arises from the loss of a β-hydrogen atom (21). The second suggested pathway

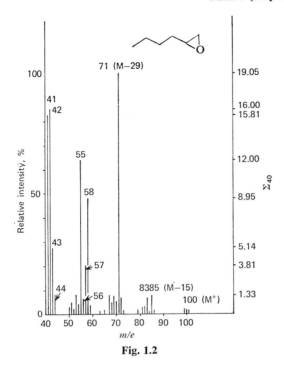

Fig. 1.2

involves a hydrogen rearrangement to give the tautomeric molecular ion **6e** which may then lose the ethyl radical by allylic fission to give **6g**, either

$$CH_3-CH_2-CH_2-\overset{..}{\underset{3}{C}H} \quad \longrightarrow \quad CH_3-CH_2-CH_2-CH=CH-CH_2\overset{+.}{O}H$$

$$\underset{6a}{} \qquad\qquad\qquad \underset{6e}{}$$

$$\underset{6g}{} \quad \xleftarrow[(?)]{} \quad \overset{.}{C}H_2-CH=CH-CH_2\overset{+.}{O}H$$

$$\underset{6f, m/e \ 71}{}$$

directly or via **6f**. The introduction of a methyl group at $C_{(3)}$ of **6a** lowers the amount of γ-cleavage, but even with two methyl groups at this position it is not totally suppressed. The third pathway has the virtue of leading to a fully conjugated ion **6h**; this pathway will be rendered inoperative by total substitution at $C_{(3)}$. γ-Cleavage is very sensitive to structural changes. This has been dramatically shown in the following pair of compounds: In

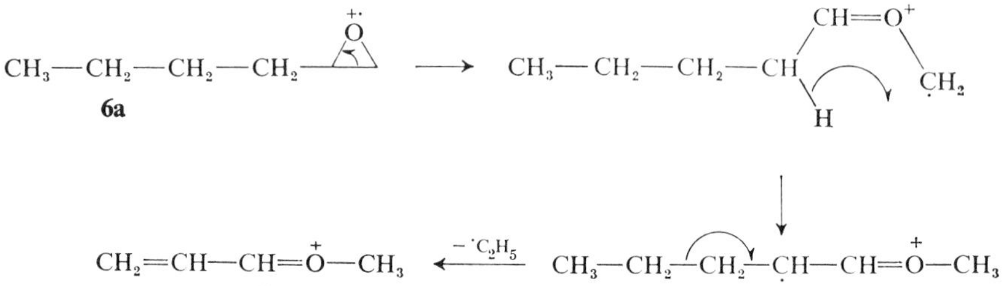

$$CH_3—CH_2—CH_2—CH_2 \overset{\overset{+\cdot}{O}}{\triangle} \longrightarrow CH_3—CH_2—CH_2—CH\overset{CH=O^+}{\underset{H}{\diagdown\ \ \dot{C}H_2}}$$

6a

$$CH_2{=}CH—CH{=}\overset{+}{O}—CH_3 \xleftarrow{-\ \cdot C_2H_5} CH_3—CH_2—CH_2—\overset{\cdot}{C}H—CH{=}\overset{+}{O}—CH_3$$

6h, m/e 71

4-methyl-1,2-epoxypentane (**7**) the percent of total ionization (Σ_{40}) for the γ-cleavage ion (m/e 85) is 17.9%, but in the steroidal epoxide **8** it is reduced to less than 1%, the fragmentation being controlled by the steroid moiety (20).

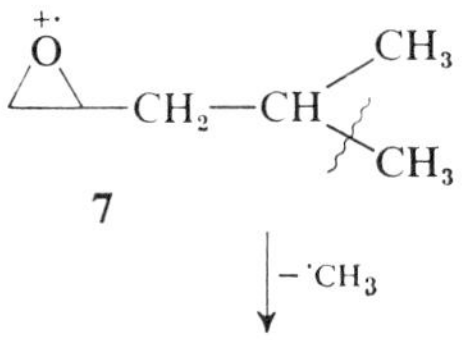

7

$$\downarrow -\cdot CH_3$$

m/e 85

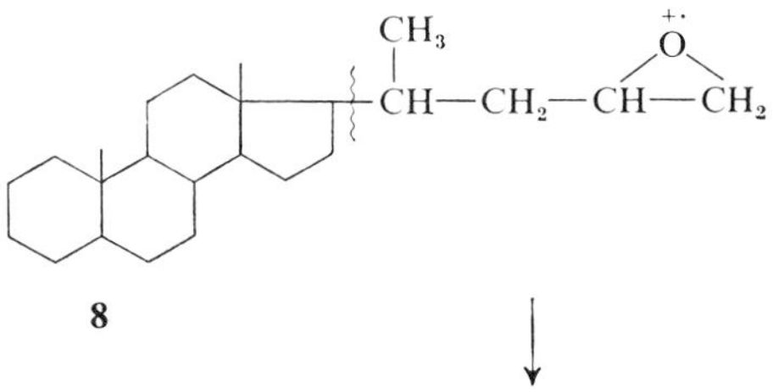

8

$$\downarrow$$

m/e 85

The cleavages and the percent of total ionization values (Σ_{40}) for 1,2-epoxyhexane (**6**) are:

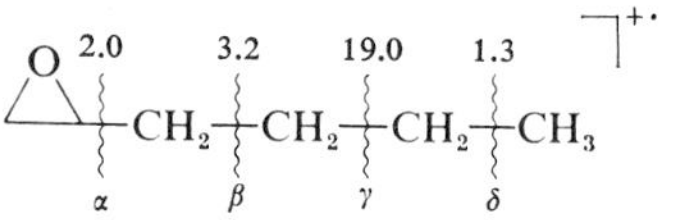

6

2. Rearrangement Reactions

Epoxides are known to rearrange on either pyrolysis or photolysis to carbonyl compounds (22), and in highly substituted compounds the mass spectra can be most easily explained as involving such rearrangements. Thus in 2,3-epoxy-2,3-dimethylbutane (**9**) an intense ion is observed at m/e 43 (21.8% Σ_{40}) and one of lesser intensity at m/e 57 (3.5% Σ_{40}), which may be readily derived from the pinacolone molecular ion **9a** formed from **9** by a methyl shift (16).

9, $M^{+\cdot}$, m/e 100 → m/e 100 → **9a**, m/e 100

$(CH_3)_3C^+$
m/e 57

$CH_3—C\equiv O^+$
m/e 43

Similarly the base peak in the mass spectrum of 2,3-epoxybutane (m/e 43) may be derived from 2-butanone. The epoxide of (−)-isopulegol (**10**) has

$CH_3—C\equiv O^+$
m/e 43

10

also been reported (23) to give a base peak at m/e 43 (although here a considerable portion may be due to $C_3H_7^+$). It has been shown, however, that in less highly substituted epoxides and especially in terminal epoxides such as 1,2-epoxyhexane rearrangement to the aldehyde or ketone is only of minor importance (16).

The operation of the McLafferty rearrangement will give rise to even-mass ions in the mass spectra of aliphatic epoxides. Such an ion was first observed (24) for methyl 9,10-epoxyoctadecanoate, and McLafferty (25) suggested it may be general for epoxides. This suggestion has been expanded by Djerassi et al. (16), and two distinct rearrangements, an "outside" and an "inside" one, have been formulated. The "inside" rearrangement involves the transfer of a hydrogen from $C_{(5)}$ and is shown for 1,2-epoxyhexane (**6**) where it leads to ions at m/e 58 and 42. The "outside" rearrangement involves transfer of the hydrogen at $C_{(4)}$ and leads to ions at m/e 44 and 56. In terminal

"inside"

m/e 58,
Σ_{40} 8.9%

m/e 42,
Σ_{40} 16.0%

6

epoxides the "inside" rearrangement is far more important than the "outside" one, as can be seen in the $\%$ Σ_{40} values for the various ions in **6**.

The possibility of rearrangement to the aldehyde or ketone prior to the McLafferty rearrangements is not considered to be an important one; as mentioned earlier, such rearrangements have been shown to be of minor importance in terminal epoxides. Further evidence is supplied by the fact that the "inside" rearrangement is not greatly affected by the environment of the transferred hydrogen. This is shown by the intensities of the M-56 peak in the spectra of the isomeric compounds **11**, **12**, and **13** where a primary, secondary, and tertiary hydrogen, respectively, is transferred. This is in

marked contrast to the McLafferty rearrangement of ketones where a secondary hydrogen atom is transferred more readily than a primary one.

3. Transannular Cleavages

Transannular cleavage with or without hydrogen transfer may be an important fragmentation process in aliphatic epoxides (16); for example, it has been shown that the base peak (m/e 28) in the mass spectrum of 1,2-epoxypropane is composed of 97% ethylene and 3% carbon monoxide, and, in effect, this arises from transannular cleavage without hydrogen transfer (16). A hydride shift after ring opening has been suggested to avoid postulation of a high-energy carbene species.

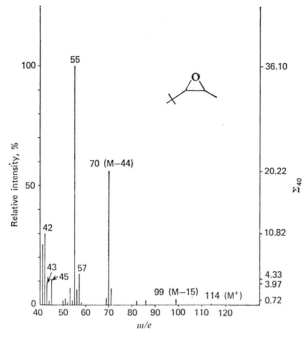

Fig. 1.3

The mass spectrum of 2,3-epoxy-4,4-dimethylpentane (**14**) (Fig. 1.3) is dominated by ions at m/e 70 ($C_5H_{10}^+$) and m/e 55 ($C_4H_7^+$) (the base peak m/e 55 is formed by loss of a methyl group from m/e 70). The ion **14b** is formed by hydrogen transfer, presumably via **14a**.

Finally it is noted that epoxidation followed by mass spectrometric analysis offers an excellent method for locating double bonds in some long-chain compounds, such as **15**, **16**, **17**, and **18** (26).

$$CH_3(CH_2)_7—CH\underset{\diagdown O \diagup}{}CH—R$$

15	*cis* and *trans* R =	$(CH_2)_7COOMe$
16	*cis* R =	$(CH_2)_{11}COOMe$
17	*cis* R =	$(CH_2)_{11}CH_2OH$
18	*cis* R =	$(CH_2)_{11}CH_3$

In contrast to epoxides with shorter alkyl chains, compounds **15–18** undergo α-cleavage as a very important process leading to intense and diagnostic ions, e.g., **16a** and **16b**. Transannular cleavage with hydrogen transfer leads

$$CH_3(CH_2)_7—CH\underset{\diagdown O \diagup}{}CH—(CH_2)_{11}—CO_2CH_3$$
$$+\cdot$$

16, M $^{+\cdot}$, *m/e* 368

$$CH_3(CH_2)_7CH\underset{\diagdown O \diagup}{}CH$$
$$+$$

16a, *m/e* 155 (32%)

$$CH\underset{\diagdown O \diagup}{}CH—(CH_2)_{11}—CO_2CH_3$$
$$+$$

16b, *m/e* 255 (31%)

to **16c** and **16d**, which further fragments to **16e** by loss of methanol.

$$H$$
$$CH_3—(CH_2)_7—CHCH—(CH_2)_{11}—CO_2CH_3$$
$$O$$
$$+\cdot\ H$$

16

$$CH_3(CH_2)_7CH{=}\overset{+}{O}H$$
16c, *m/e* 143 (20%)

$$\overset{+}{HO}{=}CH—(CH_2)_{11}CO_2CH_3$$
16d, *m/e* 243 (13%)

$$-CH_3OH$$

$$\overset{+}{HO}{=}CH—(CH_2)_{10}—CH{=}C{=}O$$
16e, *m/e* 211 (25%)

The position of the double bond in the original alkene may be readily ascertained by considering the m/e values for the two α-cleavage and trans-annular cleavage ions. It is interesting to note that the intensities of these oxygen-containing fragments increase at lower ionizing voltages. Thus at an ionizing voltage of 16 eV, **16a** becomes the base peak and **16b** is of 81% relative intensity, and similar increases are noted for ions **16c**, **16d**, and **16e**.

B. Alicyclic Epoxides

The high-resolution mass spectrum of cyclohexene oxide (**19**) is available (6). Its most notable feature is the loss of a methyl radical from the parent ion to give the base peak at m/e 83. Rearrangement of **19** to cyclohexanone does not appear to be important as the (M-CH$_3$·) ion is relatively weak in the spectrum of cyclohexanone. A possible fragmentation pathway leading to the fully conjugated oxonium ion **19a** is shown, but deuterium-labeling studies are required to ascertain the origin of the expelled methyl radical.

19a, m/e 83 (100%)

Rearrangement seems to be an important process in substituted alicyclic epoxides. Thus the base peak in the spectrum of 1,2-dimethyl-1,2-epoxy-cyclobutane (**20**) is at m/e 43; if it is assumed to be mainly due to $C_2H_3O^+$, it may be explained by ring contraction of **20** to **20a** (27).

$CH_3C\equiv O^+$
m/e 43 (100%)

A study of the mass spectra of compounds of general formula (**21**) and of 3,4-epoxymenthane (**22**) has led to the conclusion that the major features of these spectra may be accounted for by assuming an acylcyclopentane intermediate (**23**) (28, 29). This would fragment to give an acyl ion **23a** and a hydrocarbon fragment **23b**, and it was suggested that, especially for R = CH$_3$, the presence of ions **23a** and **23b** (R = CH$_3$) would be diagnostic of a methyloxiran grouping. A recent study of the mass spectrum of 1-methyl-1,2-epoxycyclohexane (**24**) with both high-resolution and deuterium-labeling techniques has indicated that a more complex situation exists (30). It is important to note that considerable variation in relative peak heights

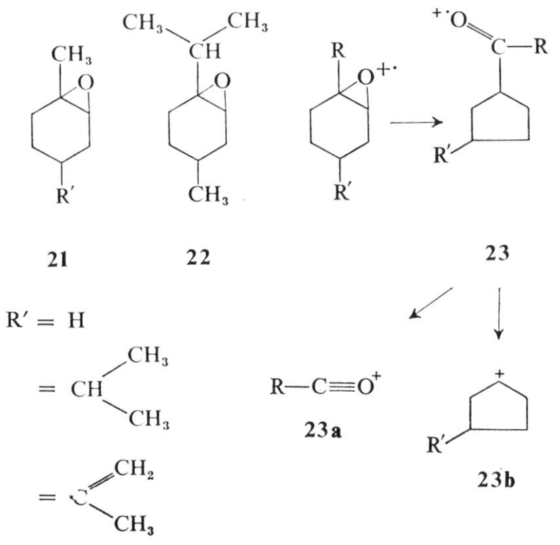

21   **22**   **23**

R′ = H

$$= CH \overset{CH_3}{\underset{CH_3}{<}}$$

$R—C≡O^+$

23a

$$= \overset{\cdot}{C} \overset{CH_2}{\underset{CH_3}{<}}$$

23b

has been observed when **24** is run on Atlas CH-4 and Hitachi RMU-6D mass spectrometers (30). The mass spectrum of **24** may be conveniently interpreted as involving a series of rearranged molecular ions **24a**, **24b**, and **24c**. The ring contraction to **24a** has been substantiated by deuterium-labeling. In the mass spectrum of 1-trideuteromethyl-1,2-epoxycyclohexane,

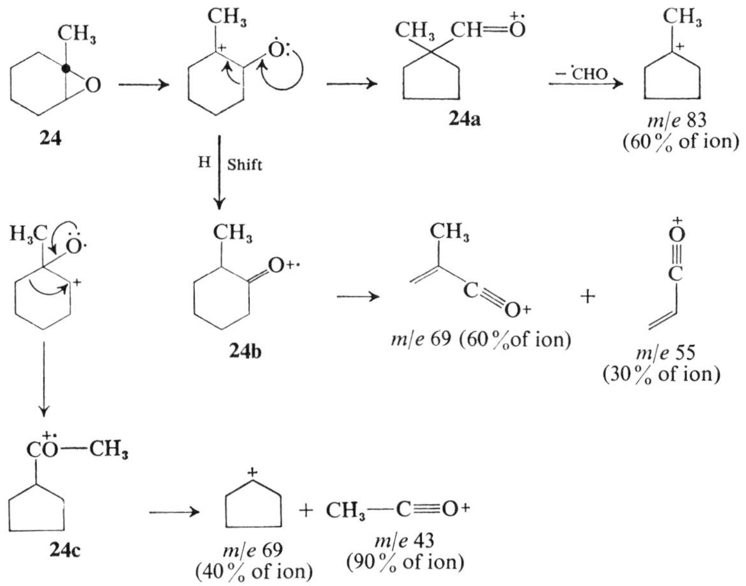

24

H | Shift

$CH_3 \; CH=\overset{+\cdot}{\overset{..}{O}}$

24a

$-\;\overset{\cdot}{C}HO$

m/e 83
(60 % of ion)

24b

$$\overset{CH_3}{\underset{}{}} \overset{}{C}≡O^+$$

m/e 69 (60 %of ion)

+

m/e 55
(30 % of ion)

24c

m/e 69
(40 % of ion)

+ $CH_3—C≡O^+$

m/e 43
(90 % of ion)

however, only a small part of the m/e 43 peak is shifted to m/e 46. This indicates that ring contraction to **24c** is not an important process leading to the acetyl ion (m/e 43), whose precise genesis is therefore not clear. Part of the ion at m/e 69 may be accounted for by rearrangement of **24** to 2-methyl-cyclohexanone (**24b**). An intense ion is also observed at (M-CH$_3$) and this probably results from α-fission (**24d**).

In general, the mass spectra of alicyclic epoxides containing the methyl-oxiran group show a very intense acetyl ion (m/e 43); thus, in alexandrofuran (**25**) the m/e 43 ion is of 98 % relative intensity and has been shown to be solely due to $C_2H_3O^+$ by exact mass measurement (31). Similarly an intense acetyl ion (78 %) is observed in the mass spectrum of caryophyllene oxide (**26**) (32). An acetyl ion (26.4 %), however, is also observed in the spectrum

of cyclohexene oxide (6). More unexpectedly, sulfolene oxide (**27**) shows an intense ion at m/e 43 (54 %, 83 %) which can only be $C_2H_3O^+$ (32, 33). The need for caution in assignment is obvious.

Mass spectrometry has proved useful in the determination of the structure of caucalol diacetate, **28** (28), and it can be used in identification of the α- and β-forms of Δ³-carene oxide, and camphene and β-pinene oxides (34). Low-resolution mass spectra of the two stereoisomeric oxides of (−)-terpinenal-(4) and of several steroidal epoxides are available in the literature (35, 36). The mass spectrum of 2,3-epoxysuccinic anhydride (**29**) has been reported (37) to give m/e 42 as the base peak (ketene) which probably arises as shown.

The mass spectra of some carotenoid epoxides of type **30** have been studied but ready rearrangement to the corresponding furanoid oxides (**31**) tends to mask any fragmentation due to the epoxide ring (38).

30, R = H or OH

31

C. Aromatic Epoxides

The mass spectra of several aromatic epoxides have been studied with high-resolution and deuterium-labeling techniques. As with aliphatic epoxides, stereoisomerism about the epoxide ring does not appear to affect the fragmentation pattern significantly and α-cleavage is again observed. Thus the loss of the benzylic hydrogen gives an intense M-1 ion (**32a**) in the mass spectra of **32**, but only a weak ion is observed at M-15 for 1-phenyl-1,2-epoxypropane (**32**, R = CH_3) and 1-phenyl-2-methyl-1,2-epoxypropane (**39, 40**). As no aromatic epoxides with side chains of greater length than methyl have been studied, no conclusions as to the importance of β- or γ-cleavage

Ph—C$\overset{\overset{+}{O}}{}$—CH—R

M-1, **32a**

Ph—C$\overset{O}{}$—CH—R

H

32, R = H, CH_3, Ph

32b,
m/e 90

32c,
m/e 89

can be drawn. Transannular cleavage without hydrogen transfer is an important process and leads to the base peak at m/e 90. This ion can be formulated as **32b** which would further lose a hydrogen atom to give the dehydrotropylium ion (**32c**, m/e 89). Alternative rationalizations postulate formation of the benzocyclopropenium ion **32d** or its open-chain isomer **32e** (41).

32

m/e 90

32d, m/e 89

or $CH{\equiv}C—CH—CH{=}CH—C{\equiv}CH$

32e, m/e 89

Evidence for rearrangement in the mass spectra of aromatic epoxides is clearly seen in the spectrum of tetraphenylethylene oxide (33), where peaks are observed at m/e 105 and 243. These may be readily explained by a phenyl migration to give the ketone 33a (39).

$$\longrightarrow \text{Ph}_3\text{CCOPh} \rceil^{+\cdot} \longrightarrow \text{Ph}_3\text{C}^+ + \text{Ph}-\text{C}{\equiv}\text{O}^+$$

33a m/e 243 m/e 105

Alkyl migration has also been shown to occur but to a much lesser extent than phenyl migration. In 1-phenyl-2-deutero-1,2-epoxypropane (34) 95% of the m/e 105 ion is due to $C_7H_5O^+$, the benzoyl ion (34a) formed by hydrogen migration, and 5% is due to $C_8H_9^+$ (34b) formed by methyl migration.

$$\longrightarrow \text{Ph}-\text{CO}\!\!\left.\right\}\!\!\text{CHD}-\text{CH}_3 \rceil^{+\cdot} \longrightarrow \text{Ph}-\text{C}{\equiv}\text{O}^+$$

34a, m/e 105

$$\longrightarrow \text{Ph}-\text{CH}\!\!\left.\right\}\!\!\text{CDO} \rceil^{+\cdot} \longrightarrow \text{Ph}-\overset{+}{\text{CH}}-\text{CH}_3$$

            $|$
            CH_3 34b, m/e 105

In isosafrole epoxide (35) the base peak arises from hydrogen migration (35a). Transannular cleavage without hydrogen transfer does not appear to be of great importance, probably because of the stabilizing of 35a by the methylenedioxy group (42).

35, m/e 178 (27%)

35a, m/e 135 (100%)

D. α,β-Epoxyketones

The mass spectra of eleven aliphatic α,β-epoxyketones have been studied by Reusch and Djerassi (43) using both deuterium-labeling and high-resolution techniques. The very similar values of the ionization potentials of 1,2-epoxypropane (9.80 eV) and acetone (9.8 to 9.9 eV) (7, 44) suggest that significant charge-localization may occur on either the epoxide or carbonyl oxygen in the molecular ion of such difunctional compounds. The predominant cleavage reaction is breakage of the bond between the epoxide and carbonyl groups, leading to formation of an acylium ion which in general gives the base peak of the spectrum. Charge retention on the oxiran moiety occurs to a much lesser extent to give, e.g., **36a**. This is clearly shown in the case of 3,4-epoxyoctan-2-one (**36**) (Fig. 1.4) where the acetyl ion accounts for 32% of the total ion current (Σ_{30}). Another intense ion observed in the mass spectrum of **36** is that at m/e 85, which results from the loss of the β-butyl substituent by α-cleavage. The ready loss of a β-alkyl substituent observed in α,β-epoxyketones is surprising in view of the fact that α-fission is relatively unimportant in aliphatic epoxides, and it has been suggested that ions of type **36c** may be better representations than type **36b**

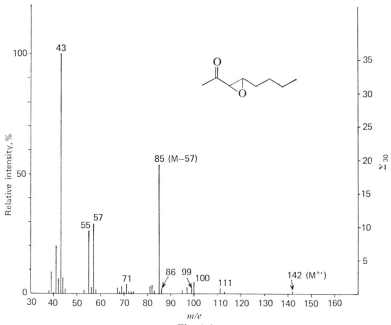

Fig. 1.4

for the results of this process. It is important to note that in **36** and in aliphatic epoxyketones in general, cleavages β and γ to the oxiran grouping appear to be completely negligible.

36, m/e 142, Σ_{30} <0.1%

$+ \; CH_3{-}C{\equiv}O^+$

m/e 43, Σ_{30} 32%

36a, m/e 99, Σ_{30} 0.8%

α,β-Epoxyketones are known to rearrange readily to dicarbonyl compounds in condensed-phase chemistry (45), but careful comparison of the mass spectra of some aliphatic α,β-epoxyketones and the possible dicarbonyl

36

36b, m/e 85, Σ_{30} 19%

$\overset{+\cdot}{O}{=}CH{-}\overset{\cdot}{C}H{-}COCH_3$

36c

compounds derived therefrom has shown that such rearrangements are of little importance in the mass spectrometer. There is little evidence of the McLafferty rearrangement (where possible) involving the epoxide oxygen by either the "outside" or the "inside" mechanism, and in **37** the McLafferty rearrangement involving the carbonyl group is not observed. This is surprising in view of the importance of this process in saturated ketones. It

37

has been suggested that competition from α-cleavage and other more facile fragmentation pathways is responsible for the lessened reactivity of the conjugated carbonyl and epoxide group in the McLafferty rearrangement (43).

An important hydrogen transfer reaction, especially in acetyl compounds (e.g., **38**, R = H), involves hydrogen transfer to the epoxide oxygen and elimination of a ketene to form the ion **38a**, which may then further lose a β-substituent to give **38b**.

From the preceding discussion it is obvious that the presence of a con-jugated carbonyl group in α,β-epoxyketones greatly affects the mass spectro-metric behavior of the epoxide group in comparison with its behavior in the

corresponding epoxyalkanes. It would be interesting to study the effect of a carbonyl group further removed, for example, in β,γ- or γ,δ-epoxyketones.

The mass spectra of a series of α,β-epoxycyclohexanones have been studied (46). The most notable feature with 3,5-dimethyl-2,3-epoxycyclohexanone (39) and isophorone oxide (40) is fission α to the carbonyl group with simul-taneous opening of the epoxide ring to give the molecular ions 39a and 40a, which then fragment, following a hydrogen transfer, to give the base peaks in the spectra (39b and 40b). It is interesting to note that in 40 the acetyl

ion (m/e 43) is of 33% relative intensity, whereas in carvone oxide (41) (which has a methyl group at $C_{(2)}$) the acetyl ion is the base peak. This may be accounted for by migration of the acyl group to give the molecular ion 41a from which an acetyl ion may be readily formed, whereas 40 (which has the methyl group at $C_{(3)}$) would not be expected to produce an acetyl ion by this pathway (although a structure analogous to 41a could result from alkyl migration).

Rearrangement of aromatic α,β-epoxyketones to dicarbonyl compounds may be expected to occur to a greater extent than in the aliphatic series,

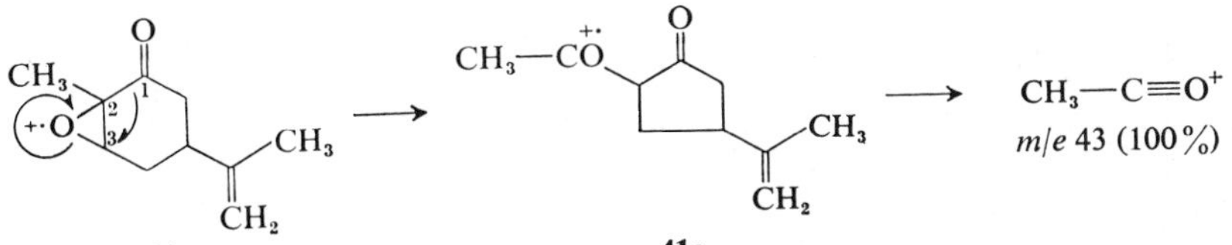

41   **41a**

parallel to the behavior of the monofunctional epoxides. Thus chalcone oxide (**42**) may rearrange to **43**, **44**, or **45** (46, 47).

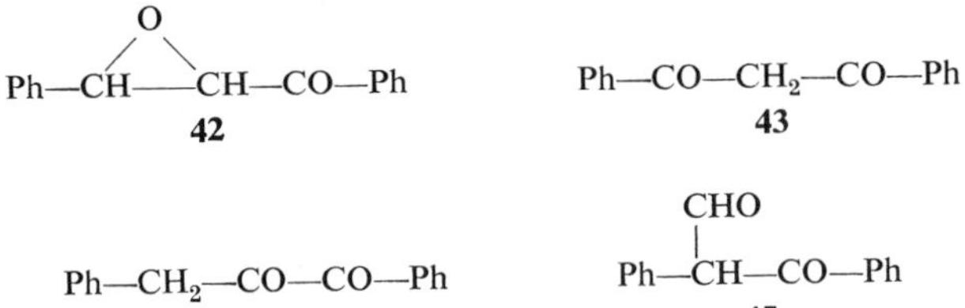

The mass spectrum of **42** is, however, markedly different to the published spectrum of dibenzoylmethane (**43**) (48), and the weak tropylium ion suggests that rearrangement to **44** is relatively unimportant. The spectra of **42** and **45** are very similar, particularly in that neither shows an appreciable loss of phenyl from the molecular ion, whereas such a loss is of considerable importance in **43**. The most notable feature in the spectra of **42** and **45** is the very

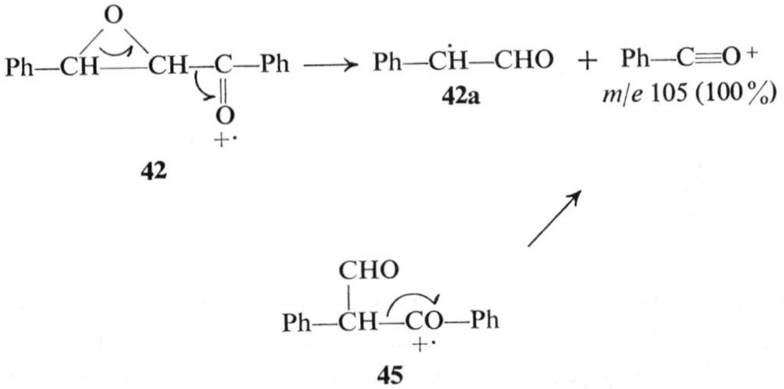

intense benzoyl ion (m/e 105). The ratio of the intensities of the benzoyl ion and the molecular ion is about 8 in **42** but only 2.3 for **45**. These figures indicate that the α-fission leading to the benzoyl ion in **42** receives a "push"

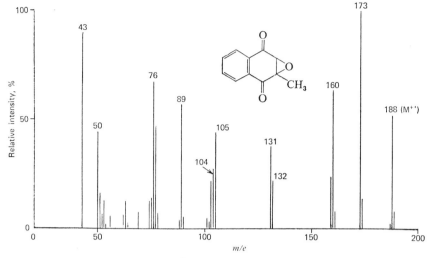

Fig. 1.5

by the simultaneous epoxide ring-opening which leads to a decrease in strain and the formation of the stable benzylic radical **42a**, all of which result in the molecular ion of **42** fragmenting more readily than that of **45** where such a "push" is not operative. This mechanism may also account for the predominant formation of acylium ions noted in aliphatic α,β-epoxyketones.

The mass spectrum of 2-methyl-1,4-naphthoquinone epoxide (**46**) is shown in Fig. 1.5. A notable feature is the intense acetyl ion at m/e 43 which can be most readily explained by rearrangement of the molecular ion to **46a**

46   **46a** m/e 43 (88%)

(**46**). The loss of two molecules of CO results in formation of the methyl-isobenzofuran (**46b**) which then further loses a hydrogen atom to form the stable isochromenyl ion (**46c**). The base peak in the spectrum is due to the loss of methyl to give the ion **46d** which surprisingly does not appear to fragment further by the loss of carbon monoxide. Another interesting feature is the intense benzyne ion at m/e 76 which may possibly arise (at least in part) directly from the parent ion.

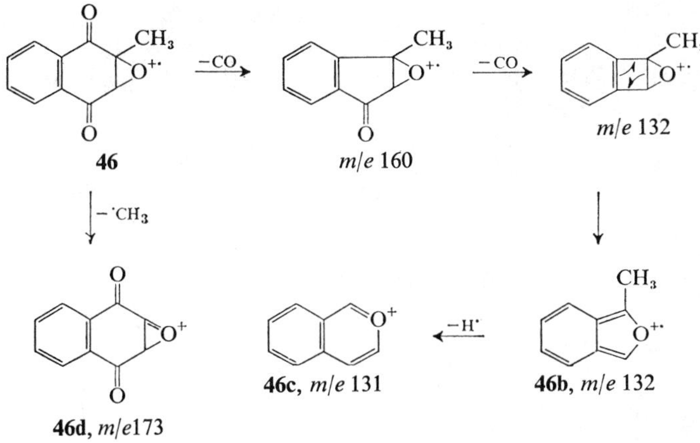

46     $-CO$ m/e 160 $-CO$ m/e 132

46d, m/e173

46c, m/e 131 $-H^{\cdot}$ 46b, m/e 132

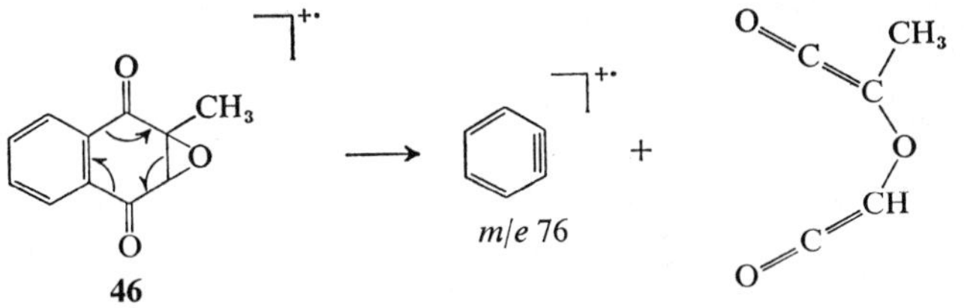

46   m/e 76

The retro-Diels-Alder reaction dominates the mass spectrum of the 2,3-epoxide of the 1,4-benzoquinone-cyclopentadiene adduct (47), with the charge residing almost entirely on the cyclopentadiene moiety (49). Such a fragmentation is in keeping with the ready thermal dissociation of 47 to cyclopentadiene and 1,4-benzoquinone epoxide (47a) (50).

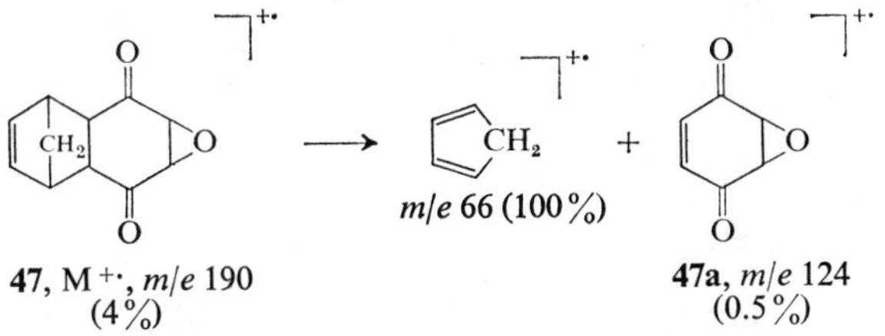

47, M$^{+\cdot}$, m/e 190
(4%)   m/e 66 (100%)   47a, m/e 124
(0.5%)

The mass spectra of 3,4-diphenyl-4,5-epoxy-2-cyclopenten-1-one (48) and 4,5-diphenyl-2-pyrone (49) are almost identical except for a somewhat less abundant molecular ion and more abundant ions at m/e 105 and 77 for the epoxyketone (51). In addition the metastable ions are very similar in all respects in both spectra. The epoxyketone may be converted to the pyrone photochemically, and the fragmentation of **48** may be explained by assuming that a similar rearrangement occurs on electron impact. The enhanced m/e 105 ion (4% for **49**, 23% for **48**) may result from the alternative pathway (**48** → **48a**) which provides an easy route to the benzoyl cation.

48, M$^{+\cdot}$, m/e 248   **49**

and

Ph—C≡O$^+$ $\xrightarrow{-CO}$ $C_6H_5^+$
m/e 105 m/e 77

48   **48a**

E. Glycidic Esters and Amides

The mass spectra of a series of β-phenylglycidic esters and amides have been examined (52, 53). The base peaks in the spectra of methyl β-phenylglycidate (**50**, R = CH$_3$), shown in Fig. 1.6, and the corresponding ethyl ester (**50**, R = C$_2$H$_5$) arise from the loss of C$_2$HO$_2$ from the molecular ions. This unusual loss may be rationalized by the migration of the alkoxy group to the β-carbon atom giving the rearranged molecular ion **50a** (rearrangement A), which fragments with the loss of the formylcarbonyl radical **50b** (57 mass units) to give the stable oxonium ion **50c**. In the case where R = C$_2$H$_5$ the last-named ion further fragments by the loss of ethylene and carbon monoxide. The rearranged molecular ion **50a** could also result from migration

50

$C_6H_7^+$ $\xleftarrow{-CO}$ Ph—CH=$\overset{+}{O}$H $\xleftarrow{-C_2H_4(R=C_2H_5)}$ Ph—CH=$\overset{+}{O}$R + CHO—ĊO
m/e 79 m/e 107   **50** m/e 121 (R = CH$_3$)
                                                      m/e 135 (R = C$_2$H$_5$)

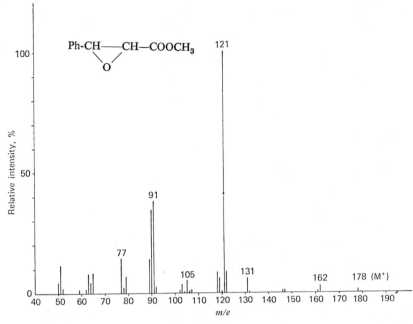

Fig. 1.6

of the ester alkyl group to the epoxide oxygen, but consideration of the spectrum of β-phenylglycidamide (**51**), Fig. 1.7, indicates that this is likely to be of minor importance. In **51** both possibilities are actually observed, rearrangement A leading to the peak at m/e 106 (**51a**) and hydrogen migration leading to m/e 107 (47% $C_7H_7O^+$), **51b**. The intensities of the two ions indicate that rearrangement A predominates in the ratio 5:1, although a strict comparison is not justified since an alkyl group would migrate in the esters and a hydrogen atom in the amide.

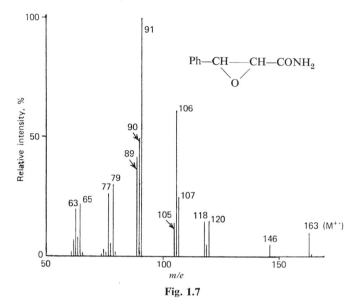

Fig. 1.7

The mass spectra of alkyl-substituted β-phenylglycidic esters are very sensitive to a change of the substituent from the α to the β position. The spectrum of ethyl α-ethyl-β-phenylglycidate is dominated by ions resulting from rearrangement A, but in the β-ethyl isomer (**52**) rearrangement A is much reduced in importance and a fragmentation path involving initial loss of ethanol and resulting finally in the isochromenyl ion (**52a**) becomes prominent.

Rearrangement to keto-esters appears to be an important process in the mass spectra of β-phenylglycidic esters. Thus, in **53** an intense ion observed at m/e 43 can be explained by rearrangement to the keto-ester **53a** (with phenyl migration). A much weaker peak arises from migration of the methyl

$$\longrightarrow CH_3\text{—}CO\text{-}\!\!\{CHPh\text{—}CO_2C_2H_5 \longrightarrow CH_3\text{—}C\!\!\equiv\!\!O^+$$
$$ ^{+\cdot}\!\} \ \textbf{53a} m/e \ 43 \ (76\%)$$

53

group to give the benzoyl ion (**53b**) of 5% relative intensity (most of the m/e 105 ion is due to $C_8H_9^+$ arising from hydrogen migration). Substituents on

$$\longrightarrow Ph\text{—}C\text{—}CH(CH_3)CO_2C_2H_5$$
$$ \underset{+\cdot}{\overset{O}{\|}} \downarrow$$

53

$$Ph\text{—}C\!\!\equiv\!\!O^+$$
$$\textbf{53b}, \ m/e \ 105 \ (5\%)$$

the α-carbon atom also give rise to acyl ions, as is seen in **54** where an intense propionyl ion **54a** is observed which may arise via ethoxycarbonyl migration.

$$\longrightarrow Ph\text{—}CH\text{-}\!\!\{\underset{+\cdot}{CO}\text{—}C_2H_5 \longrightarrow C_2H_5\text{—}C\!\!\equiv\!\!O^+$$
$$ \textbf{54a}, \ m/e \ 57 \ (95\%)$$

54

There is also, however, the possibility that it may arise from the transannular cleavage of **54b**.

$$\longrightarrow Ph\text{—}CH \overset{}{\underset{O_+}{\diagdown}} C_2H_5 \longrightarrow C_2H_5C\!\!\equiv\!\!O^+ \ +$$
$$ \textbf{54a}, \ m/e \ 57$$

54 **54b**

Another important mode of fragmentation observed in both amides and esters is transannular cleavage with or without α-hydrogen transfer. A small peak arises from loss of the epoxide oxygen in the esters.

$$Ph\text{—}\overset{+\cdot}{CR} \longleftarrow \left[\underset{Ph}{\overset{R}{\diagdown}}C\!\!\{\!\!\overset{O}{\diagup}\!\!\diagdown\!\!C\underset{H}{\overset{CO\text{—}X}{\diagup}} \right]^{+\cdot} \longrightarrow Ph\text{—}\overset{+}{CHR}$$

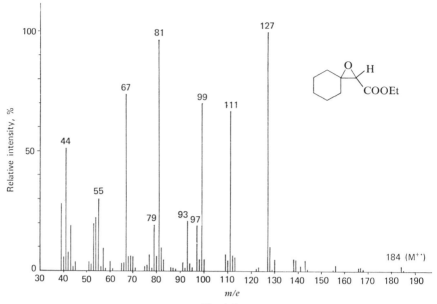

Fig. 1.8

The mass spectra of glycidic esters derived from cyclic ketones may be exemplified by that of ethyl epoxycyclohexylideneacetate **55** (Fig. 1.8) (46). Here two distinct fragmentation pathways account for nearly all of the principal ions in the spectrum. Rearrangement A is observed and leads to the base peak at m/e 127 (**55a**); the ion further fragments with loss of ethylene and water to give the intense cyclohexenyl ion **55b**. The second fragmentation pathway involves the initial loss of the ethoxycarbonyl group to give an ion at m/e 111 which may be variously formulated as **55c**, **55d**, or **55e**. Structure **55e** is considered unlikely, however, since no significant loss of

| **55** | **55c**, m/e 111 | **55d** | **55e** |

carbon monoxide is observed, but fragmentation proceeds with the loss of acetaldehyde (mass-measured), which can be most easily accommodated by structure **55c**. The ion **55c** may be expected to undergo ring contraction to **55g** which may then lose acetaldehyde with hydrogen migration to give the stable pentadienyl cation **55h** as the end product. The ion **55c** may also lose

55f, m/e 93 (21%) **55c**, m/e 111 (67%)

55g

$- CH_3CHO$

55h, m/e 67 (74%)

water to give **55f** (although other formulations are possible). In the α-methyl-substituted ester **56**, loss of the ethoxy-carbonyl group and resultant fragmentations are only of minor importance, possibly because of a competing fragmentation path which leads to the acetyl ion **56a** as the base peak in the spectrum; however, rearrangement A is prominent.

56

56a, m/e 43 (100%)

Little information is available on the mass spectra of simple alkyl-substituted glycidic esters and amides. The mass spectrum of β,β-dimethylglycidamide (**57**) has the base peak at m/e 59 (46). Its shift to m/e 60 on deuteration of the amide hydrogens indicates rearrangement to **57a** followed by α-fission to give **57b**. Rearrangement A leading to **57c** also occurs but to a much lesser extent, in marked contrast to β-phenylglycidamide (**51**) where the two processes are in the reverse order of importance. Hydrogen migration with

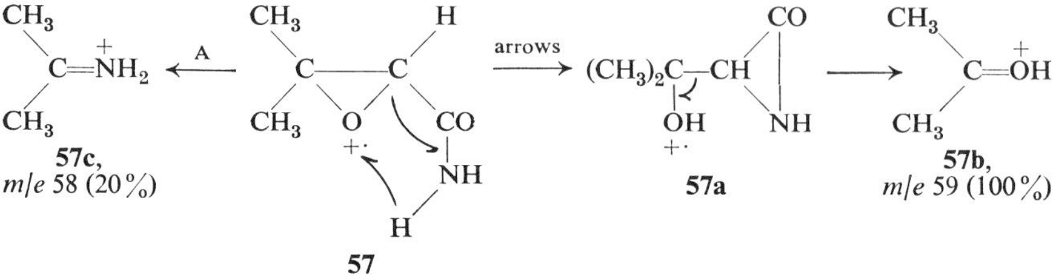

loss of isocyanic acid, leading to **57d**, is also observed, with this fragmentation being analogous to the loss of ketene in methyl α,β-epoxyketones (43).

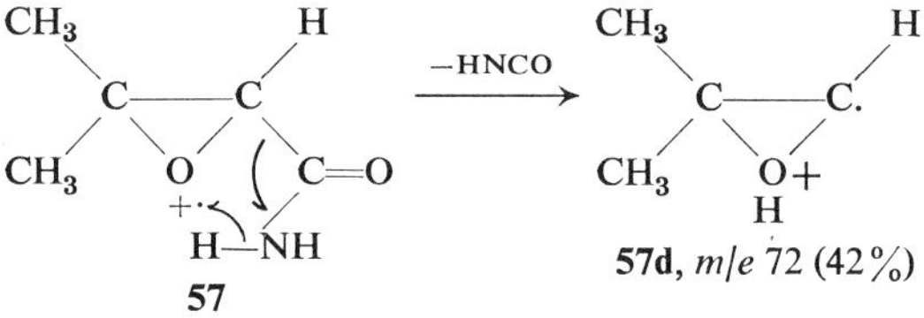

The mass spectra of ethyl β,β-dimethylglycidate (**58**, R = OC$_2$H$_5$) and β,β-dimethylglycidaldehyde (**58**, R = H) have been reported to show peaks at m/e 59 presumably due to **57b**, and it seems that both ester and aldehydo groups may be the source of a migrating hydrogen atom (54). It would be

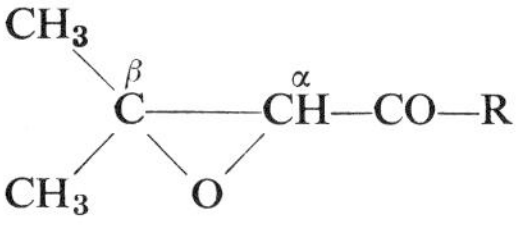

interesting to study a series of aliphatic glycidic esters and amides with longer side chains at the α and β positions and compare these with the corresponding epoxyalkanes and α,β-epoxyketones.

F. Complex Epoxy Compounds

The mass spectra of several natural products containing epoxide functions have been examined. In tutin (**59**, R = H) and mellitoxin (**59**, R = OH) the two epoxide groups do not appear to be important fragmentation-directing centers (55). An interesting (although minor) rearrangement involving the loss of a hydroxymethylene radical is observed, however, and in the case of tutin deuterium-labeling has shown that it incorporates one of the hydroxyl hydrogen atoms. The mechanism leading to ion **59a** has been suggested (55). The molecular ion is very weak (0.3% of the base peak)

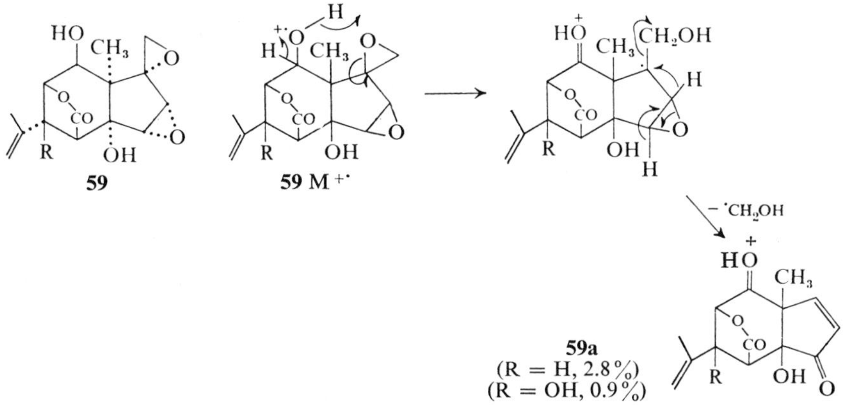

59

59 M$^{+\cdot}$

$- \cdot CH_2OH$

59a
(R = H, 2.8%)
(R = OH, 0.9%)

in the spectrum of tutin and not observed at all for mellitoxin. For further details of the fragmentation of these compounds see Chap. 6.

With dihydro-desoxy-jaborosalactone-A (**60**) the first step is homolysis of the 9-10 bond to give **60a** followed by hydrogen migration to give **60b**. Ion

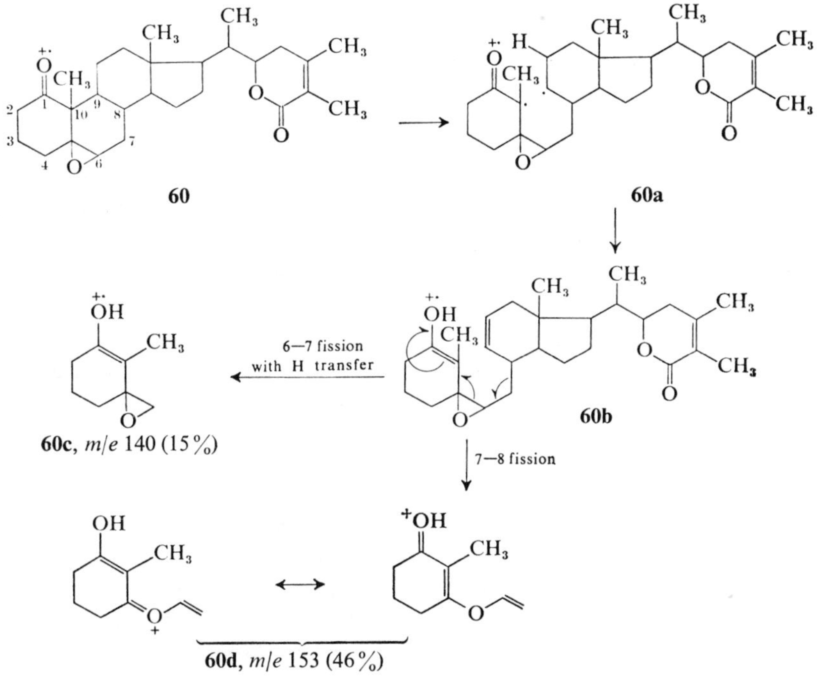

60

60a

60c, m/e 140 (15%)

6—7 fission
with H transfer

60b

7—8 fission

60d, m/e 153 (46%)

60b may further fragment by breakage of the 6-7 bond with hydrogen transfer (probably from $C_{(8)}$) to give **60c**, or breakage of the 7-8 bond to give the resonance-stabilized oxonium ion **60d** (56).

An interesting fragmentation has been observed in gedunin (**61**, R = Ac) and related compounds. Migration of the $C_{(17)}$ hydrogen atom to the epoxide oxygen leading to ion **61a** has been postulated. Ion **61a** then further loses acetic acid (gedunin) or water (deacetylgedunin, **61**, R = H) to give the base peak at m/e 299 in both cases; the expected shift to m/e 301 is observed for dihydrogedunin and deacetyldihydrogedunin (57).

61

m/e 299 (100%)

61a

In conclusion it may be mentioned that considerable loss of water is sometimes observed from either the parent or fragment ions of some epoxides (58–60). Epoxides are also useful as intermediates in the mass-spectrometric location of double bonds (26), and the application of mass spectrometry to residue analysis of some pesticides, including "heptachlor epoxide," has been reported (61). The epoxide group cannot be considered as a powerful or specific fragmentation-directing center, and its mass-spectrometric behavior is considerably altered by structural changes and the presence of other fragmentation-directing groups in the molecule.

II. OXETANS

Compared with the data available for epoxides, there is very little in the literature on the mass spectrometry of oxetans. The ionization potential of oxetan itself (**62**) has been measured as 9.85 ± 0.15 eV (62) and a high-resolution study is available (6). The mass spectrum of oxetan (or trimethylene

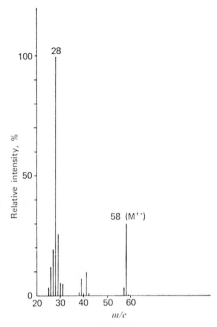

Fig. 1.9

oxide) is shown in Fig. 1.9 (62). The most important fragmentation involves cleavage of the ring to ethylene and formaldehyde with charge retention predominantly on ethylene. Ring cleavage is also the important fragmentation

$$
\boxed{}^{+\cdot}\text{O} \longrightarrow \text{CH}_2\!=\!\text{CH}_2^{+\cdot} + \text{CH}_2\text{O}
$$

62, M$^{+\cdot}$,
m/e 58   *m/e* 28

in the few substituted oxetans that have been studied (63, 64). In the bicyclic oxetan **63** the charge is distributed more evenly and in favor of the carbonyl fragment (65).

$$
\longrightarrow \text{CH}_3\!-\!\overset{+\cdot}{\underset{\parallel}{\text{C}}}\!-\!\text{CH}_3 +
$$

63, M$^{+\cdot}$, *m/e* 126   *m/e* 58   *m/e* 68
                              (Σ_{14} 4.45%)   (Σ_{14} 3.25%)

The mass spectra of some benzophenone-ketenimine adducts have been reported (66). Here also ring cleavage is the predominant reaction, as shown for the β-adduct of benzophenone and dimethyl-N-phenylketenimine (**64**), and the α-adduct **65**.

$$(CH_3)_2C{=\!=}C{=\!=}NPh$$
m/e 145 (100%)

64

$$Ph{-\!-}N{=\!=}C{=\!=}O \quad + \quad (CH_3)_2C{=\!=}C(Ph)_2$$
m/e 119 (75%) m/e 208 (100%)

65

Addenda

A detailed study of the spectrum of cyclohexene oxide (Section I-B) has appeared, and a detailed rationale of the spectrum has been advanced (67). Some discussion of the spectra of α-nitro-epoxides has also appeared (68).

REFERENCES

1. American Petroleum Institute, Research Project 44. Spectrum 760.
2. Ref. 1. Spectrum 768.
3. Ref. 1. Spectrum 777.
4. Ref. 1. Spectrum 778.
5. Ref. 1. Spectrum 772.
6. J. H. Beynon in *Advances in Mass Spectrometry*, Vol. 1, (J. D. Waldron, Ed.) Pergamon, London, 1959, pp. 328–354.
7. E. J. Gallegos and R. W. Kiser, *J. Amer. Chem. Soc.*, **83**, 773 (1961).
8. Y. Wada and R. W. Kiser, *J. Phys. Chem.*, **66**, 1652 (1962).
9. S. J. Hurst and J. M. Bruce, *J. Chem. Soc.*, **1963**, 1321.
10. K. Watanabe, *J. Chem. Phys.*, **26**, 542 (1957).
11. A. Lowrey and K. Watanabe, *J. Chem. Phys.*, **28**, 208 (1958).
12. K. H. Mueller and W. D. Walters, *J. Amer. Chem. Soc.*, **73**, 1458 (1951).
13. R. Gomer and W. A. Noyes, *J. Amer. Chem. Soc.*, **72**, 101 (1950).
14. R. J. Cvetanovic, *Can. J. Chem.*, **33**, 1684 (1955).
15. J. H. Prager, *J. Org. Chem.*, **31**, 392 (1966).
16. P. Brown, J. Kossanyi, and C. Djerassi, *Tetrahedron*, Supplement 8, Part I, 241 (1966).
17. D. Lavie, M. K. Jain, and I. Kirson, *Tetrahedron Lett.*, **1966**, 2049.
18. D. Lavie, M. K. Jain, and I. Kirson, *J. Chem. Soc.*, *C*, **1967**, 1347.
19. C. W. L. Bevan, D. E. U. Ekong, T. G. Halsall, and P. Taft, *J. Chem. Soc.*, *C*, **1967**, 820.
20. H. Budzikiewicz, C. Djerassi, and D. H. Williams, *Mass Spectrometry of Organic Compounds*, Holden-Day, San Francisco, 1967, pp. 227–258.

21. A. M. Duffield, H. Budzikiewicz, and C. Djerassi, *J. Amer. Chem. Soc.*, **87**, 2920 (1965).
22. A. Rosowsky in *Heterocyclic Compounds with Three- and Four-Membered Rings* (A. Weissberger, Ed.), Wiley-Interscience, New York, 1964, pp. 231–261.
23. K. H. Schulte-Elte and G. Ohloff, *Helv. Chim. Acta*, **50**, 153 (1967).
24. R. Ryhage and E. Stenhagen in *Mass Spectrometry of Organic Ions*, (F. W. McLafferty, Ed.), Academic Press, New York, 1963, pp. 438–440.
25. F. W. McLafferty in Ref. 24, p. 337.
26. R. T. Aplin and L. Coles, *Chem. Commun.*, **1967**, 858.
27. J.-L. Ripoll and J.-M. Conia, *Bull. Soc. Chim. Fr.*, **1965**, 2755.
28. S. Sasaki, Y. Itagaki, H. Moriyama, K. Nakanishi, E. Watanabe, and T. Aoyama, *Tetrahedron Lett.*, **1966**, 623.
29. S. Sasaki, Y. Itagaki, T. Kurokawa, and Y. Watanabe, *Chem. Ind.* (London), **1965**, 1654.
30. Ref. 20, pp. 459–460.
31. H. Budzikiewicz, C. Djerassi, and D. H. Williams, *Structure Elucidation of Natural Products by Mass Spectrometry*, Vol. II, *Steroids, Terpenoids, Sugars and Miscellaneous Classes*, Holden-Day, San Francisco, 1964, p. 151.
32. J. Baldas, Unpublished results.
33. D. S. Weinberg, C. Stafford, and N. W. Scoggins, *Tetrahedron*, **24**, 5409 (1968).
34. A. Arbuzov, Yu. Ya. Efremov, and V. L. Tal'roze, *Dokl. Akad. Nauk SSSR*, **158**, 872 (1964).
35. G. Ohlof and G. Uhde, *Helv. Chim. Acta*, **48**, 10 (1965).
36. L. Peterson, *Anal. Chem.*, **34**, 1781 (1962).
37. S. M. Creighton and D. L. Mitchell, *Can. J. Chem.*, **45**, 1304 (1967).
38. J. Baldas, Q. N. Porter, L. Cholnoky, J. Szabolcs, and B. C. L. Weedon, *Chem. Commun.*, **1966**, 852.
39. H. E. Audier, J. F. Dupin, M. Fétizon, and Y. Hoppilliard, *Tetrahedron Lett.*, **1966**, 2077.
40. I. I. Ryabinkin, I. P. Stepanov, and T. I. Temnikova, *Zh. Org. Khim.*, **3**, 216 (1967).
41. J. L. Occolowitz and G. L. White, *Aust. J. Chem.*, **21**, 997 (1968).
42. B. Willhalm, A. F. Thomas, and F. Gautschi, *Tetrahedron*, **20**, 1185 (1964).
43. W. Reusch and C. Djerassi, *Tetrahedron*, **23**, 2893 (1967).
44. J. D. Morrison and A. J. C. Nicholson, *J. Chem. Phys.*, **20**, 1021 (1952).
45. Ref. 22, p. 254.
46. J. Baldas and Q. N. Porter, Unpublished results.
47. S. Sasaki, H. Abe, K. Nakanishi, Y. Itagaki, and E. Watanabe, *Bull. Chem. Soc. Japan*, **41**, 522 (1968).
48. J. H. Bowie, D. H. Williams, S.-O. Lawesson, and G. Schroll, *J. Org. Chem.*, **31**, 1384 (1966).
49. D. F. O'Brien and J. W. Gates, *J. Org. Chem.*, **30**, 2593 (1965).
50. K. Alder, F. H. Flock, and H. Beumling, *Chem. Ber.*, **93**, 1896 (1960).
51. M. M. Bursey, L. R. Dusold, and A. Padwa, *Tetrahedron Lett.*, **1967**, 2649.
52. J. Baldas and Q. N. Porter, *Chem. Commun.*, **1966**, 571.
53. J. Baldas and Q. N. Porter, *Aust. J. Chem.*, **20**, 2655 (1967).
54. H. C. Volger, W. Brackman, and J. W. F. M. Lemmers, *Rec. Trav. Chim.*, **84**, 1203 (1965).
55. R. Hodges, E. P. White, and J. S. Shannon, *Tetrahedron Lett.*, **1964**, 371; Errata, **1964**, 541.
56. R. Tschesche, H. Schwang, H.-W. Fehlhaber, and G. Snatzke, *Tetrahedron*, **22**, 1129 (1966).

57. M. A. Baldwin, A. G. Loudon, A. Maccoll, and C. W. L. Bevan, *J. Chem. Soc., C,* **1967,** 1026.
58. H. Kanno, W. H. Schuller, and R. V. Lawrence, *J. Org. Chem.,* **31,** 4138 (1966).
59. P. J. Garratt and F. Sondheimer, *J. Chem. Soc. C.,* **1967** 565.
60. T. J. Batterham, N. K. Hart, and J. A. Lamberton, *Aust. J. Chem.,* **19,** 143 (1966).
61. T. R. Kantner and R. O. Mumma, *Residue Rev.,* **16,** 138 (1966).
62. E. J. Gallegos and R. W. Kiser, *J. Phys. Chem.,* **66,** 136 (1962).
63. J. J. Beereboom and M. S. Von Wittenau, *J. Org. Chem.,* **30,** 1231 (1965).
64. J. S. Bradshaw, *J. Org. Chem.,* **31,** 237 (1966).
65. N. C. Yang, M. Nussim, and D. R. Coulson, *Tetrahedron Lett.,* **1965,** 1525.
66. L. A. Singer and G. A. Davis, *J. Amer. Chem. Soc.,* **89,** 598 (1967).
67. M. K. Strong, P. Brown, and C. Djerassi, *Org. Mass Spec.,* **2,** 1201 (1969).
68. H. Newman and R. B. Angier, *Tetrahedron,* **26,** 825 (1970)

2 REDUCED FURANS

I. TETRAHYDROFURANS

The mass spectrum (1) of tetrahydrofuran (1) (Fig. 2.1) has been examined in detail, and both high-resolution and deuterium-labeling studies are available (2, 3). The ionization potential has been measured as 9.45 ± 0.15 eV (4). A strong molecular ion is observed, and the most important fragmentation process is elimination of formaldehyde (cf. oxetan, Chap. 1, Sec. II) to form the cyclopropane radical ion (**1a**) (base peak), which further loses a hydrogen atom to give an ion at m/e 41 which may be formulated as the cyclopropyl (**1b**) or allyl (**1c**) cation. The only other fragment of importance is the M-1 ion. A study of the deuterium-labeled compound **2** shows that 70% of the hydrogen lost originates from the α-positions (**1 → 1d**) and the remaining 30% from the β-positions to give an M-1 ion which may be formulated as the bicyclic oxonium ion **1e** or the rearranged open-chain ion **1f** (3).

1f, m/e 71   **1e**, m/e 71   **1**, M$^{+\cdot}$, m/e 72   **1d**, m/e 71

2

$\downarrow -CH_2O$

1a, m/e 42 (100%) $\xrightarrow{-H\cdot}$ **1b**, m/e 41   or $CH_2=CH-\overset{+}{C}H_2$ **1c**

The loss of formaldehyde (**3 → 3a**) and the alternative loss of acetaldehyde (**3 → 3b**) occur to a lesser extent in 2-methyltetrahydrofuran (**3**), the spectrum of which is shown in Fig. 2.2 (5). α-Cleavage now becomes a very important process, and loss of the methyl group gives the cyclic oxonium ion **3c** (m/e 71), as the base peak. This is also observed as the base peak in the spectrum of

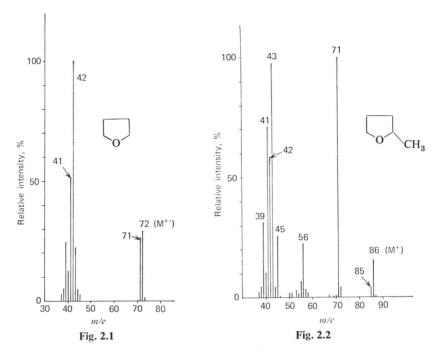

Fig. 2.1   Fig. 2.2

tetrahydrofurfuryl alcohol (**4**) (6). An M-1 ion is also observed (**3d**) but it is only about one-twentieth the intensity of **3c** and considerably weaker than the parent ion. The ion at *m/e* 45, corresponding to composition $C_2H_5O^+$,

is of some interest, and a reasonable mechanism for its formation is α-cleavage with ring-opening to **3e** followed by hydrogen transfer and elimination of the allyl radical to give protonated acetaldehyde, **3f** (*m/e* 45). The step **3e** → **3f** is analogous to the elimination of a molecule of alkene from oxonium ions in the spectra of aliphatic ethers (7).

If larger alkyl groups are substituted at $C_{(2)}$ of tetrahydrofuran, α-cleavage becomes the predominant fragmentation process and the molecular ion may be rather weak. Thus the spectrum of **5** shows a very weak parent ion and an intense base peak **5a** due to α-cleavage which accounts for 29% of the

3 **3e** **3f**, *m/e* 45

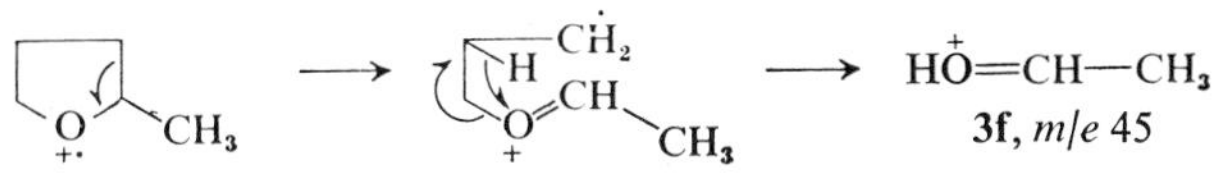

total ion current (8). The spectrum of 3-methoxycarbonylmethyl-2-pentyl-tetrahydrofuran (**6**) also shows a weak parent ion, and loss of the pentyl group gives the base peak due to ion **6a** which further fragments to a smaller

5 **5a**, *m/e* 85

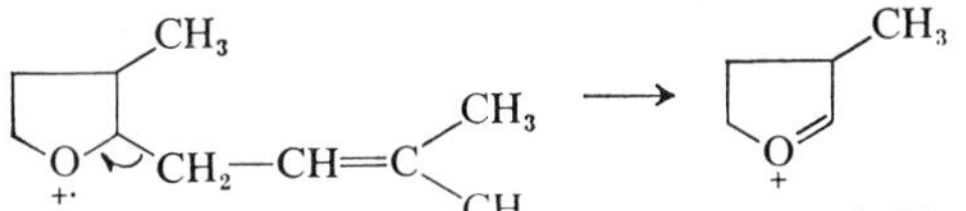

extent by loss of methanol (**6a** → **6b**) (9). Another important fragmentation pathway involves the McLafferty rearrangement with loss of methyl acetate to give the dihydrofuran radical ion **6c** which undergoes cleavage of the allylic bond to give the abundant ion **6d**.

6 **6a** **6b** **6** **6c** **6d**

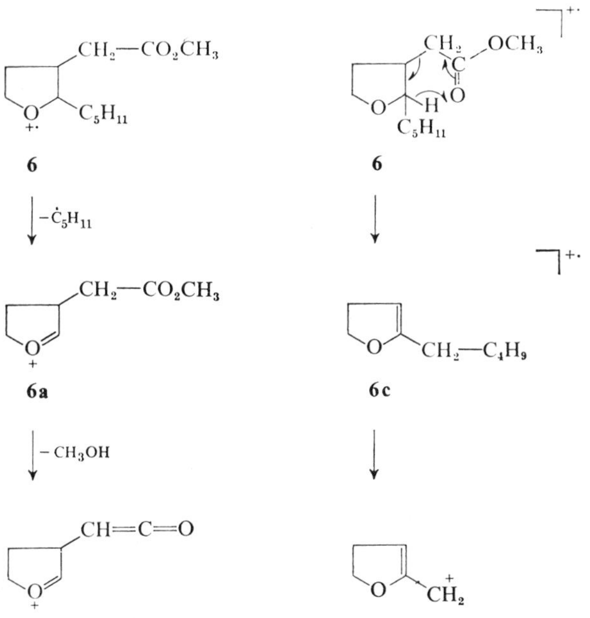

The characteristic α-cleavage is also observed in more complex 2-substituted tetrahydrofurans; for example, in gratiogenin (**7**) it accounts for the base peak (**7a**) (10).

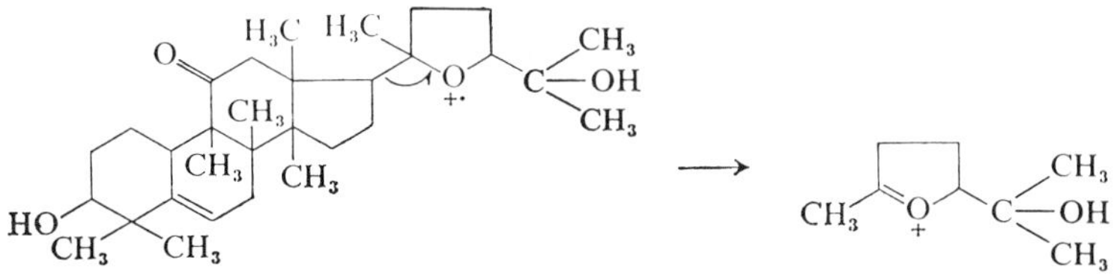

7   7a, *m/e* 143 (100%)

Mass spectrometry has proved useful in the structure elucidation of a degradation product (**8**) of the mold metabolite erythroskyrine (where α-cleavage gives the oxonium ion **8a**) (11), and in the structure determination

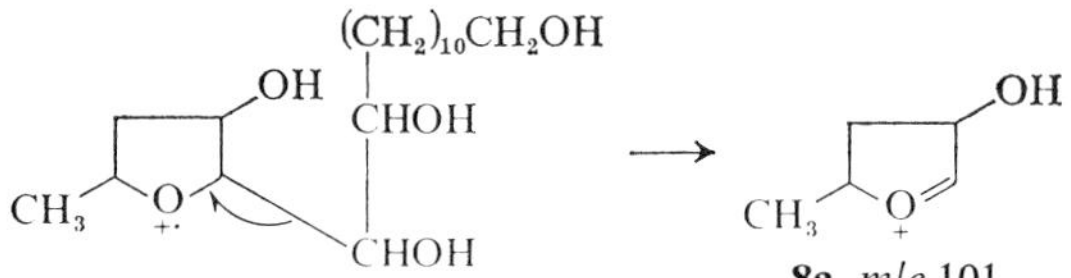

8   8a, *m/e* 101

of some high molecular weight 2,5-disubstituted tetrahydrofuran esters such as **9**, derived from the macrolide antibiotic nystatin (12). The observation of prominent peaks due to the two α-cleavage ions allows determination

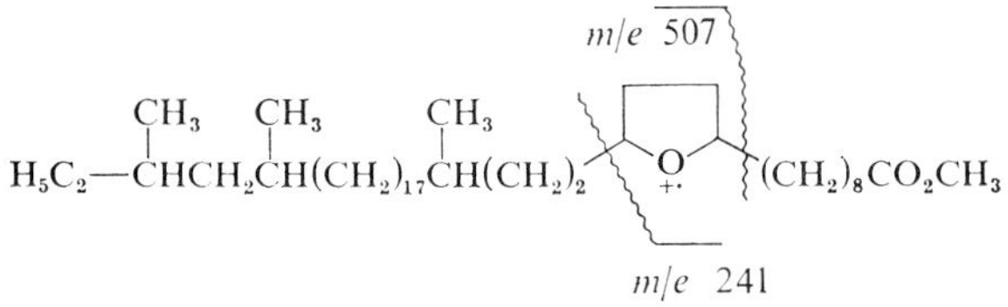

m/e 507

H_5C_2—$CHCH_2CH(CH_2)_{17}CH(CH_2)_2$ ... $(CH_2)_8CO_2CH_3$

m/e 241

9, M⁺·, *m/e* 678

of the position of the tetrahydrofuran ring in the aliphatic chain. The spectra of the eight stereoisomeric tetrahydromenthofurans **10** exhibit an intense ion at *m/e* 97, which has been rationalized as involving initial internal α-cleavage

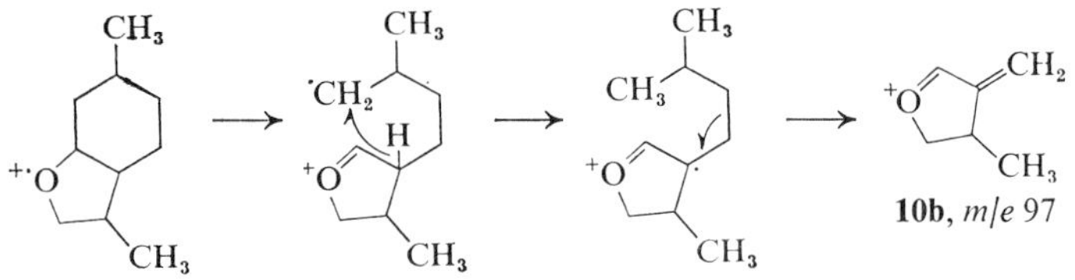

10   10a   10b, *m/e* 97

to **10a** followed by transfer of the allylic hydrogen and then fragmentation to **10b** (m/e 97) (13).

A similar mechanism has been postulated for the tricyclic tetrahydrofuran **11**, but here the methylene bridge rather than a hydrogen atom is transferred (**11a → 11b**) and fragmentation of the molecular ion **11b** then gives **11c** as the base peak (14).

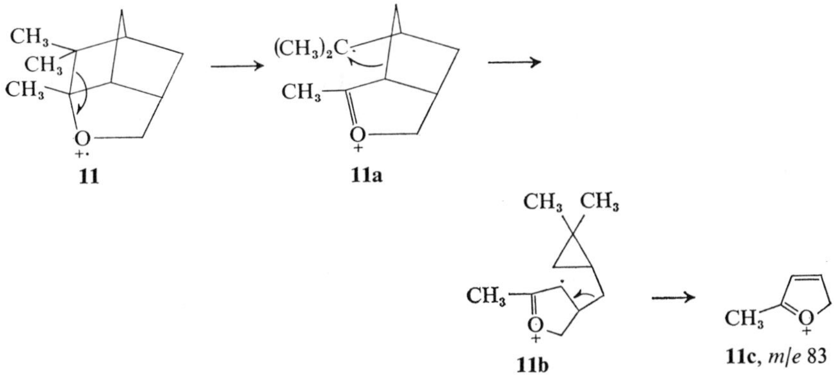

Two stereoisomeric bicyclic tetrahydrofurans (**12**) with a β-methyl group have been observed to give the base peak at M-15 (13). This may be due in part to the ion **12a**, but it is probable that ring cleavage to give the conjugated oxonium ion **12b** (which may then readily lose the β-methyl group giving the ion **12c**) represents an important pathway to an M-15 ion.

The loss of the elements of acetone (to give M-58) is an important process in the identical spectra of three stereoisomeric α,α-dimethyltetrahydrofurans

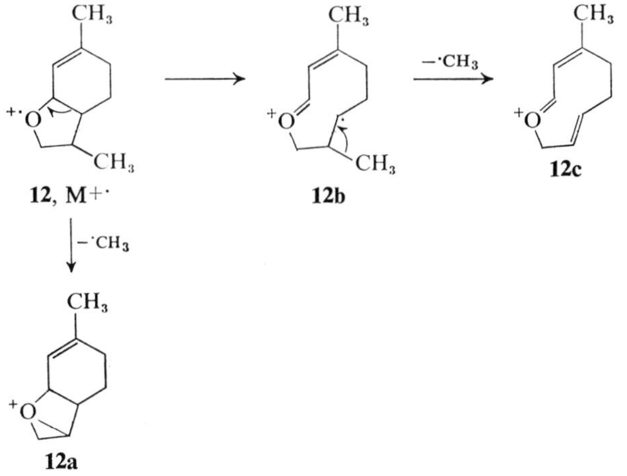

(**13**), as is the loss of one of the methyl groups (to give M-15) (15). Intense M-58 ions have also been observed in the spectra of 1,8-epoxyiridans (16, 17).

M-15 **13** M-58

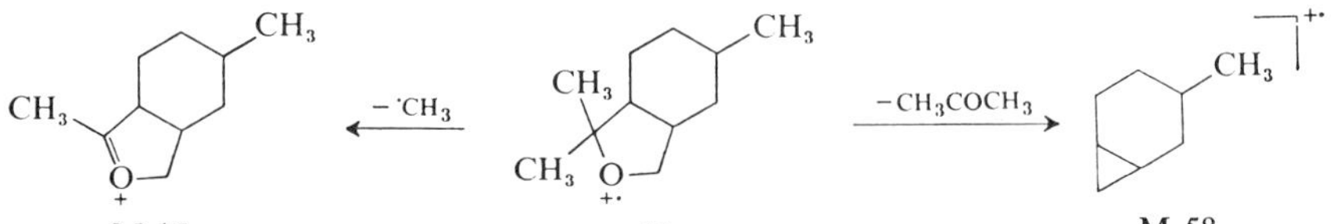

α-Cleavage is an important fragmentation for the ketone **14**, giving *m/e* 71 as the base peak (18). Cleavage β to the tetrahydrofuran ring and α to the carbonyl group gives the abundant ion **14a**, and loss of dihydrofuran with formation of the decaldehyde radical ion **14b** occurs to a lesser extent. The McLafferty rearrangement involving the carbonyl group and the alkyl chain,

m/e 71 (100%) **14**, M+· **14a**, *m/e* 99 **14b**, *m/e* 156

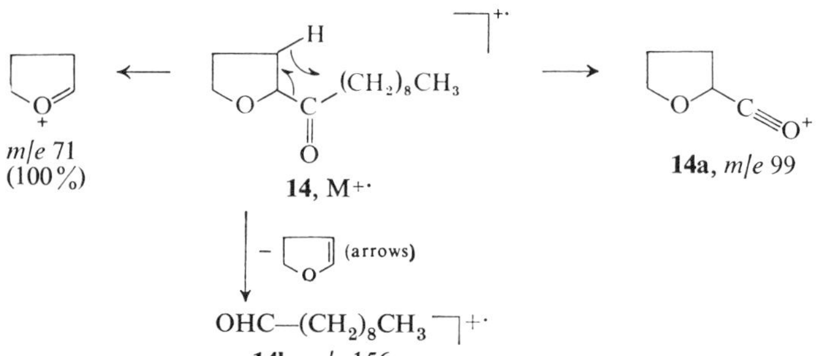

however, does not appear to be of any importance. 2-Hydroxytetrahydrofurans may show very ready loss of water; thus, in 2,4,4,5,5-pentamethyltetrahydro-2-furanol (**15**) an M-H_2O ion, but no parent ion, is observed (19).

15

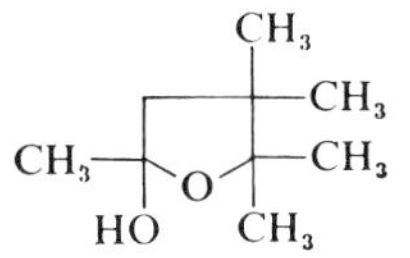

Similarly, the loss of methanol has been observed in some 2-methoxytetrahydrofurans (20).

The mass spectra of 3-substituted tetrahydrofurans (with no substituents at $C_{(2)}$ and $C_{(5)}$) do not appear to have been studied, and it would be of interest to see to what extent the spectra of isomeric 2- and 3-substituted compounds differ.

The mass spectra of a few tetrahydrofuran-3-ones have been briefly reported (21) and some high-resolution data for 2-methyltetrahydrofuran-3-one (**16**) (22) (a component of coffee aroma) is available. The loss of carbon monoxide is an important process for **16**, and the base peak at m/e 43 is mainly due to $C_2H_3O^+$, probably the acetyl ion (**16b**) which may be formed in part from the M-CO ion **16a**. α-Cleavage is of much less importance for

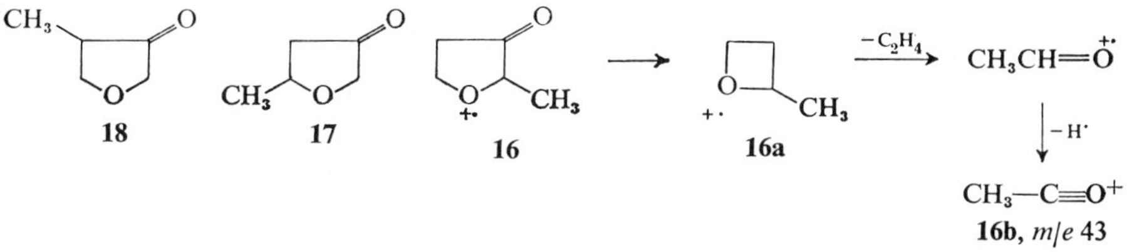

tetrahydrofuran-3-ones than for α-substituted tetrahydrofurans; thus both **16** and **17** (which have an α-methyl group) as well as **18** (with a β-methyl group) show an M-·CH_3 peak of less than 10% of the intensity of the parent ion (**21**).

Mass spectrometry has proved useful in the structure elucidation of the fungal metabolite altenin (**19**) (23).

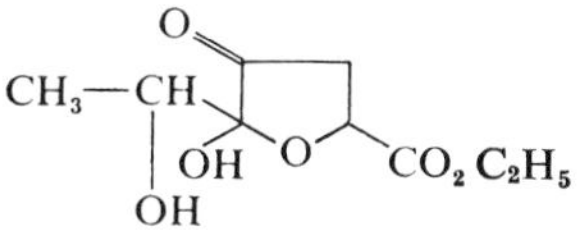

II.   DIHYDROFURANS

The major peaks in the spectra of a number of simple dihydrofurans have been reported (24, 25), and it is of interest to compare the spectra of the three double-bond isomers **20**, **21**, and **22** (25). The base peak in **20** and **21**

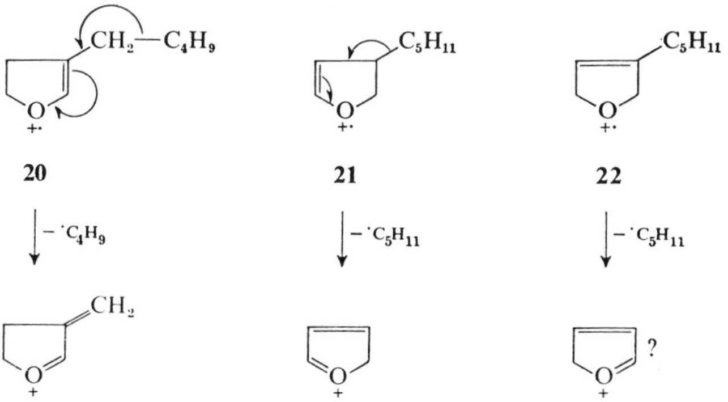

20a, m/e 83 (100%)   21a, m/e 69 (100%)   22a, m/e 69 (100%)

(20a and 21a, respectively) may be readily rationalized by the mechanisms shown, but the pathway is less obvious in 22 and here the ion may not have the structure 22a.

α-Cleavage is an important process in 2-substituted 2,5-dihydrofurans as it leads to conjugated oxonium ions such as 23a which is the base peak in the spectrum of actinidol (23) (26).

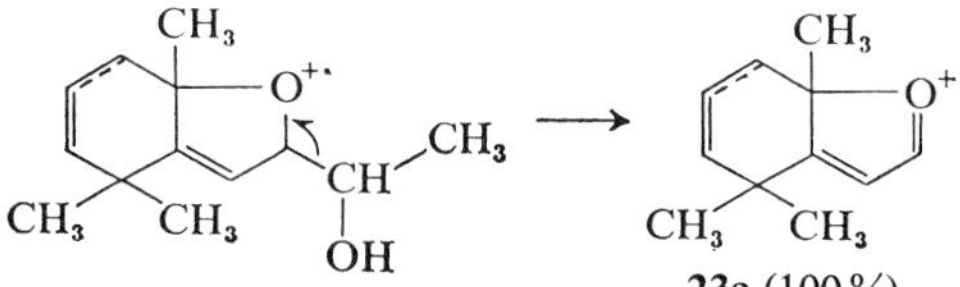

23 → 23a (100%)

The spectrum of 24 shows an abundant molecular ion and an almost equally intense M-1 peak (27), most probably arising from the loss of the allylic hydrogen atom to give the conjugated oxonium ion 24a. Loss of the bridgehead methyl group (which would also give a conjugated oxonium ion) occurs to a much lesser extent. The major portion of the spectrum of 24 can be accounted for by the loss of neutral acetylene to give the 2,5-dimethylfuran radical ion (24b) and ions derived from its further fragmentation.

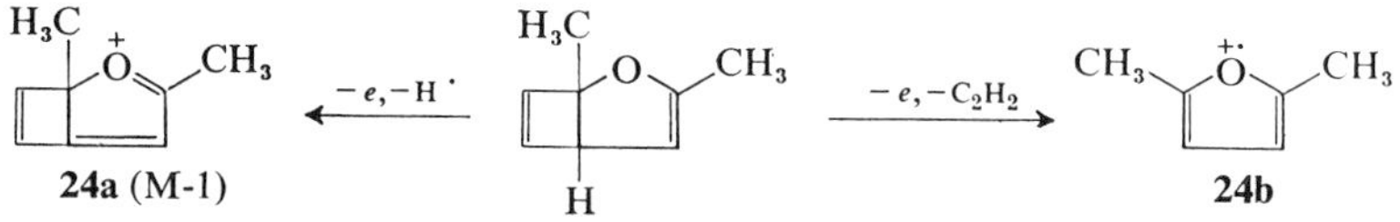

24a (M-1)   ←$-e,-H^{\cdot}$—   24   —$-e,-C_2H_2$→   24b

The retro-Diels-Alder reaction is an important fragmentation process for the bicyclic ester 25 (25 → 25a) (28) and for the steroid derivative 26 (26 → 26a) (29), both of which contain the dihydrofuran grouping.

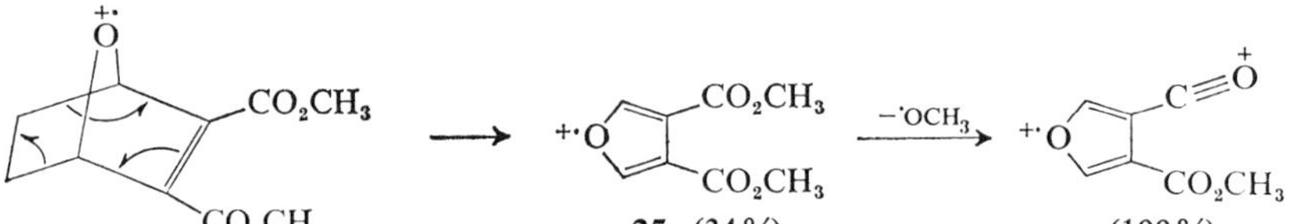

25, M⁺·   →   25a (34%)   —$-^{\cdot}OCH_3$→   (100%)

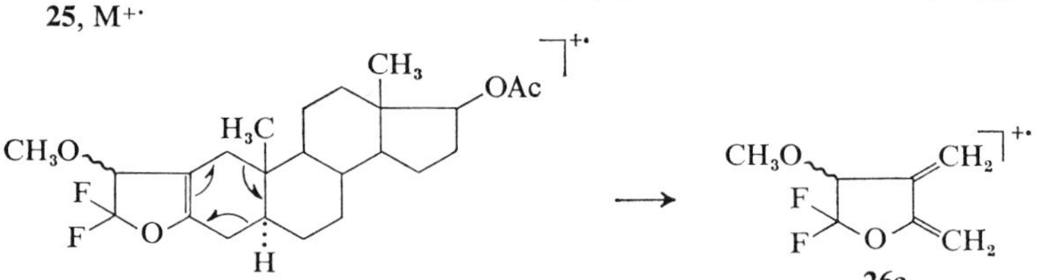

26   →   26a

In 2-oxabicyclo(3.3.0)octa-3,7-diene (**27**) expulsion of CHO· is an important process and leads to an abundant ion at m/e 79, which has been formulated as the phenonium (or benzonium) ion **27a** (30). This further loses a hydrogen molecule to form the phenyl ion **27b**.

27, M⁺·, m/e 108 m/e 108 m/e 79

C₆H₅⁺ ⟵ᴴ² **27b**, m/e 77   **27a**, m/e 79

The spectrum of 2-acetyl-2,5-dimethyl-3(2*H*)-furanone (**28**) exhibits a weak molecular ion and a ready loss of ketene, probably via a six-membered transition state, to give the abundant 3-hydroxy-2,5-dimethylfuran ion (**28a**) (31).

28   **28a**

Mass spectrometry has proved useful in the identification of 2,5-dimethyl-4-hydroxy-3(2*H*)-furanone as a component of pineapple flavor (32).

III. COMPLEX TETRAHYDRO- AND DIHYDROFURANS

A. Lignans

The mass spectra of some simple lignans containing the tetrahydrofuran nucleus, of which galgravin (**29**) (Fig. 2.3) is a typical example, have been examined (33). Stereoisomerism has been shown to have little effect on the fragmentation pattern and the intensity; differences are usually too slight to be useful in distinguishing between stereoisomers. α-Cleavage with loss of Ar· is an unimportant process; the two major fragmentation pathways of the molecular ion correspond to cleavage of the tetrahydrofuran ring along the lines *a* and *b* in **29** (33). Path *a* may lead to charge retention on the aldehyde moiety **29a**; this fragments further by loss of a hydrogen atom to give the

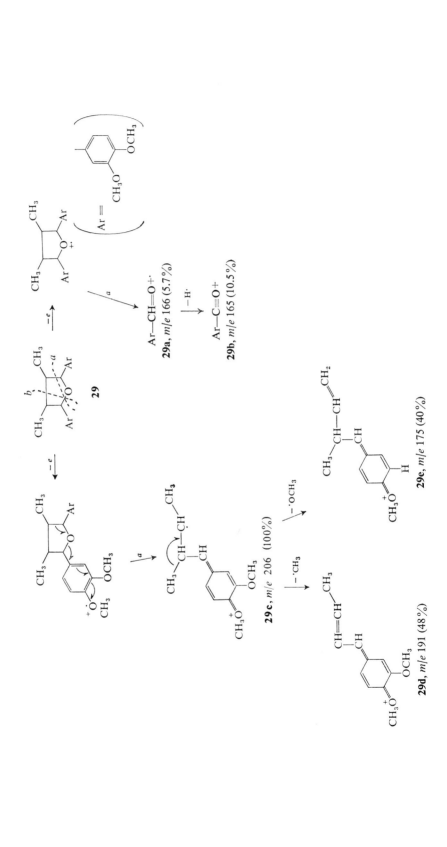

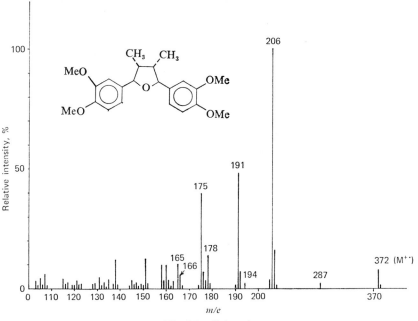

Fig. 2.3 Galgravin.

substituted benzoyl ion **29b**, which in turn also further fragments. Charge retention on **29c** is much more important and results in the base peak of the spectrum. The ion **29c** further fragments in a variety of ways, two of the more important ones being by loss of a methyl group from the side chain to give **29d** and by loss of a methoxy group to give the ion formulated as **29e**. Its intensity suggests that transfer of a hydrogen from the side chain to the site of the lost methoxyl group is involved.

For fragmentation by path *b* several mechanisms may be written, including the concerted one shown. This gives the ions **29f** and **29g**, and the latter fragments further to the benzoyl ion **29b**.

Galbacin (**30**) shows a fragmentation pattern (**33**) very similar to galgravin and its stereoisomer galgelbin but in this case the ions produced by path *b* are of considerably higher intensity.

The mass spectra of dihydrogmelinol (**31**) and its stereoisomer di-*O*-methylolivil are extremely similar (33, 34). Fragmentation by paths *a* and *b* is still observed; however, the most important fragmentation path involves displacement of the benzyl side chain at $C_{(4)}$ either with hydrogen transfer (**31** → **31a** and **31b**) or without (**31** → **31c** and **31d**).

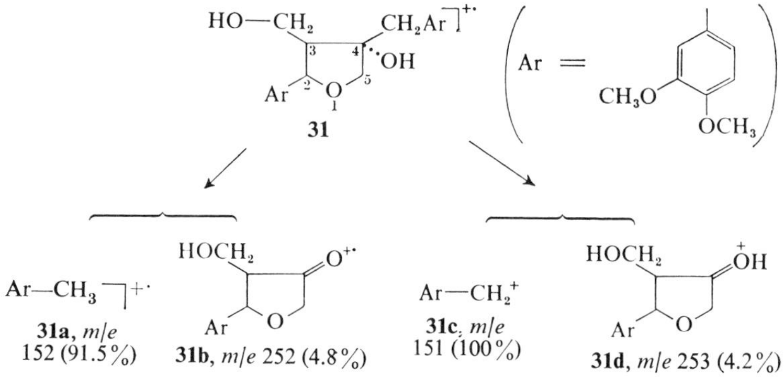

Stereochemistry also has little effect on the spectra of lignans containing two fused tetrahydrofuran rings, such as di-*O*-methylpinoresinol (**32**) and its stereoisomer epieudesmin (33, 35). The two spectra may be largely rationalized in terms of the fragmentation paths *a* and *b* observed in the simpler lignans. Thus for **32**, path *a* results in formation of both the aldehyde fragment **32a** and the corresponding ion **32b** as well as in loss of formaldehyde to give **32c**. Path *b* does not result in fragmentation but in the ring-opened

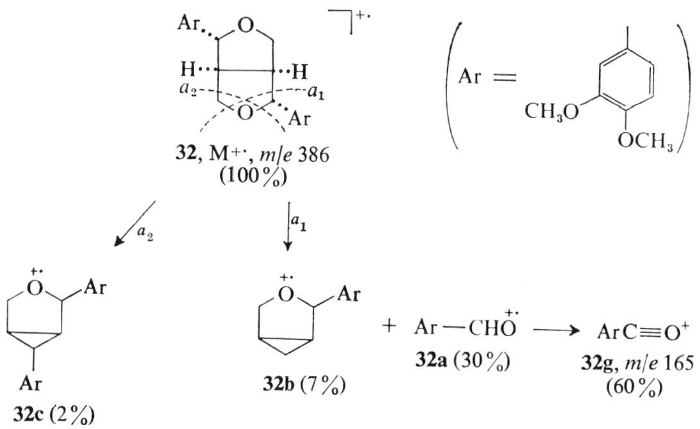

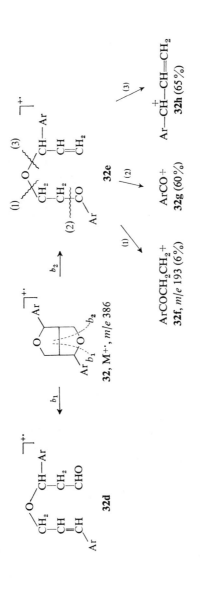

molecular ions **32d** and **32e**. Fragmentation of **32e** leads directly to **32f**, **g**, and **h**. Another fragmentation of some importance is the production of the ion $ArCH_2^+$ (**32i**) from the parent ion via **32j**. The literature should be consulted for further examples and more detailed discussion (33, 35).

32 **32j** **32i**, *m/e* 151

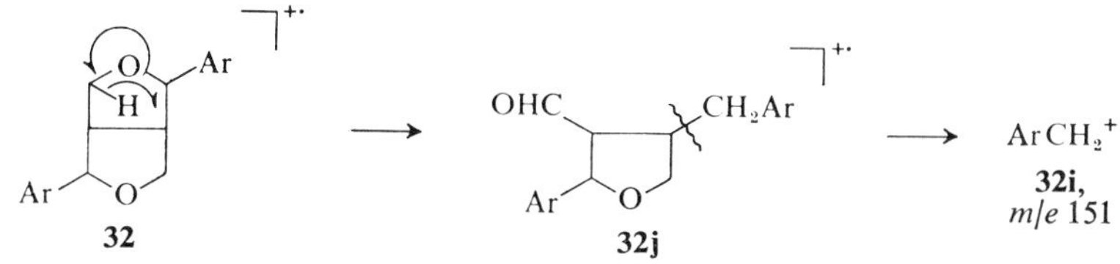

B. Terpenes

The spectra of a number of terpenes containing the spiro-tetrahydrofuran group have been examined. That of grindelane (**33**) is dominated by a very intense base peak at M-139, whose formation may be rationalized as involving initial α-cleavage to **33a** followed by hydrogen transfer and fragmentation

33, M$^{+\cdot}$ **33a**

33b, *m/e* 153 (100%) (M-139)

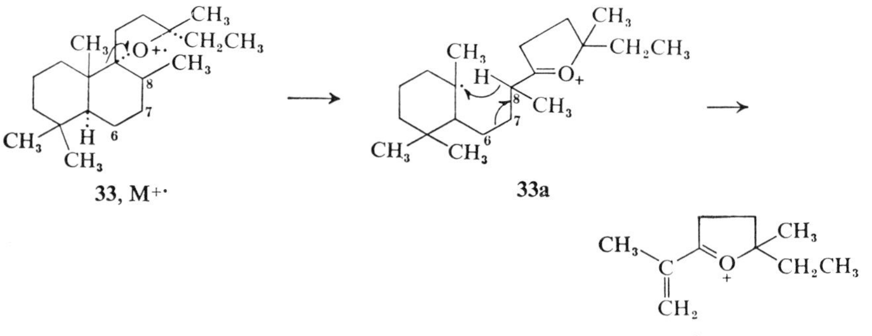

to the conjugated oxonium ion **33b** (M-139) (36). A much weaker ion at M-125 corresponds to hydrogen transfer (probably from $C_{(7)}$) and cleavage of the $C_{(5)}$—$C_{(6)}$ bond (**33a** → **33c**). A weak ion at M-152 corresponds to

33a

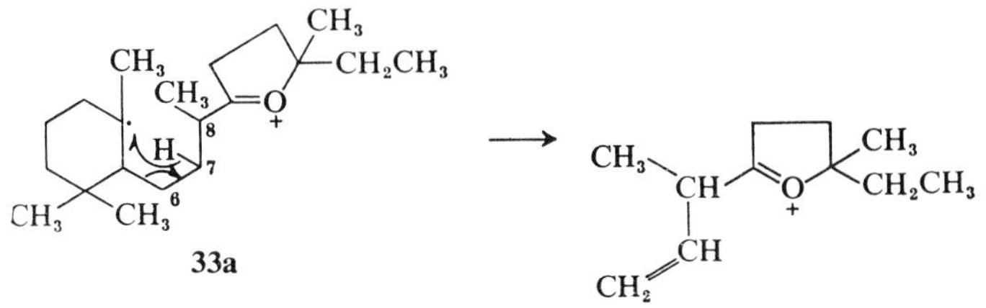

33c, *m/e* 167 (M-125)

cleavage of the $C_{(7)}$—$C_{(8)}$ bond in **33a** with no hydrogen rearrangement, possibly to give **33d**.

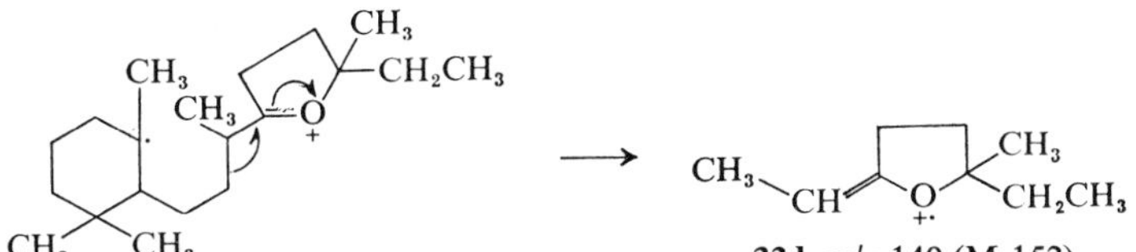

33d, *m/e* 140 (M-152)

33a

The retro-Diels-Alder fragmentation is very important in the spectra of diterpenes with a $C_{(7)}$—$C_{(8)}$ double bond such as grindelene (**34**, R = CH_3), methyl grindelate (**34**, R = $COOCH_3$), and grindelic acid (**34**, R = COOH) (in whose structure elucidation mass spectrometry played a useful role), leading to the base peak **34a** in all three cases (36, 37). Further fragmentation

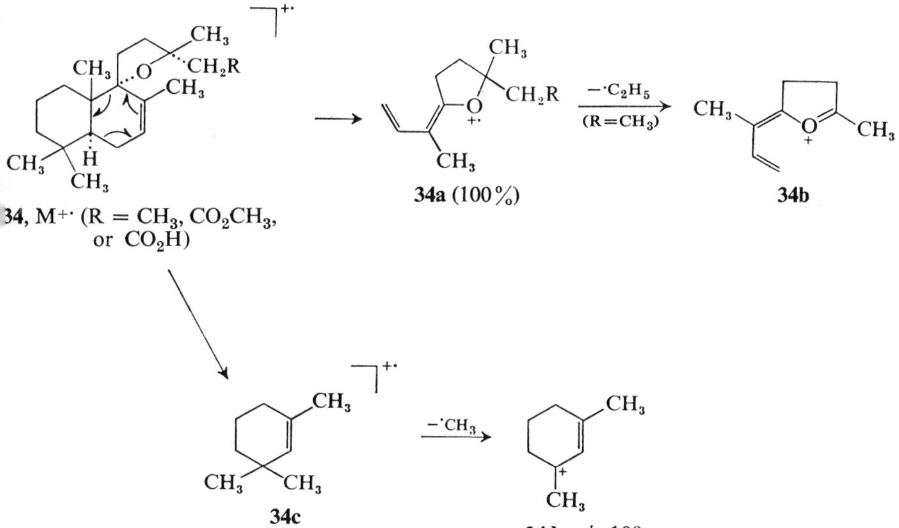

34, M$^{+\cdot}$ (R = CH$_3$, CO$_2$CH$_3$, or CO$_2$H)

34a (100%)

34b

34c

34d, *m/e* 109

of **34a** depends on the nature of the R group. Thus for grindelene (R = CH_3) either the ethyl group at $C_{(13)}$ (**34a** → **34b**) or to a lesser extent the methyl group at $C_{(13)}$ may be lost, while in methyl grindelate (R = $COOCH_3$), ions due to the loss of methanol, acetic acid (or methyl formate), and methyl acetate are observed. Charge retention on the cyclohexene fragment **34c** with further loss of a methyl radical gives the ion **34d** which is very prominent in the spectrum of grindelene (**34**, R = CH_3).

The retro-Diels-Alder fragmentation is also important in the spectrum of methyl 6-oxogrindelate (**35**) where it gives the intense ion **35a**, but in methyl 6-hydroxygrindelate (**36**) the molecular ion loses water very readily to give an M-18 ion which cannot undergo the usual retro-Diels-Alder

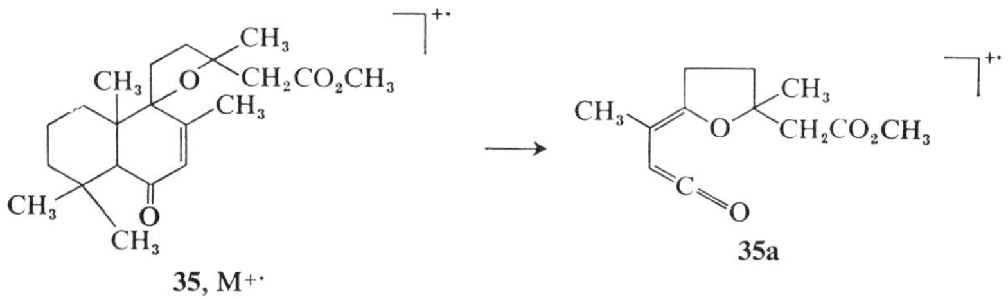

35, M+·   35a

fragmentation. Instead it leads to a single peak at m/e 187 which may be formulated as **36a** or the ring-expanded tropylium ion **36b** (38).

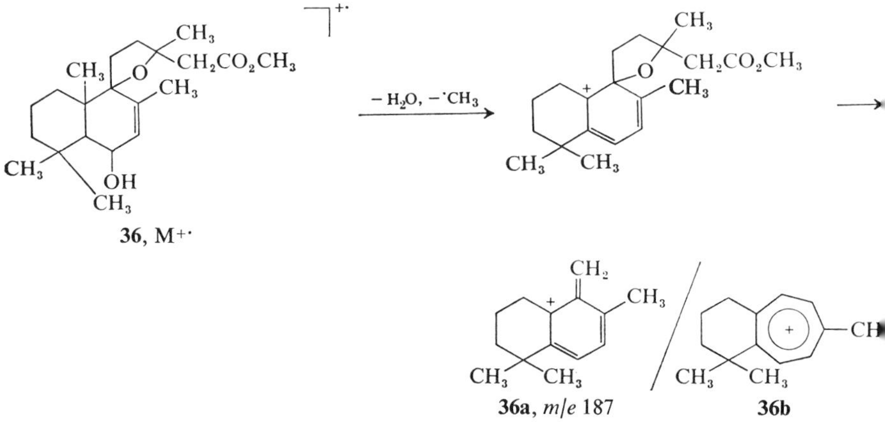

36, M+·

36a, m/e 187   36b

Mass spectrometry has proved very useful in the structure elucidation and the study of the biosynthesis of the C_{25}-terpenoid fungal metabolite ophiobolin A (**37**) (39, 40) (also referred to as cochliobolin, see Ref. 41 for suggested nomenclature). The distinction between the two possible structures **37** and **38** could readily be made from the fact that the base peak in the spectrum of ophiobolin A (cochliobolin) was at m/e 165 and at m/e 167 in its derivatives which have an isopropyl rather than an isopropylidene group. The origin of the m/e 165 ($C_{11}H_{17}O$) peak may readily be accounted for with structure **37** by α-cleavage of the $C_{(10)}-C_{(14)}$ bond to **37a**, followed by transfer of a hydrogen atom from $C_{(13)}$ to $C_{(10)}$ and then fragmentation to give the conjugated oxonium ion **37b** (m/e 165). It cannot easily be explained by structure **38**.

37, $M^{+\cdot}$

38

37a   37b, m/e 165

The spectrum of the complex bistetrahydrofuran derivative **39**, produced by acid-catalyzed isomerization of farnesol diepoxide, has been rationalized by the sequence shown (42). An extremely weak (0.04%) parent ion is observed and α-cleavage is the predominant fragmentation process, leading directly to fragments **39a, b, c**, and **d**. The base peak ion **39d** further fragments in a variety of ways giving **39e, f**, and **g**. The intense peak at m/e 43 may be assigned to the acetyl ion **39h** which may be formed from **39g** (and possibly from **39f** as well). The ions **39i** (base peak), **39a, 39h**, and **39e** are also observed as important ions in the spectrum of linalöol oxide (**40**) (43).

The spectra of a number of triterpenes containing a tetrahydrofuran ring as part of the structure have been reported and it appears that fragmentation is largely determined by the triterpenoid framework. For example, in the spectrum of **41** (44) the base peak **41a** results from the retro-Diels-Alder fragmentation typical of Δ^{12}-unsaturated pentacyclic triterpenes (45). The effect of a tetrahydrofuran ring may be more pronounced if it is suitably sited to trigger a favored fragmentation. Thus the spectrum of the steroid derivative **42** (46) shows a very weak parent ion (1.5%) and a base peak at M-15 most probably resulting from α-cleavage to produce the oxonium ion **42a**. (Part of the M-15 ion may possibly also result from loss of a methyl from the ethoxy group.)

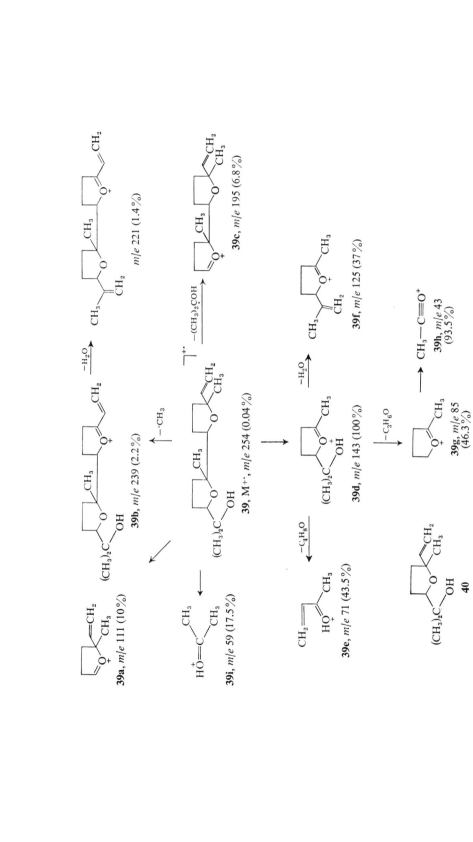

Mass spectrometry has been utilized in the investigation of holadysone (**43**) and the derived dione **44** (47, 48); the cleavages proposed to account for the major ions are shown.

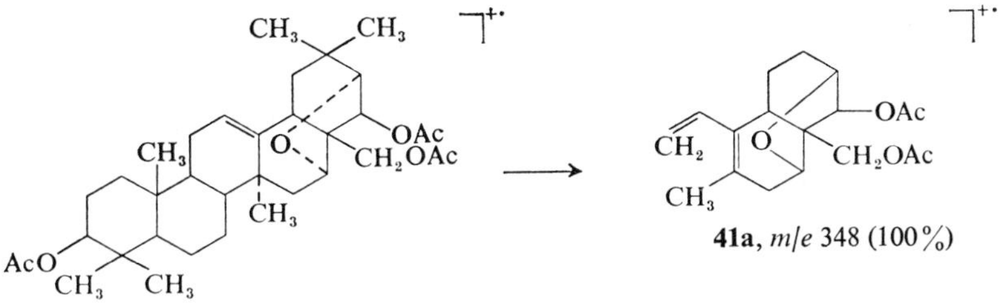

41

41a, m/e 348 (100%)

Further examples of the use of mass spectrometry in investigation of complex compounds containing a tetrahydrofuran ring may be found in the literature (49–54).

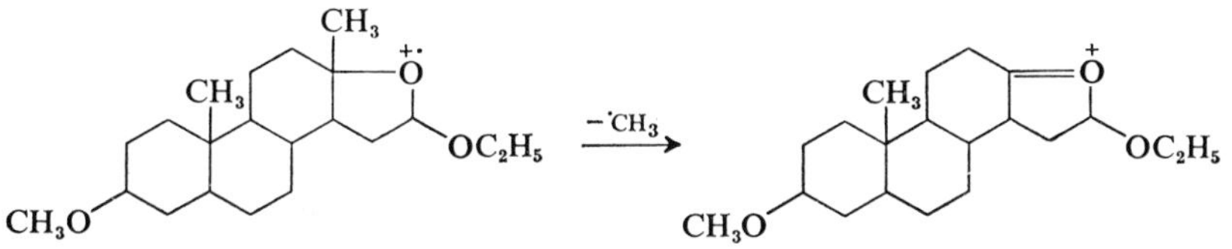

42

42a, M-15

The spectra of a number of carotenoid furanoid oxides, such as aurochrome (**45**), containing the 2,5-dihydrofuran ring system, have been examined (55). α-Cleavage is observed to a considerable extent, resulting in the conjugated

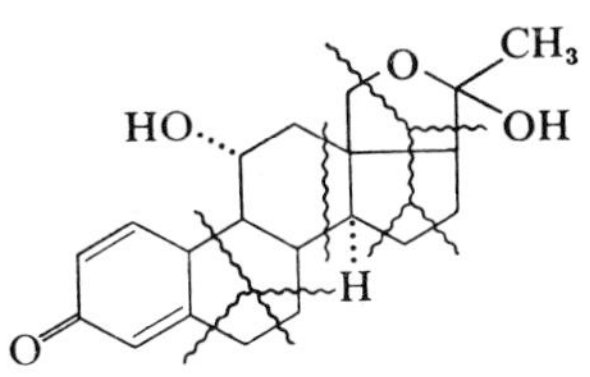

43

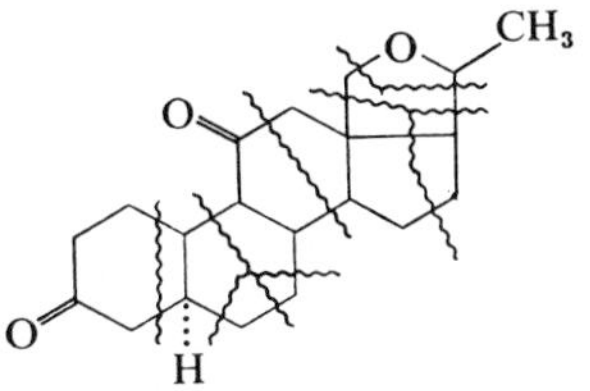

44

oxonium ion **45a**. The base peak is observed at m/e 205 ($C_{14}H_{21}O$ by high resolution) and has been rationalized by rearrangement of the parent ion to **45b** followed by α-cleavage to yield the homopyrilium-type ion **45c** (*m/e* 205). The expected mass shifts in **45a** and **45c** are observed in the spectrum of auroxanthin (**46**). It is important to note that both the cleavages **45** → **45a** and **45b** → **45c** result in breaking-off of a vinylic substituent, a

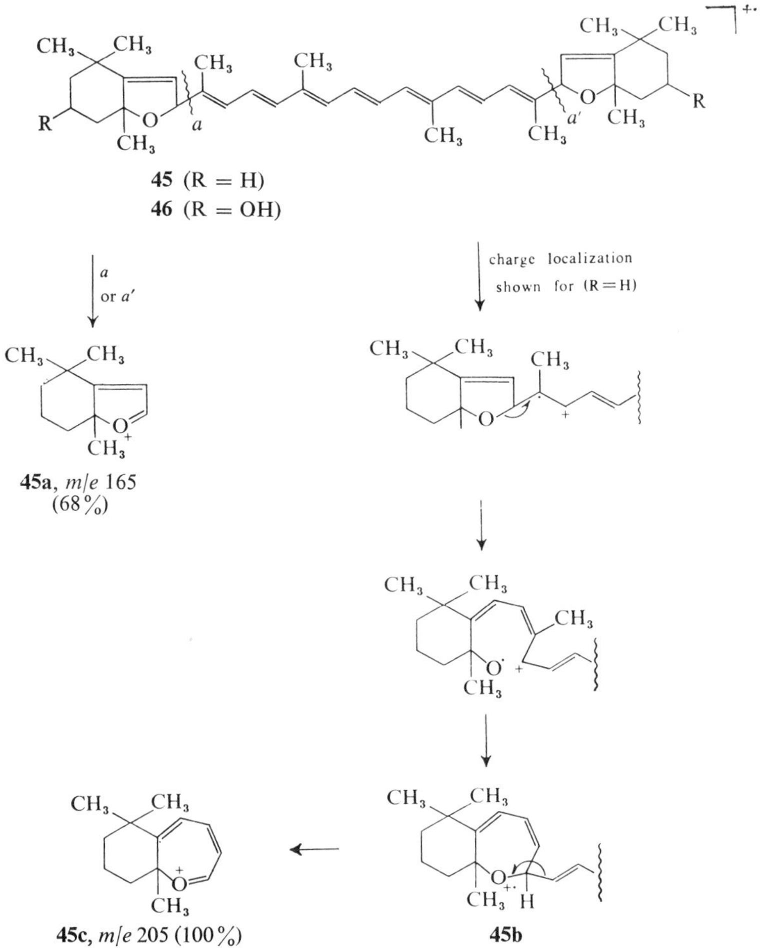

45 (R = H)
46 (R = OH)

45a, *m/e* 165
(68%)

charge localization
shown for (R=H)

45c, *m/e* 205 (100%)

45b

process which is normally unfavored, but here the vinylic radical may well be stabilized by cyclization onto the polyene chain.

IV. DIHYDROBENZOFURANS

The mass spectrum of dihydrobenzofuran (**47**) (Fig. 2.4) is characterized by an intense molecular ion (base peak), and the major fragmentation pathway

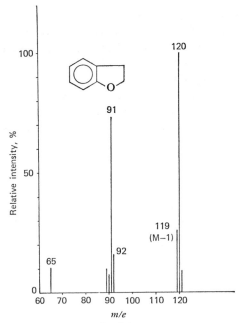

Fig. 2.4

involves loss of a hydrogen atom followed by expulsion of carbon monoxide and rearrangement to give an intense peak due to the tropylium ion **47c** (56).

Although no deuterium-labeling studies are available, both $C_{(2)}$ (**47** → **47a**) and $C_{(3)}$ (**47** → **47b**) are likely to be important sources of the lost hydrogen

47, M⁺˙, m/e 120   **47a**, m/e 119   **47b**, m/e 119

↓ → C_2H_4 ↓ − CO   − ˙CH₃

47d, m/e 92   **47c**, m/e 91   **48**, m/e 134 (100%)   **48a**, m/e 133 (42%)

↓ − CO

$C_5H_4^{+˙}$
47e,
m/e 64

atom in formation of the M-1 ion. Loss of a hydrogen atom from $C_{(3)}$ may be expected to be more favored than the corresponding loss in tetrahydrofuran (1) as now it involves the loss of a benzylic hydrogen and results in the resonance-stabilized ion **47b**. A minor fragmentation path is shown in the sequence **47** → **47d** → **47e**.

With alkyl substituents at $C_{(2)}$ α-cleavage becomes an important process. Thus in 2-methyldihydrobenzofuran (**48**) loss of the methyl group results in the cyclic oxonium ion **47a** which in this spectrum is equal in intensity to the parent ion. Loss of a hydrogen atom from $C_{(2)}$ or $C_{(3}$ occurs to a lesser extent (57).

49, M+·, *m/e* 270 (81%)

49b, *m/e* 121 (96%)

49c, *m/e* 148 (61%)

49a, *m/e* 149 (100%)

49d, *m/e* 268 (6%)

In dihydrobenzofurans with a 2-benzylic substituent such as **49**, α-cleavage is an important process, leading in this example to **49a** as the base peak (58). The ion **49a** fragments further by loss of CO to form **49b**, which may also arise directly from the parent ion. Aromatization of the dihydrofuran ring by loss of *p*-methoxytoluene to form the 6-methoxybenzofuran radical ion **49c** is also an important process. Aromatization by loss of hydrogen (**49** → **49d**) occurs to a lesser extent. A benzylic substituent, however, is not essential for aromatization to occur, and the spectrum of debromoaplysin (**50**) contains a considerable ion attributable to the benzofuran **50a** (59).

The spectra of a number of dihydrobenzofuran-3-ones have been examined (56, 60, 61). The parent compound **51** exhibits an intense molecular ion (base peak), and two distinct fragmentation paths are observed. The first involves α-cleavage with loss of a $C_{(2)}$ hydrogen atom to give **51a** which then expels CO to give an ion at m/e 105. This has been formulated as the benzoyl

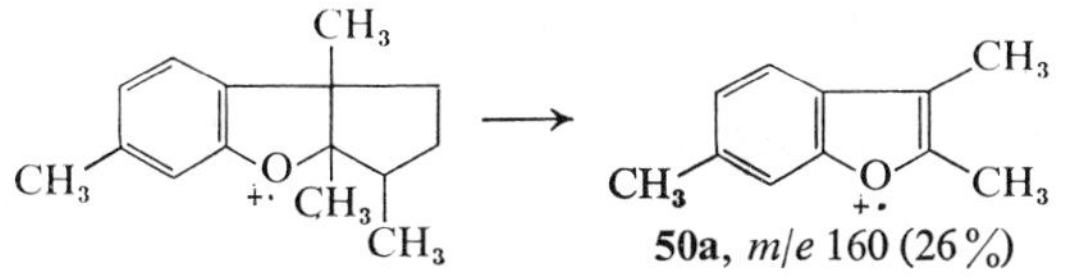

50, $M^{+\cdot}$ m/e 216 (34%) **50a**, m/e 160 (26%)

ion **51b** which in turn loses a molecule of CO to form the phenyl ion **51c**. The other fragmentation path involves loss of formaldehyde (**51 → 51d**) followed by loss of carbon monoxide to form the benzyne radical ion **51e**, which then loses

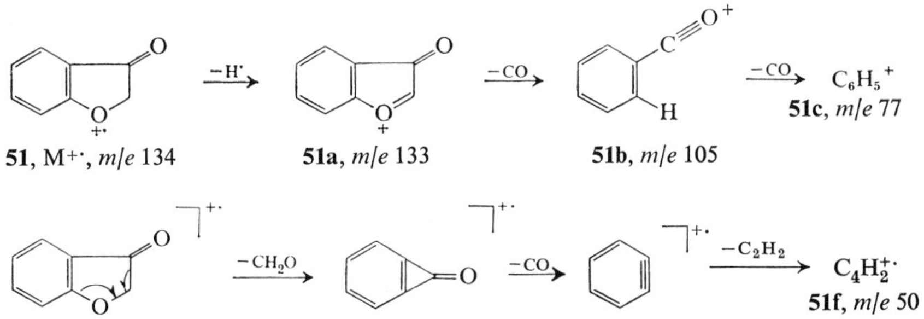

51, $M^{+\cdot}$, m/e 134   **51a**, m/e 133   **51b**, m/e 105   **51c**, m/e 77

51   **51d**, m/e 104   **51e**, m/e 76   **51f**, m/e 50

acetylene to give **51f**. A similar fragmentation pattern is observed for 2-methyldihydrobenzofuran-3-one, but in the spectrum of 4,6-dimethoxydihydrobenzofuran-3-one no significant ion corresponding to **51e** is observed, although ions corresponding to the sequence **51 → 51a → 51b → 51c** are prominent. In the spectra of the aurone derivatives **52** and **53** the predominant cleavage is α to the ring ether-oxygen.

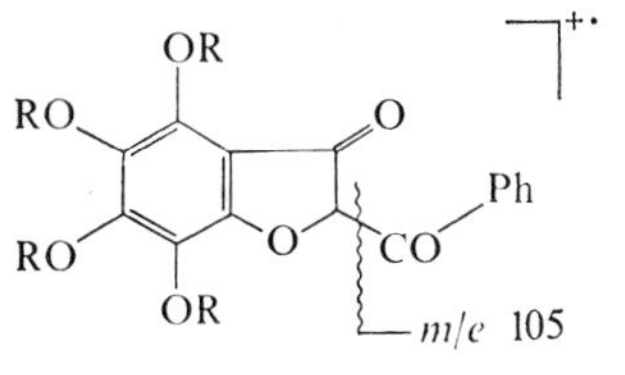

52

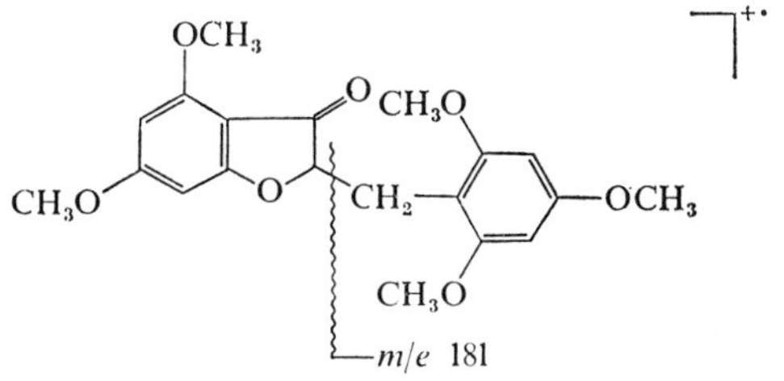

53

V. PHTHALANS (DIHYDROISOBENZOFURANS)

The mass spectra of a number of phthalan derivatives have been reported (62, 63). Curvulol (**54**) and *O,O*-dimethylcurvulol (**55**) show strong molecular ions, and cleavage α to the ring ether-oxygen atom (a very favored process, as the positive charge is stabilized by the oxygenated benzene ring) leads to **54a** and **55a** which dominate the spectra (63). Further slight loss of CO from **54a** and **55a** is the only other notable fragmentation observed.

54 (*R* = H)	**54a** (100%)	
55 (*R* = CH$_3$)	**55a** (100%)	

The retro-Diels-Alder reaction with a resultant loss of acetylene, although observed, is not the major fragmentation pathway for 1,4-dihydro-1,4-epoxynaphthalene (**56**) (56). The intense peak at *m/e* 116 (**56b**) is probably due to indene, which may be formed by rearrangement of the molecular ion **56** to **56c** by a 1,4 hydrogen-transfer followed by loss of CO. Further loss of a hydrogen atom from **56b** would lead to the indenyl ion (**56d**), which

56a, *m/e* 118 (18%)

56, M$^{+ \cdot}$ *m/e* 144 (63%)

56c

56e, *m/e* 128 (5.2%)

C$_9$H$_7$$^{+ \cdot}$
56d, *m/e* 115 (100%)

56b, *m/e* 116 (67%)

may also be formed by direct loss of CHO· from the parent ion; however, no metastable-ion data has been reported. Loss of oxygen occurs to a small extent, leading to the stable naphthalene radical ion **56e**.

Addenda

A further study of the mass spectrum of tetrahydrofuran (Section I) supports the conclusions already reached (64).

REFERENCES

1. American Petroleum Institute, Research Project 44. Spectrum 780.
2. J. H. Beynon in *Advances in Mass Spectrometry*, Vol. 1 (J. D. Waldron, Ed.), Pergamon, London, 1959, p. 348.
3. A. M. Duffield, H. Budzikiewicz, and C. Djerassi, *J. Amer. Chem. Soc.*, **87,** 2920 (1965).
4. E. J. Gallegos and R. W. Kiser, *J. Phys. Chem.*, **66,** 136 (1962).
5. Ref. 1. Spectrum 798.
6. J. Collin, *Bull. Soc. Chim. Belges*, **69,** 575 (1960).
7. H. Budzikiewicz, C. Djerassi, and D. H. Williams, *Mass Spectrometry of Organic Compounds*, Holden-Day, San Francisco, 1967, pp. 228–229.
8. K. Biemann, *Mass Spectrometry*, McGraw-Hill, New York, 1962, p. 98.
9. R. E. Wolff and M. Lenfant, *Bull. Soc. Chim. Fr.*, **1965,** 2470.
10. R. Tschesche, G. Biernoth, and G. Snatzke, *Justus Liebigs Ann. Chem.*, **674,** 196 (1967).
11. J. Shoji and S. Shibata, *Chem. Ind.* (London), **1964,** 419.
12. M. Ikeda, M. Suzuki, and C. Djerassi, *Tetrahedron Lett.*, **1967,** 3745.
13. G. Ohloff, K. H. Schulte-Elte, and B. Willhalm, *Helv. Chim. Acta*, **49,** 2135 (1966).
14. D. J. Goldsmith, B. C. Clark, and R. C. Jaines, *Tetrahedron Lett.*, **1966,** 1149.
15. D. J. Goldsmith, B. C. Clark, and R. C. Jaines, *Tetrahedron Lett.*, **1967,** 1211.
16. H. Strickler, G. Ohloff, and E. Sz. Kováts, *Tetrahedron Lett.*, **1964,** 649.
17. H. Strickler, G. Ohloff, and E. Sz. Kováts, *Helv. Chim. Acta*, **50,** 759 (1967).
18. S. C. Cascon, W. B. Mors, B. M. Tursch, R. T. Aplin, and L. J. Durham, *J. Amer. Chem. Soc.*, **87,** 5237 (1965).
19. N. C. Yang and Do-Minh Thap, *J. Org. Chem.*, **32,** 2462 (1967).
20. N. J. Turro and R. M. Southam, *Tetrahedron Lett.*, **1967,** 545.
21. M. A. Gianturco, P. Friedel, and A. S. Giammarino, *Tetrahedron*, **20,** 1763 (1964).
22. R. Viani, F. Müggler-Chavan, D. Reymond, and R. H. Egli, *Helv. Chim. Acta*, **48,** 1809 (1965).
23. N. Sugiyama, C. Kashima, Y. Hosoi, T. Ikeda, and R. Mohri, *Bull. Chem. Soc. Japan*, **39,** 2470 (1966).
24. M. A. Gianturco, P. Friedel, and V. Flanagan, *Tetrahedron Lett.*, **1965,** 1847.
25. M. A. Gianturco and P. Friedel, *Can. J. Chem.*, **44,** 1083 (1966).
26. T. Sakan, S. Isoe, and S. B. Hyeon, *Tetrahedron Lett.*, **1967,** 1623.
27. L. A. Paquette and J. H. Barrett, *J. Amer. Chem. Soc.*, **88,** 1718 (1966).
28. P. Deslongchamps and J. Kallos, *Can. J. Chem.*, **45,** 2235 (1967).
29. P. Hodge, J. A. Edwards, and J. H. Fried, *Tetrahedron Lett.*, **1966,** 5175.
30. J. E. Fraz, M. Dietrich, and A. Henshall, *Chem. Ind.* (London), **1966,** 1177.
31. R. Shapiro, J. Hachmann, and R. Wahl, *J. Org. Chem.*, **31,** 2710 (1966).
32. J. O. Rodin, C. M. Himel, R. M. Silverstein, R. W. Leeper, and W. A. Gartner, *J. Food Sci.*, **30,** 280 (1965).
33. A. Pelter, A. P. Stainton, and M. Barber, *J. Heterocycl. Chem.*, **3,** 191 (1965).
34. A. J. Birch, P. L. Macdonald, and A. Pelter, *J. Chem. Soc.*, C, **1967,** 1968.
35. A. M. Duffield, *J. Heterocycl. Chem.*, **4,** 16 (1967).
36. C. R. Enzell and R. Ryhage, *Ark. Kemi*, **23,** 367 (1965).
37. T. Bruun, L. M. Jackman, and E. Stenhagen, *Acta Chem. Scand.*, **16,** 1675 (1962).
38. H. E. Audier, S. Bory, M. Fétizon, and N.-T. Anh, *Bull. Soc. Chim. Fr.*, **1966,** 4002.
39. L. Canonica, A. Fiecchi, M. Galli Kienle, and A. Scala, *Tetrahedron Lett.*, **1966,** 1211.
40. S. Nozoe, K. Hirai, and K. Tsuda, *Tetrahedron Lett.*, **1966,** 2211.

41. K. Tsuda, S. Nozoe, M. Morisaki, K. Hirai, A. Itai, S. Okuda, L. Canonica, A. Fiecchi, M. Galli Kienle, and A. Scala, *Tetrahedron Lett.*, **1966**, 3369.
42. G. Ohloff, K.-H. Schulte-Elte, and B. Willhalm, *Helv. Chim. Acta*, **47**, 602 (1964).
43. B. Willhalm, A. F. Thomas, and M. Stoll, *Acta Chem. Scand.*, **18**, 1573 (1964).
44. R. Tschesche and G. Wulff, *Tetrahedron Lett.*, **1965**, 1569.
45. H. Budzikiewicz, J. M. Wilson, and C. Djerassi, *J. Amer. Chem. Soc.*, **85**, 3688 (1963).
46. G. Quinkert, G. Cimbollek, and G. Buhr, *Tetrahedron Lett.*, **1966**, 4573.
47. R. Tschesche, I. Mörner, and G. Snatzke, *Justus Liebigs Ann. Chem.*, **670**, 103 (1963).
48. R. Tschesche, *Bull. Soc. Chim. Fr.*, **1965**, 1219.
49. E. Fujita, T. Fujita, and M. Shibuya, *Tetrahedron Lett.*, **1966**, 3153.
50. M. Anteunis, M. Bracke, F. Alderweireldt, and M. Verzele, *Bull. Soc. Chim. Belges*, **73**, 910 (1964).
51. C. Giannotti, B. C. Das, and E. Lederer, *Bull. Soc. Chim. Fr.*, **1966**, 3299.
52. I. Yosioka, T. Nishimura, A. Matsuda, and I. Kitagawa, *Tetrahedron Lett.*, **1966**, 5979.
53. R. O. Dórchaí and J. B. Thomson, *Tetrahedron Lett.*, **1965**, 2223.
54. M. H. A. Elgamal, M. B. E. Fayez, and G. Snatzke, *Tetrahedron*, **21**, 2109 (1965).
55. J. Baldas, Q. N. Porter, L. Cholnoky, J. Szabolcs, and B. C. L. Weedon, *Chem. Commun.*, **1966**, 852.
56. B. Willhalm, A. F. Thomas, and F. Gautschi, *Tetrahedron*, **20**, 1185 (1964).
57. G. Fráter and H. Schmid, *Helv. Chim. Acta*, **50**, 255 (1967).
58. J. G. Down and R. G. Cooke, Unpublished results.
59. S. Yamamura and Y. Hirata, *Tetrahedron*, **19**, 1485 (1963).
60. K. Heyns, H.-F. Grützmacher, W. Mayer, F. Merger, and G. Frank, *Justus Liebigs Ann. Chem.*, **675**, 134 (1964).
61. R. Hänsel, H. Rimpler, and R. Schwarz, *Tetrahedron Lett.*, **1965**, 1545.
62. L. Pavlova and A. A. Polyyakova, *Zh. Org. Khim.* **3**, 601 (1967).
63. A. Ali Qureshi, R. W. Rickards, and A. Kamal, *Tetrahedron*, **23**, 3801 (1967).
64. R. Smakman and Th. J. De Boer, *Org. Mass Spec.*, **1**, 403 (1968).

3 REDUCED PYRANS

I. TETRAHYDROPYRANS

A. Tetrahydropyran

The mass spectrum of tetrahydropyran (**1**) is shown in Fig. 3.1 (1). High-resolution data for the major ions (2), together with the spectra of the $2,2,6,6-d_4$-, $3,3,5,5-d_4$-, and $4,4-d_2$-derivatives (**3**, **4**), have led to the following rationale for the observed spectrum. The fragmentation is similar to that observed for piperidine (Chap. 11, Sec. I). The labeling results also show that

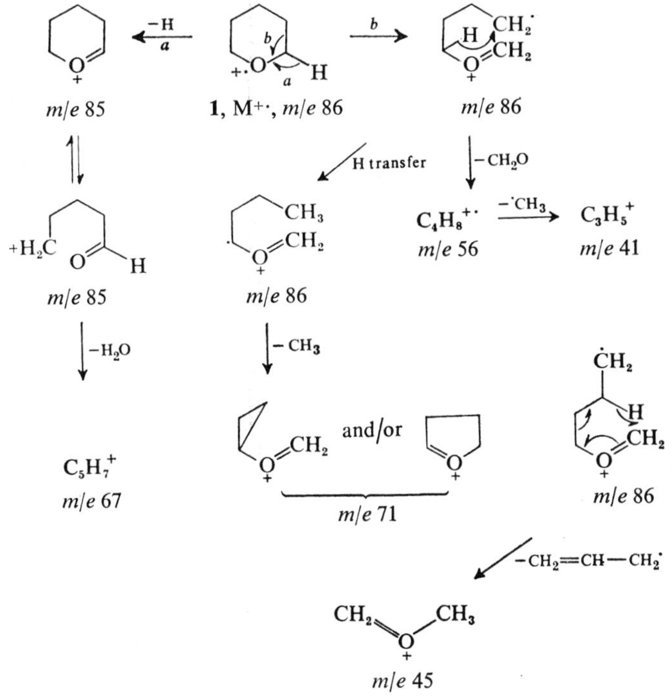

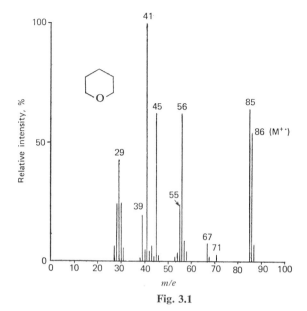

Fig. 3.1

some 10% of the m/e 85 ion results from the loss of hydrogen atoms from the 3- and 4-positions, whereas the loss of water in the formation of m/e 67 from m/e 85 involves comparable amounts of α-, β-, and γ-hydrogen atoms. Presumably the oxygen atom abstracts hydrogen at random after the initial fission of the CH_2—O bond.

B.   2-Substituted Tetrahydropyrans with a $C_{(2)}$—C Link

As was observed for the corresponding tetrahydrofuran derivatives, α-cleavage with resulting loss of a 2-substituent and formation of a cyclic oxonium ion is the predominant fragmentation process in the spectra of 2-substituted tetrahydropyrans. For example, the spectrum of **2** exhibits a weak parent ion, and α-cleavage (**2 → 2a**) results in loss of the allylic side chain to give the cyclic oxonium ion **2a** as the base peak at m/e 99, accounting for 22.3% of the total ion current (5). Similarly, in 2-(2,4,4-trimethyltetra-hydropyranyl)ethanol (**3**) loss of the larger 2-substituent results in the base peak **3a** (6). In the spectrum of **4**, a double-bond isomer of **2** found to occur naturally in Bulgarian rose oil (7), only a very weak peak is observed at m/e 99 as now α-cleavage would result in the unfavored loss of a vinylic radical. The base peak is now due to the loss of a methyl group from the parent ion. (Only a weak M-$\cdot CH_3$ peak is observed for **2**.) This loss may be explained by formation of the bicyclic oxonium ion **4a**, a fragmentation process which may become important if α-cleavage is inhibited. In the absence of deuterium-labeling, however, the above suggestion is only tentative, and it is possible that part of the M-·CH_3 ion may originate from the ring

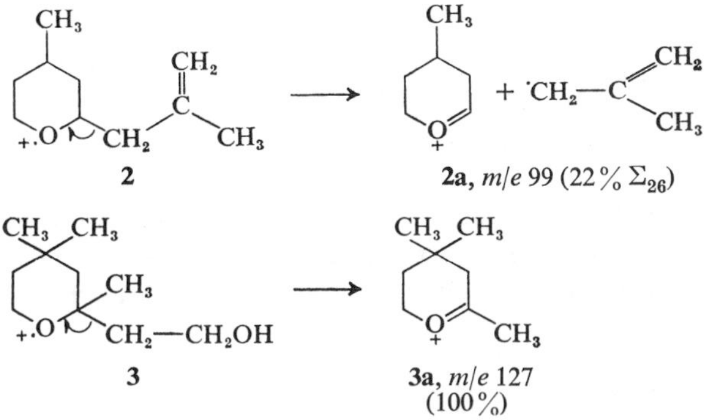

2   **2a**, m/e 99 (22 % Σ_{26})

3   **3a**, m/e 127
                                              (100 %)

methylene groups. A 2-carboxy group may be readily lost, as shown by the prominent m/e 99 ion in the spectrum of 4-methyltetrahydropyran-2-carboxylic acid (**5**) (7).

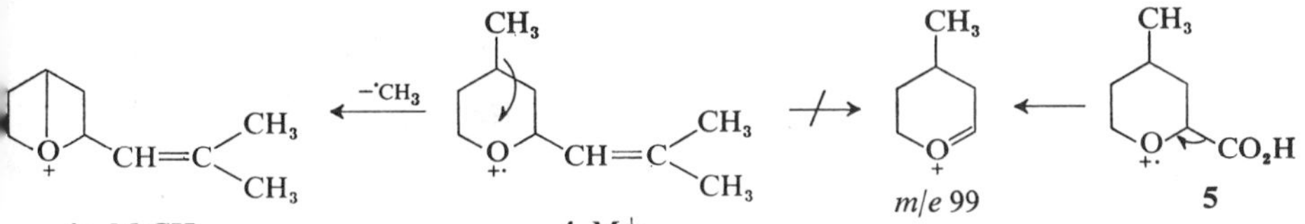

4a, M-CH_3 **4**, M$^{+·}$ m/e 99   **5**

Apparently no studies have yet been made of 3- or 4-alkly-substituted tetrahydropyrans lacking 2-substituents. It would be of interest to see to what extent the spectra of isomeric 2-, 3-, and 4-alkyl-substituted tetrahydropyrans differ, as useful correlations between structure and fragmentation pattern might result.

Ready loss of water in the polysubstituted tetrahydropyran pederol (**6**) results in no molecular ion being observed; however, α-cleavage of the molecular ion occurs to give **6a** (8). α-Cleavage appears to be of little importance in perfluoro-2-propyltetrahydropyran (**7**), resulting

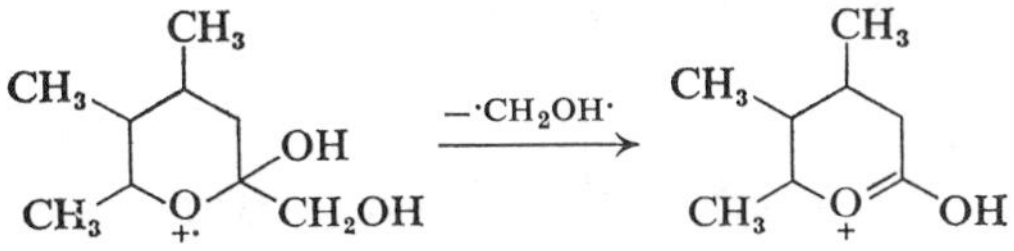

6, M$^{+·}$ (not observed)   **6a**, m/e 143

in only a weak $(M-\cdot C_3F_7)$ ion (**7a**). The parent ion is not observed and the spectrum is dominated by fluorocarbon ions (9).

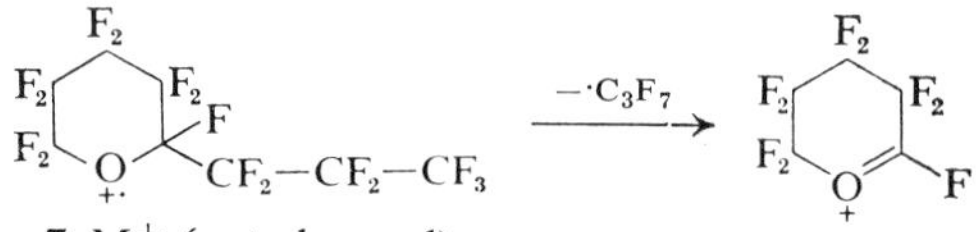

7, M$^{+\cdot}$ (not observed)   **7a** (1.4%)

The characteristic α-cleavage of 2-substituted tetrahydropyran derivatives has been used to advantage in the structure elucidation of panaxadiol (**8**) and the related panaxatriol (**9**) (10–13). The base peak (m/e 127) in the

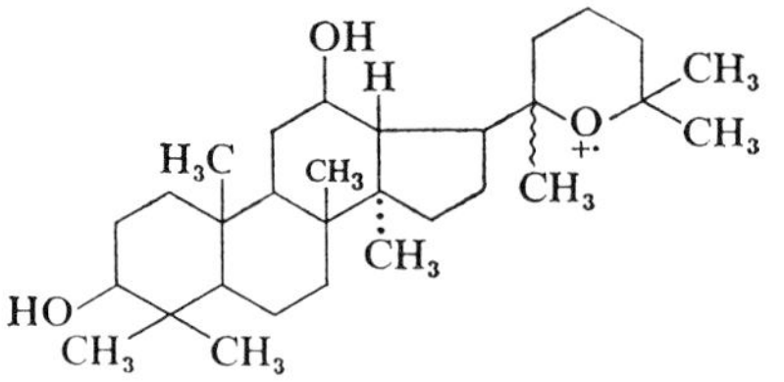

8, M$^{+\cdot}$

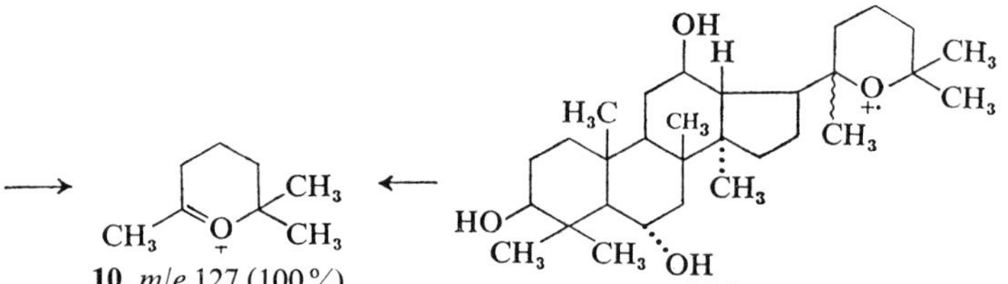

10, m/e 127 (100%)   **9**, M$^{+\cdot}$

spectrum of panaxadiol is due to the α-cleavage ion **10**, thus confirming the presence of a trimethyltetrahydropyranyl ring. Panaxatriol (**9**) differs from panaxadiol (**8**) in the presence of an extra hydroxyl group, but the base peak is also at m/e 127, showing that the third hydroxyl group is not located on the tetrahydropyranyl ring.

Mass spectrometry played an important role in the structure elucidation of 5,9-epoxyhexacosanoic acid and some other degradation products of the macrolide antibiotic pimaricin (14). The methyl ester (**11**) of 5,9-epoxy-hexacosanoic acid shows two prominent α-cleavage ions **11a** and **11b**, confirming the position of the ether link.

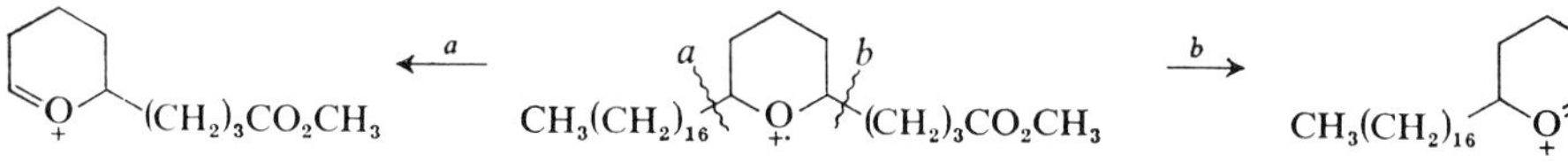

11a, m/e 185   **11**, M$^{+\cdot}$ m/e 424   **11b**, m/e 323

Similarly the structure of 5,9-epoxynonacosanoic acid, a degradation product of lucensomycin, has been determined largely on the basis of the mass spectrum of its methyl ester (**12**) (15). The base peaks in the spectra of **11** and **12** appear at m/e 144 and have been explained by hydrogen transfer and

11, R $= (CH_2)_{16}CH_3$
12, R $= (CH_2)_{19}CH_3$

11c

11d, m/e 144 (100%)

ring opening to form the ketone **11c** which then undergoes the McLafferty rearrangement to give **11d** (m/e 144).

C. 2- and 3-Oxygenated Tetrahydropyrans

The mass spectrum of 2-acetoxytetrahydropyran (**13**) is unexceptional. The base peak is due to the acetyl ion (shown by deuterium labeling to arise from the acetoxy group), loss of the acetoxy radical by α-cleavage gives **13a**, and loss of acetic acid leads to the dihydropyran radical ion (**13b**) whose further fragmentation accounts for most of the other major peaks in the spectrum (16).

13a, m/e 85   **13**, M+·   **13b**, m/e 84
(11%)   (10%)

The acetyl ion is also the base peak in the spectrum of 3-acetoxytetra-hydropyran (**14**). Loss of acetic acid leads to an abundant dihydropyran radical ion (**14a**) (17). An important ion, hardly observed (0.2% of the base peak) in the 2-acetoxy isomer but characteristically associated with 2-substituted tetrahydrofurans, is observed at m/e 71 (**14b**) and is the result of an important rearrangement of 3-oxygenated tetrahydropyrans.

14a, m/e 84   **14**, M+·   **14b**, m/e 71
(31%)   (34%)

The spectra of *cis*- and *trans*-2,3-diacetoxytetrahydropyran (**15**) and the isomeric tetrahydrofuran-2-aldehyde diacetate (**16**) are extremely similar (except for minor variations in peak intensities). The base peak is due to the acetyl ion (*m/e* 43) and an ion at *m/e* 71 is observed as the second most intense peak in all three spectra (16, 18). The ring contraction **15 → 15a** has been postulated to account for the *m/e* 71 peak in the spectrum of **15** (18), but this mechanism has been questioned because the step **15a → 15b** involves the loss of a carbene species (CH₃COOCH:), a reaction hardly ever observed in mass spectrometry (19).

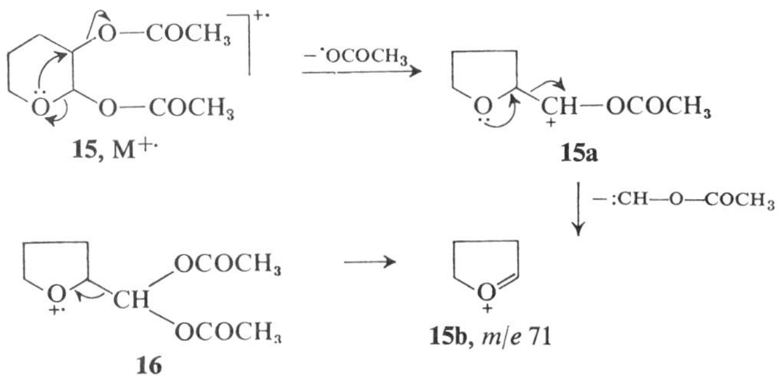

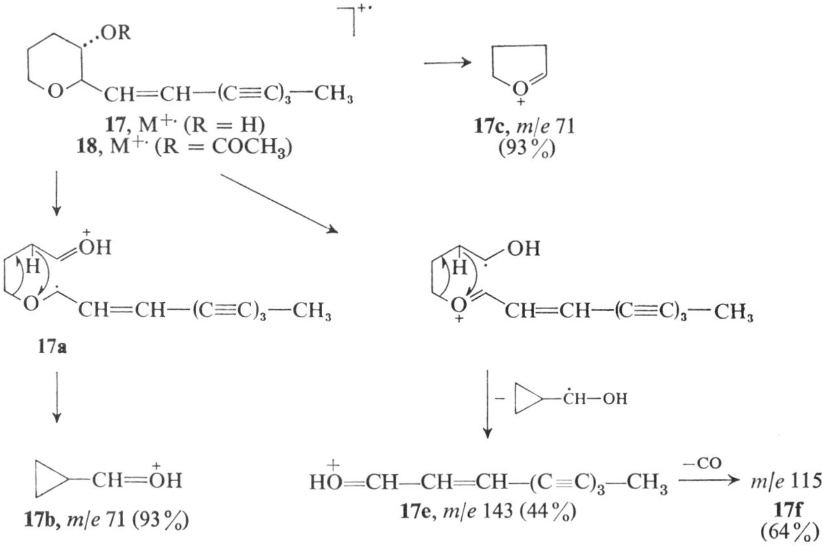

A situation similar in some respects exists in the spectra of the two fish poisons ichthyothereol (**17**) and its acetate (**18**) (20). In both cases a prominent *m/e* 71 ion is observed, and high resolution has shown its composition to be $C_4H_7O^+$. The origin of the *m/e* 71 ion in the spectrum of ichthyothereol has been explained by cleavage of the $C_{(2)}$—$C_{(3)}$ bond to give the ion **17a**, followed by transfer of a $C_{(4)}$ hydrogen atom to the ethereal oxygen with resultant fragmentation. The mechanism of type **15** → **15a** → **15b** is also considered as a possible route to m/e 71 (**17c**), especially for the acetate (**18**). Cleavage of the $C_{(2)}$—$C_{(3)}$ bond with formation of the oxonium ion **17d**, followed by hydrogen transfer and fragmentation, leads to **17e** which readily loses CO to give an ion **17f** of uncertain structure. Another important ion in the spectrum of **17** is at m/e 100 (mainly $C_5H_8O_2$) and has been assigned the structure **17g**. It is of interest to note that **17** loses only a trace of water, while loss of acetic acid results in the base peak of the acetate (**18**).

17   **17g**, m/e 100

A possible rationale has been suggested for the genesis of the m/e 71 ion in 3-hydroxytetrahydropyrans (**19**) (with or without 2-substituents) which does not require expulsion of a carbene (**19**). It is shown in the sequence **19** → **19a** → **19b**. Differentiation between this and a mechanism like that

19, M+·   **19**, M+·

19b, m/e 71

19a

suggested for formation of the m/e 71 ion in **17** (i.e., **19** → **19c**) could readily be achieved by deuterium labeling of the hydroxyl hydrogen, which would result in deuterium incorporation in **19c** but not in **19b**.

19   **19c**, m/e 71

The spectrum of the ketone **20** derived from ichthyothereol also shows a base peak at m/e 71 (20), and the spectrum of the deuterium-labeled

20b, m/e 99

20, M+· (R = H)
21, M+· (R = D)

20a, m/e 71 (R = H)
21a, m/e 74 (R = D)

ketone is consistent with a mechanism involving initial loss of CO (although an M-CO peak is not observed) followed by α-cleavage with loss of the *n*-decyl side chain to give **20a** and **21a**, respectively. Only a very weak peak is observed at m/e 99 for **20**, indicating that α-cleavage to give **20b**, although it would involve the loss of a comparatively stable *n*-decyl radical, is not an important reaction of the parent ion. In **22**, however, which contains a large 2-substituent at $C_{(6)}$, α-cleavage results in the base peak of the spectrum **22a** (21, 22). Another abundant ion in the spectrum of **20**, observed at m/e

22

22a, m/e 141 (100%)

100, is due to the McLafferty rearrangement (**20** → **20c**). 2-Tetrahydropyranyl

20, M+·

20c, m/e 100

ethers such as **23** behave normally. Its base peak is observed at m/e 85 (**23a**) and is due to α-cleavage (23). There is no evidence of a m/e 71 peak in the spectrum of **23**.

23, M+·

23a, m/e 85 (100%)

The spectra of a number of diterpenes containing a tetrahydropyran ring system, such as manoyl oxide (**24**), dihydromanoyl oxide (**25**), and related

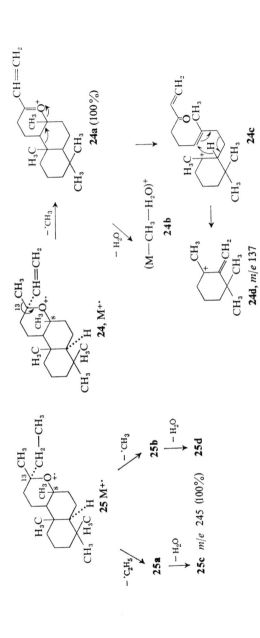

compounds, have been examined and show some interesting features (24–26). α-Cleavage initiated by the ether oxygen is an important process, and molecular ions are either very weak or not observed at all. The spectrum of **25** exhibits an abundant M-·C_2H_5 ion (**25a**) and a weaker M-·CH_3 ion which may be due to loss of either the methyl group at $C_{(8)}$ or that at $C_{(13)}$. In **24**, however, loss of the vinyl group at $C_{(13)}$ is very unfavored and hence the M-·CH_3 ion gives the base peak **24a**. A notable feature of the oxonium ions **25a**, **25b**, and **24a** is the large extent to which they lose a further molecule of water (substantiated by a metastable ion for **24a → 24b**). An abundant ion at m/e 137 in both spectra has been rationalized as arising from the initially formed oxonium ions by the sequence **24a → 24c → 24d** (m/e 137) shown for **24**. Another feature of the spectra is the concerted cyclic fragmentation of the tetrahydropyran ring leading to ions at m/e 192 and m/e 177. In general, when the ion due to the loss of the larger substituent at

24 (R = CH=CH$_2$)
25 (R = C$_2$H$_5$)

24e, m/e 192

24f, m/e 177

$C_{(13)}$ is weak (as with manoyl oxide), ions corresponding to **24e** and **24f** are intense; however, if the ion due to the loss of the larger $C_{(13)}$ substituent is intense, the ions produced by this fragmentation are weak.

An interesting example of the use of mass spectrometry in solving structural problems is provided by ketomanoyl oxide (**26**). Incorporation of four deuterium atoms into dihydroketomanoyl oxide on base-catalyzed deuterium exchange shows that only structure **26** with the carbonyl group at $C_{(2)}$ is possible for ketomanoyl oxide itself (27, 28).

26

The mass spectrum of the naturally occurring sugiresinol (**27**) has been studied in detail (29), and the fragmentation scheme illustrated (supported by spectra of deuterium-labeled derivatives and by observations of metastable ions) has been put forward to account for the major ions in the spectrum.

No ion corresponding to **27a** is observed; one might have been expected from the rearrangement of 3-hydroxytetrahydropyrans. Cyclic fragmentation of the tetrahydropyran ring is an important process which may occur by either a concerted or stepwise mechanism and which leads to **27b** as the base peak

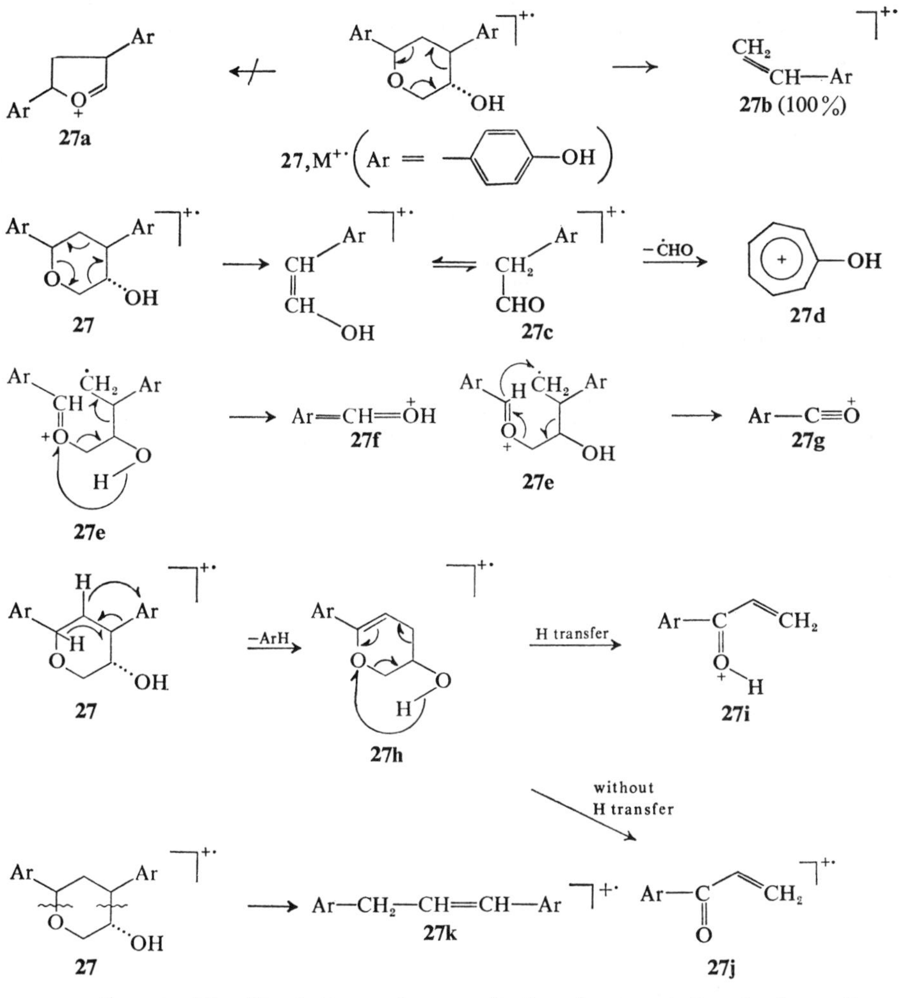

as well as to **27c**. The latter undergoes further fragmentation by loss of a formyl radical to the hydroxytropylium ion **27d** (also formed in part by other routes). Internal α-cleavage of **27** results in the ring-opened molecular ion **27e** which fragments with hydrogen transfer to give **27f** and **27g**. The loss of arene (**27 → 27h**) gives rise to the second-most intense ion in the spectrum, **27h**, which further fragments by the retro-Diels-Alder process, both with

(**27h → 27i**) and without (**27h → 27j**) transfer of the hydroxyl hydrogen. Another important ion corresponds to the cleavage of two benzylic bonds in **27** to give **27k**.

Mass spectrometry has proved valuable in structure elucidation of the insect metabolite pederone (**28**), where comparison of the spectra of pederone and pederin (**29**) provided an easy method of determining which of the two hydroxy groups in pederin remained unaltered in pederone (30).

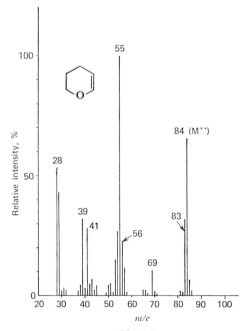

28 (R = O)
29 (R = OH(H))

II. DIHYDROPYRANS

The mass spectrum of dihydropyran (**30**) (Fig. 3.2) (16) shows an intense molecular ion and a relatively abundant M-1 ion which may arise by loss of a hydrogen atom from either $C_{(4)}$ or $C_{(2)}$ to give the conjugated oxonium

Fig. 3.2

ions **30a** and **30b**; no deuterium-labeling results are available to evaluate their relative importance. The retro-Diels-Alder reaction is observed but

30d, *m/e* 28   **30**, M⁺·,   **30c**, *m/e* 56
(90% C₂H₄)   *m/e* 84   (25% C₃H₄O)

C₄H₇⁺
30e, *m/e* 55   **30a**, *m/e* 83   **30b**, *m/e* 83

is not of great importance, only 25% of the *m/e* 56 ion being due to **30c**. The alkene fragment, ethylene, is more intense, but in the absence of deuterium labeling it cannot be said with certainty that **30d** is due exclusively to the retro-Diels-Alder process. The base peak is due to the hydrocarbon fragment C₄H₇⁺ (**30e**) (31), corresponding to loss of CHO· from the parent ion.

Polysubstitution of the dihydropyran ring favors the retro-Diels-Alder reaction and in 2,2,6-trimethyl-2,3-dihydropyran (**31**) the alkene fragment **31a** accounts for 10.3% of the total ion current (32, 33). However, formation of the carbonyl fragment **31b** is almost completely suppressed. The base

31a, *m/e* 56   **31**, M⁺·   **31b**, *m/e* 70

peak in the spectrum is at *m/e* 71, and a reasonable mechanism for its formation would be cleavage of the allylic C₍₃₎—C₍₄₎ bond with formation of the conjugated oxonium ion **31c**, followed by transfer of a β-hydrogen atom from one of the methyl groups at C₍₂₎ to the oxygen atom with simultaneous fragmentation (a reaction typical of oxonium ions derived from aliphatic ethers), to give **31d** and the stable allylic radical **31e**. There is no evidence to substantiate the origin of the transferred hydrogen atom, however.

The retro-Diels-Alder reaction is similarly very important in the spectrum of dihydroneroloxide (**32**) (34). Here the diene fragment **32a** gives the base

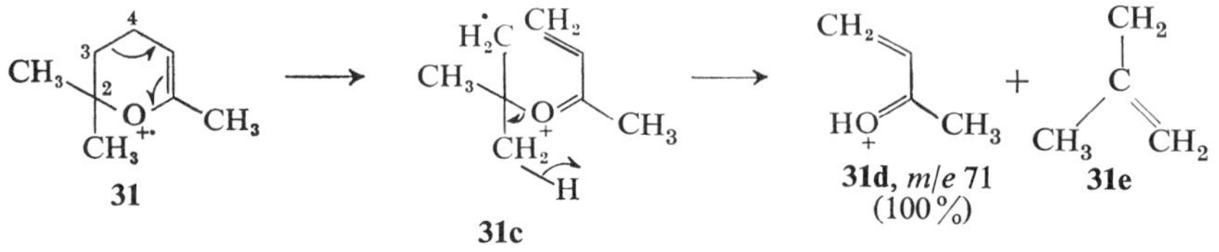

31   **31c**   **31d**, *m/e* 71   **31e**
                                                   (100%)

peak. α-Cleavage, which might be expected to be very favored in this case, results in the relatively weak peak **32b**. The other retro-Diels-Alder product **32c** readily loses a hydrogen atom to give the ion **32d** (*m/e* 85). Another

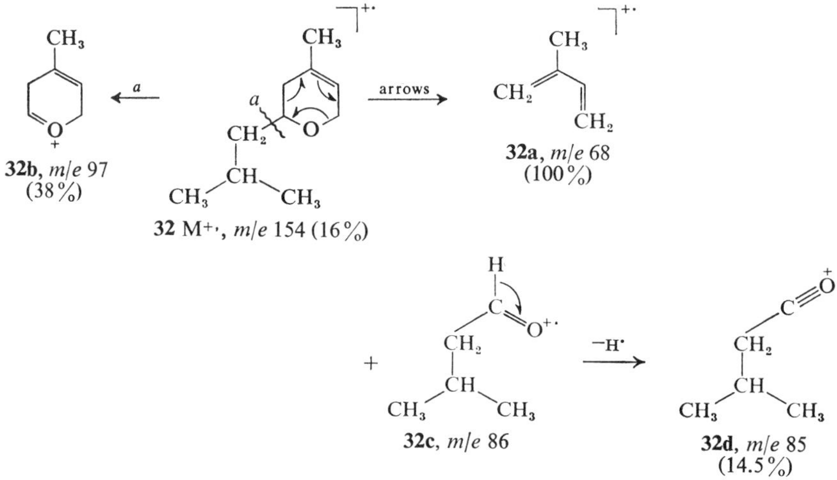

32b, *m/e* 97   **32a**, *m/e* 68
(38%)   (100%)

32 M+·, *m/e* 154 (16%)

32c, *m/e* 86   **32d**, *m/e* 85
                          (14.5%)

example is **33** where the charge is spread comparably over both the retro-Diels-Alder fragments (35).

The spectra of *cis-* and *trans-*2,5,5,1a-tetramethyl-1a,4a,5,6,7,8-hexa-hydro-γ-chromene (**34**) show prominent molecular ions (more intense in the

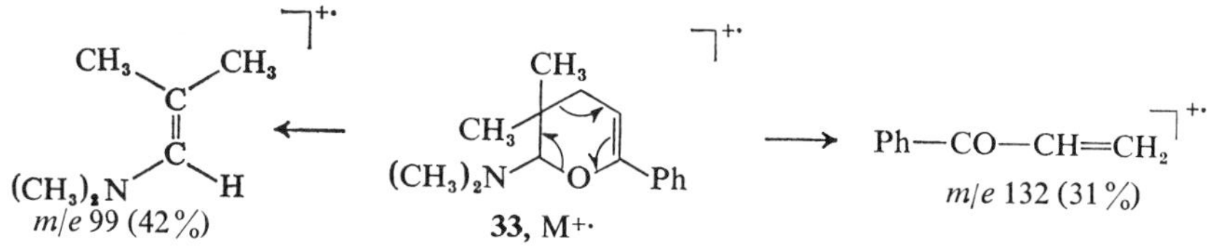

m/e 99 (42%)   **33**, M+·   *m/e* 132 (31%)

trans isomer) and peaks due to M-·CH$_3$ and M-·CH$_3$-H$_2$O (36). It has been suggested that the M-·CH$_3$ peak (**34a**) is due to loss of the C$_{(5)}$ methyl group, and this view is supported by the observation of an intense M-·C$_2$H$_5$ ion, but practically no M-·CH$_3$ peak, in the C$_{(5)}$-ethyl compound **35**. The retro-Diels-Alder reaction results in **34b** (which further fragments by loss of a

hydrogen atom, or by loss of a methyl group to give **34c**) and (with hydrogen transfer) in the ion **34d**, *m/e* 71 (cf. **31d**). The base peak in both the isomers **34** is at *m/e* 109 and deuterium labeling shows that it is predominantly due to **34c**. It has been suggested that **34c** arises from the M-15 ion **34a** since no shift of the *m/e* 109 ion is observed in the spectrum of **35** (36). (This, however, could be explained by a readier loss of the ethyl radical in **35a**; cf. the much readier loss of the C$_{(5)}$ ethyl rather than the C$_{(5)}$ methyl group in **35**.) The minor component of the *m/e* 109 peak has been shown to be due to the oxygen-containing fragment **34f**. This is probably formed by α-cleavage of the

C$_{(8)}$—C$_{(1a)}$ bond to give **34e**, followed by a hydrogen transfer and fragmentation to the 2,6-dimethylpyrilium ion **34f**. A notable feature in the spectra of **34** is the considerable ion due to loss of acetone which has been rationalized by the sequence **34** → **34g** (36). The spectra of the tricyclic

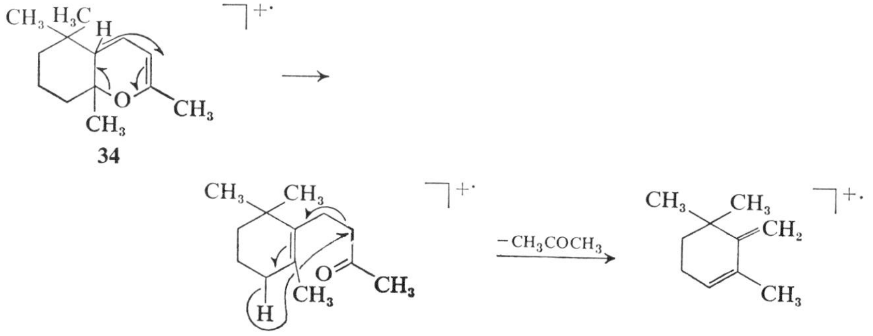

34

sclareol oxides (**36**) have also been discussed and show many features in common with **34** (26, 36).

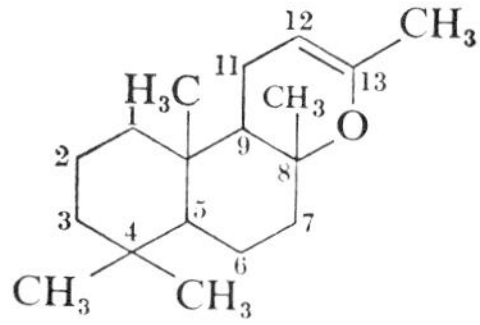

36

Mass spectrometry proved valuable in investigation of the structure of the crotonaldehyde- (**37**) and cinnamaldehyde-dimedone (**38**) condensation products (37). The initial loss of dimedone (**37** → **37a** and **38** → **38a**) and subsequent fragmentation by loss of a hydrogen atom or the ·R group can best be accounted for by the cyclic ether structures **37** and **38**.

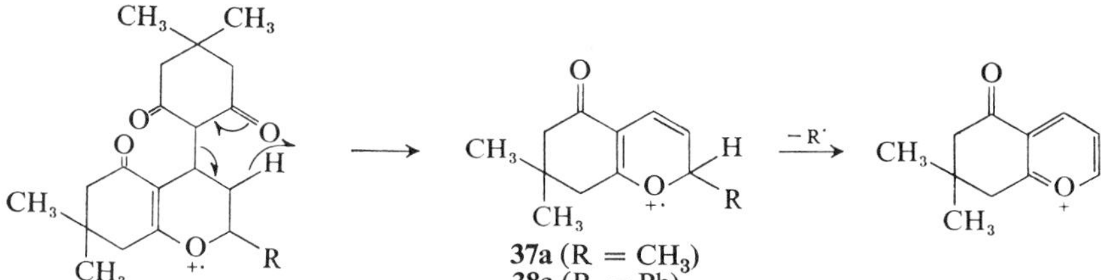

37, M⁺· (R = CH₃)
38, M⁺· (R = Ph)

37a (R = CH₃)
38a (R = Ph)

III. CHROMANS

Two important fragmentation paths are observed in the spectrum of 2,2-dimethylchroman (**39**) (32, 38, 39). Localization of the positive charge on the ether oxygen in the molecular ion results in α-cleavage with loss of a methyl group to form the cyclic oxonium ion **39a** which undergoes little further fragmentation. The retro-Diels-Alder reaction to give **39b** is insignificant.

39, M$^{+\cdot}$, m/e 162   **39a**, m/e 147 m/e 119

The base peak is one mass unit higher, at m/e 107, and the following two-step mechanism is suggested. Cleavage of the benzylic $C_{(3)}$—$C_{(4)}$ bond (β to the benzene ring) would result in the ring-opened molecular ion **39c**, which could then fragment by transfer of a β hydrogen atom to the oxygen atom and elimination of the stable isobutenyl radical to form the m/e 107 ion, which may be formulated as **39d** or the hydroxytropylium ion **39e**. There is as yet no deuterium-labeling evidence to determine the origin of the transferred hydrogen atom. The step **39c** → **39d** is analogous to the transfer of a

39b, m/e 106   **39**, M$^{+\cdot}$ **39c**

39e, m/e 107   **39d**, m/e 107 (100%)

β-hydrogen atom and loss of a molecule of alkene observed for the oxonium ions in the spectra of acyclic aliphatic ethers. The fragmentation sequence **39** → **39d** amounts to a retro-Diels-Alder reaction with hydrogen transfer and will henceforth be referred to as such.

In chroman itself (**40**) the retro-Diels-Alder reaction with hydrogen transfer (**40** → **40a**) is suppressed because it would result in the unfavored elimination of a vinyl radical (**40b**). The normal retro-Diels-Alder reaction, which results in loss of ethylene and formation of an ion at m/e 106 (formulated as **40c** or the troponium radical ion **40d**), is now favored. Loss of a hydrogen atom results in **40e** which further loses CO to form the indenyl ion **40f**. A notable feature is the considerable M-·CH₃ ion **40g**, and it has been suggested that this ion arises both from the parent ion directly and the

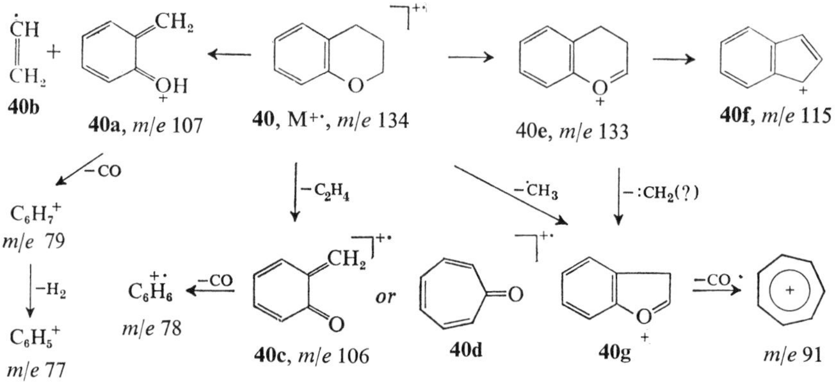

M-1 ion (**40e → 40g**), metastable ions being observed for both transitions (39). Later workers, however, have not been able to detect the metastable ion corresponding to **40e → 40g** (40), and since the very unfavored loss of methylene is involved it seems unlikely that a significant portion of **40g** arises from **40e**.

The most important fragment ion in the spectrum of 6-methoxy-chroman (**41**) is formed by loss of a methyl group, quite possibly from the methoxy group as the major source (39). This would result in the stable quinonoid oxonium ion **41a**.

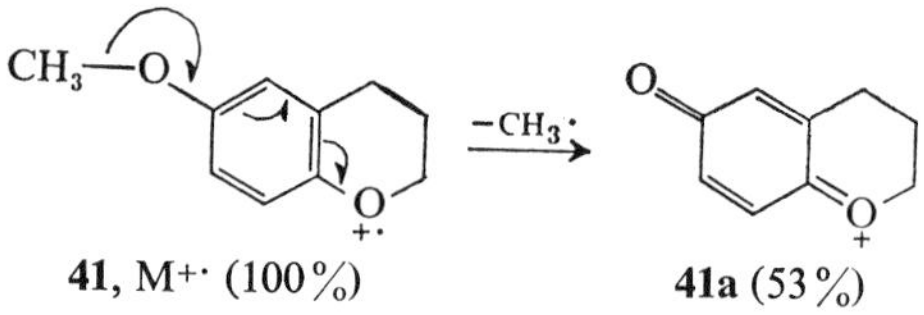

Dihydroevodionol (**42**) behaves similarly to 2,2-dimethylchroman, showing an intense M-·CH₃ ion (**42a**) and base peak **42b** resulting from the retro-Diels-Alder reaction with hydrogen transfer. Loss of the substituents on the benzene ring is of little importance (38).

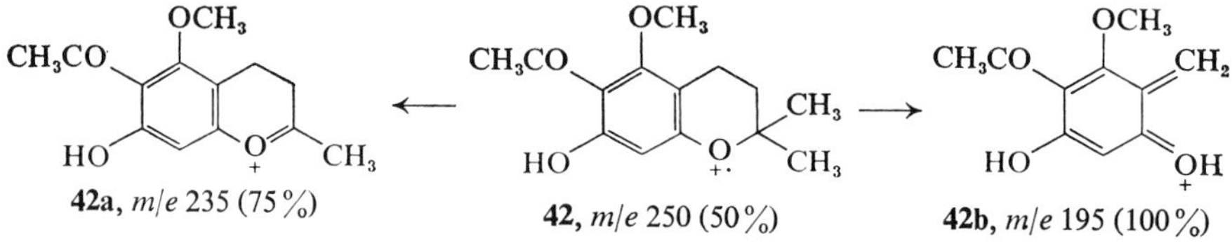

The spectra of a number of 4-oxygenated chromans (**43**) have been studied, and several distinct fragmentation paths are observed (32, 39). Loss of ROH results in the formation of chromen (**43a**), which further loses a hydrogen atom to form the intense chromenyl ion **43b**. The extent of loss of OR·

43, R = H
   R = C$_2$H$_5$
   R = COCH$_3$

43a, *m/e* 132

43b, *m/e* 131

43g, R = H (48%)
    R = C$_2$H$_5$ (9%)

43c, *m/e* 133

43d

or

−C$_2$H$_4$ (R = H, 18%)
(R = C$_2$H$_5$, 100%)
(R = COCH$_3$, 28%)

43e, *m/e* 105   =O (?) $\xrightarrow{-CO}$ C$_6$H$_5{}^+$

43f, *m/e* 77

depends on the nature of R, and leads to an ion at *m/e* 133 (**43c**) which is likely to have the tropylium ion structure **43d**. This then loses ethylene to form the troponium ion **43e** which further loses a molecule of CO to form the phenyl ion (**43f**). Loss of ethylene from the molecular ion is also observed, and appears to be most important in chroman-4-ol (**43,** R = H). The retro-Diels-Alder reaction becomes the predominant fragmentation process for chroman-4-ones; thus, for the parent compound **44** the ion due to loss of ethylene (**44a**) together with its further fragmentation product (by loss of CO) account for 56% of Σ_{51} (32). A similar extensive loss of ethylene is observed for 6-hydroxy- and 6-methoxychroman-4-one (39).

44, M$^{+\cdot}$, *m/e* 148

44a, *m/e* 120

44b, *m/e* 92

An interesting rearrangement shown to involve phenyl migration and resulting in a considerable ion through loss of a benzyl radical is observed in the spectrum of 2,2-diphenylchroman (**45**) (41). A plausible mechanism leading to **45a** has been suggested (42).

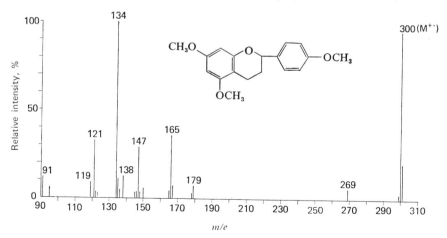

45, M$^+$·, m/e 286 (86%)

45a, m/e 195 (45%)

IV. FLAVANS AND ISOFLAVANS

The retro-Diels-Alder reaction is the major fragmentation pathway in the spectra of flavans (43–45). In the naturally occurring flavan **46** (Fig. 3.3) it leads to the styrene **46a** (ring C fragment) as the base peak (44) and to **46b** (ring A fragment) of a little over one-third the intensity. The further fragmentations of **46a** and **46b** are as shown. The charge distribution over the two retro-Diels-Alder fragments is very dependent on the substitution of rings A and C. Thus the ring A fragment is not observed if ring A is unsubstituted (as in flavan itself) or if ring C is polysubstituted. In **46**, where

Fig. 3.3

ring A is more highly substituted than ring C, the charge still remains predominantly with the ring C fragment. The other important fragmentation

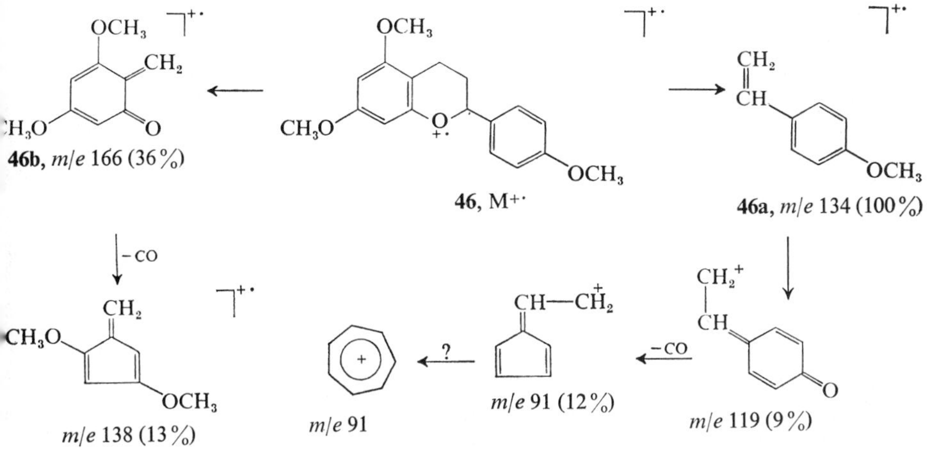

46b, m/e 166 (36%)

46, $M^{+\cdot}$

46a, m/e 134 (100%)

m/e 138 (13%)

m/e 91

m/e 91 (12%)

m/e 119 (9%)

pathway in flavans involves hydrogen transfer, possibly from $C_{(4)}$ to $C_{(2)}$. In **46** it leads to the ions **46c** and **46d**. A further ion of interest in the spectrum

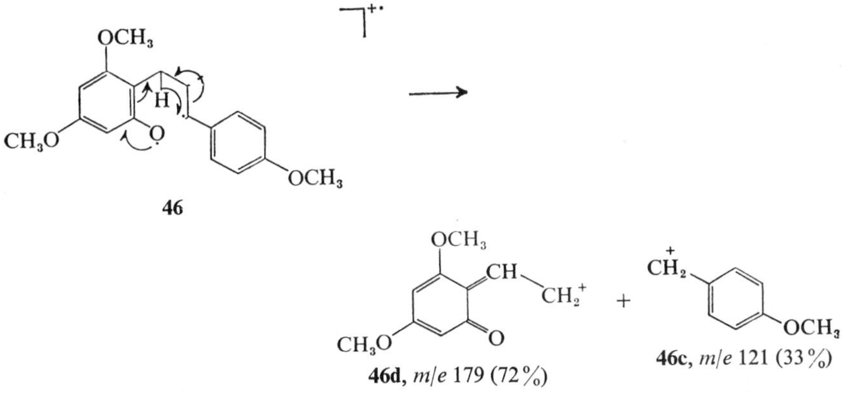

46

46d, m/e 179 (72%)

46c, m/e 121 (33%)

of **46** is observed at m/e 147 (**46e**). Two possible mechanisms have been put forward for its formation, but deuterium labeling is required to distinguish between them (43).

With 2′-hydroxyflavans an intense ion due to loss of water is observed, which may or may not be the base peak depending on the temperature of the probe (46). The mechanism of the water loss is uncertain, but it appears from deuterium-labeling studies that the 2′-hydroxy group is not directly involved.

The spectra of isoflavans resemble those of flavans in that the retro-Diels-Alder reaction is the major fragmentation process, and in 7-methoxyisoflavan

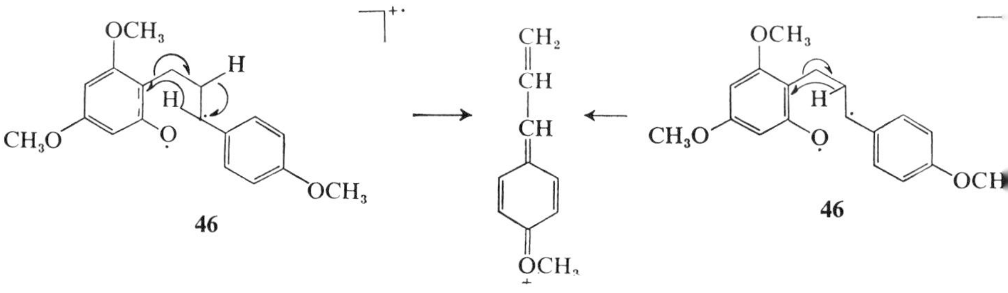

46e, *m/e* 147 (29%)

(47) it leads to the ring A fragment 47b as the base peak with the ring C fragment 47a about one-third the intensity (45). In 48, however, which has a methoxy group on the C ring as well, the ring C fragment is the base peak

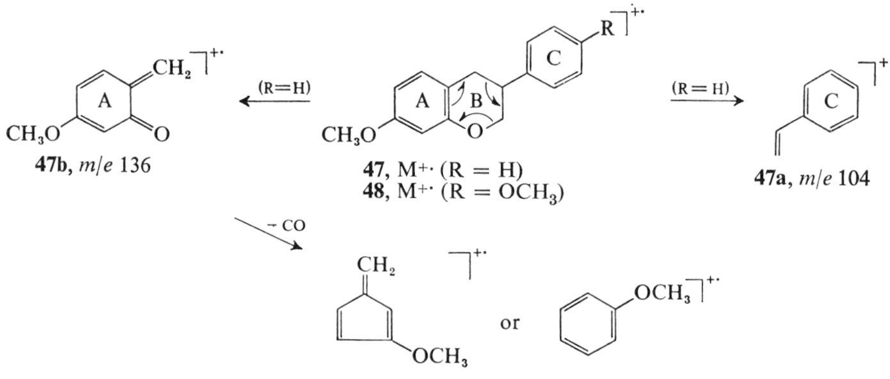

47b, *m/e* 136

47, M⁺· (R = H)
48, M⁺· (R = OCH₃)

47a, *m/e* 104

m/e 108

and no peak due to the ring A fragment is observed (45). Notable peaks in the spectrum of 47 are at *m/e* 91 and 149, and their origin may be rationalized by hydrogen migration from $C_{(2)}$ to $C_{(3)}$ with resultant ring contraction to form the 6-methoxy-2-benzyldihydrobenzofuran radical ion 47c, from which 47d and 47e may readily be derived by simple α-cleavage.

An alternative route to 47c may also be envisaged, involving phenyl migration from $C_{(3)}$ to $C_{(2)}$ and resultant ring contraction, but it is much less likely on general grounds. Suitable deuterium-labeling experiments, however, would distinguish between the two possibilities.

The retro-Diels-Alder reaction with hydrogen transfer, which is insignificant for **47** and **48**, becomes important in the isoflavan **49**. Charge retention

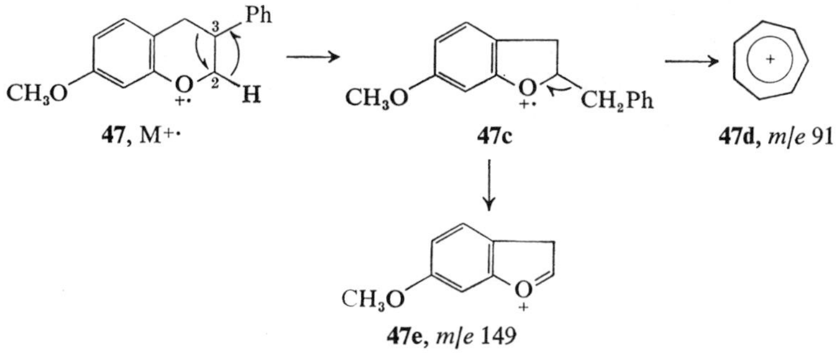

47, M+·   **47c**   **47d,** *m/e* 91

47e, *m/e* 149

on both the fragments **49a** and **49b** is observed (43). The ion **49b** may be stabilized by interaction with the adjacent hydroxyl group (**49c**). The suggested mechanism involves homolysis of the $C_{(3)}$—$C_{(4)}$ bond to form a stable dibenzylic diradical **49d**, followed by hydrogen transfer and fragmentation. The base peak **49e** results from the retro-Diels-Alder reaction with charge

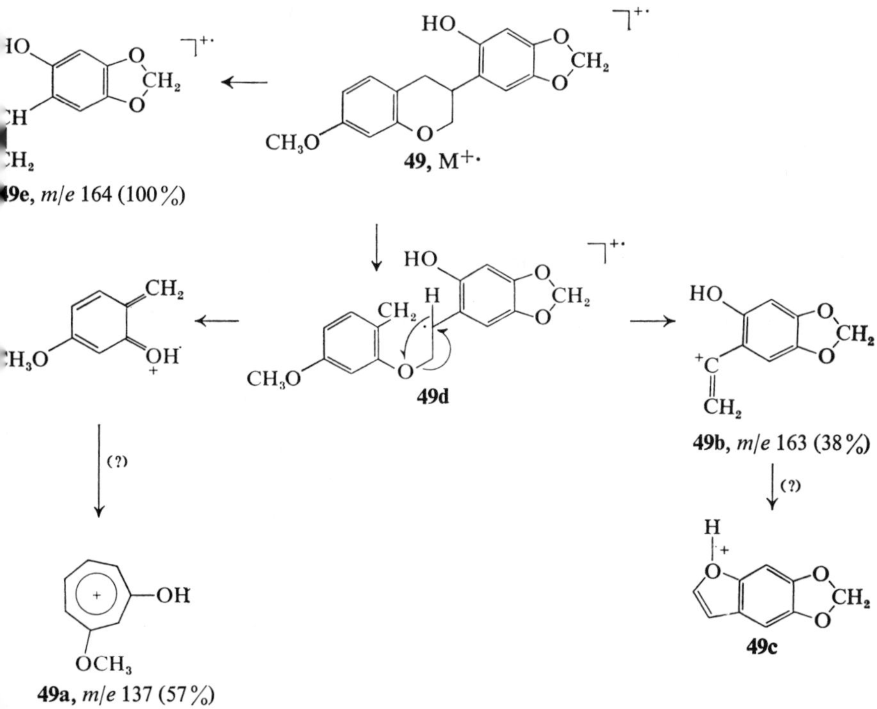

49e, *m/e* 164 (100%)   **49, M+·**

49d

49b, *m/e* 163 (38%)

49a, *m/e* 137 (57%)   **49c**

retention on the ring C fragment. It is noteworthy that no loss of water from the molecular ion is observed for **49** despite the 2′-hydroxy group. Such a loss is very important in the 2′-hydroxy-flavans.

The retro-Diels-Alder reaction is also the major fragmentation pathway for flavan-4-ols (43). In the parent compound **50** it leads to the ring C fragment as the base peak (**50a**) and the ring A fragment (**50b**) of one-half the intensity. With **51**, which has a methoxy group on ring C, the charge

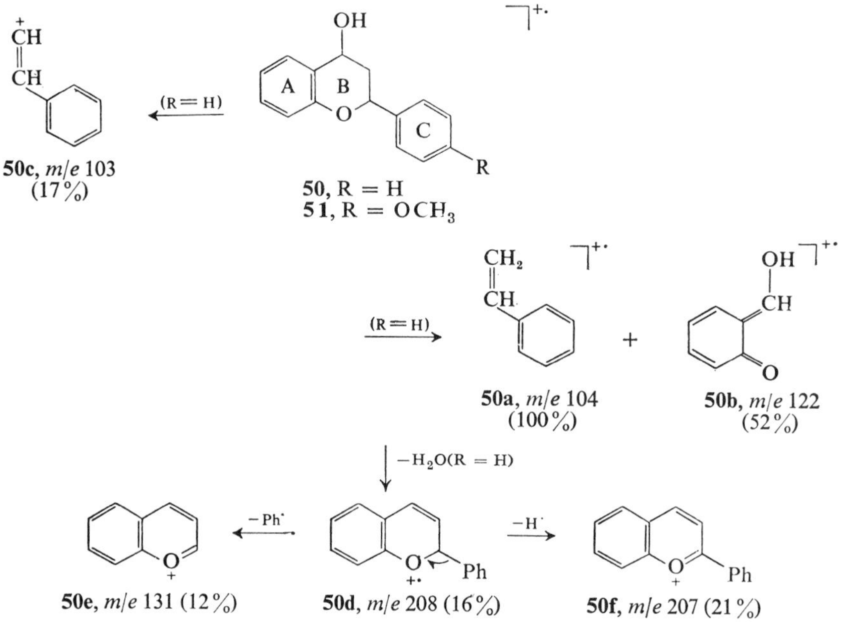

50c, m/e 103
(17%)

50, R = H
51, R = OCH$_3$

50a, m/e 104
(100%)

50b, m/e 122
(52%)

$-H_2O(R = H)$

50e, m/e 131 (12%)

50d, m/e 208 (16%)

50f, m/e 207 (21%)

is retained almost entirely on the ring C fragment, as with the corresponding flavans and isoflavans. The retro-Diels-Alder reaction with hydrogen transfer also occurs to a small extent, leading to the hydrogen-deficient ion **50c.**

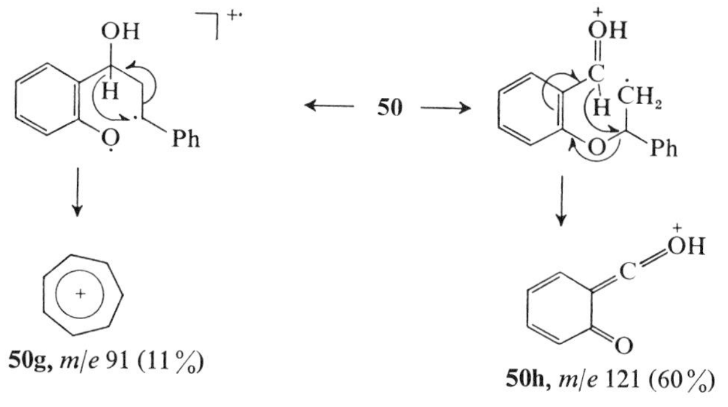

50g, m/e 91 (11%)

50h, m/e 121 (60%)

Loss of water leads to **50d**, which then loses either a phenyl radical or a hydrogen atom to form the stable chromenyl ions **50e** and **50f**. Hydrogen transference is also of importance, and two paths can be distinguished. The first (already observed with flavans) leads to the tropylium ion **50g**, and the other leads to the intense ion **50h**.

Mass spectrometry has proved valuable in the structure elucidation of theaflavin (**52**), a natural product isolated from black tea (47, 48). The mass

53c, m/e 191

spectra of derivatives of **52**, including the heptamethyl ether **53**, provided excellent support for the proposed structure. The ions **53a**, **53b**, and **53c** clearly indicate the presence of a chroman system; **53a** arises from the retro-Diels-Alder reaction without hydrogen transfer, **53b** results from hydrogen transfer, and **53c** is formed by dehydration to a chromen followed by α-cleavage. The successive loss of two neutral molecules of **53a** from the parent ion further showed that two chroman groups must be present.

V. FLAVANONES, ISOFLAVANONES, AND FLAVANONOLS

The mass spectrum of flavanone (**54**) is typical of the class; three distinct fragmentation paths are observed (38). The retro-Diels-Alder reaction leads to the ring A ion **54a** as the base peak, and a considerable ion due to the ring

C fragment is also observed. In 4'-methoxyflavanone (**55**), however, the charge resides almost entirely on the ring C ion **55b** (43). The retro-Diels-Alder

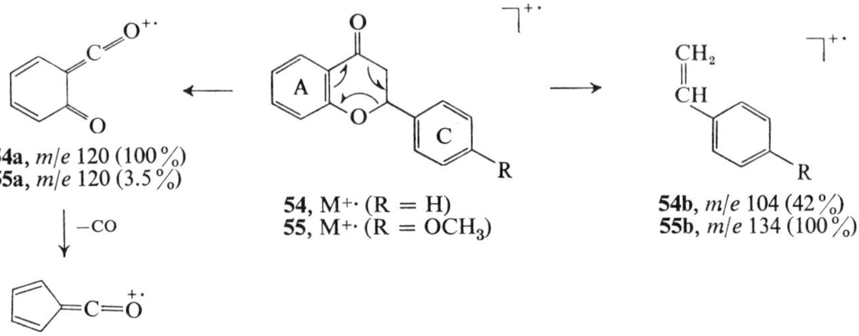

54a, m/e 120 (100%)
55a, m/e 120 (3.5%)

$\downarrow$ —CO

m/e 92

54, M$^{+\cdot}$ (R = H)
55, M$^{+\cdot}$ (R = OCH$_3$)

54b, m/e 104 (42%)
55b, m/e 134 (100%)

fragmentation with hydrogen transfer is also observed and leads to the ions **54c** and **54d**, and **55c** and **55d**. A characteristic fragmentation of flavanones

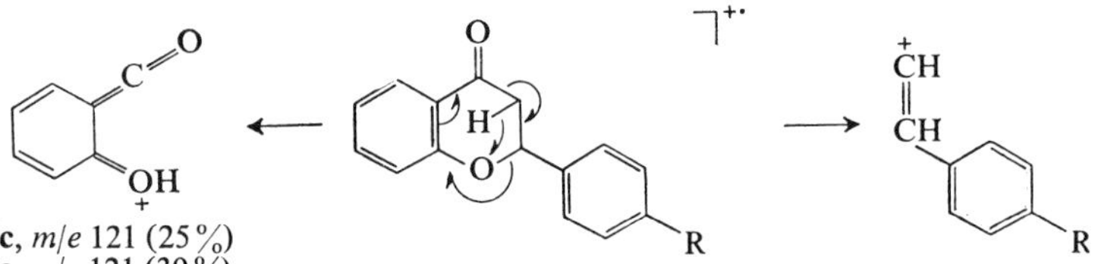

54c, m/e 121 (25%)
55c, m/e 121 (30%)

54, M$^{+\cdot}$ (R = H)
55, M$^{+\cdot}$ (R = OCH$_3$)

54d, m/e 103
55d, m/e 133 (3%)

is their favored cleavage α to the ring oxygen atom with loss of either the C ring (a) or a hydrogen atom (b) to form stable 4-hydroxychromenyl ions

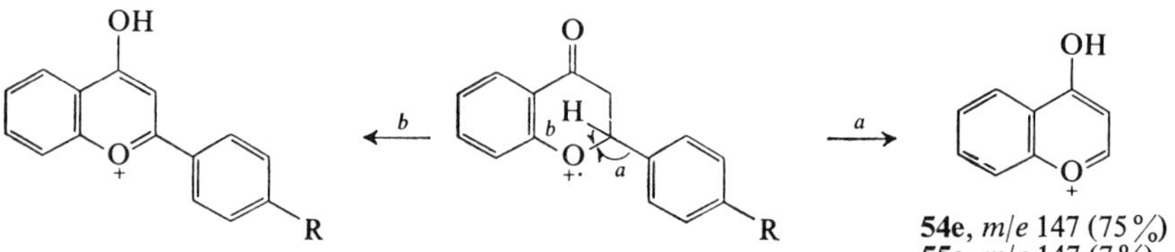

54f, m/e 223
55f, m/e 253 (30%)

54, M$^{+\cdot}$
55, M$^{+\cdot}$

54e, m/e 147 (75%)
55e, m/e 147 (7%)

(**54e** and **54f**, and **55e** and **55f**). An interesting ion in the spectrum of **55** resulting from hydrogen transfer and elimination of chromone (43) is due to anisole (**55g**).

Other simple flavanones follow the fragmentation pattern outlined for **54** and **55**. With 2'-hydroxyflavanone (**56**) and 2'-hydroxy-4',5,7-trimethoxy-flavanone, the normal fragmentation pattern is observed but the base

55g, m/e 108
(10%)

55

peak is at M-H$_2$O and the third most intense peak in the spectra is due to the further loss of a hydrogen atom (M-19). It has been suggested (46) that these ions arise from ring-opening of the molecular ion (shown for **56**) followed by ring closure on the 2'-hydroxy group to form **56a**, which may

56, M$^{+\cdot}$ m/e 240 (22%)

− H$_2$O

56b, m/e 222 (100%)

56a

56c

− H$^{\cdot}$

56d, m/e 221 (60%)

then readily lose water to form **56b** as the base peak. The ion **56b** cyclizes to form **56c** which then readily loses a hydrogen atom to form the stable chromenyl ion **56d**. It is important to note that the spectra of flavanones

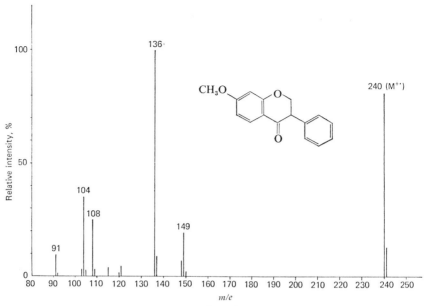

Fig. 3.4

and the corresponding 2′-hydroxychalcones are very similar, and it is probable that an equilibrium exists between the two ionized forms (45).

The spectra of isoflavanones are much simpler than those of flavanones. Thus the fragmentation of 7-methoxyisoflavanone (**57**) (see Fig. 3.4) originates mainly in the retro-Diels-Alder reaction and leads to ions **57a** and **57b**. The ions **57c** and **57d** are formed by further fragmentation of **57b** by loss of CO followed by loss of $CH_3\cdot$. The retro-Diels-Alder reaction with hydrogen transfer occurs only to a small extent (45). As is to be expected, no loss of the phenyl group from the parent ion is observed [contrast flavanone (**54**)]. This, along with the absence of an M-1 ion, may provide a basis for distinguishing between flavanones and isoflavanones by mass spectrometry.

The spectra of flavanonols differ from the flavanoids already studied inasmuch as other fragmentation processes compete very favorably with the retro-Diels-Alder reaction. Thus in the spectrum of dihydroquercetin (**58**) (Fig. 3.5) the base peak is due to the retro-Diels-Alder reaction with hydrogen transfer (**58a**) (45). The retro-Diels-Alder reaction leads to the ions **58b** and **58c** (which are isobaric in this case). A characteristic fragmentation of the ring C retro-Diels-Alder fragments **58c** and **59c** [from aromadendrine (**59**)] is the further loss of 29 mass units (CHO·). This is probably preceded by rearrangement to the aldehydes **58d** and **59d** and results in the ions **58e** and **59e**. Another characteristic of flavanonols is the loss of 29 mass units from the molecular ion (resulting in the abundant m/e 275 ion in the spectrum of **58**) (45).

In 3,5,7-trihydroxy-4′-methoxyflavan-4-one (**60**) the retro-Diels-Alder reaction with hydrogen transfer (**60** → **60a**) and without it (**60** → **60b** + **60c**) are both important processes (43). The ion **60c** fragments further, mainly by loss of a methyl group (**60c** → **60d**). A weaker ion at m/e 121 (**60g**), however, may be attributed to loss of CHO· from **60c**. A notable feature of the spectrum of **60** is the intense ion (base peak) at m/e 137 (**60e**) which is very weak in the spectrum of **58**. Two mechanisms involving migration of either the $C_{(2)}$ hydrogen or the $C_{(3)}$ hydroxyl hydrogen in **60f** have been suggested, but deuterium studies proved inconclusive (43). The M-29 ion is very weak in the spectrum of **60** (contrast **58**), but a considerable (26%) M-1 ion is observed.

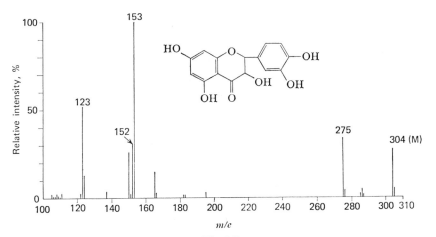

Fig. 3.5

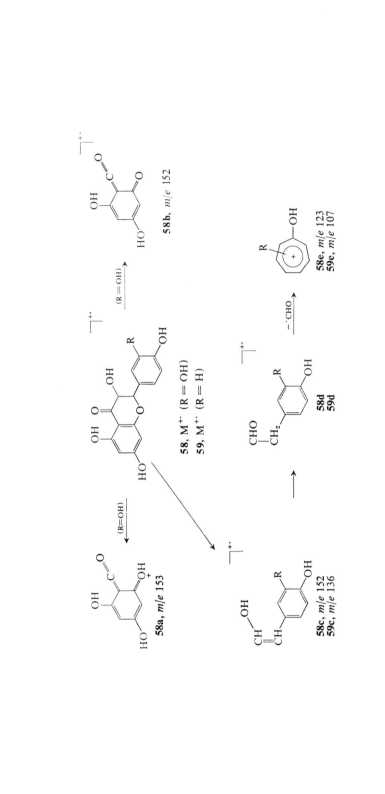

58b, *m/e* 152

(R = OH)

58, M⁺· (R = OH)
59, M⁺· (R = H)

(R=OH)

58a, *m/e* 153

58c, *m/e* 152
59c, *m/e* 136

58d
59d

−·CHO

58e, *m/e* 123
59e, *m/e* 107

Eucomin (**61**) and eucomol (**62**) are the first members of a new class of natural products, the *homo-isoflavanones*. Ions due to the retro-Diels-Alder

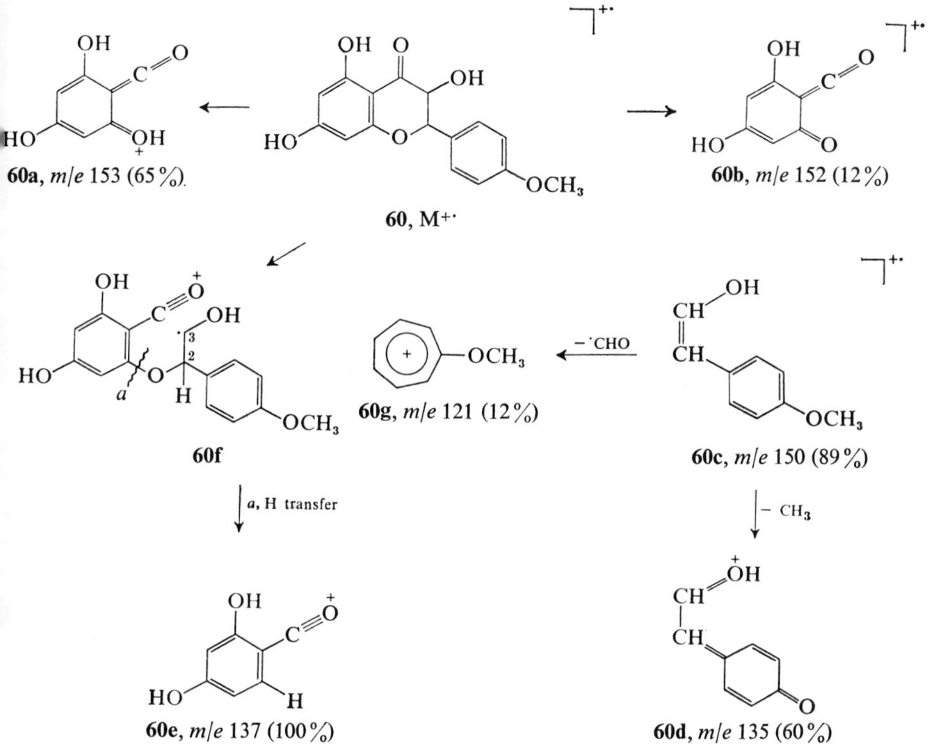

60a, *m/e* 153 (65%).

60, M+·

60b, *m/e* 152 (12%)

60f

60g, *m/e* 121 (12%)

60c, *m/e* 150 (89%)

a, H transfer

– CH₃

60e, *m/e* 137 (100%)

60d, *m/e* 135 (60%)

reaction with hydrogen transfer (**61a**) and without it (**61b** + **61c**) are observed in the spectrum of eucomin (49), whereas in eucomol (**62**) cleavage α to the

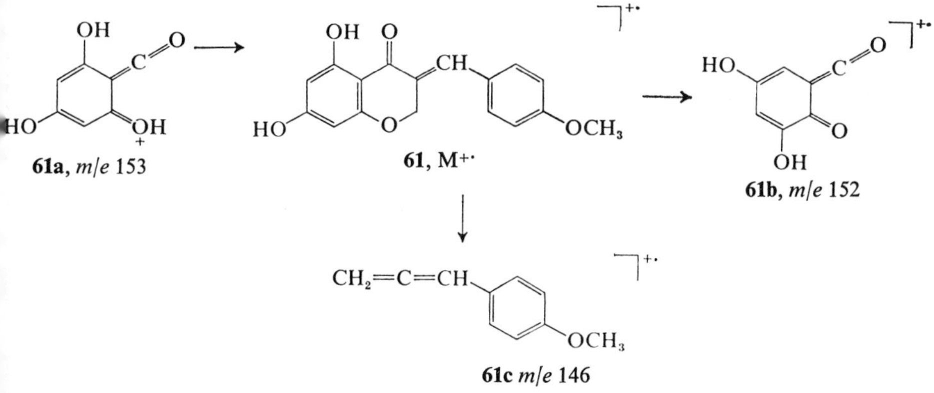

61a, *m/e* 153

61, M+·

61b, *m/e* 152

$CH_2{=}C{=}CH$

61c *m/e* 146

B-ring hydroxyl group appears to be important, leading to ions **62a** and **62b** (49).

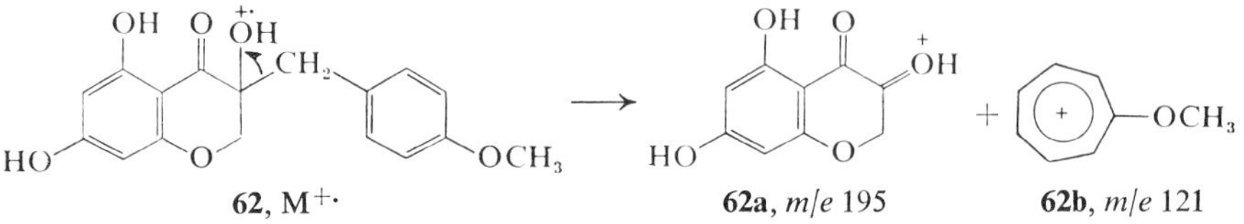

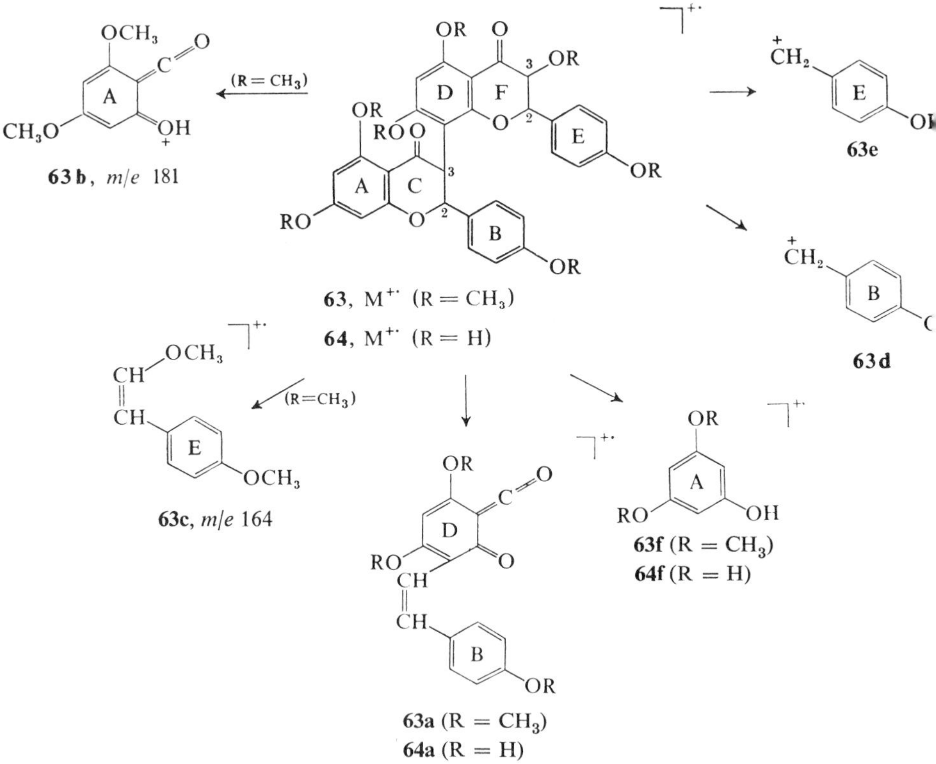

Mass spectrometry has proved very valuable in the structure elucidation of a number of biflavanoids and a flavonylflavanone (50–55). The fragmentation of biflavanones may be rationalized in terms of processes similar to those observed in monoflavanones. Thus with GB-1 heptamethyl ether (**63**) two successive retro-Diels-Alder reactions (substantiated by metastable peaks) are observed and lead to the base peak **63a** (52). The ion **63b** results from the retro-Diels-Alder fragmentation of ring C with hydrogen transfer to ring A, and the ion **63c** by the retro-Diels-Alder fragmentation of ring F with charge retention on ring E. The ion at m/e 121 in the spectrum of **63** (**63d** and **63e**, R = CH$_3$) may be accounted for by hydrogen migration from C$_{(3)}$ to C$_{(2)}$ in rings C and F. The formation of phloroglucinol (**64f**) from GB-1 (**64**) and of phloroglucinol dimethyl ether (**63f**) from **63** is probably due to thermal decomposition. Phloroglucinol gives the base peak of the spectrum of **64** and becomes so intense at higher ion-chamber temperatures that there is difficulty in detecting the molecular ion. However the greater volatility of **63** results in phloroglucinol dimethyl ether (**63f**) being of only 10% intensity relative to the base peak in the spectrum of **63**.

VI. ROTENOIDS AND PTEROCARPANS

The mass spectra of a considerable number of rotenoids have been studied. The retro-Diels-Alder reaction with charge retention wholly on the chromen fragment is the major fragmentation pathway, resulting in the base peak (**65a**) of the spectrum of the simplest rotenoid **65** (Fig. 3.6) (45). Further loss of a hydrogen atom is highly favored for it results in the stable chromenyl ion **65b**. The retro-Diels-Alder reaction with hydrogen transfer also occurs to a considerable extent, resulting in **65c**.

In the spectra of substituted rotenoids, ions due to the retro-Diels-Alder reaction with or without hydrogen transfer and charge retention on ring

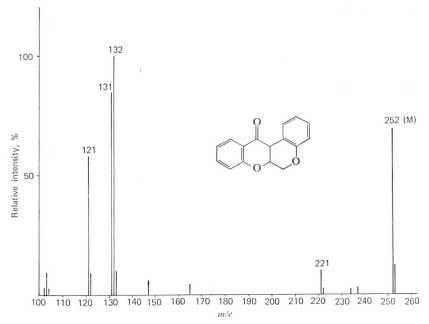

Fig. 3.6

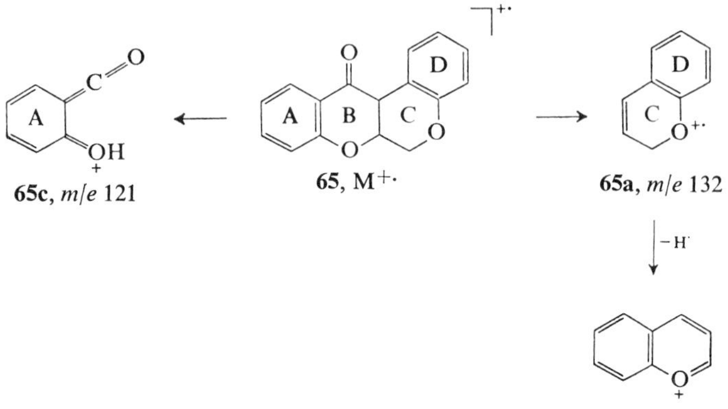

65c, *m/e* 121   65, M$^{+\cdot}$ 65a, *m/e* 132

65b, *m/e* 131

A are generally very weak, the base peak being due to the chromen fragment (rings C and D). Thus with sermundone (**66**) the base peak is **66a**, which then loses either a hydrogen atom to form the chromenyl ion **66b** or a methyl group to form the quinonoid oxonium ion **66c** (45, 56). That the lost methyl group originates from one of the methoxyl groups in **66a** is supported by the absence of this fragmentation in the related rotenoid pachyrrhizone, which has a methylenedioxy group in place of two methoxyls in ring D (56). Further fragmentation of the ions **66b** and **66c** (by losses of CH_2O, CO, or $CH_3\cdot$) is observed, but is only of very minor importance. The other retro-Diels-Alder fragment (**66d**) is very weak (1.6%) and is accompanied by a hydrogen-transfer ion at *m/e* 167 (2.5%).

In pterocarpans the retro-Diels-Alder reaction does not lead to fragmentation, and this is reflected in the intensity of the molecular ion and

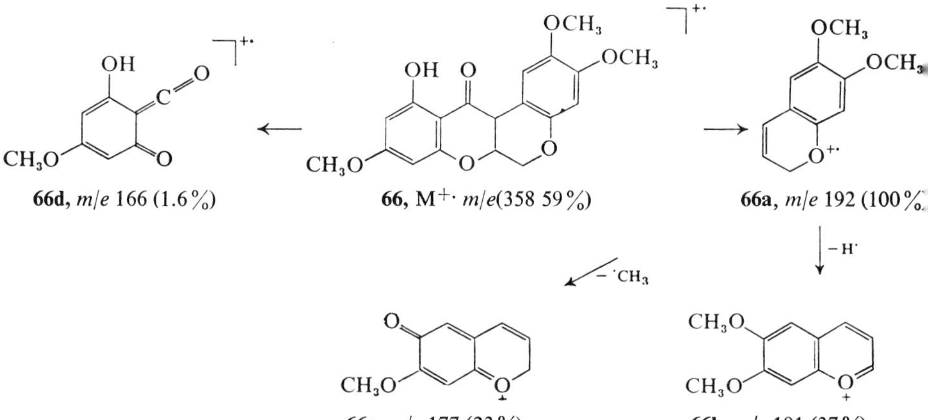

66d, *m/e* 166 (1.6%)   66, M$^{+\cdot}$ *m/e*(358 59%)   66a, *m/e* 192 (100%)

66c, *m/e* 177 (23%)   66b, *m/e* 191 (37%)

the small degree of fragmentation observed in the spectrum of pterocarpin (**67**) (57). Benzofuran radical ions such as **67a** and chromenyl ions such as **67b** and **67c**, formed as shown, are of general importance in the spectra of

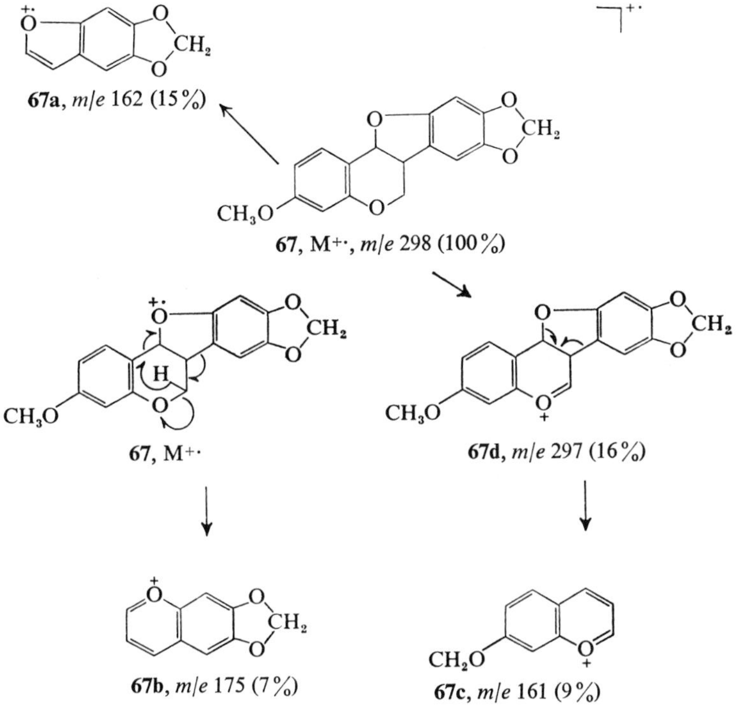

67a, *m/e* 162 (15%)

67, M+·, *m/e* 298 (100%)

67, M+·

67d, *m/e* 297 (16%)

67b, *m/e* 175 (7%)

67c, *m/e* 161 (9%)

pterocarpans. Other ions of importance in the spectrum of pterocarpin (**67**) are the M-1 ion (**67d**), the M-·CH₃ (9%) ion, and an ion at *m/e* 148 (**67e**) which probably arises from the M-1 ion as shown.

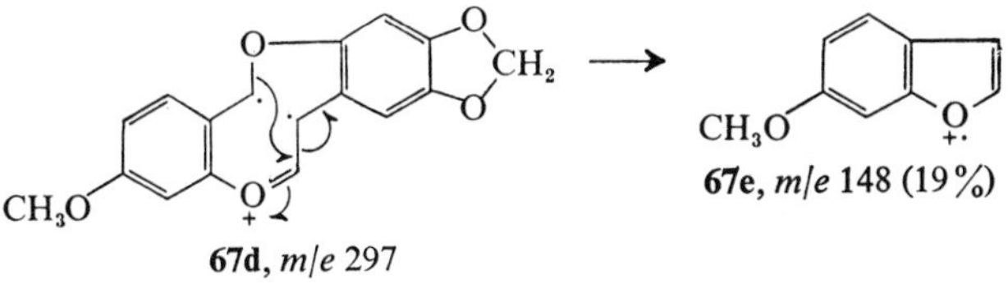

67d, *m/e* 297

67e, *m/e* 148 (19%)

VII.   COMPLEX CHROMANS

The fragmentation processes postulated for simple chromans may be readily extended to more complex systems.

The spectra of γ-tocotrienol (**68**) and plastochromanol-8 (**69**) show prominent molecular ions (58) and intense ions (**68a** and **69a**) due to the

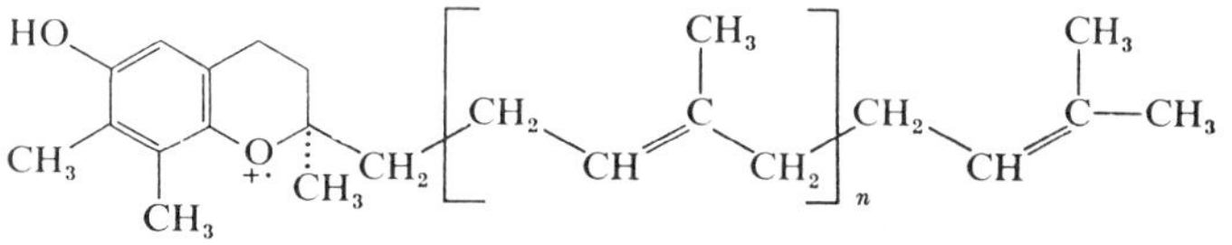

68 M$^+$, $n = 2$, m/e 410 (53%)
69 M$^+$, $n = 7$, m/e 750 (14%)

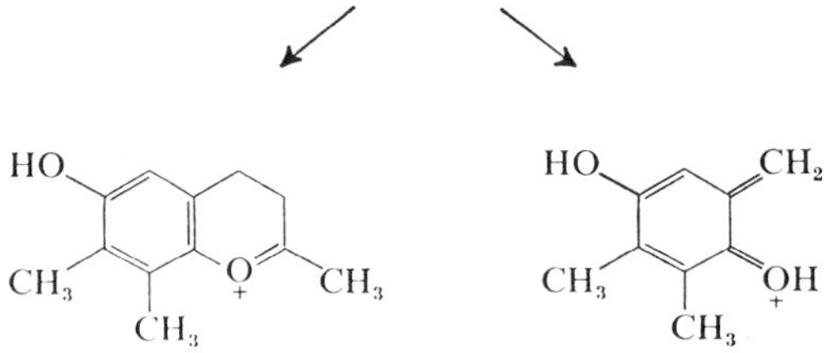

m/e 191
68b (21%)
69b (24.5%)

m/e 151
68a (100%)
69a (83%)

retro-Diels-Alder reaction with hydrogen transfer; prominent ions due to α-cleavage (**68b** and **69b**) are also observed [cf. the spectrum of 2,2-dimethyl-chroman (**39**) where α-cleavage and the retro-Diels-Alder reaction with hydrogen transfer are the two most important fragmentation processes]. Weak ions due to successive losses of isoprene from the parent ion, and also

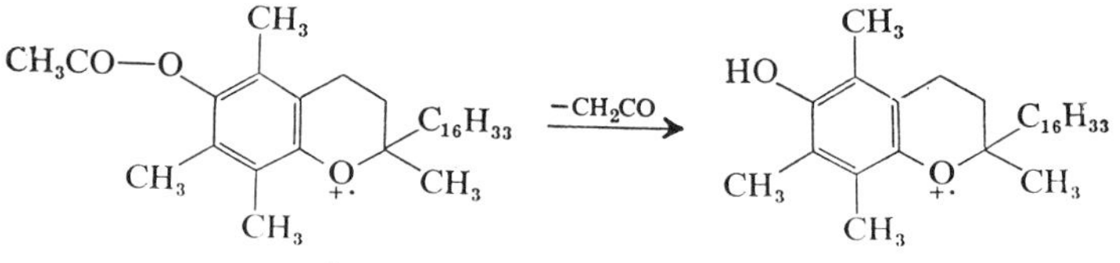

70, M$^{+\cdot}$

$-CH_2CO$

70a

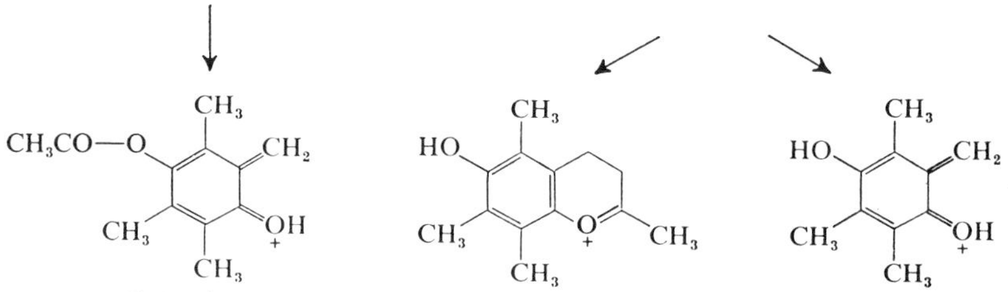

70d, *m/e* 207

70c, *m/e* 205

70b, *m/e* 165

intense ions at m/e 81 (C_6H_9), m/e 69 (C_5H_9), and m/e 41 (C_3H_5) are observed and are characteristic of the isoprenoid side chain.

A simpler fragmentation pattern is observed in O-acetyl-α-tocopherol (**70**) in which the side chain is saturated (59). The base peak is due to loss of ketene from the acetyl group, resulting in **70a**. This undergoes the retro-Diels-Alder reaction with hydrogen transfer to give an intense ion at m/e 165 (**70b**) and also α-cleavage to give a weaker ion at m/e 205 (**70c**). The retro-Diels-Alder reaction with hydrogen transfer also occurs (to a lesser extent) directly from the molecular ion to give **70d**.

The retro-Diels-Alder reaction with or without hydrogen transfer is quite unimportant in selinidin (**71**) which shows a very simple mass spectrum with

71, M$^{+\cdot}$, m/e 328

71a, m/e 228

71b, m/e 213 (100%)

71c, m/e 83

71d, m/e 55

only five significant peaks (60). Loss of the side chain from the intense molecular ion gives the chromen **71a** which then loses a methyl group to form the stable ion (base peak) **71b**. The only other important ions, **71c** and **71d**, are derived from the side chain. It is interesting to note that there is no ion corresponding to loss of CO from the parent ion although such a loss is characteristic of compounds containing the coumarin grouping.

Mass spectrometry has proved useful in the structure investigation of some quinone dimers containing the chroman grouping. Thus it has given definite confirmation to structure **72** for the acid-catalyzed dimerization product from 1-tocoquinone (61). In addition to the retro-Diels-Alder ion **72a** and its further fragmentation products **72b** and **72c** [also observed in the spectrum of the dimer **73** (62)], an ion (**72d**) that confirms structure **72** and is due to loss of a duroquinonyl radical from the parent ion is observed.

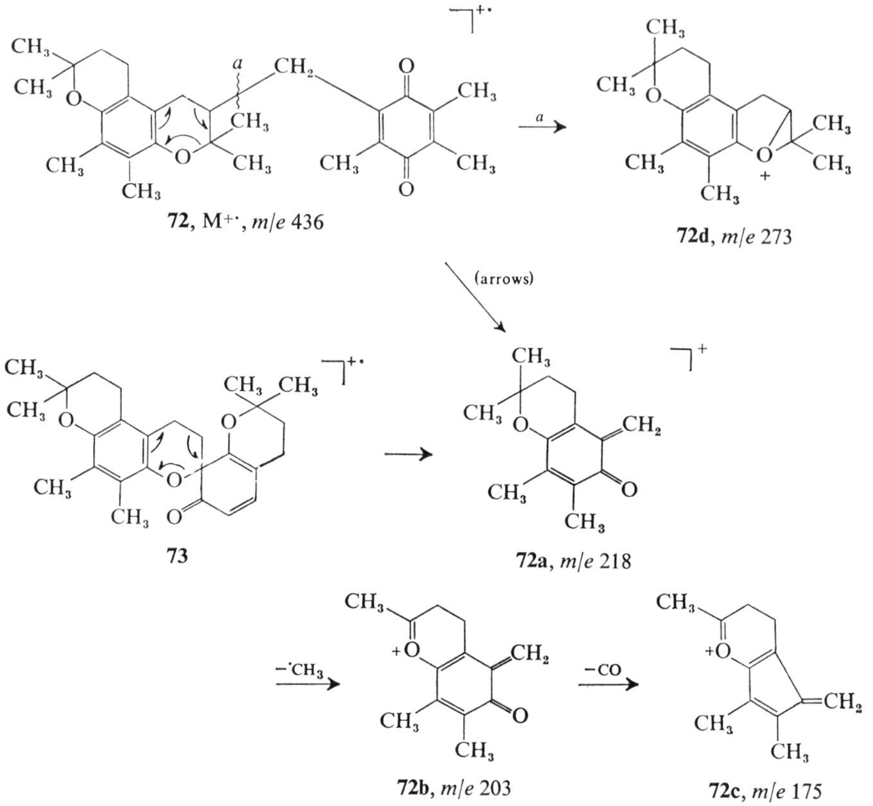

72, M$^{+\cdot}$, m/e 436

72d, m/e 273

(arrows)

73

72a, m/e 218

$\xrightarrow{-\cdot CH_3}$

72b, m/e 203

$\xrightarrow{-CO}$

72c, m/e 175

Little data is available on the mass spectra of derivatives of isochroman, but it appears that the retro-Diels-Alder reaction may be an important fragmentation process as is cleavage α to the ring oxygen atom. Thus for **74** the loss of acetaldehyde from the parent ion by a retro-Diels-Alder process results in the ion **74a** as the base peak (63). This ion further fragments either by loss of CO (**74a** → **74b**) or by loss of a methyl radical (**74a** → **74c**). The other important fragmentation pathway involves successive losses of a methyl group and water, leading to the relatively stable isochromenyl ion **74d** which fragments further only by loss of CO.

The spectrum of the aphid pigment erythroaphin-*fb* (**75**) is rather simple and has been rationalized as shown (63). The molecular ion fragments by loss of a methyl radical to form the stable ion **75a** which does not appear to fragment further. It has been suggested that a "benzylic" methyl group is probably lost as this would lead to extended conjugation; however, the other possible route (to **75f**) has not been discounted. The major fragmentation

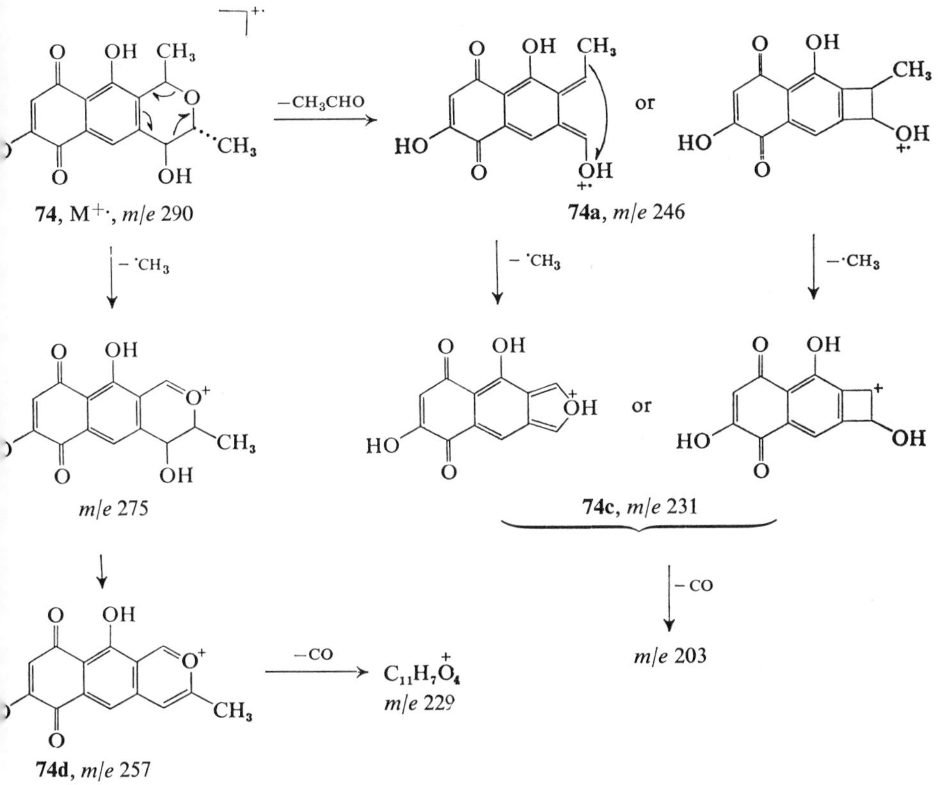

74, M$^{+\cdot}$, m/e 290   **74a, m/e 246**

m/e 275   **74c, m/e 231**

$-CO$ $C_{11}H_7O_4^+$ m/e 203
m/e 229

74d, m/e 257

pathway involves successive loss of two molecules of acetaldehyde (**75 →
75b → 75c**) followed by loss of an electron to give the doubly-charged ion
75d as the base peak of the spectrum. Another important ion is at m/e 451
(**75e**) arising by loss of a methyl group from **75b**. The mass spectra of a
number of other aphid pigments have been studied. The spectra vary in
complexity but significant loss of acetaldehyde from the parent ion is not
always observed (64–66).

VIII. LARGER RINGS WITH ONE OXYGEN ATOM

There appear to be no examples of the mass spectra of simple seven-membered
and larger oxygen-containing ring compounds. The spectrum of 2-methyl-
2,3,4,5-tetrahydro-4′-bromo-3′-methyl-benzo[b]oxepin-3,5-dione (**76**) exhibits an intense molecular ion, and a peak at M-72 due to loss of COCH$_2$CO
is consistent with the formulation of **76** as containing a seven-membered
ring (67).

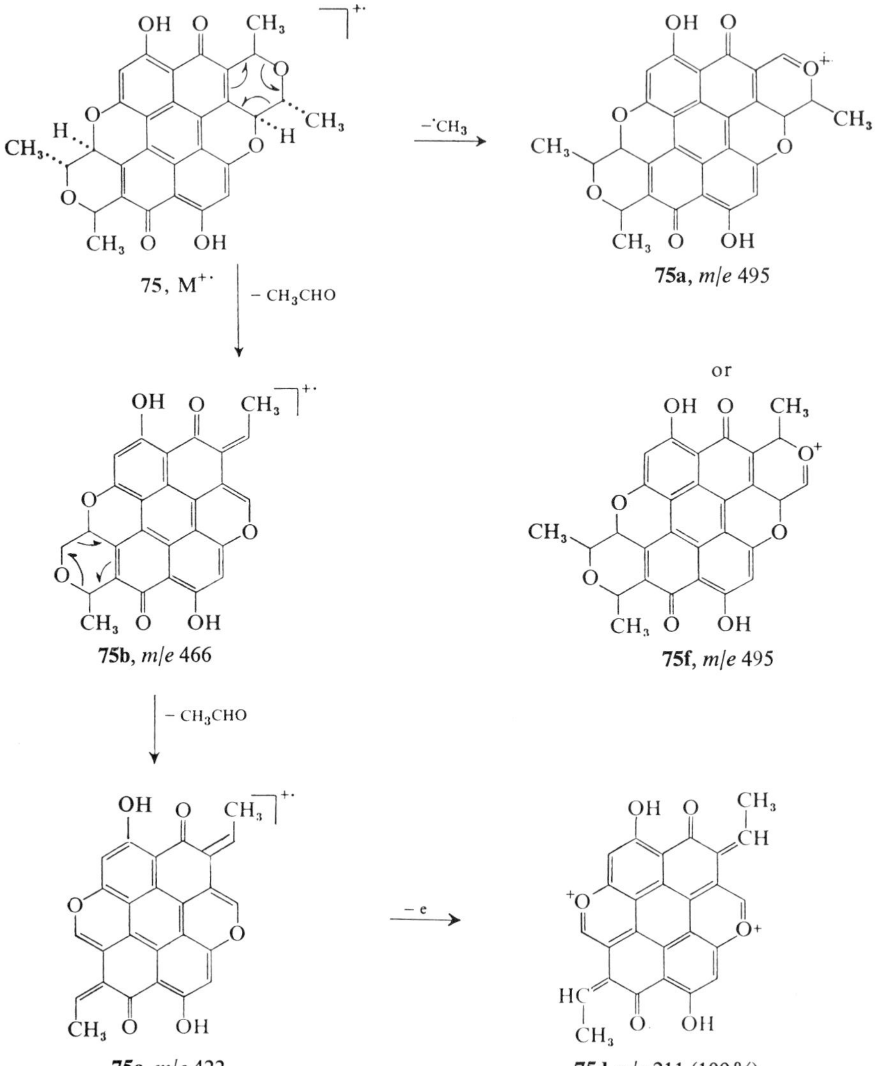

75, M$^{+\cdot}$

75a, m/e 495

75b, m/e 466

or

75f, m/e 495

75c, m/e 422

75d m/e 211 (100%)

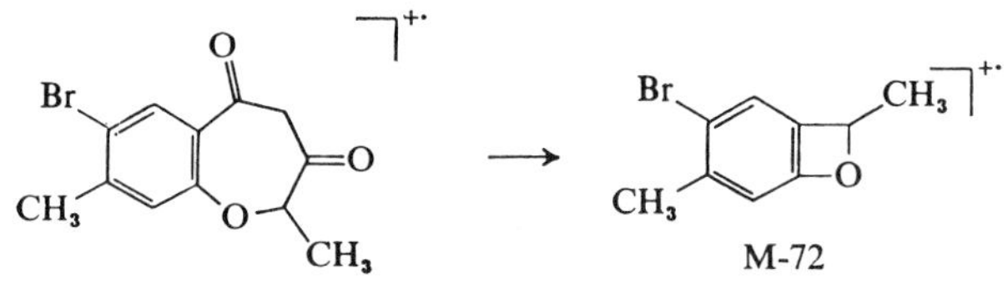

76 M$^{+\cdot}$

M-72

Addenda

Further work has appeared on the fragmentation of flavan-3-ols and -3,4-diols (Section IV) (68, 69), 3-hydroxyflavanones (Section V) (69) and bis-2,2-dimethylchromenes and -chromanes (Sections II and VII) (70). A detailed high resolution study has led to elucidation of the fragmentation pathways of alkyl tetrahydropyranyl ethers and thioethers (Section I-C) (71).

REFERENCES

1. American Petroleum Institute, Research Project 44. Spectrum 1828.
2. J. H. Beynon in *Advances in Mass Spectrometry*, Vol. 1, (J. D. Waldron, Ed.), Pergamon, London, 1959, p. 328.
3. R. Smakman and Th. J. De Boer, *Org. Mass. Spectrosc.*, **1**, 403 (1968).
4. Q. N. Porter and C. M. Richards, Unpublished observations.
5. K. Biemann, *Mass Spectrometry*, McGraw-Hill, New York, 1962, p. 16.
6. R. C. Blume, D. Plant, and W. P. O'Neill, *J. Org. Chem.*, **30**, 1553 (1965).
7. C. F. Seidel, D. Felix, A. Eschenmoser, K. Biemann, and M. Stoll, *Helv. Chim. Acta*, **44**, 598 (1961).
8. C. Cardani, D. Ghiringhelli, R. Mondelli, and A. Quilico, *Gazz. Chim. Ital.*, **96**, 3, (1966).
9. J. von Hoene and W. M. Hickman, *J. Chem. Phys.*, **32**, 876 (1960).
10. S. Shibata, M. Fujita, H. Itokawa, O. Tanaka, and T. Ishii, *Tetrahedron Lett.*, **1962**, 419.
11. S. Shibata, M. Fujita, H. Itokawa, O. Tanaka, and T. Ishii, *Chem. Pharm. Bull. Japan*, **11**, 759 (1963).
12. G. B. Elyakov, A. K. Dzizenko, and Yu. N. Elkin, *Tetrahedron Lett.*, **1966**, 141.
13. S. Shibata, O. Tanaka, K. Sôma, Y. Iida, T. Ando, and H. Nakamura, *Tetrahedron Lett.*, **1965**, 207.
14. O. Ceder, J. M. Waisvisz, M. G. Van Der Hoeven, and R. Ryhage, *Acta Chem. Scand.*, **18**, 83 (1964).
15. G. Gaudiano, P. Bravo, and A. Quilico, *Gazz. Chim. Ital.*, **96**, 1351 (1966).
16. M. Venugopalan and C. B. Anderson, *Indian J. Chem.*, **3**, 20 (1965).
17. S. Moon and J. M. Lodge, *J. Org. Chem.*, **29**, 3453 (1964).
18. M. Venugopalan and C. B. Anderson, *Chem. Ind.* (London), **1964**, 370.
19. H. Budzikiewicz, C. Djerassi, and D. H. Williams, *Mass Spectrometry of Organic Compounds*, Holden-Day, San Francisco, 1967, pp. 255–257.
20. S. C. Cascow, W. B. Mors, B. M. Tursch, R. T. Aplin, and L. J. Durham, *J. Amer. Chem. Soc.*, **87**, 5237 (1965).
21. J.-F. Biellmann, *Tetrahedron Lett.*, **1966**, 4803.
22. J.-F. Biellmann, *Bull. Soc. Chim. Fr.*, **1967**, 3459.
23. Ref. 19, pp. 478–479.
24. R. Hodges and R. I. Reed, *Tetrahedron*, **10**, 71 (1960).
25. C. R. Enzell and R. Ryhage, *Ark. Kemi*, **23**, 367 (1965).
26. H. Audier, S. Bory, and M. Fétizon, *Bull. Soc. Chim. Fr.*, **1964**, 1381.
27. P. K. Grant and R. Hodges, *Chem. Ind.* (London), **1960**, 1300.
28. C. R. Enzell, *Acta Chem. Scand.*, **14**, 2053 (1960).
29. C. R. Enzell, B. R. Thomas, and I. Wahlberg, *Tetrahedron Lett.*, **1967**, 2211.
30. C. Cardani, D. Ghiringhelli, A. Quilico, and A. Selva, *Tetrahedron Lett.*, **1967**, 4023.

31. Ref. 19, p. 254.
32. H. Budzikiewicz, J. I. Brauman, and C. Djerassi, *Tetrahedron*, **21**, 1855 (1965).
33. A. F. Thomas and M. Stoll, *Chem. Ind.* (London), **1963**, 1491.
34. G. Ohloff, K.-H. Schulte-Elte, and B. Willhalm, *Helv. Chim. Acta*, **47**, 602 (1964).
35. I. Fleming and M. H. Karger, *J. Chem. Soc.*, *C*, **1967**, 226.
36. N. S. Wulfson, V. I. Zaretskii, V. L. Sadovskaya, A. V. Semenovsky, W. A. Smit, and V. F. Kucherov, *Tetrahedron*, **22**, 603 (1966).
37. J. Baldas and Q. N. Porter, *Tetrahedron Lett.*, **1968**, 1351.
38. C. S. Barnes and J. L. Occolowitz, *Aust. J. Chem.*, **17**, 975 (1964).
39. B. Willhalm, A. F. Thomas, and F. Gautschi, *Tetrahedron*, **20**, 1185 (1964).
40. Ref. 19, p. 23.
41. C. S. Barnes, M. I. Strong, and J. L. Occolowitz, *Tetrahedron*, **19**, 839 (1963).
42. P. Brown and C. Djerassi, *Angew. Chem.*, *Int. Ed.*, **6**, 477 (1967).
43. A. Pelter, P. Stainton, and M. Barber, *J. Heterocycl. Chem.*, **2**, 262 (1965).
44. A. J. Birch and M. Salahuddin, *Tetrahedron Lett.*, **1964**, 2211.
45. H. Audier, *Bull. Soc. Chim. Fr.*, **1966**, 2892.
46. A. Pelter and P. Stainton, *J. Chem. Soc.*, *C*, **1967**, 1933.
47. Y. Taniko, A. Ferretti, V. Flanagan, M. Gianturco, and M. Vogel, *Tetrahedron Lett.*, **1965**, 4019.
48. A. G. Brown, C. P. Falshaw, E. Haslam, A. Holmes, and W. D. Ollis, *Tetrahedron Lett.*, **1966**, 1193.
49. P. Böhler and Ch. Tamm, *Tetrahedron Lett.*, **1967**, 3479.
50. S. E. Drewes, D. G. Roux, S. H. Eggers, and J. Feeney, *J. Chem. Soc.*, *C*, **1967**, 1217.
51. A. J. Birch, C. J. Dahl, and A. Pelter, *Tetrahedron Lett.*, **1967**, 481.
52. B. Jackson, H. D. Locksley, F. Scheinmann, and W. A. Wolstenholme, *Tetrahedron Lett.*, **1967**, 787.
53. A. Pelter, *Tetrahedron Lett.*, **1967**, 1767.
54. B. Jackson, H. D. Locksley, F. Scheinmann, and W. A. Wolstenholme, *Tetrahedron Lett.*, **1967**, 3049.
55. C. G. Karanjgaokar, P. V. Radhakrishnan, and K. Venkataraman, *Tetrahedron Lett.*, **1967**, 3195.
56. R. I. Reed and J. M. Wilson, *J. Chem. Soc.*, **1963**, 5949.
57. W. D. Ollis, *Experientia*, **22**, 777 (1966).
58. H. Mayer, J. Metzger, and O. Isler, *Helv. Chim. Acta*, **50**, 1376 (1967).
59. H. Mayer, W. Vetter, J. Metzger, R. Rüegg, and O. Isler, *Helv. Chim. Acta*, **50**, 1168 (1967).
60. T. R. Seshadri, M. S. Sood, K. L. Handa, and Vishwapaul, *Tetrahedron*, **23**, 1883 (1967).
61. P. Mamont, P. Cohen, and R. Azerad, *Bull. Soc. Chim. Fr.*, **1967**, 1485.
62. P. Cohen and P. Mamont, *Bull. Soc. Chim. Fr.*, **1967**, 1164.
63. J. H. Bowie and D. W. Cameron, *J. Chem. Soc.*, *B*, **1966**, 684.
64. J. H. Bowie and D. W. Cameron, *J. Chem. Soc.*, *C*, **1967**, 704.
65. D. W. Cameron and H. W.-S. Chan, *J. Chem. Soc.*, *C*, **1966**, 1825.
66. J. H. Bowie and D. W. Cameron, *J. Chem. Soc.*, *C*, **1967**, 708.
67. J. H. P. Tyman and R. Pickles, *Tetrahedron Lett.*, **1966**, 4993.
68. S. E. Drews, *J. Chem. Soc.* [*C*], 1140 (1968).
69. J. W. Clark-Lewis, *Austral. J. Chem.*, **21**, 3025 (1968).
70. A. M. Duffield, *Org. Mass Spec.*, **2**, 965 (1969).
71. S. J. Isser, A. M. Duffield and C. Djerassi, *J. Org. Chem.*, **33**, 2266 (1968).

4 FURANS AND BENZOFURANS

I. FURANS

A. Furan

Furan (1) is an example of a 6π-electron heteroaromatic system. Its stability is evidenced by a resonance energy five times that of the acyclic analog, divinylether (2) (1), and by an intense molecular ion in the mass spectrum (Fig. 4.1) accounting for 25% of the total ion current (2). The ionization potential of furan is 9.00 ± 0.10 eV (3), considerably lower than that of benzene (9.56 eV) (4). Theoretical considerations indicate that the most energetically favored bond-cleavage in the furan molecular ion is that of a carbon–oxygen bond, and it results in the ring-opened molecular ion **1a**, which may then undergo electronic rearrangement to **1b** (5). Homolytic cleavage of the $C_{(4)}$—$C_{(5)}$ bond in **1b** results in elimination of a formyl radical and formation of an ion at m/e 39, which accounts for 39% of the total ion current and which is best formulated as the cyclopropenyl ion (**1c**), a stable 2π-electron aromatic system. A metastable ion at m/e 38.1 (m/e 40 → m/e 39) indicates that another pathway to **1c** involves loss of CO from the parent ion followed by loss of H· from the cyclopropene radical ion (m/e 40) so formed. Heterolytic cleavage of the $C_{(4)}$—$C_{(5)}$ bond in **1b** would result in elimination of the cyclopropenyl radical and formation of the formyl ion **1d** (m/e 29). Alternatively the formyl ion may be envisaged as arising from the formylcyclopropene ion **1e** formed by electronic rearrangement of **1a** as shown. The loss of neutral acetylene from the parent ion is not an important process, as is indicated by the relatively weak peak at m/e 42.

Recent experiments on the fragmentation of 2-deutero- and 2,5-dideuterofuran show that hydrogen scrambling in this heterocycle does not occur to a significant extent before fragment ion formation (6). This may be contrasted with the situation observed with thiophen (Chap. 8, Sec. I.H.1) in which scrambling is essentially complete before the major fragmentation reaction.

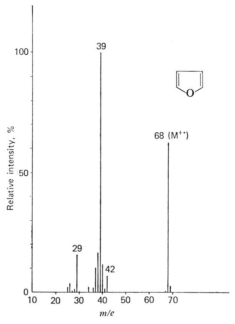

Fig. 4.1

This may simply indicate that much slower randomization occurs with furan or, alternatively, there may in fact be no randomization, quite a likely hypothesis since oxygen cannot expand its valence shell in the way suggested for sulfur. Thus a mechanism for randomization involving interchange of carbon atoms and their attached hydrogen atoms is not available for furan.

It is noteworthy that, in contrast to their thiophen analogs, diphenylfurans do not undergo photochemically induced phenyl-group migrations (7).

It is interesting to note that the photosensitized photolysis of furan yields carbon monoxide and cyclopropene (1f) and that the suggested mechanism involves initial breakage of a carbon–oxygen bond (8).

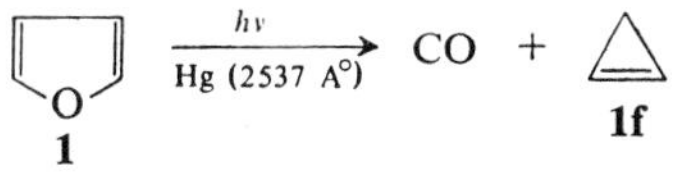

Tetrafluorofuran (3) resembles furan in that it shows major peaks due to the molecular ion and the trifluorocyclopropenyl ion (3a) (9).

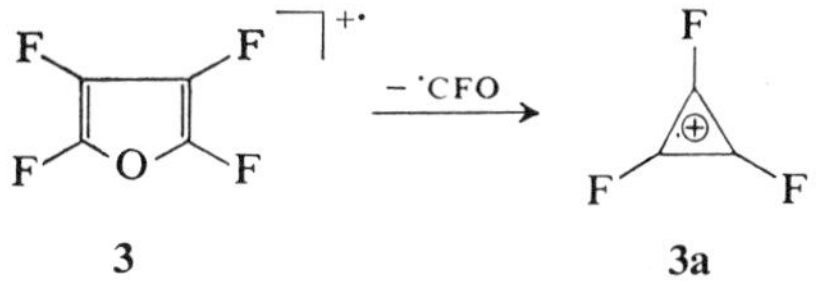

B. Alkylfurans

Methylfurans show an abundant molecular ion and ring fragmentation remains an important process. In 2-methylfuran (4) cleavage of the $O-C_{(2)}$ or the $O-C_{(5)}$ bond may occur, resulting in two different ring-opened molecular ions (4a and 4b, respectively). These fragment by the processes described for furan, giving the intense cyclopropenyl (4d, M-CH$_3$CO·) and methylcyclopropenyl (4c, M-HCO·) ions as well as a weaker acetyl ion 4e

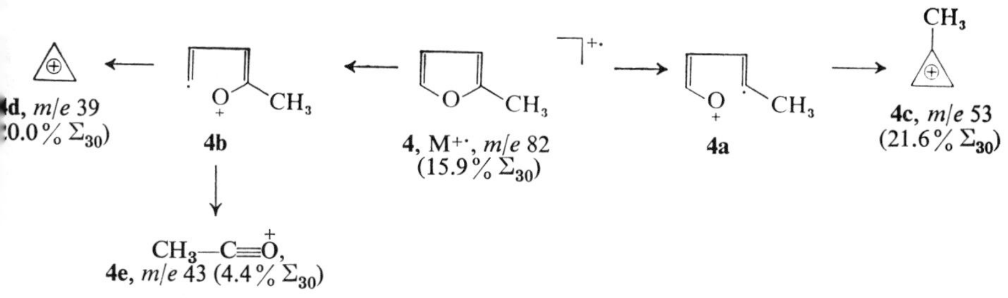

(10). In the spectrum of 2,5-dimethylfuran, which has two α-methyl groups, the M-CH$_3$CO· ion (m/e 53) is still intense but the base peak is due to the acetyl ion (m/e 43) (10). However, an M-CH$_3$CO· ion is not specific for α-methylfurans as it is also observed for 3-methylfuran (5) where it arises by separate elimination of CO and the β-methyl group (5 → 5a → 5b). The base peak **5c** may arise by loss of HCO· from the parent ion (**5** → **5c**) and probably to some extent also by loss of H· from **5a**. The intensity of the acetyl

ion (m/e 43) is, however, characteristic of α-methyl substituents and increases markedly in any one isomeric pair with an increasing number of α-methyl substituents as is shown by the intensity of the acetyl ion (m/e 43) as a percentage of the total ion current (Σ_{30}) for three isomeric pairs of methyl furans (11):

2-Methyl	4.4	2,5-Dimethyl	24.6	2,3,5-Trimethyl	11.6
3-Methyl	0.1	2,4-Dimethyl	4.1	2,3,4-Trimethyl	5.5

The other important fragmentation process for methylfurans is cleavage β to the furan ring, resulting in an intense M-1 ion characteristic of methyl-furans. The M-1 ion may be represented as the fully conjugated oxonium ions **4f** and **5d** for 2- and 3-methylfuran, respectively, but is probably better formulated as a ring-enlarged aromatic pyrilium ion (**6**) by analogy with the well-established occurrence of tropylium ions in the spectra of alkylbenzenes

(12). Cleavage α to the furan ring with loss of the methyl group is insignificant in 2- and 3-methylfuran as it leads to an unfavored vinyl or diradical ion [**4h** and **4g** in the case of 2-methylfuran (**4**)]. Di- and tri-methylfurans (**7**), however, show significant M-CH$_3$· peaks (10, 13), and it is probable that elimination of the methyl group occurs during or after rearrangement of the molecular ion to a ring-expanded aromatic pyrilium ion **7a**, in agreement with the behavior of analogous benzene derivatives such as the xylenes (12).

7, R = H or CH$_3$ **7a, R = H or CH$_3$**

In the spectrum of 2,3,4,5-tetramethylfuran the M-CH$_3$· ion is second in intensity only to the M-1 ion and accounts for 16.9% of the total ion current (Σ_{30}) (10).

With larger 2-substituents, ring fragmentation with resultant formation of cyclopropenyl and acyl ions is unimportant, and β-fission becomes the dominant fragmentation process. Thus for 2-n-propylfuran (**8**) the intense base peak at m/e 81 (M-29) has been shown by high resolution to be entirely due to the loss of C$_2$H$_5$ (**8 → 8a**), the alternative loss of HCO· (**8 → 8b**) with formation of a propylcyclopropenyl ion not being observed (10).

8c, m/e 109   **8**, m/e 110 (11.9% Σ_{30})   **8a**, m/e 81
(0.4% Σ_{30})   (43.1% Σ_{30})

8b, m/e 81

β-Fission with loss of a hydrogen atom (**8 → 8c**) occurs only to a slight extent in keeping with the greater stability and hence much readier loss of an ethyl radical compared to a hydrogen atom in electron-impact-induced bond-cleavage reactions. γ-Cleavage is unimportant as is shown by the weak M-CH$_3$· ion observed for **8** (0.6% of Σ_{30}). If the 2-side-chain is n-propyl or longer, a McLafferty rearrangement can occur (10, 14). Thus with 2-n-butyl-(**9**) and 2-n-pentylfuran (**10**) the loss of propene and butene, respectively, results in m/e 82 as the most intense ion (after β-fission) in both spectra.

Mass spectrometry provides a ready means of detecting branching on the carbon atom adjacent to the ring of a 2-alkylfuran; thus the isomeric 2-n-propyl- (**8**) and 2-isopropylfuran (**11**) have base peaks at M-29 (loss of an ethyl radical) and M-15 (loss of methyl), respectively (10).

β-Fission is also the dominant fragmentation reaction for 3-alkyl-furans, and in 2,5-dimethyl-3-ethylfuran (12) loss of a methyl group results in 12a

9, M$^{+\cdot}$, R = CH$_3$
10, M$^{+\cdot}$, R = C$_2$H$_5$

m/e 82
9a (10.6 % Σ_{30})
10a (11.9 % Σ_{30})

as the base peak, accounting for 37.9 % of the total ion current (10). On the other hand the acetyl ion (m/e 43), which gives the base peak for 2,5-dimethylfuran, is comparatively weak (3.1 % of Σ_{30}) even though there are two α-methyl groups in 12.

11, M$^{+\cdot}$, m/e 110

m/e 95 (27.0 % Σ_{30})

Spectra have been reported for a number of alkenylfurans with the double bond in conjugation with the ring. In 2-vinylfuran (13) the parent ion dominates the spectrum as β-cleavage is now not possible and only those ions resulting from ring fragmentation are significant (10). Loss of CO with

12, M$^{+\cdot}$

12a, m/e 109 (37.9 % Σ_{30})

recyclization results in the cyclopentadiene radical ion (13a) which further loses a hydrogen atom forming the stable cyclopentadienyl ion (13b). Alternatively, loss of CHO$\cdot$ from the parent ion leads to 13b directly.

With 2-n-propenylfuran (14) loss of H$\cdot$ is favored relative to ring-opening since it gives the fully conjugated oxonium ion 14a. Loss of CO occurs as the second step, forming the intense benzonium ion 14b which further loses a molecule of hydrogen to give the phenyl ion (14c). Cleavage α to the furan ring (14 → 14d) is negligible, but a significant ion is observed at m/e

81 (**14e**) (10). Its formation must involve hydrogen rearrangement if it arises directly from the parent ion. Alternatively, it may be formed by loss of

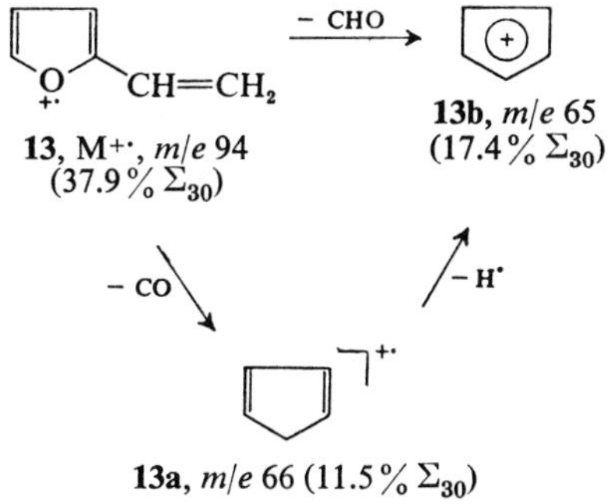

- CHO
CH=CH₂
13, M+·, m/e 94
(37.9% Σ30)
13b, m/e 65
(17.4% Σ30)
- CO
- H·
13a, m/e 66 (11.5% Σ30)

neutral acetylene from the M-1 ion (**14a**), but no metastable ion data has been reported to allow a choice between these two paths. 3-Phenylfuran (**15**)

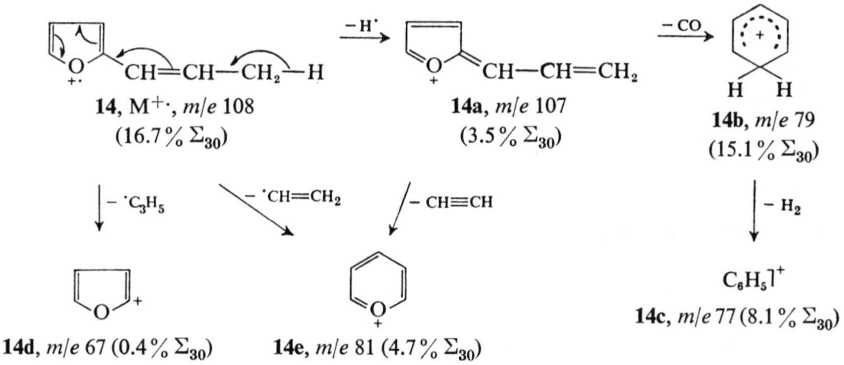

- H·
- CO
CH=CH—CH₂—H
14, M+·, m/e 108
(16.7% Σ30)
CH—CH=CH₂
14a, m/e 107
(3.5% Σ30)
14b, m/e 79
(15.1% Σ30)
- ·C₃H₅
- ·CH=CH₂
- CH≡CH
- H₂
C₆H₅⁺
14c, m/e 77 (8.1% Σ30)
14d, m/e 67 (0.4% Σ30)
14e, m/e 81 (4.7% Σ30)

behaves similarly to 2-vinylfuran (**13**), the major fragmentation process being loss of CHO· and formation of the phenylcyclopropenyl ion (**15a**) which may further lose acetylene to give an ion at *m/e* 89 (15).

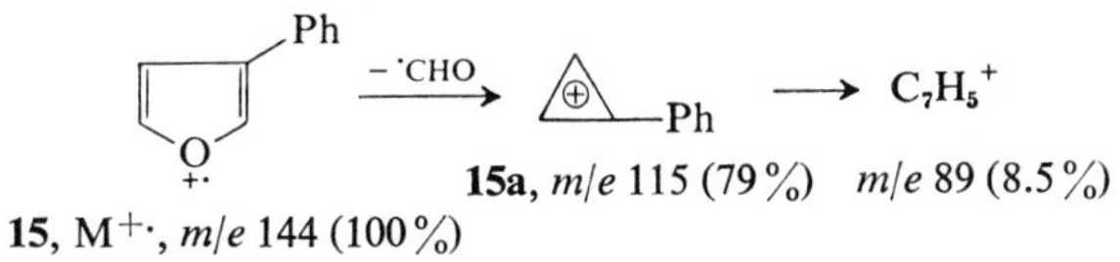

Ph
- ·CHO
Ph
C₇H₅⁺
15a, m/e 115 (79%)
m/e 89 (8.5%)
15, M+·, m/e 144 (100%)

C. Fragmentation of Some Furfuryl Esters and Ketones

β-Cleavage remains an important fragmentation process for these furfuryl compounds. In the simple furfuryl esters **16** it results in the base peak,

m/e 81 (**16a**) (15, 16). Other peaks (whose intensity varies depending on R) are due to furfuryl alcohol (**16b**), formed by loss of a ketene, and the ions R^+ and R—$C\equiv O^+$.

16a, m/e 81 (100%)   **16**, $M^{+\cdot}$ (R = CH_3, nC_3H_7,   **16b**, m/e 98
                          isoC_3H_7, C_4H_9, isoC_4H_9,
                          $-CH\begin{smallmatrix}CH_3\\C_2H_5\end{smallmatrix}$)

With 1-(2-furyl)propan-2-one (**17**) and 1-(5-methyl-2-furyl)propan-2-one (**18**) intense ions are observed for β-cleavage with charge retention on the pyrilium (**17a** and **18a**) and acetyl ions (**17b** and **18b**) (15).

17, $M^{+\cdot}$ (R = H)   **17a**, m/e 81 (98.6%)   **17b**, m/e 43 (100%)
18, $M^{+\cdot}$ (R = CH_3)   **18a**, m/e 95 (100%)   **18b**, m/e 43 (34%)

D.   Furanaldehydes and Furyl Ketones

The mass spectra of 2-furanaldehydes of general formula **19** are characterized by an abundant parent ion and an abundant M-1 ion, the resonance-stabilized furoyl cation **19a**. This further fragments by loss of two molecules of carbon monoxide, forming a cyclopropenyl ion (**19b**) (10). A considerable cyclopropenyl ion (m/e 39) is also observed for 5-bromo- and 5-iodofurfural

19, $M^{+\cdot}$,
R = H, m/e 96 (21.8% Σ_{30})
R = CH_3, m/e 110 (14.5% Σ_{30})

19a, m/e 95 (R = H, 21.2% Σ_{30})
     m/e 109 (R = CH_3, 12.9% Σ_{30})

19b
m/e 39 (R = H, 27.6% Σ_{30})
m/e 53 (R = CH_3, 19.0% Σ_{30})

(**19**, R = Br, R = I), indicating that migration of the aldehydic hydrogen occurs during the elimination of R· and two molecules of CO (17). The spectrum of furan-3-aldehyde resembles that of its 2-isomer (**19**, R = H) but there is some variation in relative peak intensities (10).

An intense furoyl ion is also observed in the spectra of 2-furyl alkyl ketones together with acyl ions formed by fission of the furyl–carbonyl bond with charge retention on the side chain. If the side chain is *n*-butyryl or longer, the McLafferty rearrangement involving the carbonyl group becomes an important process. Thus it gives the base peak (**20a**) of the spectrum of 2-*n*-valerylfuran (**20**), competing favorably with formation of the furoyl ion (**20b**) (10).

20b, *m/e* 95
(16.6% Σ_{30})

20, M$^{+\cdot}$, *m/e* 152 (1.1% Σ_{30})

20a, *m/e* 110
(22.5% Σ_{30})

In 2-furyl phenyl ketone (**21**) an appreciable M-CO ion (**21a**) is observed and the presence of a peak at *m/e* 115 due to the phenylcyclopropenyl ion (**21b**) shows that phenyl migration has occurred (17). The major fragments in the spectrum of **21** are the furoyl (**21c**) and benzoyl (**21d**) ions together with their further fragmentation products.

21c, *m/e* 95 (60%)

21, M$^{+\cdot}$, *m/e* 172
(100%)

21a, *m/e* 144
(11%)

21b,
m/e 115
(10%)

21d, *m/e* 105 (100%)

E. Furancarboxylic Acids

2-Furoic acid (**22**) exhibits an abundant parent ion. β-Cleavage results in loss of OH· to give a substantial furoyl ion (**22a**), while α-cleavage leads to charge retention mainly on the ion **22b**. The base peak is due to the cyclopropenyl ion (**22c**), which may be formed after ring-opening by cleavage of the $C_{(5)}$—O bond ($22d \rightarrow 22c$) and in part also by further fragmentation of $22a$ (18, 19). Both 2- and 3-furoic acids show only weak ions due to furan

m/e 67
(4.9%)

22, $M^{+\cdot}$, m/e 112
(40.7%)

$22a$, m/e 95
(36.5%)

$\overset{+}{C}O_2H$

$22b$, m/e 45
(62.2%)

$22e$ m/e 68
(2.8%)

$22d$

$22c$, m/e 39
(100%)

($22e$), indicating that decarboxylation of the parent ion is not an important process (18, 19).

The base peak for 3-furoic acid (23) is again the cyclopropenyl ion ($22c$) but no significant ion is observed for the expected carboxycyclopropenyl ion

$23b$, m/e 95 (92.5%)

23, $M^{+\cdot}$, m/e 112
(58.2%)

$23a$, m/e 83

$22c$, m/e 39
(100%)

m/e 95

$22e$ m/e 68 (4.1%)

($23a$) so that $23a$, if formed, decarboxylates very readily. The loss of OH·
($23 \rightarrow 23b$) is much more marked than in the 2-isomer.

Furan-2,5- and 3,4-dicarboxylic acids and furantetracarboxylic acid show ready decarboxylation (the parent ion of the tetracarboxylic acid is very weak) which appears to end at the monocarboxylic acid stage and may be due in part to thermal decomposition prior to ionization (18). The spectrum of 2-methylfuran-3,4-dicarboxylic acid (24) is of interest as the base peak is due to loss of the methyl group giving an ion formulated as **24a** (18). It is possible that some stabilization of the positive charge in **24a** may result by interaction with the hydroxyl or carbonyl oxygen but its intensity is surprising. Loss of OH· or H_2O from the parent ion (**24**) does not appear

24, M$^{+\cdot}$, m/e 170 (17.7%)   24a, M-15, (100%)

to be significant, in contrast to the corresponding dimethyl ester, **29**, where the base peak is due to loss of methanol via the "ortho-effect" mechanism (17).

F. Furoic Esters (Furancarboxylates)

Loss of OR· is a highly favored process for furancarboxylates (e.g., methyl and ethyl furan-2-carboxylate, **25**, R = CH_3, R = C_2H_5) and leads to the intense furoyl ion (**22a**) which further fragments by loss of two molecules of

22   25, M$^{+\cdot}$ (R = CH_3 or C_2H_5)   22a, m/e 95 (100%)   22c, m/e 39

CO to the cyclopropenyl ion (**22c**) (16, 17). In the ethyl ester (**25**, R = C_2H_5) a considerable M-28 peak is observed and is due to loss of ethylene from the alkoxy group (**25** → **22**). Loss of the R· group from the parent ion (**25**) does not occur to a significant extent. A methyl group adjacent to the ester grouping allows the operation of the "ortho-effect" and a considerable loss of methanol is observed in the two isomeric esters **26** and **27**; the base peak is due to loss of OCH$_3$· in both cases (17). The two esters differ in that a

26   27 m/e 108 (15%)

prominent acetyl ion (m/e 43) is observed only for **27** which has a 2-methyl substituent (cf. methylfurans). Loss of a methyl group occurs to a considerable extent in both the esters **26** and **27**, and it has been suggested that an "ortho-effect," e.g., **27 → 27a**, is operative as an isolated methyl ester does not show an appreciable M-CH$_3$· ion (17). That the lost methyl group arises from the ester group rather than the furan ring is shown by the spectrum of **28** which exhibits a prominent M-C$_2$H$_5$· ion but a negligible M-CH$_3$· ion. A similar "ortho-effect" has also been postulated for pyrrolecarboxylates (20). A very pronounced "ortho-effect" resulting in M-CH$_3$OH as the base

27, M$^+$·, m/e 140   27a, m/e 125 (23%)   28

peak (**29a**) is observed for the diester **29** (R = H). Negligible loss of CH$_2$DOD is observed for the deuterium-labeled derivative **29** (R = D), establishing that the nuclear methyl group is involved (17).

m/e 166   29, M$^+$·   29a, m/e 166

The mass spectra of a number of bromofurancarboxylates have been studied, and the fragmentation scheme for methyl 5-bromofuran-2-carboxylate (**30**) (Fig. 4.2) is shown (17). As usual, the base peak (**30a**) arises from loss of the methoxyl group. Weaker ions are observed which are due to loss of ·COOCH$_3$ (**30 → 30b**) and of ·COOCH$_2$· with hydrogen rearrangement to **30c**. The ions at m/e 117/119 are due to the bromocyclopropenyl cation (**30d**) which further loses a bromine atom to give an abundant m/e 38 ion, C$_3$H$_2$$^+$·. Ions due to the loss of formaldehyde and a bromine atom appear to be characteristic of methyl 5-bromofuran-2-carboxylates, and a concerted loss of (CH$_2$O + Br·) from the hydrogen-rearranged molecular ion **30f** has been postulated to account for the m/e 95 ion (**30e**) in the spectrum of **30**. The weak ions at m/e 133/135 (**30g**) are of interest as they arise by the formal loss of :CCOOCH$_3$ from the parent ion. Migration of the methoxyl group to give **30h** may be involved as this would result in elimination of a neutral

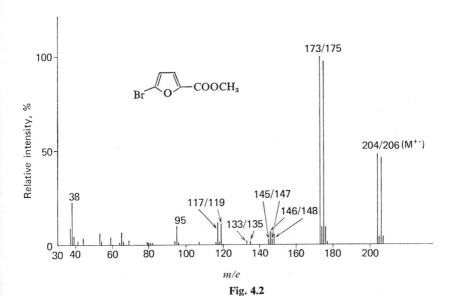

30c, *m/e* 146/148

30d, *m/e* 117/119 $\xrightarrow{-Br^{\cdot}}$ $C_3H_2^{+\cdot}$ *m/e* 38

$-CO_2CH_2$

$-^{\cdot}OCH_3$

30a, *m/e* 173/175

30, M$^{+\cdot}$

$-^{\cdot}CO_2H$

30b *m/e* 145/147

$-^{\cdot\cdot}CCO_2CH_3^{\cdot\cdot}$

30g, *m/e* 133/135

H rearrangement

$-CH_2O, Br^{\cdot}$
(arrows)

30e, *m/e* 95

30f

Fig. 4.2

fragment, $O=C=\cdot C—OCH_3 \leftrightarrow \cdot O—C\equiv C—OCH_3$, more stable than $:CCOOCH_3$.

30, M⁺·   30h   30g, m/e 133/135

It is interesting to compare the mass spectra of the two isomeric esters **31** and **32** (17). Ions due to M-CH₃· and M-CH₃OH are observed for **31**, where the methyl group is adjacent to the ester function, but are absent in **32**. Both spectra show prominent methylbromocyclopropenyl ions (**33**),

31   32   33, m/e 131/133

confirming that the ester group must be present at $C_{(2)}$ (or $C_{(5)}$) in both cases. The prominent M-(CH₂O + Br·) ion in **31** is characteristic of methyl 5-bromofuran-2-carboxylates, as mentioned above, and hence the structure of **31** could have been deduced by mass spectrometric evidence alone. A pronounced ion at m/e 43 for **32** indicates that an α-methyl group is present and hence only the position of the bromo group would have to be determined.

G. Other Substituted Furans

Spectra have been reported for a few furans with other simple functional groups. 2-Cyanofuran (**34**) behaves as expected, giving a cyclopropenyl ion at m/e 39 and a considerably more intense cyanocyclopropenyl ion (**34a**) (17). This further loses HCN, giving the ion C_3H^+ (m/e 37).

m/e 39   34, M⁺·, m/e 93   34a, m/e 64 m/e 37
(16%)   (100%)   (42%)   (12%)

5-Bromo-2-nitrofuran (**35**) (17) differs somewhat from nitrobenzene in that the M-NO₂ fragment is relatively weak and of comparable intensity to the M-NO ion. The most abundant fragment is the bromocyclopropenyl ion at m/e 117/119 which further characteristically loses a bromine atom to give the ion $C_3H_2^{+\cdot}$, m/e 38. An abundant fragment at m/e 133/135 is also observed, indicating loss of the nitro group together with $C_{(2)}$, but metastable ions have not been reported to distinguish between a two-step process (M-NO-CO) and a concerted one (possibly via **35a**). Another important ion occurs at m/e 96 (M-·OBr) but the mechanism of its formation is not obvious.

m/e 117/119   **35, M$^{+\cdot}$**   **35a** m/e 133/135

H. Bifurans and Difurylmethane

The mass spectra of bifurans are of considerable interest, as they may be complicated by rearrangement processes made possible by the proximity of the two furan rings (21). The parent ion is the base peak in the spectrum of 2,2'-bifuran (**36**), and the most abundant fragment ion is due to concerted loss of two molecules of carbon monoxide resulting in the stable benzene radical ion **36a**. The loss of the two molecules of carbon monoxide is shown

36, M$^{+\cdot}$ **36a**, m/e 78

m/e 95   **36b**, m/e 105

to be concerted by the appropriate metastable ion, and therefore suggests that the 2,2'- rather than the 5,5'-carbon atoms are involved, probably with simultaneous formation of the 3-3' carbon–carbon bond. The loss of two stable fragments (CO) and the formation of the stable benzene ion would then provide the driving force. Another important fragmentation pathway involves the loss of CHO· from the parent ion followed by rearrangement to the benzoyl ion (36b), while loss of a cyclopropenyl radical from the parent ion results in a weak 2-furoyl ion at m/e 95. Loss of a methoxyl group is first observed with the bifuran ester 37, followed by loss of two molecules of CO (most probably from the same ring) and rearrangement to the 4-methoxycarbonylbenzoyl ion (37a) (21).

37, M$^{+\cdot}$

$-2CO$, Rearrangement

37a, m/e 163

The spectrum of difurylmethane (38) resembles that of a benzylic compound as the base peak is at m/e 91, most probably due to the tropylium ion (38a). β-Cleavage with formation of the pyrilium ion m/e 81 (38b) is relatively unimportant. Other notable ions are due to loss of CO and CHO· from the

38b, m/e 81
(18%)

38, M$^{+\cdot}$

$-CO, -\dot{C}HO$

38a, m/e 91
(100%)

parent ion (m/e 120 and 119, respectively), and a weak ion is observed at m/e 105 (benzoyl?) corresponding to the loss of [CH$_3$ + CO]. Its formation would imply considerable hydrogen and skeletal rearrangement (22).

I.   Complex Furanoid Compounds

As the furan ring is of common occurrence in natural products and important in general organic chemistry, it is useful to determine to what extent mass spectrometry may be used in its detection and in elucidating its environment.

The characteristic β-cleavage of alkylfurans was used to advantage in the structure elucidation of a furanoid fatty acid 39 isolated from *Exocarpus* seed oil (23). An abundant molecular ion was observed, and consideration of the two prominent β-cleavage ions 39a (base peak) and 39b allowed the position of the furan ring in the alkyl chain to be unambiguously fixed. Weaker ions due to γ-cleavage were also observed.

β-Cleavage is also of great importance in 3-alkylfuran derivatives and results in the pyrilium ion (m/e 81) as the base peak dominating the spectra

$$CH_3—(CH_2)_3\overset{\gamma}{\underset{}{\vert}}CH_2\overset{\beta}{\underset{}{\vert}}CH_2\quad \text{[furan]}\quad CH_2\overset{\beta'}{\underset{}{\vert}}CH_2\overset{\gamma'}{\underset{}{\vert}}(CH_2)_5—CO_2CH_3$$

39, M$^{+\cdot}$, m/e 308 (78%)

39b, m/e 237 (23%)   39a, m/e 165 (100%)

of torreyal (**40**, R = CHO) and neotorreyol (**40**, R = CH$_2$OH) (24). In the spectrum of dendrolasin (**40**, R = CH$_3$) the m/e 81 ion is also intense, but the base peak is now due to the highly favored allylic cleavage with formation of the dimethylallyl ion (**40a**) (25); however, the corresponding fragmentation in **40** (R = CHO and CH$_2$OH) appears to be unimportant. Common features in the three spectra are weak peaks at m/e 67 and 95 due to cleavage α and γ to the furan ring, respectively, an ion at m/e 53 most probably due to loss of CO from the pyrilium ion, and a weak cyclopropenyl ion (m/e 39) as well as ions due to fragmentation of the side chain.

The combined gas–liquid chromatography (GLC)/mass spectrometric technique was used for the isolation and structure elucidation of perillen (**41**), a compound present in small amount in the mandibular glands of an ant (25). The molecular ion for **41** was at m/e 150 (C$_{10}$H$_{14}$O). A comparison of its spectrum with that of dendrolasin (**40**, R = CH$_3$) showed that ions at m/e 82, 81, 69, 53, and 39 were present and of the same relative intensity in both spectra, indicating that dendrolasin and perillen are very similar structurally. This allowed structure **41** to be assigned to perillen. Similarly β-cleavage results in an intense m/e 81 ion in the spectra of a number of diterpenes containing the 3-furylmethyl grouping such as daniellic acid (**42**) (26), marrubiin (**43**) (18), and related compounds (27, 28).

However an intense m/e 81 ion does not necessarily indicate the presence of a 3-furylmethyl group as this ion is also observed in the spectrum of columbin (**44**) where it must be formed by cleavage of two bonds β to the furan ring together with a hydrogen transfer to the furan moiety (18, 29, 30). This group of compounds may, however, be distinguished by an intense peak due to the stable vinylfuran ion (**44a**), possibly formed by a cyclic concerted mechanism and corresponding to the base peak for columbin and

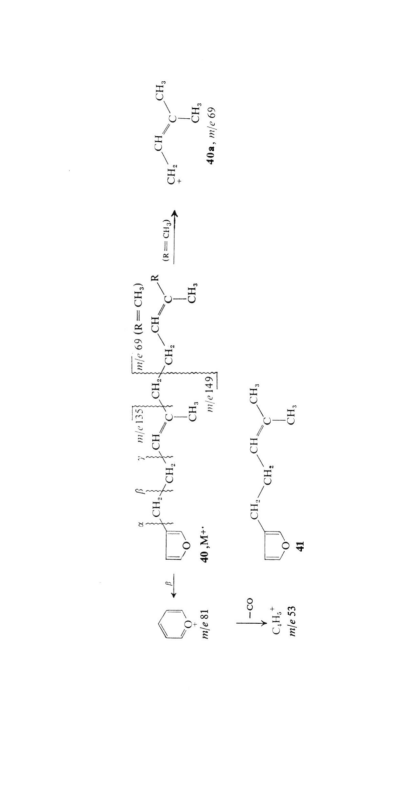

42, $M^{+\cdot}$ → m/e 81 (100% from **42**) (62% from **43**) ← **43**, $M^{+\cdot}$

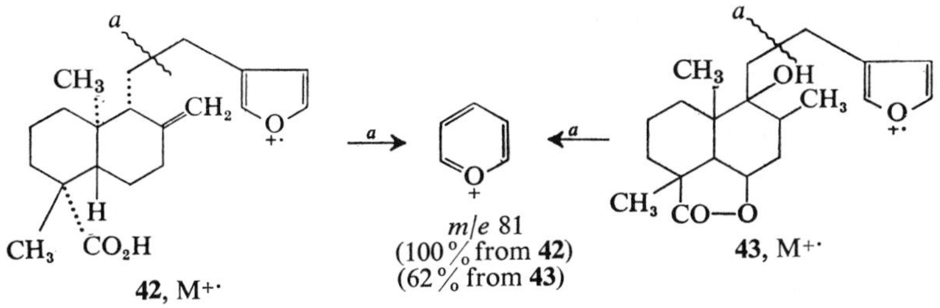

tinophyllone (**45**) (31). Charge retention on the alicyclic moiety **44b** occurs only to a minor extent as **44b** is less able to stabilize the positive charge.

In tinophyllone (**45**) the alternative fragmentation of the lactone ring, resulting in the loss of neutral 3-furanaldehyde and CO, is also observed

44 —fission at *a* with H transfer→ m/e 81 (51%)

44 —$-CO_2$ (arrows)→ **44a**, m/e 94 (100%) + **44b**, m/e 220 (3%)

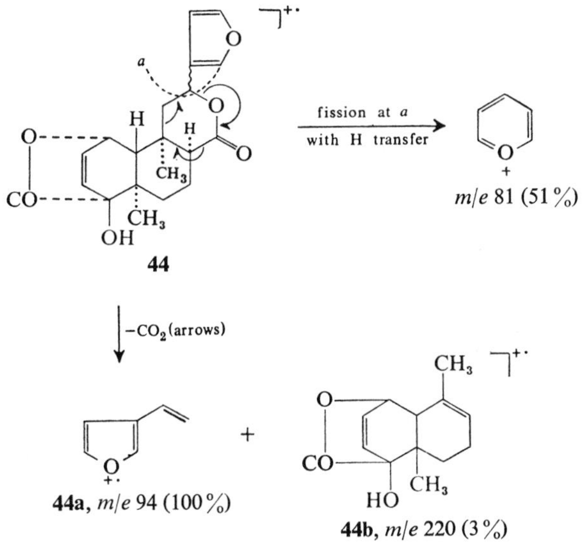

and leads to the ion **45a** which further fragments by loss of methanol (31). An ion corresponding to **45a**, however, does not appear to be significant in the spectrum of columbin.

In lactones with a 3-furyl group adjacent to the lactone oxygen atom, loss of 3-furanaldehyde may become an important fragmentation process if formation of the vinylfuran ion is not possible as in fraxinellon (**46**). Here the favored cleavage of the allylic bond *a–b* results in the ring-opened molecular ion **46a**, which may then readily eliminate 3-furanaldehyde leading to m/e 136

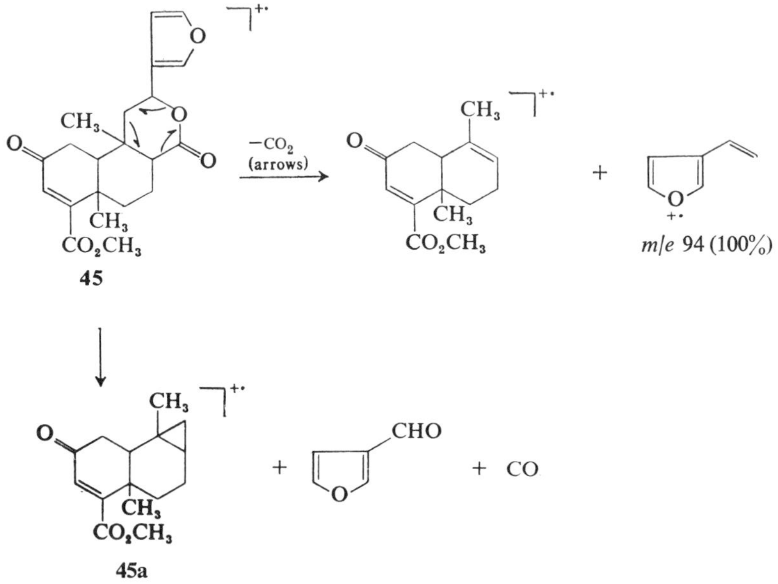

45 → **45a**

(**46b**) as the base peak of the spectrum (32). The dihydro-compound **47** fragments in a different manner. Here the carbonyl–carbon bond is probably broken first, resulting in loss of CO and formation of the ion **47a**(M-28). Loss of 3-furanaldehyde now readily occurs (**47a** → **47b**) and the ion **47b** further fragments as expected by loss of either a hydrogen atom (**47b** → **47c**) or a methyl group [**47b** → **47d** (base peak)].

Loss of 3-furanaldehyde (**48** → **48a**) is similarly observed from the molecular ion of methyl angolensate (**48**) where the furyl group is attached to a six-membered (δ) lactone ring (33). Considerable loss of 3-furanaldehyde is also observed from fragment ions and its loss from **48b** leads to **48c** as

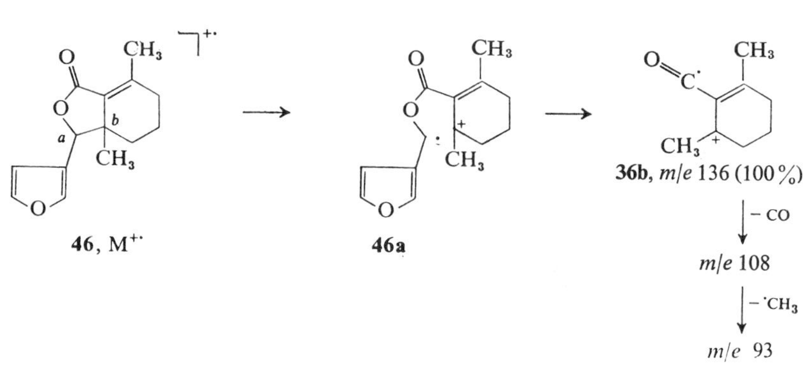

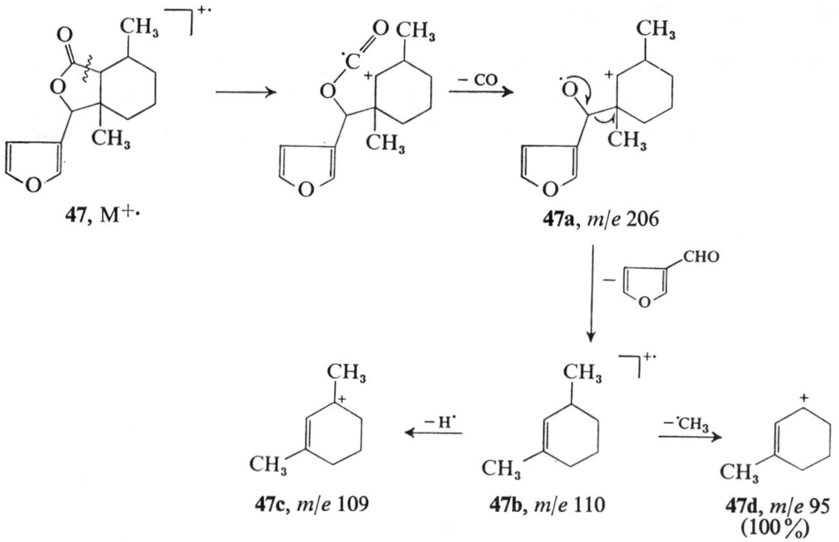

47, M+·
– CO
47a, m/e 206
– CHO
– H·
– ·CH3
47c, m/e 109
47b, m/e 110
47d, m/e 95 (100%)

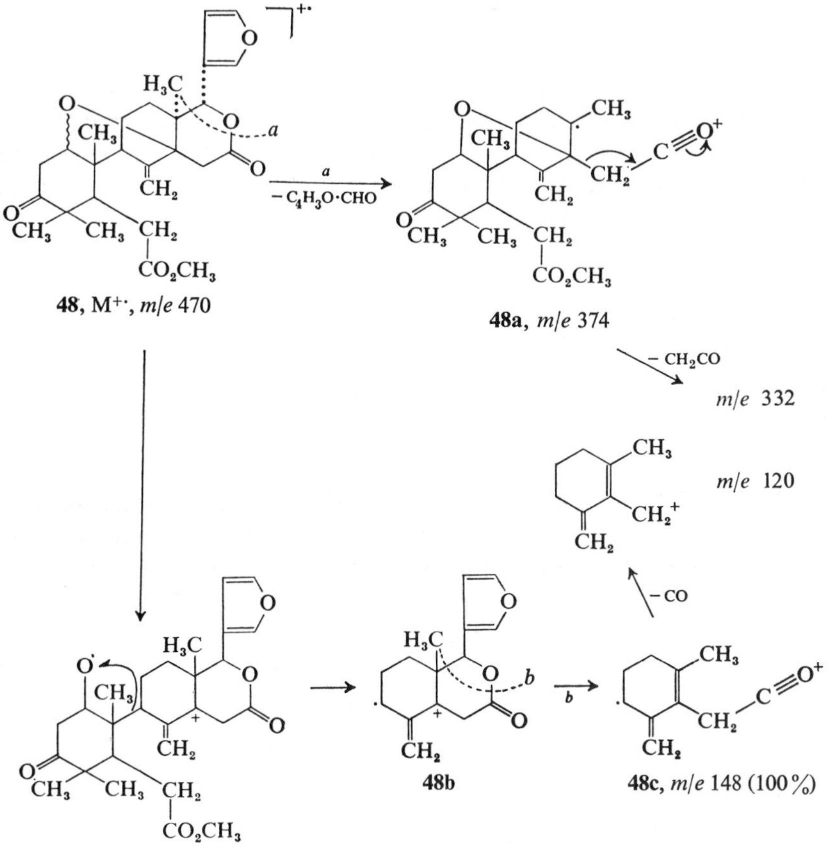

48, M+·, m/e 470
a
– C4H3O·CHO
48a, m/e 374
– CH2CO
m/e 332
m/e 120
– CO
48b
b
48c, m/e 148 (100%)

the base peak in the spectrum. Mass spectrometry may thus provide a means of distinguishing between the two isomeric groupings **49** and **50** in a compound, but further examples are necessary before generalizations can be made.

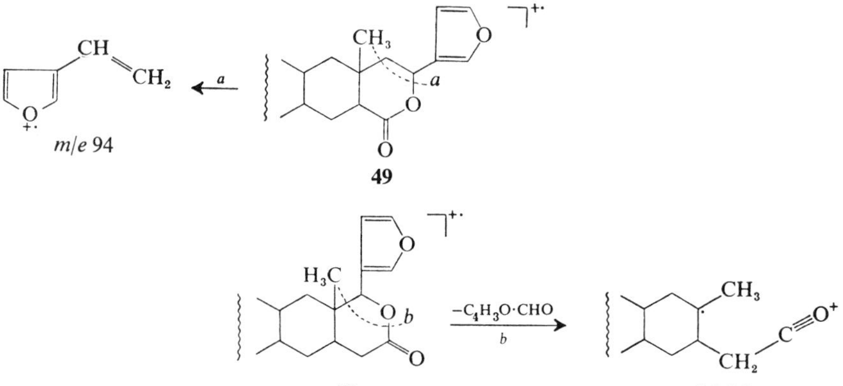

49

50   M-96

If a furan ring is fused at the 2- and 3-positions to a saturated six-membered carbocyclic ring, the retro-Diels-Alder reaction becomes the predominant fragmentation process. Thus for menthofuran (**51**) it results in the resonance-stabilized ion **51a** as the base peak, dominating the spectrum and accounting

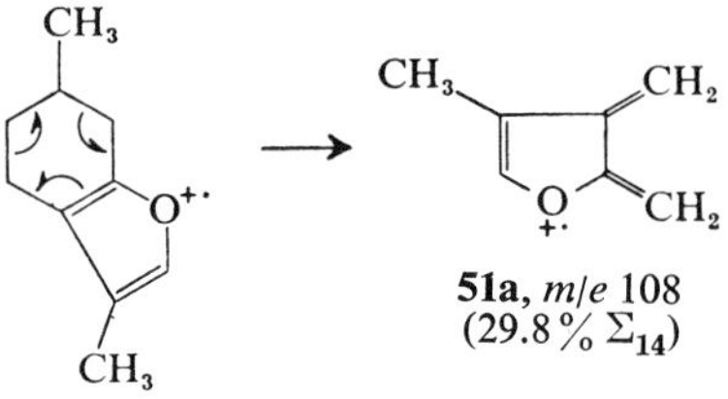

51a, m/e 108
(29.8% Σ_{14})

51

for 29.8% of the total ion current (Σ_{14}) (34). Little other fragmentation is observed, including only a very weak M-1 ion. Similarly, with the more complex methyl vouacapenate (**52**), the retro-Diels-Alder reaction leads to a

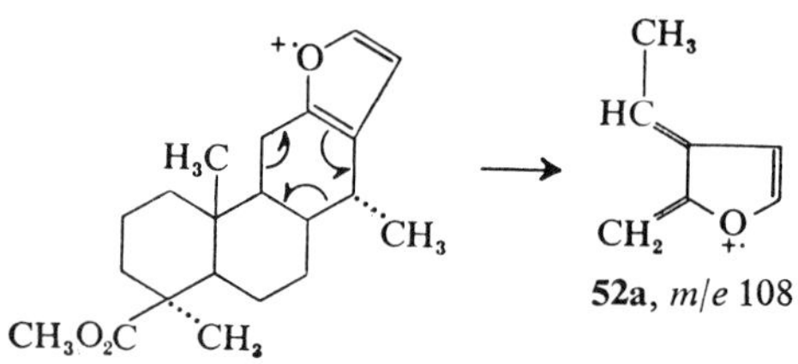

52a, m/e 108

52

very intense ion **52a** (*m/e* 108) (35), and in the spectrum of furanoligularenone (**53**) intense ions are observed for both the retro-Diels-Alder fragments (**53a** and **53b**) (36). If a carbonyl group is present α to the furan ring as in

53a *m/e* 122   **53**   **53b**, *m/e* 108 (100%)

euryopsonol (**54**), the retro-Diels-Alder reaction remains important and results in a characteristic mass shift in the ion **54a** (37). An intense ion at *m/e* 108, however, does not necessarily indicate a methylfuran fused at the

54   **54a**, *m/e* 122 (100%)

2- and 3-positions to a reduced six-membered ring as is shown by alexandrofuran (**55**) where the base peak is also *m/e* 108 (**55a**) (38) and the retro-Diels-Alder reaction as such cannot be operative.

55   **55a**, *m/e* 108 (100%)

   The most favored cleavage for **56** [condensation product of (+)-menthofuran with 4-methylcyclohexanone] and some related compounds is of a bond β to both of the furan rings, resulting in the fully conjugated molecular ion **56a**. Hydrogen transfer and loss of a *sec*-butyl radical results in the well-stabilized ion **56b** as the base peak (39). Loss of propene via the retro-Diels-Alder fragmentation of the two menthofuran moieties does not appear to be of major importance (contrast menthofuran itself).

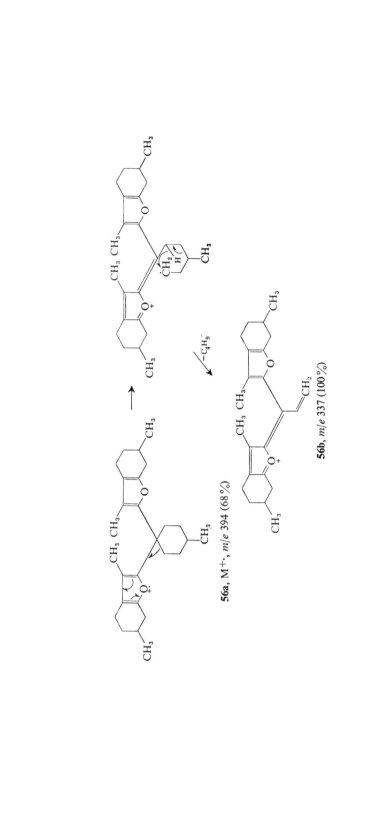

56a, M⁺·, *m/e* 394 (68%)

$-C_4H_9\cdot$

56b, *m/e* 337 (100%)

Finally it may be added that the combined GLC/mass spectrometric technique has been valuable in the identification of a considerable number of simple furan compounds as constituents of coffee aroma (15, 40), and of 2-*n*-pentylfuran and furfural as constituents of the aroma of cranberries (41).

II. BENZOFURANS

The mass spectrum of benzofuran (**57**) is characterized by an intense molecular ion as the base peak (42), its high stability being due to extension of the conjugation of the furan ring to the benzofuran system and to lack of easily cleaved bonds. The two major fragments *m/e* 89 and 90 are due to loss of CHO˙ and CO from the parent ion, respectively. Loss of CO may be envisaged as resulting in the benzocyclopropene radical ion **57a** (or its open chain isomer **57b**) which would then lose a hydrogen atom to give the *m/e*

57, M⁺˙, *m/e* 118 (100%)

| −CO

−˙CHO

57a, *m/e* 90 (33%)   **57c**, *m/e* 89   **57e**, *m/e* 89
(30%)   or

or   CH≡C—CH=CH—CH₂—C≡CH ⎤⁺˙   CH≡C—CH=CH—CH—C≡CH
57b, *m/e* 90   **57d**, *m/e* 89

|−C₂H₂

C₅H₃⁺
m/e 63 (16%)

89 ion (**57c**, **57d**, or perhaps **57e**). Elimination of acetylene from *m/e* 89 gives C₅H₃⁺ (*m/e* 63), the only other significant ion in the spectrum. The *m/e* 89 ion may also arise directly from the parent ion as shown, by loss of ˙CHO in a mechanism analogous to that for the formation of C₃H₃⁺ from furan (Sec. I.A), but this direct path is not substantiated by a metastable peak.

The benzodifuran **58** also shows an intense molecular ion, and fragments by two successive losses of CO (**58** → **58a** → **58b**) (43). The ion **58a** further fragments by the loss of H˙ to a small extent. Naphtho[2,1-*b*]furan (**59**) also follows the same general pattern but here the M-(CO + H˙) ion is considerably more intense than M-CO (43).

58, M⁺˙, *m/e* 158   —CO→   **58a**, *m/e* 130   —CO→   **58b**, *m/e* 102
(100%)

|—H

59   **58c**, *m/e* 129   or   or noncyclic isomer

The mass spectra of 2-methyl- and 2-ethylbenzofuran (**60**, R = H and CH₃, respectively) are very similar to their corresponding 3-isomers **61** (42). As in alkylfurans, cleavage β to the furan ring is the dominant fragmentation process, resulting in the stable ring-expanded aromatic chromenyl ion **62** as the base peak in all four spectra. Weak ions at M-1 (*m/e* 145) due to β-cleavage with loss of H˙ are also observed for the two ethyl isomers.

60, R = H, *m/e* 132 (79%)   **62**, *m/e* 131   **61**, R = H, *m/e* 132 (94%)
R = CH₃, *m/e* 146 (39%)   (100%)   R = CH₃, *m/e* 146 (45%)

2,3-Dimethylbenzofuran (**63**) is of interest as the M-CH₃˙ ion **63a** is more intense than M-1 (**63b**) (42), and it is most probable that loss of the methyl group occurs during or after the rearrangement and ring-expansion of the molecular ion [cf. dimethylfurans (Sec. I.B) and xylenes (12)]. The M-1 ion (**63b**) fragments further to a small extent by loss of CO (**63b** → **63c**) followed by loss of a molecule of hydrogen to the stable indenyl ion, **63d**, *m/e* 115.

Abundant M-1 ions are also observed in the spectra of benzofurans with methyl substituents on the benzene ring, such as 4,7-dimethylbenzofuran (**64**)

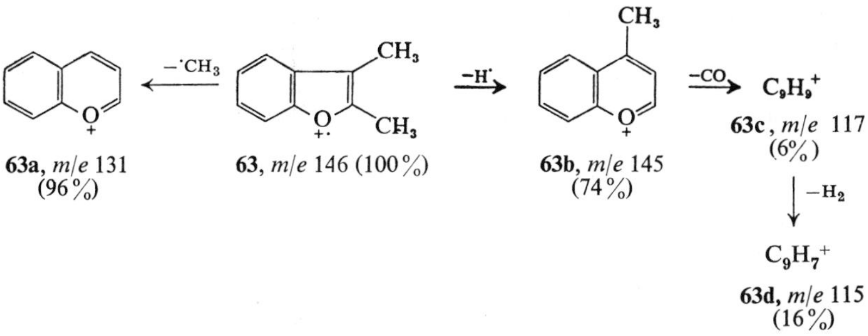

−·CH₃
CH₃
CH₃
−H·
CH₃
−CO
C₉H₉⁺
63a, m/e 131 (96%)
63, m/e 146 (100%)
63b, m/e 145 (74%)
63c, m/e 117 (6%)
−H₂
C₉H₇⁺
63d, m/e 115 (16%)

(44). It has been suggested (44) that the conjugated oxonium ions **64a** and **64b**, formed by loss of H˙ from the $C_{(4)}$ and $C_{(7)}$ methyl groups, respectively, are a better representation of the M-1 ion than the tropylium ion structure **64c**. The grounds for this suggestion are that, while in *p*-xylene the loss of $CH_3^{\cdot}$ and formation of the tropylium ion (*m/e* 91) results in the base peak and M-1 is only 30% the intensity, the reverse is true for **64**, M-1 being much more intense than M-$CH_3^{\cdot}$. Alternatively the M-1 ion may be formulated as the ring-expanded 10π-electron system **64d**.

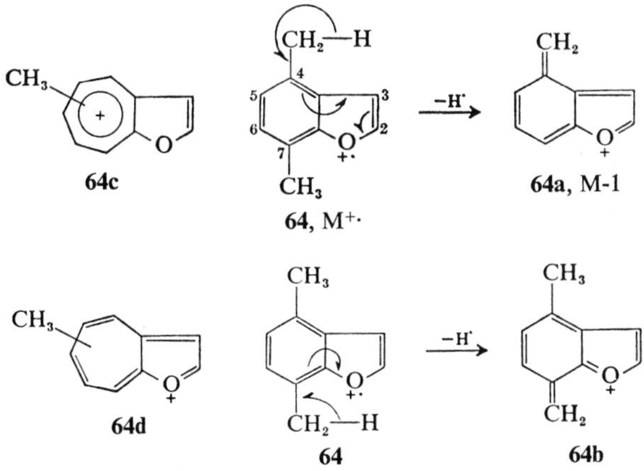

CH₃
64c
CH₂—H
−H·
CH₂
CH₃
64, M⁺·
64a, M-1
CH₃
CH₃
CH₃
−H·
64d
CH₂—H
64
CH₂
64b

The mass spectra of a few benzofurans with substituents other than alkyl groups have been reported. With benzofurans having a carbonyl substituent at $C_{(2)}$ two distinct fragmentation paths are observed. Cleavage β to the furan ring with loss of R˙ is the most important fragmentation for 2-formyl- and 2-acetylbenzofuran (42) (**65**, R = H and CH_3, respectively), giving the ion **65a** (base peak in the latter case) which further fragments by a concerted loss of two molecules of carbon monoxide to the m/e 89 ion. The minor fragmentation pathway involves loss of CO in the 2-formyl compound (**65**, R = H) and loss of ketene for **65**, R = CH$_3$, with hydrogen rearrangements in both cases, giving the benzofuran radical ion **65c** which further fragments

by loss of CO followed by H$^{\cdot}$ to m/e 89. With benzofuran-2-carboxylic acid (**66**) (45) a third fragmentation pathway as well as the two pathways described for **65** is observed which involves three successive losses of carbon monoxide, possibly as outlined in the scheme **66** → **66a** → **66b** → **66c** resulting in ionized benzene as the stable final product.

Nitrobenzofurans substituted in the benzene ring resemble other aromatic nitrocompounds in that they readily lose NO and NO$_2$ from the molecular ion (18), while acetamidobenzofurans fragment very readily by loss of ketene and less readily by a further elimination of CHO$^{\cdot}$ from the furan ring (18).

A very pronounced "ortho-effect" is operative in the suitably 2,3-disubstituted benzofuran derivative **67**, loss of ethanol resulting in the base peak (**67a**) which further loses CO giving **67b**, while loss of $^{\cdot}$OC$_2$H$_5$, a normally favored process, is of considerably lesser importance (46). In the N-acetyl compound **68** loss of ethanol directly from the parent ion (**68** → **68a**) is of minor importance. Loss of ketene gives **67** and from then on the spectrum is very similar to that of **67** (46).

Discussion of the spectra of furanocoumarins and furanochromones is reserved for Chap. 5.

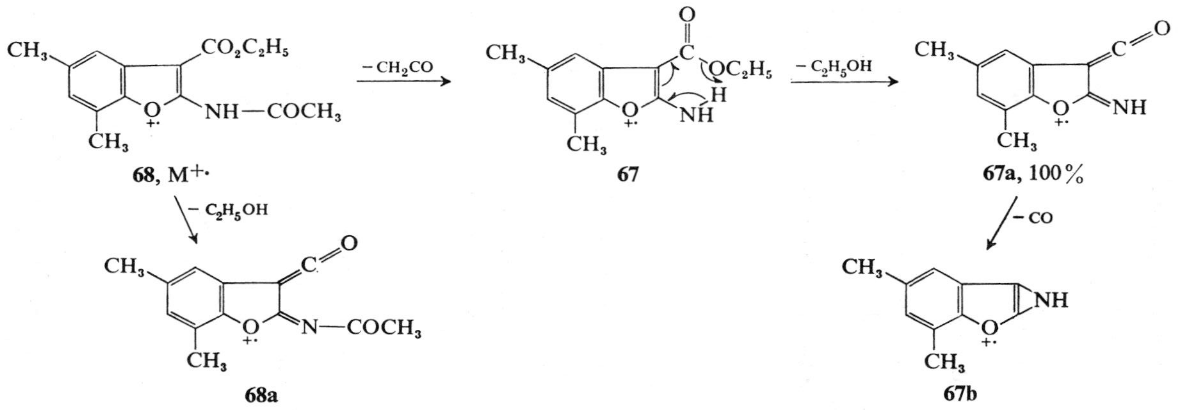
CH3
CO2C2H5
NH—COCH3
CH3
68, M+·
− CH2CO
− C2H5OH
67
67a, 100%
− CO
68a
67b

The spectrum of the hetero-pseudoaromatic cycloocta[*c*]furan (**69**) appears to be rather complex (47). Ions at *m*/*e* 118, 115, and 89 may be assigned the structures **69a**, **69b**, and **69c**, respectively.

$-C_2H_2$
$-{}^{\cdot}CHO$
$-C_2H_2$
$C_7H_5^+$

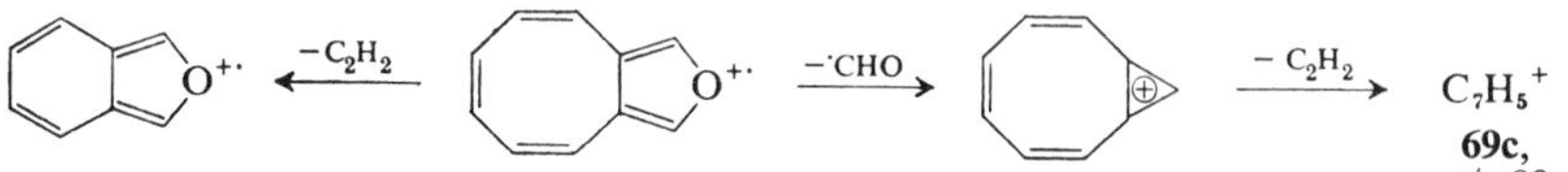

69a, *m*/*e* 118   **69**, $M^{+\cdot}$, *m*/*e* 144   **69b**, *m*/*e* 115   **69c**, *m*/*e* 89

III. DIBENZOFURANS

Dibenzofuran (**70**) is a highly stable aromatic system, and as might be expected the intense molecular ion dominates the spectrum (48, 49). The only important fragmentation pathway is loss of CO followed by loss of $H^{\cdot}$ giving *m*/*e* 139 which may possibly be the benzodehydrotropylium ion (**70a**).

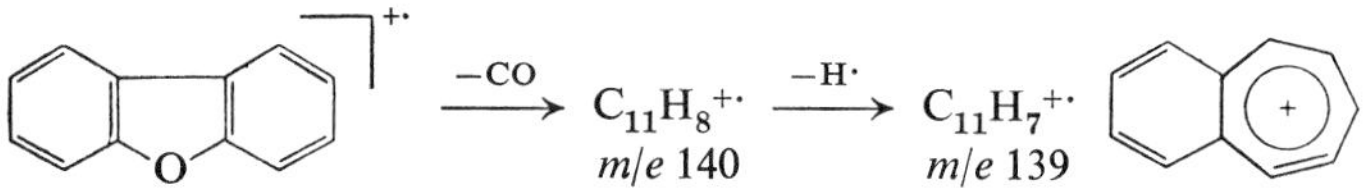

$-CO$
$C_{11}H_8^{+\cdot}$
m/*e* 140
$-H\cdot$
$C_{11}H_7^{+\cdot}$
m/*e* 139

70, $M^{+\cdot}$, *m*/*e* 168 (100%)   **70a**

Mass spectrometry has been useful in the structure elucidation of "dihydroisothelephoric acid hexaacetate" (**71**) (50). Apart from establishing the correct molecular formula, the major fragmentation path involving five stepwise losses of ketene followed by the loss of an acetoxyl radical (with formation of the conjugated oxonium ion **71a**) showed that one of the acetoxyl groups was different to the others.

CH_2—O—$COCH_3$
CH_3CO—O
O—$COCH_3$
CH_3CO—O
O—$COCH_3$
O—$COCH_3$
$-5\,CH_2CO,$
$-CH_3CO_2^{\cdot}$
CH_2
HO
OH
HO
OH
OH

71, $M^{+\cdot}$

71a

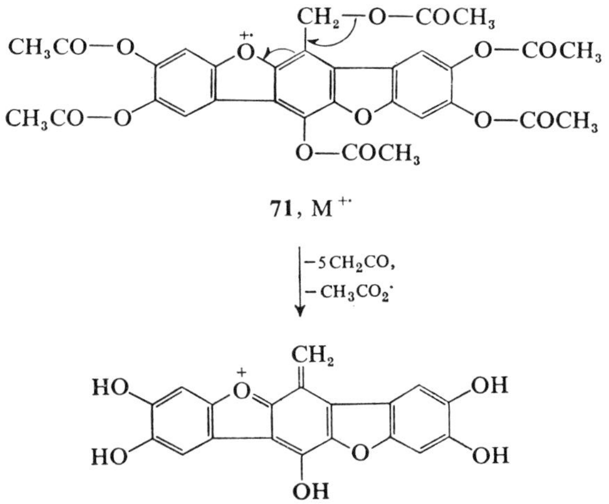

Addenda

A study of the fragmentation of a range of substituted benzofurans (Section II) has appeared (51). The spectra of the complex benzofuran usnetol (Section II) and the naturally occurring dibenzofuranoid compound usnic acid (Section III) have been analysed (52).

REFERENCES

1. J. Collin, *Bull. Soc. Chim. Belges*, **69**, 449 (1960).
2. American Petroleum Institute, Research Project 44. Spectrum 545.
3. I. Omura, K. Higasi, and H. Baba, *Bull. Chem. Soc. Japan*, **29**, 501 (1956).
4. K. Biemann, *Mass Spectrometry*, McGraw-Hill, New York, 1962, p. 202.
5. H. Budzikiewicz, C. Djerassi, and D. H. Williams, *Interpretation of Mass Spectra of Organic Compounds*, Holden-Day, San Francisco, 1964, p. 226.
6. D. H. Williams, R. G. Cooks, J. Ronayne, and S. W. Tam, *Tetrahedron Lett.*, **1968**, 1777.
7. A. Padwa and R. Hartman, *J. Amer. Chem. Soc.*, **88**, 3759 (1966).
8. R. Srinivasan, *J. Amer. Chem. Soc.*, **89**, 1758 (1967).
9. J. Burdon, J. C. Tatlow, and D. F. Thomas, *Chem. Commun.*, **1966**, 48.
10. K. Heyns, R. Stute, and H. Scharmann, *Tetrahedron*, **22**, 2223 (1966).
11. H. Budzikiewicz, C. Djerassi, and D. H. Williams, *Mass Spectrometry of Organic Compounds*, Holden-Day, San Francisco, 1967, p. 616.
12. H. M. Grubb and S. Meyerson in *Mass Spectrometry of Organic Ions* (F. W. McLafferty, Ed.), Academic Press, New York, 1963, pp. 516–519.
13. G. Spiteller in *Advances in Heterocyclic Chemistry*, Vol. 7 (A. R. Katritzky and A. J. Boulton, Eds.), Academic Press, New York, 1966, p. 306.
14. S. S. Chang, T. H. Smouse, R. G. Krishnamurthy, B. P. Mookhertee, and R. B. Reddy, *Chem. Ind.* (London), **1966**, 1926.
15. M. Stoll, M. Winter, F. Gautschi, I. Flament, and B. Willhalm, *Helv. Chim. Acta*, **50**, 628 (1967).
16. Ref. 4, pp. 111–112.
17. R. Grigg, M. V. Sargent, D. H. Williams, and J. A. Knight, *Tetrahedron*, **21**, 3441 (1965).
18. R. I. Reed and W. K. Reid, *J. Chem. Soc.*, **1963**, 5933.
19. R. I. Reed, W. K. Reid, and J. M. Wilson in *Advances in Mass Spectrometry*, Vol. II, (R. M. Elliot, Ed.), Pergamon, London, 1963, p. 416.
20. H. Budzikiewicz, C. Djerassi, A. H. Jackson, G. W. Kenner, D. J. Newman, and J. M. Wilson, *J. Chem. Soc.*, **1964**, 1949.
21. R. Grigg, J. A. Knight, and M. V. Sargent, *J. Chem. Soc.*, C, **1966**, 976.
22. Ref. 4, p. 139.
23. L. J. Morris, M. O. Marshall, and W. Kelly, *Tetrahedron Lett.*, **1966**, 4249.
24. T. Sakai, K. Nishimura, and Y. Hirose, *Bull. Chem. Soc. Japan*, **38**, 381 (1965).
25. R. Bernardi, C. Cardani, D. Ghiringhelli, A. Selva, A. Baggini, and M. Pavan, *Tetrahedron Lett.*, **1967**, 3893.
26. C. R. Enzell and R. Ryhage, *Ark. Kemi*, **23**, 367 (1965).
27. V. P. Arya and R. Ryhage, *J. Sci. Ind. Res.*, **21B**, 283 (1962).
28. L. Canonica, B. Rindone, C. Scolastico, G. Ferrari, and C. Casagrande, *Tetrahedron Lett.*, **1967**, 2639.

29. R. I. Reed in *Mass Spectrometry of Organic Ions* (F. W. McLafferty, Ed.), Academic Press, New York, 1963, Chap. 13.
30. Ref. 13, p. 325.
31. G. Anguilar-Santos, *Chem. Ind.* (London), **1965**, 1074.
32. M. Pailer, G. Schaden, G. Spiteller, and W. Fenzl, *Monatsh. Chem.*, **96**, 1324 (1965).
33. C. W. L. Bevan, J. W. Powell, D. A. H. Taylor, T. G. Halsall, P. Tuft, and M. Welford, *J. Chem. Soc., C*, **1967**, 163.
34. Ref. 4, p. 106.
35. H. E. Audier, S. Bory, M. Fétizon, and N.-T. Anh, *Bull. Soc. Chim. Fr.*, **1966**, 4002.
36. F. Patil, J.-M. Lehn, G. Ourisson, Y. Tanahashi, and T. Takahashi, *Bull. Soc. Chim. Fr.*, **1965**, 3085.
37. D. E. A. Rivett and G. R. Woolard, *Tetrahedron*, **23**, 2431 (1967).
38. H. Budzikiewicz, C. Djerassi, and D. H. Williams, *Structure Elucidation of Natural Products by Mass Spectrometry*, Vol. II, Holden-Day, San Francisco, 1964, p. 152.
39. P. H. Boyle, W. Cocker, T. B. H. McMurry, and A. C. Pratt, *J. Chem. Soc., C*, **1967**, 1993.
40. M. A. Gianturco, A. S. Giammarino, P. Friedel, and V. Flanagan, *Tetrahedron*, **20**, 2951 (1964).
41. K. Anjou and E. von Sydow, *Acta. Chem. Scand.*, **21**, 945 (1967).
42. B. Willhalm, A. F. Thomas, and F. Gautschi, *Tetrahedron*, **20**, 1185 (1964).
43. C. E. Loader and C. J. Timmons, *J. Chem. Soc., C*, **1967**, 1677.
44. Ref. 11, p. 622.
45. N. S. Vulf'son, V. I. Zaretskii, and V. G. Zaikin, *Isv. Akad. Nauk SSSR, Ser. Khim.*, **1963**, 2215.
46. G. Spiteller, *Monatsh. Chem.*, **92**, 1142 (1961).
47. E. Le Goff and R. B. La Count, *Tetrahedron Lett.*, **1965**, 2787.
48. Ref. 2. Spectrum 633.
49. C. S. Barnes and J. L. Occolowitz, *Aust. J. Chem.*, **17**, 975 (1964).
50. J. Gripenberg and M. Lounasmaa, *Acta. Chem. Scand.*, **20**, 2202 (1966).
51. E. N. Givens, L. G. Alexakos, and P. B. Venuto, *Tetrahedron*, **25**, 2407 (1969).
52. R. M. Letcher, *Org. Mass Spec.*, **1**, 551 (1968).

5 PYRONES AND BENZOPYRONES

I. 2-PYRONES

The spectrum of 2-pyrone (**1**) exhibits an abundant molecular ion (Fig. 5.1), and the two major fragments are the M-CO ion ($C_4H_4O^+$) at m/e 68 and $C_3H_3^+$ (m/e 39, base peak) (1). The subsequent fragmentation of the M-CO ion is very similar to that of furan (Chap. 4), the major fragment being m/e 39 (most probably the cyclopropenyl ion **1b**) formed both by direct loss of CHO· and by sequential loss of CO and H·. On this basis it would be reasonable to infer that the M-CO ion (m/e 68) was the furan radical ion **1a**. If the only (or at least the major) pathway to m/e 39 (**1b**) were via the furan radical ion (**1a**), equal incorporation of deuterium into **1b** (and all other ions derived from **1a**) would be expected for 3-deutero-2-pyrone (**2**) and 6-deutero-2-pyrone (**3**) as loss of CO from the molecular ion in both cases would result in $C_{(3)}$ and $C_{(6)}$ becoming equivalent α-positions in **1a**. However, the incorporation of deuterium into **1b** (i.e., shift to m/e 40) was found to be much greater in the 3-deutero-compound **2** than in **3**, and it was concluded

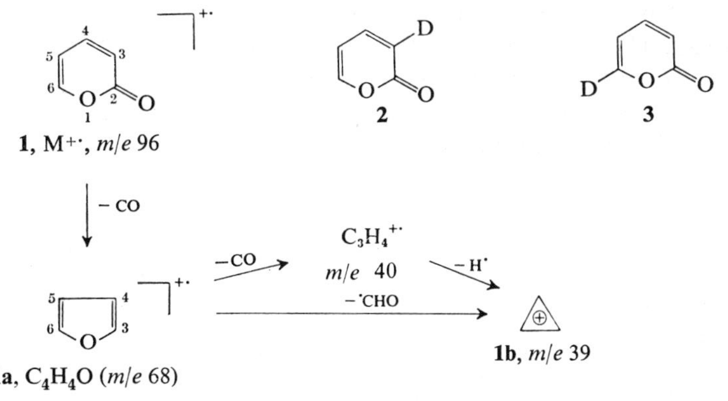

1, M⁺·, m/e 96

$-$ CO

1a, C_4H_4O (m/e 68)

$-CO \rightarrow C_3H_4^{+·}$ m/e 40 $-H·$; $-·CHO \rightarrow$ **1b**, m/e 39

2

3

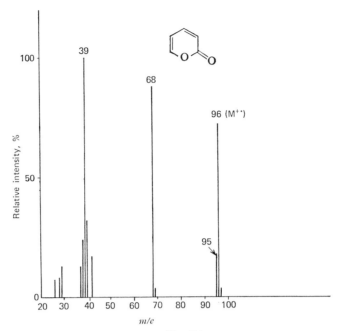

Fig. 5.1

that the M-CO ion ($C_4H_4O^{+\cdot}$) did not have the furan structure **1a** (1). It has been pointed out (2) that the intermediacy of the furan radical ion (**1a**) in the fragmentation of 2-pyrone is still an open question as the differences in isotopic distribution in the spectra of **2** and **3** can be equally well explained by multiple pathways to m/e 39 (**1b**). Strong evidence for a second pathway to m/e 39 (**1b**), involving initial specific loss of the $C_{(6)}$ hydrogen atom (**1 →** **1c**) followed by two successive losses of carbon monoxide (**1c → 1d → 1b**), is provided by the observation that while **2** exhibits only an M-1 ion, **3** (which has a $C_{(6)}$ deuterium atom) shows only an M-2 ion. This accounts for the lesser extent to which the m/e 39 ion **1b** is shifted in the case of **3**. This second pathway also occurs in a number of variously substituted 2-pyrones as evidenced by the appropriate metastable ions.

Evidence supporting the formation of a $C_{(3)}$—O bond after the expulsion of CO in some 2-pyrones is provided by the spectrum of 3-bromo-4,6-dimethyl-2-pyrone (**4**) (3). Here the appropriate metastable ions (*) show that an important fragmentation pathway involves initial loss of CO from the parent ion (**4 → 4a**) followed by sequential loss of the bromine atom (**4a → 4b**) and a second molecule of CO (**4b → 4c**) to give the dimethyl-cyclopropenyl ion **4c** (m/e 67) whose composition ($C_5H_7^+$) has been established by high-resolution measurement. As **4c** contains the two methyl groups and the ring hydrogen of **4** it also contains $C_{(4)}$, $C_{(5)}$, and $C_{(6)}$; and as $C_{(2)}$ has already been lost as CO in the first step (**4 → 4a**), the second loss of CO

4, M⁺·   **4a**, m/e 174 / 176   **4c**, m/e 67

m/e 95

4b

(**4b → 4c**) must involve $C_{(3)}$ and ring O. (It is assumed, of course, that the loss of Br' (**4a → 4b**) is not accompanied by the unlikely simultaneous migration of the $C_{(6)}$ methyl group to $C_{(3)}$.) This indicates that formation of the $C_{(3)}$—O bond has occurred at some earlier stage of the fragmentation sequence. While the most attractive representation (which will be used henceforth) of the M-CO ion from 2-pyrones would be a furan radical ion, the considerations above do not constitute proof; it would have to be shown that both the $C_{(3)}$—O and the $C_{(6)}$—O bonds are present simultaneously in the M-CO ion. It is, however, also conceivable that a number of structurally different ions of the same composition, including a furan radical ion, contribute to M-CO, and it must be remembered that the structure of the molecular ion from furan itself is as yet unknown.

The mass spectra of a number of other substituted monocyclic 2-pyrones have been reported. The expulsion of CO and the formation of substituted cyclopropenyl ions are processes of general importance together with the initial loss of the $C_{(6)}$ substituent. A typical example is 4-methoxy-6-methyl-2-pyrone (**5**) (Fig. 5.2) where loss of CO results in the base peak, the ion being formulated as **5a** (4, 5). This further fragments in a number of ways typical of similarly substituted furan derivatives. Thus loss of CHO· and CH₃CO· result in the cyclopropenyl ions **5b** (m/e 83) and **5c** (m/e 69), respectively, and the intense acetyl ion (**5e**, m/e 43) is typical of 2-methylfurans (Chap. 4,

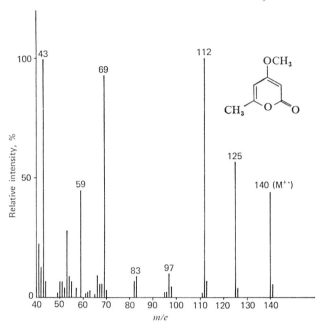

Fig. 5.2

Sec. I.B), while cleavage β to the furan ring results in the loss of the methoxyl methyl group with the formation of **5d**. An intense M-CH$_3$· ion (**5f**) is also observed which fragments by concerted loss of two molecules of CO to give **5c** (m/e 69). This suggests that the formulation **5g** is involved as it is unlikely that two isolated CO groupings would be lost simultaneously (6).

An "ortho-effect" is observed in suitably substituted 2-pyrone derivatives such as 5-carboxy-4,6-dimethyl-2-pyrone (**6**) where loss of water from the molecular ion results in **6a** (5). Expulsion of CO gives **6b** and this too fragments further by elimination of water (**6b → 6c**). An interesting further fragmentation of the M-CH$_3$· ion, **7a**, of 5-methoxycarbonyl-6-methyl-2-pyrone (**7**) (5) is the loss of the methoxycarbonyl group with hydrogen transfer, probably resulting in the β-furoyl ion (**7b**).

The spectra of 4-hydroxy-2-pyrones such as 4-hydroxy-6-methyl-2-pyrone (**8**) and dehydroacetic acid (**9**) are consistent with the molecular ion being mainly in the enol form as the fragmentation pattern resembles that of other 2-pyrones and the retro-Diels-Alder reaction does not appear to be important (5). However, an important ion (**10**) at m/e 85 in both spectra is probably best envisaged as arising from the keto form (e.g., **8a** in the case of **8**) by a mechanism of the "retro-Diels-Alder reaction with hydrogen transfer" type which has been observed in 4-hydroxycoumarins (Sec. III.C).

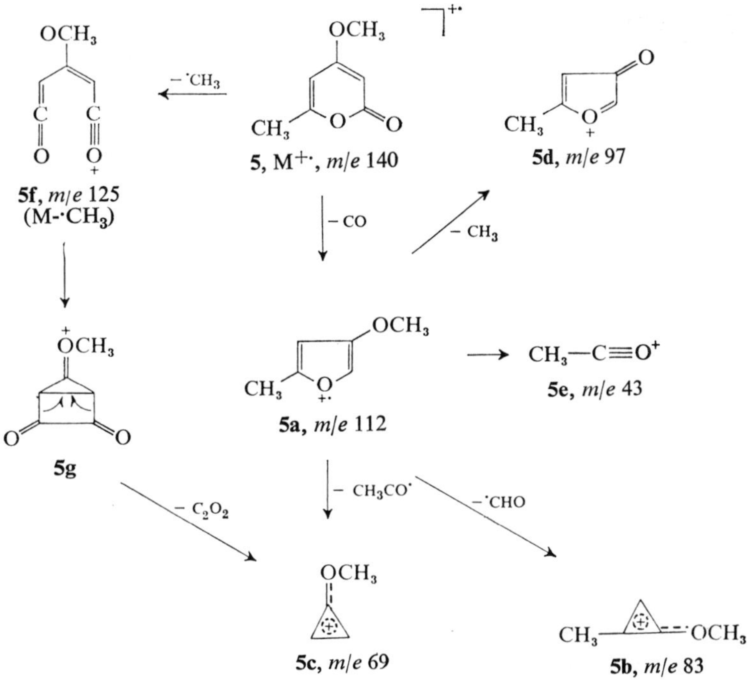
OCH3
– ·CH3
C
O
C
O
+
5f, m/e 125
(M-·CH3)
OCH3
CH3
O
O
5, M+·, m/e 140
– CO
– CH3
O
CH3
O
+
5d, m/e 97
OCH3
CH3
O
+·
5a, m/e 112
CH3—C≡O+
5e, m/e 43
+
OCH3
O
O
5g
– C2O2
– CH3CO·
– ·CHO
OCH3
5c, m/e 69
CH3
·OCH3
5b, m/e 83

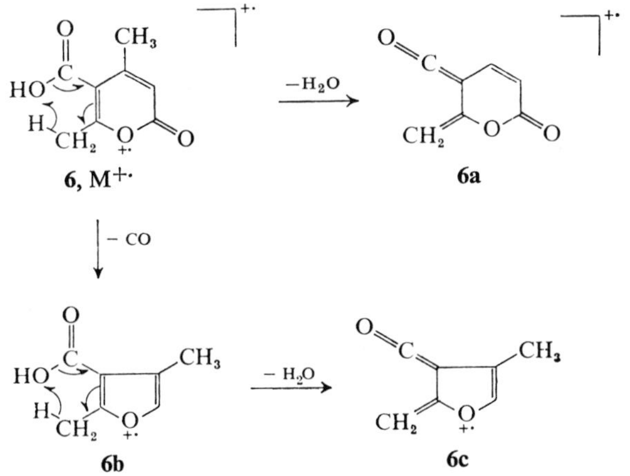
O
CH3
HO
C
H
CH2
O
+·
O
6, M+·
–H2O
O
C
CH2
O
O
6a
– CO
O
C
CH3
HO
H
CH2
O
+·
6b
– H2O
O
C
CH3
CH2
O
+·
6c

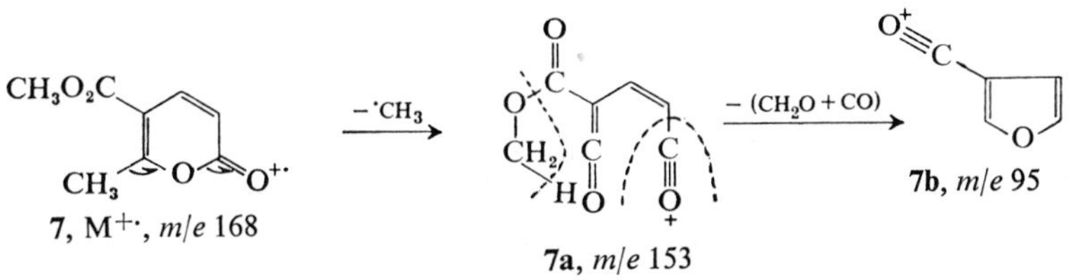
CH3O2C
CH3
O
O+·
7, M+·, m/e 168
– ·CH3
O
C
O
CH2
H
C
O
C
O
+
7a, m/e 153
– (CH2O + CO)
O+
C
O
7b, m/e 95

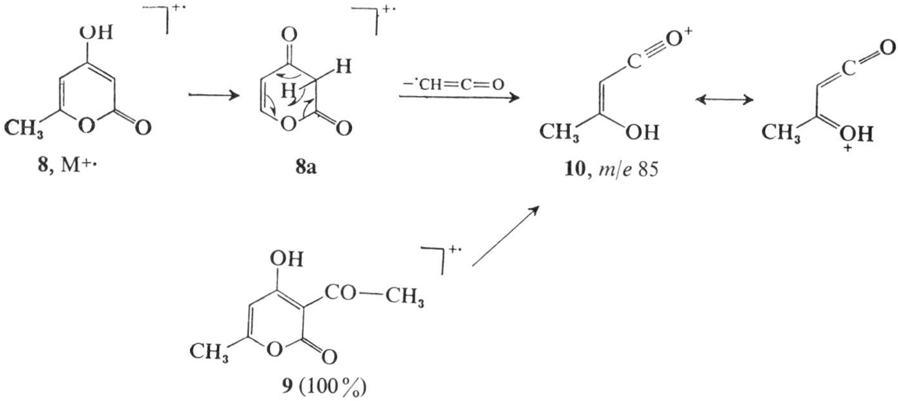

8, M+·   8a   −·CH=C=O   10, m/e 85

9 (100%)

II. 4-PYRONES

The mass spectrum of 4-pyrone (11) has been studied in detail with the use of both deuterium- and ^{18}O-labeling techniques (7). It is of interest to compare the spectrum with that of 2-pyrone (1). The M-CO ion (m/e 68, and most probably the furan radical ion 11a) and the ion at m/e 39 ($C_3H_3^+$) are now much reduced in importance, being only about 16% and 9%, respectively, of the intensity of the base peak which is due to the parent ion (m/e 96). The important fragmentation process is the retro-Diels-Alder

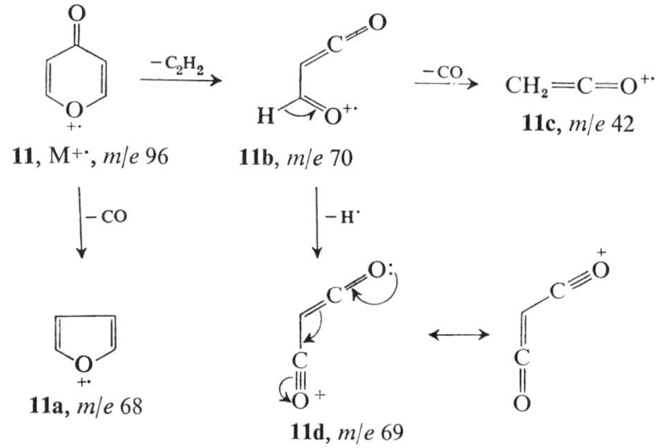

11, M+·, m/e 96   −C_2H_2 11b, m/e 70   −CO   11c, m/e 42

−CO   −H·

11a, m/e 68   11d, m/e 69

reaction with loss of neutral acetylene and charge retention on the ketene fragment 11b, which then loses CO to give the ketene radical ion (11c), m/e 42. The only other important ion is 11d (m/e 69) which low ionization-voltage studies suggest arises by loss of H· from m/e 70 (11b).

The spectra of the isomeric 3,5-dimethyl-4-pyrone (**12**) (7) and 2,6-dimethyl-4-pyrone (**13**) (7) provide an interesting comparison. In both cases loss of propyne via the retro-Diels-Alder reaction results in relatively weak ions at m/e 84 (**12a** and **13a**) whose further fragmentations differ markedly

12, M$^{+\cdot}$, m/e 124

13, M$^{+\cdot}$, m/e 124

$-CH_3C\equiv CH$

$-CH_3C\equiv CH$

12a, m/e 84

13a, m/e 84

$-CO$

$-H\cdot$

$-\,^{\cdot}CH_3$

$CH_3-CH=C=O$
12b, m/e 56

12c, m/e 83

13b, m/e 69

and consistently with the two different structures. Thus **12a**, which has the methyl group at $C_{(5)}$ (or $C_{(3)}$), fragments preferentially by loss of CO giving methylketene, **12b** (m/e 56), as the most important fragment ion in the spectrum while only relatively minor loss of H· from $C_{(6)}$ (or $C_{(2)}$) to give **12c** (m/e 83) is observed; on the other hand **13a**, with a methyl group at $C_{(6)}$ (or $C_{(2)}$), fragments preferentially by loss of a methyl radical (**13a** → **13b**) in keeping with the greater stability and hence much easier loss of a methyl group compared with a hydrogen atom, while loss of CO is here relatively unimportant. The fragmentation of the M-CO ions (m/e 96) is also consistent with their formulation as the 3,4-dimethyl- (**12d**) and the 2,5-dimethylfuran radical ion (**13d**), respectively. Thus in both cases loss of H· and CH₃· is

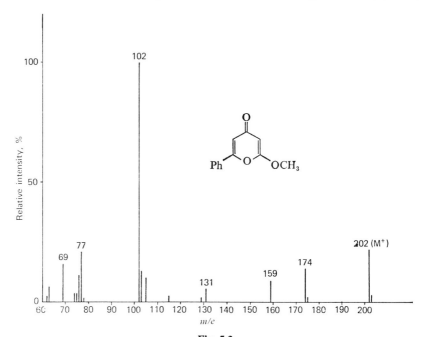

observed to give ions formulated as **14** and **15**, but only from **13d** with its two 2-methyl groups is an intense acetyl ion (**13c**, m/e 43) produced.

With phenyl-substituted 4-pyrones the retro-Diels-Alder reaction results in preferential charge-retention on the phenylacetylene fragment; in 2-methoxy-6-phenyl-4-pyrone (**16**) (Fig. 5.3) the retro-Diels-Alder reaction is the most important fragmentation process, resulting in ionized phenyl-acetylene (**16a**) as the base peak whereas the ketene fragment **16b** (m/e 100)

Fig. 5.3

is not observed (4). Loss of CO from the molecular ion gives **16c** which may further fragment as shown (**16c** → **16d** → **16e**). Loss of CO from the molecular ion competes favorably with the retro-Diels-Alder reaction (**17** → **17a**)

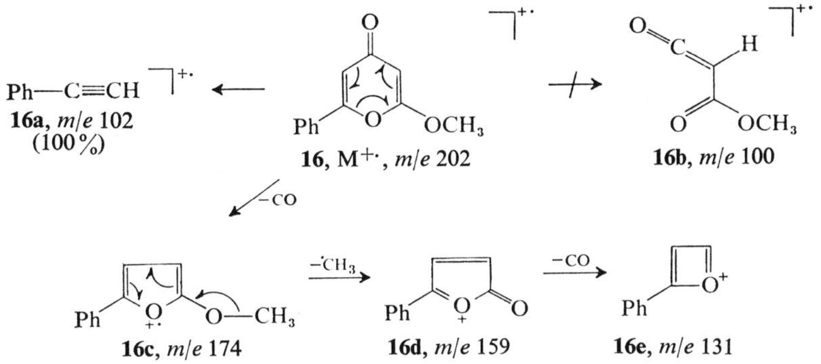

Ph—C≡CH
16a, m/e 102
(100%)
Ph
OCH3
16, M+·, m/e 202
16b, m/e 100
−CO
−CH3
−CO
Ph
O—CH3
16c, m/e 174
Ph
16d, m/e 159
Ph
16e, m/e 131

in 2,6-diphenyl-4-pyrone (**17**) giving **17b** as the base peak (8). Other important ions in the spectrum are at *m/e* 105 (38%) (the benzoyl ion, probably formed from **17b**) and the phenyl ion, *m/e* 77 (74%).

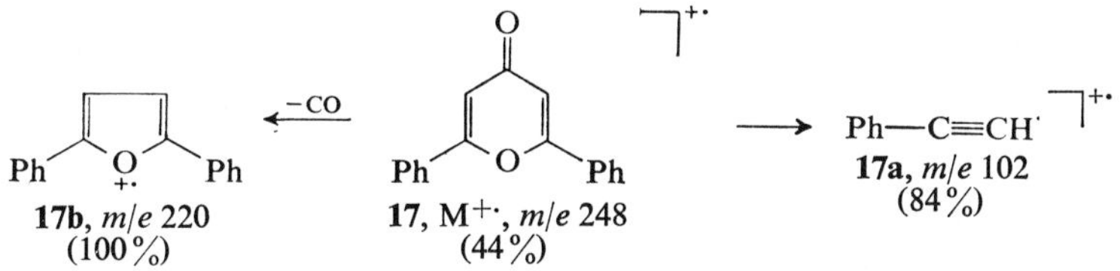

Ph
Ph
17b, m/e 220
(100%)
−CO
Ph
Ph
17, M+·, m/e 248
(44%)
Ph—C≡CH
17a, m/e 102
(84%)

It has been suggested that the characteristic retro-Diels-Alder fragmentation of 4-pyrones and its absence in the isomeric 2-pyrones may serve as the basis for the mass-spectrometric differentiation of these two classes of compounds (4).

III. COUMARINS AND DERIVATIVES

A. Coumarins

The coumarin ring system is of common occurrence in natural product chemistry, and studies of the general fragmentation pattern of coumarins and of the effect of various substituents in various positions are likely to be of considerable use in structure elucidation work in this field as well as being of use in the investigation of the general chemistry of the coumarin system.

The parent compound, coumarin (18), exhibits an abundant molecular ion, and as there are no other very facile bond-cleavages available the loss of a stable neutral molecule of carbon monoxide is a very important fragmentation process, resulting in the base peak, m/e 118 (9). The subsequent fragmentation of the M-CO ion (m/e 118), i.e., by expulsion of a second molecule of carbon monoxide (18a → 18b) followed by loss of a hydrogen atom to give m/e 89 (18c), is very similar to that of benzofuran itself and suggests that the M-CO ion may be formulated as the benzofuran radical ion 18a. There is evidence (10, and cf. Sec. I) to suggest that the M-CO ion of coumarins may be better represented as an open-chain ion rather than

as an ionized benzofuran system, but the latter representation is much more convenient for purposes of discussion and will be used here. Other ions of lower intensity in the spectrum of coumarin (18) are the hydrocarbon fragments m/e 63 and 51 due to the further fragmentation of 18c.

The fragmentation pattern of umbelliferone (19), which has a $C_{(7)}$ hydroxyl group, resembles that of coumarin itself (9). An abundant molecular ion (m/e 162, 80%) is observed and the base peak is due to the M-CO ion (m/e 134, 100%). The appearance potential of the M-CO (m/e 134) ion in 19 is closer to the appearance potential of the M-CO ion (18a) in coumarin than to the corresponding ion in phenol, indicating that the initial loss of carbon monoxide involves the lactone carbonyl group rather than the phenolic hydroxyl group of 19. As with coumarin, further loss of CO and H· from the M-CO ion of 19 is important, and loss of the third oxygen atom (also as CO) finally results in the benzene radical ion, m/e 78 (32%).

Loss of carbon monoxide from the molecular ion is also an important process in 7-methoxycoumarin (herniarin, **20**) (9) but further loss of carbon monoxide from **20a** to give m/e 120 (3%) is unimportant. Instead, loss of the methyl group becomes very much favored, probably as a result of the relative stability of the conjugated quinonoid oxonium ion **20b**. This species further fragments to a small extent by two consecutive losses of carbon monoxide (**20b** → **20c** → **20d**), forming the phenyl ion, m/e 77 (**20d**). Loss of the methoxyl group from the parent ion (M-31) is unimportant. Increased substitution as in 6,7-dimethoxycoumarin (dimethylaesculetin, **21**) results in a considerably more complex fragmentation pattern (9). The base peak is

21, M+·, m/e 206 (100%)   **21b**, m/e 178   **21c**, m/e 163

21a, m/e 191   **21d**, m/e 163   **22**

21f, m/e 120   **21e**, m/e 135 m/e 107 $\xrightarrow{-CO}$ $C_6H_7^+$ **21g**, m/e 79

due to the molecular ion, and loss of a methyl radical results in M-15 as the most abundant fragment ion. A study of the mass spectrum of 6-methoxy-7-deuteromethoxycoumarin (**22**) has shown that only the $C_{(6)}$-methoxyl group is involved, establishing structure **21a** for the M-15 ion (11). The spectrum of **22** also shows that both the $C_{(5)}$- and $C_{(6)}$-methoxyl groups are involved in the further loss of a methyl radical from the M-CO ion. This is not unexpected as a conjugated quinonoid oxonium ion (**21c** for loss of $C_{(6)}$-Me and **21d** for loss of $C_{(5)}$-Me) may now result in both cases. The ions at m/e 163 (**21c** and **21d**) further fragment by loss of CO to **21e**, which then loses either the methyl group (**21e** → **21f**) or two successive molecules of carbon monoxide (**21e** → **21g**).

The above result indicates that the M-15 ion will be important in coumarins with methoxyl groups at $C_{(6)}$ and (most probably) at $C_{(8)}$ as loss of a methyl group will result in the formation of a quinonoid oxonium ion (e.g., **21d**), but it will be much less important for $C_{(5)}$- and $C_{(7)}$-methoxycoumarins where formation of such ions is not possible. The absence of an $M\text{-}CH_3 \cdot$ ion for 7-methoxycoumarin (**20**) is in keeping with this generalization (9).

The allylic-ether side chain in 7-(6′,7′-dihydroxy-3′,7′-dimethyl-2′-octenyloxy)coumarin (**23**) provides sites for facile fragmentation and as a result only a weak parent ion is observed (5%) (12, 13). Loss of the allylic

23, M⁺·, *m/e* 332 (5%)

23e, *m/e* 59 (46%)

23a, *m/e* 162 (100%)

23c, *m/e* 171 (21%)

23b, *m/e* 134 (26%)

23d, *m/e* 153 (91%)

side chain with hydrogen transfer, most probably from the allylic $C_{(3')}$-methyl group via a six-membered ring transition-state as shown, results in the 7-hydroxycoumarin radical ion (**23a**) as the base peak, which further undergoes the usual expulsion of carbon monoxide to give the benzofuran **23b**. Cleavage of the allylically activated $C_{(1')}$—O ether bond with charge retention on the side chain gives the allylic carbonium ion **23c**, which further fragments in a variety of ways, among the more important being (*1*) loss of water from the $C_{(6')}$—$C_{(7')}$ diol system to give **23d**, and (*2*) cleavage of the

$C_{(6')}$—$C_{(7')}$ bond and formation of the oxonium ion **23e** which may also arise directly from the parent ion.

Cleavage β to the benzene ring, typical of alkylbenzenes, is the most important fragmentation process in coumarins with a $C_{(8)}$-alkyl side chain such as dihydro-osthol (**24**) and 7-methoxy-8-(2-formyl-2-methylpropyl)-coumarin (**25**) (9, 14). The resulting ion, the base peak in both cases, has been assigned the ring-expanded tropylium structure **26a**. An alternative, and possibly preferable, representation is the conjugated oxonium ion **26b**. The further fragmentation of **26a** (or **26b**) is by loss of the methoxy group as formaldehyde with hydrogen rearrangement to **26c** which then undergoes

24, M$^{+\cdot}$

26a, m/e 189 (100%)

CH_3O

25, M$^{+\cdot}$

and/or

26b, m/e 189

$-CH_2O$

26c, m/e 159

$-CO$

26d, m/e 131

the usual loss of carbon monoxide to give **26d**. Cleavage β to the benzene ring is also observed for mexoticin (**27**) where it occurs both with hydrogen transfer from the side chain (**27** → **27a**) and without (**27** → **27b**) (15). The expected cleavage of the $C_{(2')}$—$C_{(3')}$ bond of the side chain gives the two oxonium ions, m/e 59 (**27c**) and M-59. The most abundant fragment ion in the spectrum is m/e 207 ($C_{11}H_{11}O_4$ by high-resolution mass-measurement), and its genesis is rather unusual as it apparently involves loss of the five-carbon $C_{(8)}$ side chain with transfer of two hydrogen atoms to the coumarin moiety. Possible structures for the ion are **27d** or the resonance-stabilized **27e**. The driving force for this important fragmentation is not clear, although the extended conjugation in **27e** may be the answer.

The mass spectrum of osthol (**28**) (9), which has a $C_{(8)}$-dimethylallyl substituent, is considerably more complex than that of its dihydro analog **24**. β-Cleavage of the side chain, although much less prominent than in **24** as it now involves cleavage of a vinylic bond, is still of considerable importance,

CH₃ / CH₃ C=⁺OH

27c, m/e 59

⟵

OCH₃

CH₃O   CH₂ 1'
HO—CH 2'
HO—C(CH₃)₂ 3'

⟶

OCH₃

CH₃⁺O H · H

27d, m/e 207

27, M⁺·, m/e 308

a

OCH₃

CH₃O CH₂

↕

⁺OCH₃

CH₃O CH₂

27b, m/e 219

a with H transfer

↓

OCH₃

H₃CO CH₃

27a, m/e 220

or

OCH₃

CH₃O H H⁺

27e, m/e 207

giving **28a** which further fragments as outlined for **26a**, i.e., **28a** → **28b** → **28c**. The most abundant fragment is now at M-15, but loss of a methyl radical from a $C_{(7)}$-methoxyl group in coumarins is most unfavored since a conjugated oxonium ion cannot be formed. [An M-CH₃· ion is not observed for 7-methoxycoumarin or dihydro-osthol (**24**).] It has been suggested, therefore, that the lost methyl group originates from the side chain with hydrogen rearrangement and formation of the conjugated oxonium ion **28d**, which then fragments further by the usual loss of carbon monoxide (**28d** → **28e**) followed by loss of (most probably) the methoxyl methyl-group giving **28f**. Loss of the methoxyl group is also considerable (contrast **24**, where only a very weak M-CH₃O· ion is observed), the resulting ion probably having the tropylium structure **28g**. The M-CO ion is very weak (2%), and it appears to be a general rule for coumarins that if the molecular ion can undergo other facile fragmentations, the M-CO ion becomes relatively unimportant although marked losses of carbon monoxide from fragment ions will still be observed.

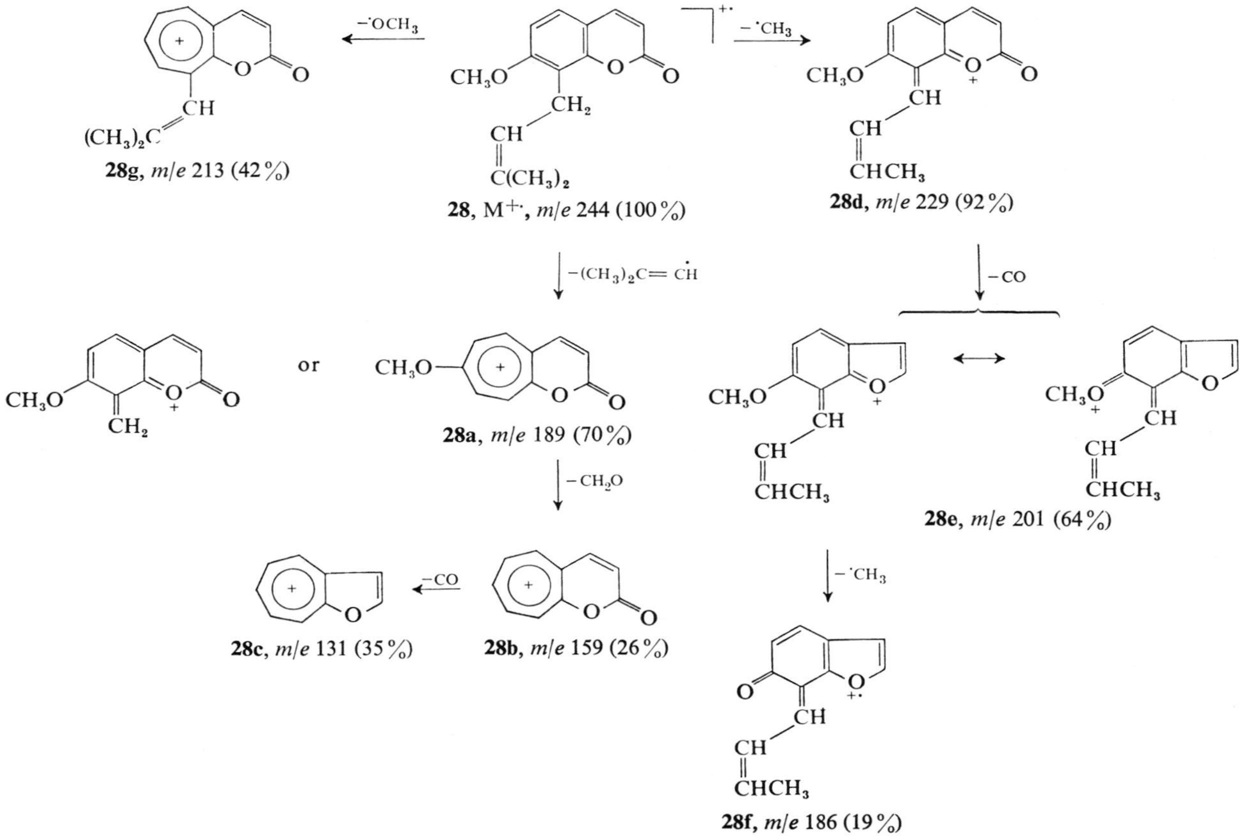

28d, m/e 229 (92%)

28, M$^{+\cdot}$, m/e 244 (100%)

28g, m/e 213 (42%)

28a, m/e 189 (70%)

28b, m/e 159 (26%)

28c, m/e 131 (35%)

28e, m/e 201 (64%)

28f, m/e 186 (19%)

$-\,^{\cdot}CH_3$

$-\,^{\cdot}OCH_3$

$-(CH_3)_2C=\overset{\cdot}{C}H$

$-CH_2O$

$-CO$

$-CO$

$-\,^{\cdot}CH_3$

or

B.   Coumarins Substituted on the Pyrone Ring

The mass spectra of 7-hydroxy- (**29**, R = H) (Fig. 5.4) and 7-methoxy-4-methylcoumarin (**29**, R = CH$_3$) (16, 17) resemble those of 7-hydroxy- and

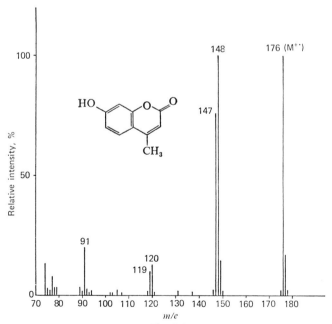

7-methoxycoumarin in that intense molecular and M-CO ions are observed in all cases. The subsequent fragmentation of the M-CO ions (**29a**) is modified

Fig. 5.4

by the $C_{(4)}$-methyl substituent as now the favored loss of a hydrogen atom probably occurs with formation of the ring-expanded chromenyl ions **29b** (cf. the β-cleavage of 2- and 3-alkylbenzofurans). In the methoxy compound (**29**, $R = CH_3$), loss of a methyl radical from the M-CO ion is more marked than loss of H·, giving the conjugated oxonium ion **29c** which further loses two molecules of carbon monoxide sequentially, as **29b** ($R = H$) also does, to form (most probably) the tropylium ion (m/e 91, **29d**). That the methyl group lost in the step **29a** ($R = CH_3$) → **29c** originates from the methoxyl group and not $C_{(4)}$ is shown by the absence of such an ion in the spectrum of the hydroxy compound **29** ($R = H$). A similar fragmentation pattern is observed for some other 4-methylcoumarin derivatives but if a 6-ethyl substituent is present, the highly favored cleavage β to the benzene ring at $C_{(6)}$ results in abundant M-CH$_3$· ions (17). The spectrum of 3-acetyl-6-methoxycoumarin (**30**) is also characterized by an abundant M-CH$_3$· ion, and the lost methyl group may originate to some extent from the acetyl group (**30** → **30b**) as well as from the 6-methoxy group (**30** → **30a**). Loss of ketene from the molecular ion is also observed, forming the 6-methoxy-coumarin ion (**30c**) which then undergoes the expected further fragmentations (17).

30c $-\,CH_2CO$ 30, M$^{+\cdot}$ $-\,^{\cdot}CH_3$

30a   or   30b

C. 4-Hydroxycoumarins and Related Compounds

The mass spectra of 4-hydroxycoumarins are quite different from those of other coumarin derivatives, and the spectra can be most easily rationalized by assuming that the molecular ion exists in the tautomeric β-keto-lactone form. Thus for 4,7-dihydroxycoumarin (**31**) (Fig. 5.5) the tautomeric form **31a** allows the operation of the retro-Diels-Alder reaction [path (i)] resulting in loss of neutral ketene to give **31b** as the base peak of the spectrum (16). The ion **31b** then further fragments by consecutive loss of two molecules of CO (**31b** → **31c** → **31d**). The retro-Diels-Alder reaction with hydrogen transfer [path (ii)] is also very important, giving an abundant ion at m/e 137

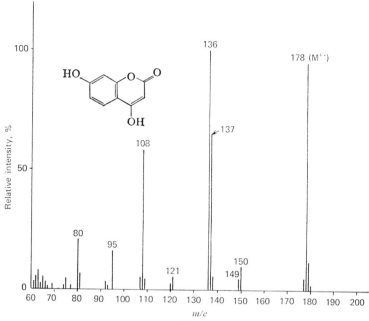

Fig. 5.5

which has been formulated as **31e**, but which is possibly better represented as the resonance-stabilized ion **31f** with the implication that the hydrogen atom is transferred from $C_{(3)}$ to the ring oxygen. Deuterium-labeling experiments support paths (i) and (ii). Loss of carbon monoxide from the molecular ion [path (iii)] is of minor importance and may involve loss of the $C_{(2)}$ carbonyl group since the further fragmentation of **31g** (M-CO), by loss of H· (**31g → 31h**) and loss of CHO· with hydrogen rearrangement (**31g → 31i**), is consistent with a benzofuran-3-one structure (see Chap. 2, Sec. IV) for this ion.

Fragmentation pathways (i) and (ii) and to a lesser extent (iii) are also important for 4-hydroxy-3-phenylcoumarins (or isoflavonols) such as 4,5,7-trihydroxy-3-phenylcoumarin (**32**) (16). Charge retention is observed on both of the retro-Diels-Alder fragments. The phenyl-ketene radical ion, **32b** (ring C fragment), is the base peak of the spectrum, and **32a** (ring A fragment) is also of considerable intensity and further fragments to a small extent by the usual loss of CO (**32a → 32c**). Pathway (ii) is also important but here charge retention is essentially on the ring A fragment **32d**, the other fragment (**32e**) being rather weak. Pathway (iii) is followed but the resulting ions (**32f** and **32g**) are of minor importance. The relative importance of pathways (i) and (ii) and the charge distribution on the various fragments is dependent

31a

(i) \ − CH₂CO

31f, m/e 137

31e, m/e 137

31 M⁺·, m/e 178

31b, m/e 136 (100%)

31c, m/e 108

31d, m/e 80

31a $\xrightarrow[\text{(iii)}]{-CO}$

31g, m/e 150

31i, m/e 121

31h, m/e 144

on the substitution of rings A and C. In **33** (16), which has a methoxyl group on ring C, pathway (i) results in charge retention entirely on the ring C fragment **33b**, and pathway (ii) is less important than in **32**, whereas if methoxyl groups are present on ring A but not ring C, pathway (ii) is favored more than pathway (i) (16). As expected, pathways (i) and (ii) are not observed for 4-methoxy-3-phenylcoumarins as tautomerism involving structures such as **32** and **33** can no longer occur (16).

The characteristic fragmentation of 4-hydroxy-3-phenylcoumarins has contributed much to the structure elucidation of the complex naturally-occurring 4-hydroxy-3-phenylcoumarins robustic acid (**34**) (18, 19) and the related scandenin (**35**) (20–22) together with the isomeric lonchocarpic acid (**36**) (21, 22). The spectrum of robustic acid (**34**) is readily rationalized in terms of the characteristic fragmentations of the 4-hydroxy-3-phenylcoumarin and 2,2-dimethylchromen moieties. Pathways (i) and (ii) are followed directly from the molecular ion giving **34a** and **34b**, respectively, which further fragment as shown (**34a** → **34c** → **34d** and **34b** → **34e** → **34f**). The very characteristic loss of a methyl radical from the 2,2-dimethylchromen system

32 (R $=$ H), m/e 270 (65%)

33 (R $=$ OCH$_3$), m/e 300 (15%)

32a, m/e 152 (40%)

32b, m/e 118 (10

33b, m/e 148 (10

32f, m/e 242 (2.2%)

32d, m/e 153 (45%)

32c, m/e 124 (11%)

Ph—C$\equiv$O$^+$
32g, m/e 105
(12%)

32e (R $=$ H), m/e 117 (2.4%)

gives the intense chromenyl ion **34g**, which then fragments via the retro-Diels-Alder reaction [pathway (i)] giving the even-electron ion **34c** as the most abundant fragment in the spectrum. Fragmentation of **34g** by pathway (ii) is less favored as it would give an odd-electron ion (**34e**). The fragmentations of **35** and **36** are very similar to those of robustic acid (**34**).

If a 3-alkoxycarbonyl group is introduced into a 4-hydroxycoumarin, the fragmentation pattern is modified by the interaction of the C$_{(3)}$- and C$_{(4)}$-substituents. Three main fragmentation paths are observed for the esters **37** (17). Loss of a neutral molecule of an aldehyde results in **37a**, while loss of a molecule of an alcohol via an "ortho-effect" gives **37b** which then readily undergoes a retro-Diels-Alder reaction with loss of C$_3$O$_2$ and formation of **37c**, the base peak in the spectrum of the ethyl ester **37** (R $=$ CH$_3$). The further fragmentation of **37c** by loss of CO is unexceptional (**37c** → **37d**). Initial loss of the alkoxy radical is also observed, giving **37e** which may then undergo the retro-Diels-Alder reaction with formation of **37f**; this

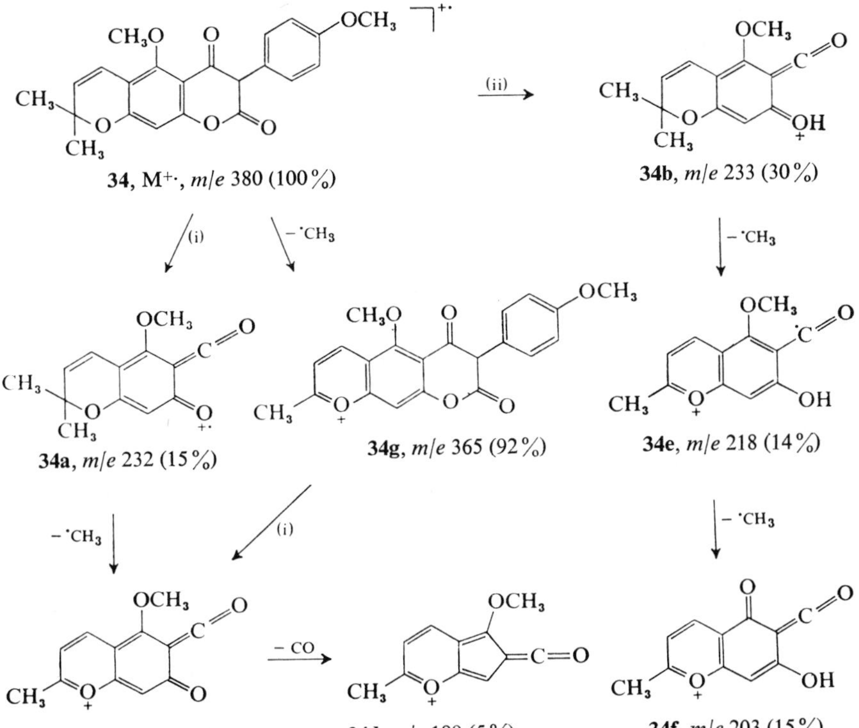

34, M$^{+\cdot}$, m/e 380 (100%)

34b, m/e 233 (30%)

34a, m/e 232 (15%)

34g, m/e 365 (92%)

34e, m/e 218 (14%)

34c, m/e 217 (95%)

34d, m/e 189 (5%)

34f, m/e 203 (15%)

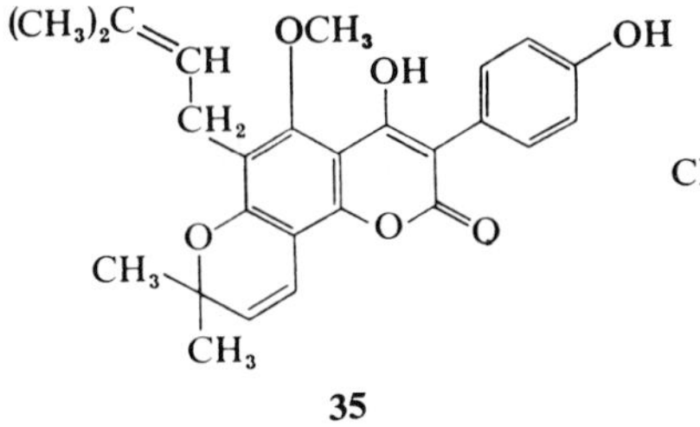

35

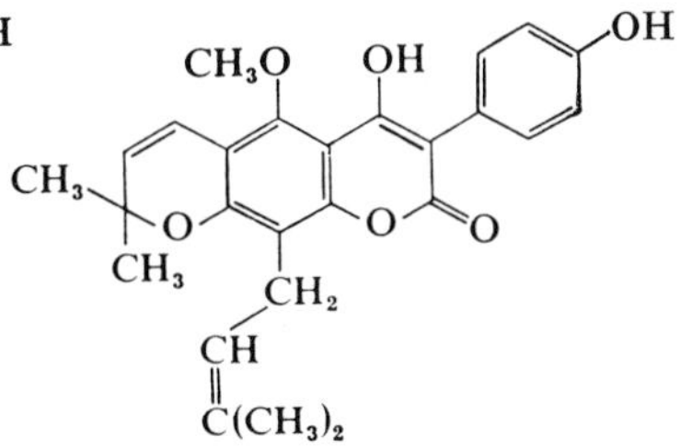

36

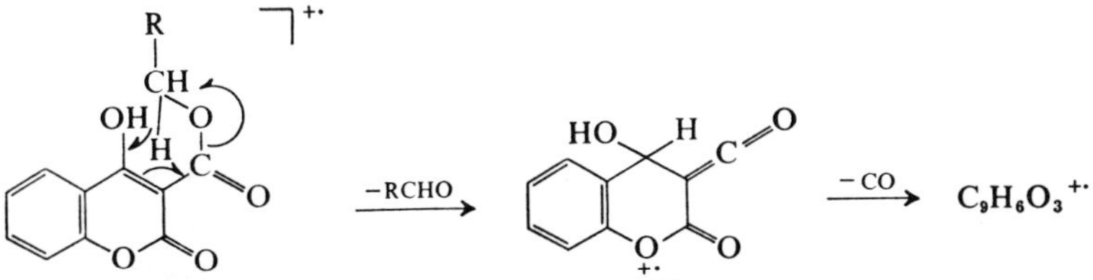

37, M$^{+\cdot}$ (R = H or CH$_3$)

37a

$-CO \rightarrow$ C$_9$H$_6$O$_3$$^{+\cdot}$

$-RCHO$

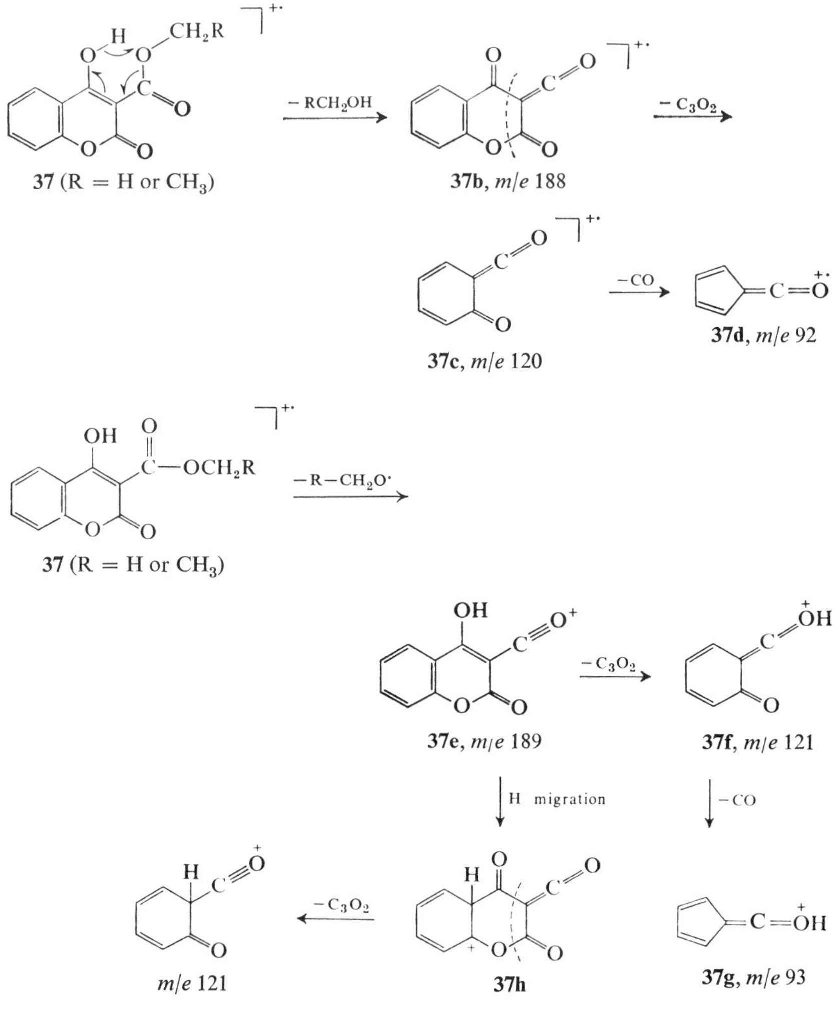
37 (R = H or CH3)
− RCH2OH
37b, m/e 188
− C3O2
− CO
37c, m/e 120
37d, m/e 92
37 (R = H or CH3)
− R−CH2O·
37e, m/e 189
− C3O2
37f, m/e 121
H migration
− CO
− C3O2
m/e 121
37h
37g, m/e 93

further loses CO (**37f** → **37g**). A mechanism involving hydrogen rearrangement (**37e** → **37h**) prior to elimination of C_3O_2 has been suggested but appears to be unnecessary in this case. This last fragmentation pathway (i.e., **37** → **37e**) decreases in importance with increasing substitution of the benzene ring.

The spectrum of 3-acetoacetyl-4-hydroxy-6-methylcoumarin (**38**) shows two losses of ketene from the $C_{(3)}$ side chain giving an ion **38a** whose further fragmentation is very similar to that of 6-methyl-4-hydroxycoumarin (17). Loss of $\cdot CH_2COCH_3$ is also observed, to give an analog to **37e**. The spectrum of 3-formyl-4-hydroxy-5,7-dimethylcoumarin (**39**) (17) is of interest as it exhibits an abundant M-CO ion (**39a**) which further fragments in a manner typical of 4-hydroxycoumarins, i.e., by the pathways (i), (ii), and

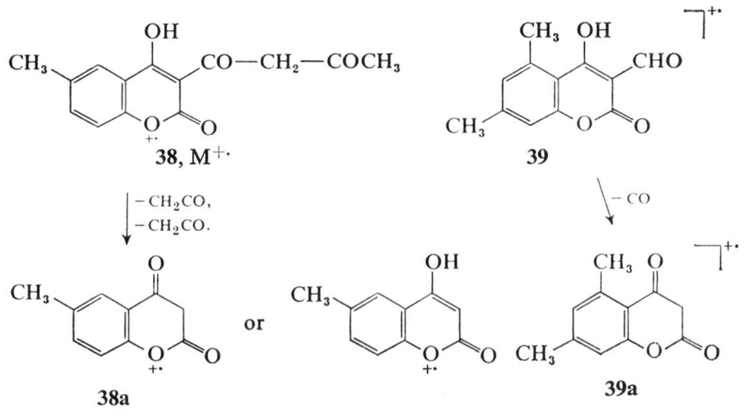

(iii) described for **31**, indicating that the initial loss of CO occurs from the 3-formyl group with hydrogen rearrangement to give **39a**.

IV. FURANO- AND PYRONOCOUMARINS AND OTHER NATURAL PRODUCTS RELATED TO COUMARIN

A. Furanocoumarins

The mass spectra of a considerable number of furanocoumarins have been reported, and in general the fragmentation patterns observed are similar to those of analogous simple coumarin derivatives. Thus angelicin (**40**) exhibits an abundant molecular ion and loss of CO results in **40a** as the base peak (9). Further loss of two molecules of CO occurs to a lesser extent giving ions at m/e 130 (25%) and m/e 102 (41%). The spectrum of the linear isomer,

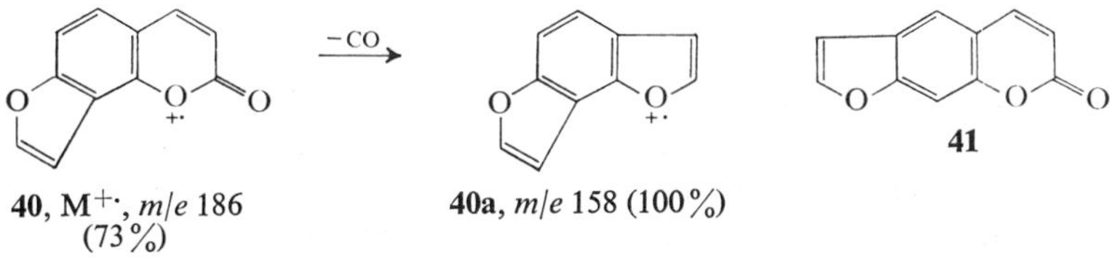

psoralin (**41**), is similar but here the molecular ion is the base peak (23). Similar spectra are also observed for the isomeric methoxyfuranocoumarins, xanthotoxin (**42**) and bergapten (**43**) (9), loss of the methoxyl methyl-group being marked as conjugated oxonium ions (**42a** and **43a**) may be formed in both cases. Other fragmentations of xanthotoxin (**42**) are as shown, the remaining three oxygen atoms in **42b** being lost sequentially as CO. The

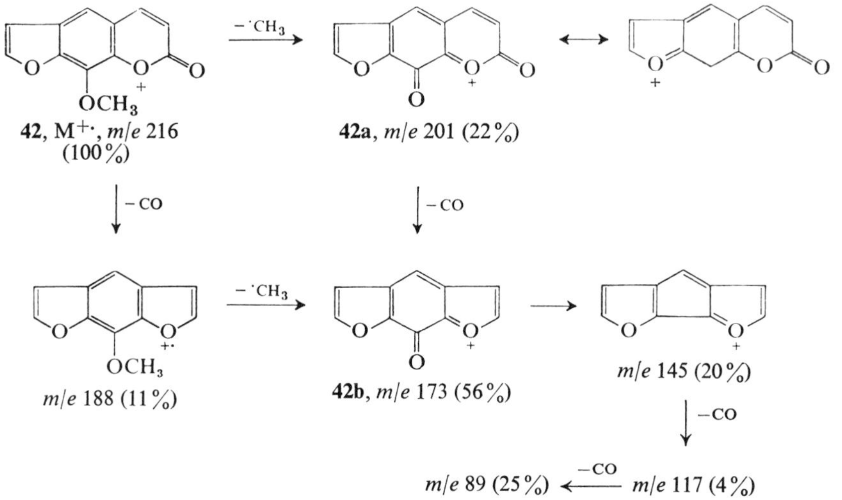

dimethoxyfuranocoumarin, isopimpinellin (**44**), fragments in a similar manner losing successively both of the methyl groups; M-15 is the base peak (9). Repeated loss of CO is also observed but only after at least one

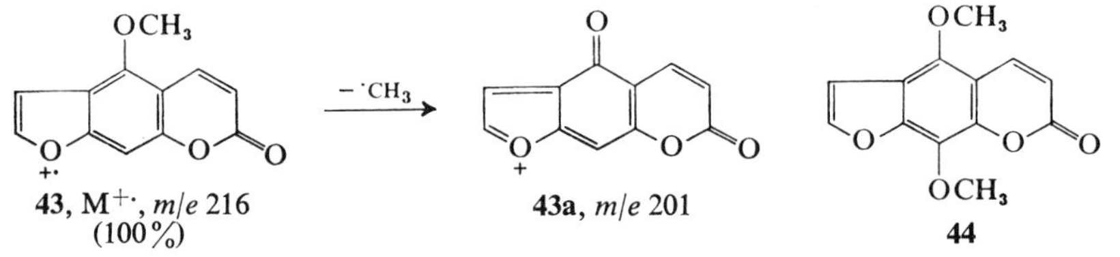

loss of a methyl radical from the molecular ion. The expected loss of a stable dimethylallyl radical to furnish an ion at m/e 201 (**42a**) is, however, not observed for imperatorin (**45**) (23); instead, loss of the side chain occurs with hydrogen transfer resulting in **47** (m/e 202) as the base peak of the spectrum. The further sequential loss of four molecules of CO from **47** is unexceptional. It was suggested that the transferred hydrogen atom originated from $C_{(1')}$, but hydrogen transfer from $C_{(4')}$ via a six-membered cyclic transition state, as shown, is far more likely (see also 23). The ion **47** (m/e 202) is also the base peak in the spectrum of the epoxy-compound prangenin (**46**) (23), and it is probable that elimination of the stable neutral diene fragments,

45a and **46a**, is an important factor contributing to the driving force for this type of fragmentation. Other ions observed in the spectra of **45** and **46** are due to fragmentation of the side chain. Thus cleavage of the O—C$_{(1')}$ bond in **45** results in an abundant ion at m/e 69 (**45b**) while the corresponding ion in the spectrum of **46** is at m/e 85 (**46b**).

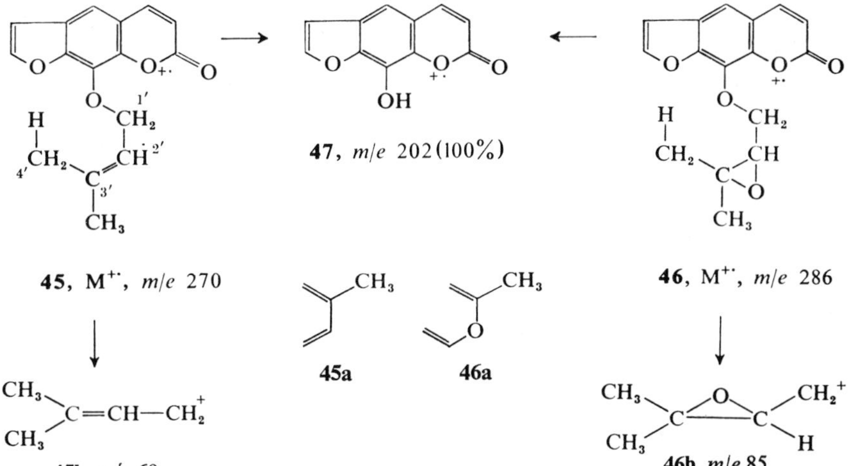

47, m/e 202 (100%)

45, M$^{+\cdot}$, m/e 270

45a   **46a**

45b, m/e 69

46, M$^{+\cdot}$, m/e 286

46b, m/e 85

The spectrum of 2-isopropylpsoralin (**48**) (24) is relatively simple, being characterized by the important β-cleavage typical of alkylbenzofurans (see Chap. 4). This results in loss of a methyl radical with the probable formation of the stable ring-expanded chromenyl ion **48a** (base peak) which further fragments only to a small extent by loss of CO (**48a** → **48b**). Loss of CO from

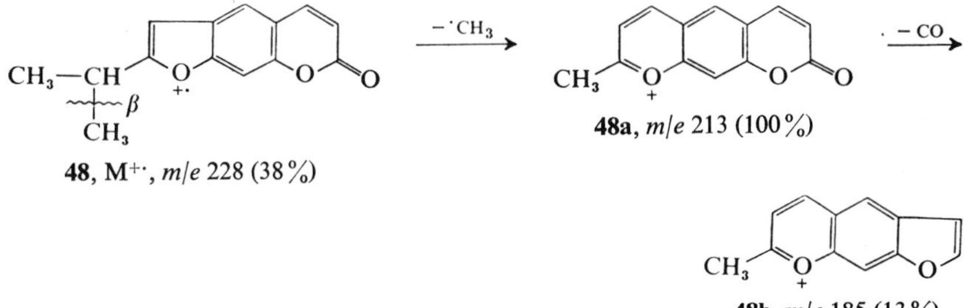

48, M$^{+\cdot}$, m/e 228 (38%)

48a, m/e 213 (100%)

48b, m/e 185 (13%)

the molecular ion of 4-bromoxanthotoxol (**49**) is observed to a small extent (m/e 252/254, 10%). The major fragmentation pathway is the ready ejection of the bromine atom, probably with hydrogen rearrangement, to form the oxonium ion **49a** (m/e 201, base peak) which then typically undergoes four consecutive losses of CO (24). If the spectrum of **49** is measured with a heated inlet system, an abundant ion is observed at m/e 202 (54%) corresponding to loss of the bromine atom and abstraction of a hydrogen atom from the surroundings. If the sample is introduced directly into the ion source the m/e 202 species is not observed so that the hydrogen abstraction is probably a reaction occurring in the heated inlet system (24). 4-Nitroxanthotoxin (**50**) behaves as a typical aromatic nitro-compound. The

49, m/e 280 /282 (53%)   **49a**, m/e 201 (100%)   **50**

molecular ion is the base peak, and considerable M-NO (10%) and M-NO$_2$ (29%) ions are observed which further fragment by loss of a methyl radical and losses of CO, several pathways being apparent (24).

The spectrum of 2,3-dihydroxanthotoxin (**51**) (24) differs from that of xanthotoxin (**42**) in that the M-CO (24%) ion is considerably more intense than the M-CH$_3$· ion (10%). A surprising feature is the absence of ions due to loss of a hydrogen atom from the dihydrobenzofuran system. The fragmentation of marmesin (**52**) is largely associated with the dihydrobenzofuran system (24); loss of the side chain with hydrogen transfer results in **52a** which, in contrast to **51**, undergoes ready loss of a hydrogen atom with formation of the oxonium ion **52b** as the base peak. Cleavage α to the dihydrobenzofuran ring results both in charge retention on the side chain with formation of the oxonium ion **52c** (m/e 59), and (most probably) in an

51   **52**, M$^+$·, m/e 246 (45%)   **52a**, m/e 188 (87%)

53   **52c**, m/e 59 (60%)   **52b**, m/e 187 (100%)

alternative route to **52b**, although this is not supported by observation of a metastable ion. Loss of CO directly from the molecular ion is not observed although it occurs to a small extent from **52a** and **52b** [m/e 160 (25%) and m/e 159 (10%), respectively]. Another fragmentation pathway is by loss of water from the molecular ion, possibly with rearrangement to give the

52

$\xrightarrow{-H_2O}$

52d, m/e 228 (7%)

$\xrightarrow{-\cdot CH_3}$

52e, m/e 213 (24%)

dimethylchromene system **52d**, this being consistent with the further ready loss of a methyl radical, **52d** → **52e**. In contrast to that in 4-bromoxanthotoxol (**49**), the bromine atom of 6-bromomarmesin (**53**) is very stable and bromine-containing analogs of the ions derived from the ring structure of marmesin are observed (24).

B. Pyronocoumarins

The mass spectrum of 4'-methyl-2'-pyrono(5',6':3,4)coumarin (**54**) is characterized by an intense loss of CO (M-CO is the base peak) followed by three smaller consecutive losses of CO (17). The order of elimination of the CO molecules has been shown by ^{13}C-labeling and by the observation of a metastable ion for the concerted loss of the second and third molecules of carbon monoxide, indicating the probable presence of two adjacent CO units in the M-CO ion (**54a**). The energy required for each elimination of CO was also measured and the fragmentation scheme illustrated was suggested. The energy required for the removal of the first molecule of CO from the parent ion (step *a*) is 2.2 eV, and it might be expected that a similar amount of energy would be required for the second loss (step *b*). This, however, requires only 0.7 eV, indicating that rearrangement to a more stable ion, perhaps **54b**, is involved. On the basis of these results it was suggested (17) that the M-CO ion, **54a**, is better represented as a ring-opened ion than as an ionized furan system. Loss of a hydrogen atom is observed from the ions **54a** (**54a** → **54c**) and **54d**, the driving force being formation of a more stable system such as **54c**. The loss of the third molecule of CO (step *c*) also requires considerable energy as extrusion of a carbon atom from the benzene ring is involved whereas the final loss of CO (step *d*) is assisted by formation of the stable ion **54e**. The isomeric 4′-pyrono(5′,6′:3,4)coumarins behave in a

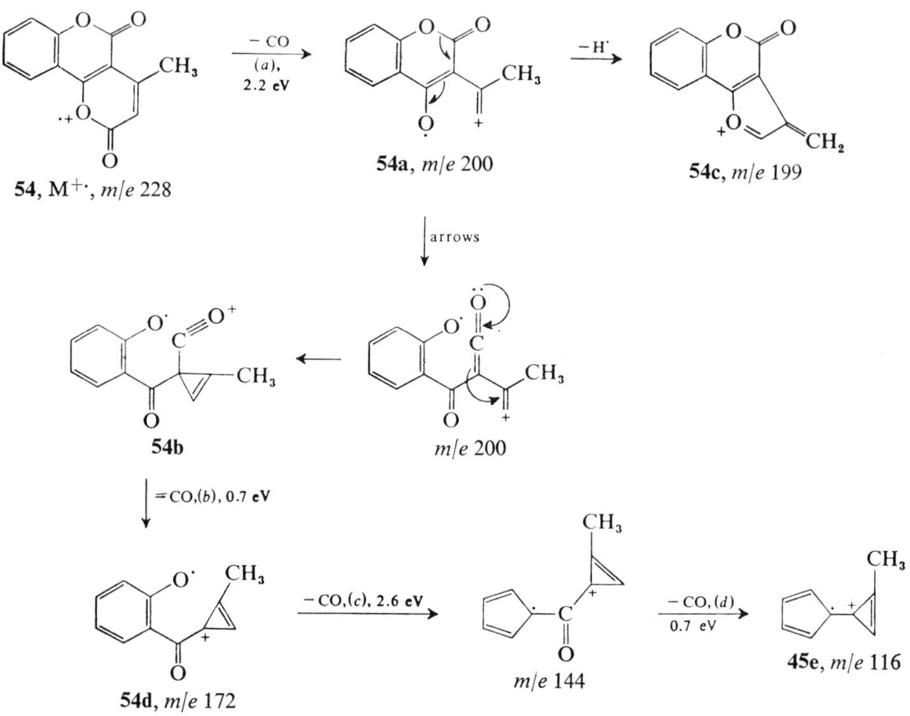
54, M+·, m/e 228
− CO (a), 2.2 eV
54a, m/e 200
− H·
54c, m/e 199
arrows
m/e 200
54b
− CO,(b), 0.7 eV
54d, m/e 172
− CO,(c), 2.6 eV
m/e 144
− CO, (d) 0.7 eV
45e, m/e 116

similar manner, four successive losses of CO being observed. The presence of a 6-methoxy-group as in 6,7-dimethoxy-2′-methyl-4′-pyrono(5′6′:3,4)-coumarin (**55**) modifies the fragmentation pattern with loss of a methyl radical giving the conjugated oxonium ion **55a** as the most abundant fragment. Loss of CO directly from the molecular ion is not observed (17).

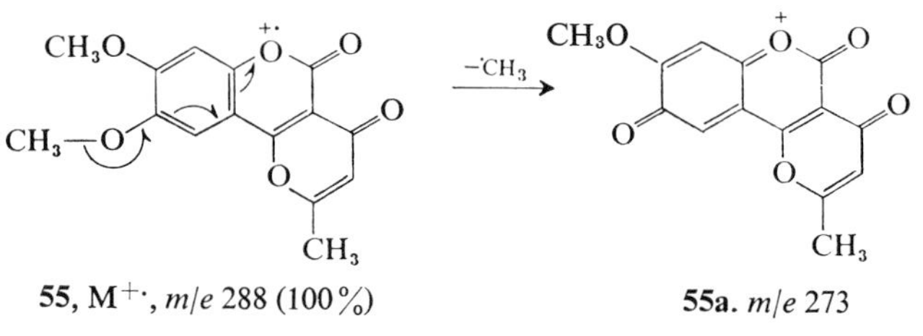
−·CH3
55, M+·, m/e 288 (100%)
55a. m/e 273

C.   Other Natural Products Related to Coumarin

The spectra of compounds containing both a 2,2-dimethylchromen and a coumarin system, such as xanthyletin (**56**) (25), the angular isomer seselin (**57**), and its methoxy-derivative brayelin (**58**) (9), are dominated by the intense M-CH$_3$· ion characteristic of 2,2-dimethylchromens, the chromenyl ion **56a** accounting for 55% of the total ion current in the spectrum of **56**. Loss of CO from the M-CH$_3$· ions is observed (e.g., **56a** → **56b**) though not

CH$_3$ ⟶ $-$·CH$_3$ ⟶

56, M$^{+·}$, m/e 228 (18%)

56a, m/e 213 (100%) $-$CO ⟶

CH$_3$

56b, m/e 185 (17%)

57, R = H
58, R = OCH$_3$

59

to a major extent, and in the spectrum of brayelin (**58**) the M-CH$_3$· ion fragments by a further loss of a methyl radical (most probably from the methoxy group) as well as by loss of CO. Similarly the major fragments in the spectrum of **59** (25) are characteristic of the 4-chromanone system (see Chap. 3, Sec. III), direct loss of CO from the molecular ion not being observed.

Loss of water is an important fragmentation process for the isocoumarin derivative **60** (26), possibly with formation of the ring-opened ion **60a**, which then successively loses three molecules of CO, finally giving a stable ion at m/e 178 which may be formulated as the phenanthrene radical ion **60b**. The base peak is m/e 89 (**60d**) which may be formed by loss of the benzylic substituent (**60** → **60c**) followed by loss of two molecules of CO. The spectrum of **61** is of interest as the base peak is due to loss of water followed by three successive losses of CO. A prominent ion at m/e 165 which may be assigned

60, M$^{+\cdot}$, m/e 280 (5%)

$\xrightarrow{-\text{H}_2\text{O}}$

60a, m/e 262 (71%)

$\xrightarrow{-3\text{CO}}$

60b, m/e 178 (40%)

60c

$\xrightarrow{-2\text{CO}}$ $\text{C}_7\text{H}_5{}^+$

60d, m/e 89 (100%)

61

61a, m/e 165 (40%)

the fluorenyl structure **61a** is formed by loss of the carboxy group followed by loss of two molecules of CO (26).

V.   CHROMONES, FLAVONES, AND RELATED COMPOUNDS

A.   Chromones

The mass spectrum of the parent compound, chromone (**62**), (Fig. 5.6), differs considerably from that of the isomeric coumarin (**4**, 27). The parent ion is the base peak and loss of CO gives an abundant ion at m/e 118. This possibly has the same structure as the M-CO ion of coumarin and may be formulated conveniently as the benzofuran radical ion **62a** (although a ring-opened form cannot be excluded). It further undergoes the typical loss of CO and H· to form ions at m/e 90 and 89 (cf. benzofuran, Chap. 4, Sec. II). The second important fragmentation process, not observed for coumarin, is the retro-Diels-Alder reaction with loss of neutral acetylene and charge retention on the fragment **62b** which further fragments by the usual sequential loss of two molecules of CO (**62b → 62c**). The same fragmentations are observed for 2-methylchromone (**63**). In this case, however, the M-CO ion (**63a**) may lose a hydrogen atom with formation of the oxonium ion **63d** or the ring-expanded chromenyl ion **63e** (27) (cf. 2-methylbenzofuran, Chap. 4, Sec. II).

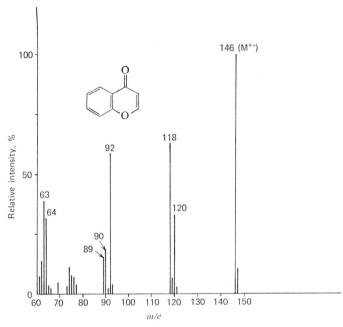

Fig. 5.6

The retro-Diels-Alder reaction may operate directly on the parent ion of a chromone or (in some cases) only on fragment ions. It may serve as a basis for mass-spectrometric differentiation between the isomeric coumarins and chromones (cf. 2- and 4-pyrones). In some cases such as **64**, however, there is no evidence of ions due to retro-Diels-Alder fragmentations whereas loss of CO is very much favored, affording **64a** as the base peak here (27).

$$\xrightarrow{-RC\equiv CH}$$

$$\xrightarrow{-CO}$$

$$\xrightarrow{-CO} C_3H_4^{+\cdot}$$

62, R = H, m/e 146
63, R = CH$_3$, m/e 160

62b

m/e 92

62c,
m/e 64

$\downarrow -CO$

$$\xrightarrow{-H^{\cdot}(R=CH_3)}$$ or

62a, R = H, m/e 118
63a, R = CH$_3$, m/e 132

63d, m/e 131

63e, m/e 131

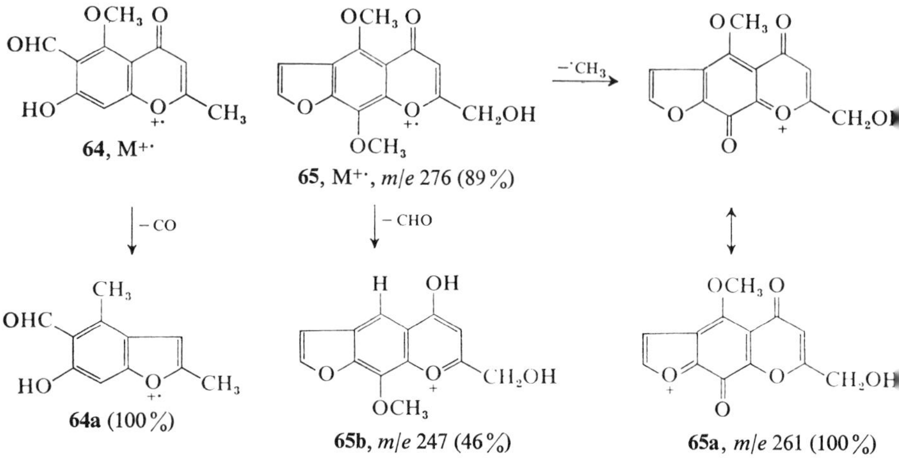

64, M+·
65, M+·, m/e 276 (89%)
−·CH3
−CO
−CHO
64a (100%)
65b, m/e 247 (46%)
65a, m/e 261 (100%)

As with furanocoumarins, loss of a methoxy methyl-group is an important fragmentation process for the furanochromone ammiol (**65**) resulting in the conjugated oxonium ion **65a** as the base peak (27). A notable feature of the spectrum of **65** (and many others) is a considerable ion due to concerted loss of CHO· from the molecular ion. It has been suggested (27) that the lost CHO· radical originates from a methoxyl group, possibly with formation of the chromenyl ion **65b**, as deuteration studies show that the hydroxyl hydrogen atoms are not involved in the loss. Cleavage β to the benzene ring of peucenin (**66**) (28) gives *m/e* 205 as the base peak which may be formulated as the conjugated oxonium ion **66a** or the tropylium ion **66b**. The retro-Diels-Alder fragmentation of *m/e* 205 can be more readily rationalized in terms of **66b**, in which case it would result in a structure such as **66c** for *m/e* 165.

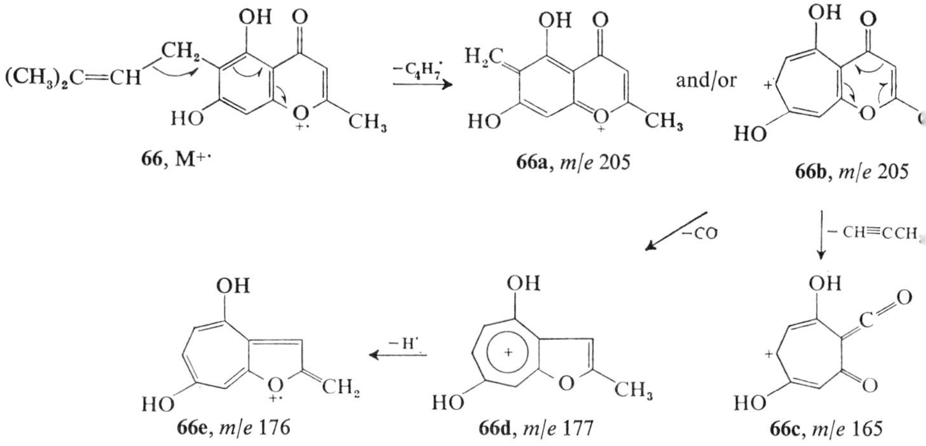

66, M+·
−C4H7·
and/or
66a, m/e 205
66b, m/e 205
−CO
−CH≡CCH3
−H·
66e, m/e 176
66d, m/e 177
66c, m/e 165

Alternatively, the m/e 205 ion may fragment by loss of CO followed by H·
(**66b** → **66d** → **66e**). Introduction of a long alkyl side chain as in 2-*n*-hepta-
decylchromone (**67**) (29) modifies the normal fragmentation pattern as now

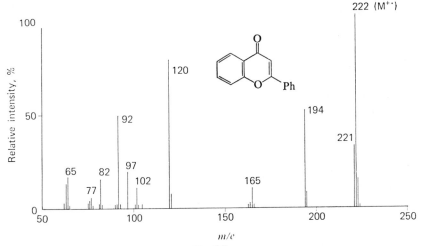

the base peak is due to cleavage of the alkyl side-chain γ to the pyrone ring,
possibly with hydrogen rearrangement and formation of the stable 4-hydroxy-
2-vinylchromenyl ion **67a**. The retro-Diels-Alder reaction is only of minor
importance, but a considerable ion is observed at m/e 121 which is due to
the retro-Diels-Alder reaction with hydrogen transfer.

B. Flavones and Isoflavones

The mass spectra of flavones are characterized by intense molecular ions
(the base peak in most cases), indicative of a stable heterocyclic system with
no facile bond-cleavages available. The major fragmentation pathway for
the parent compound, flavone (**68**, Fig. 5.7), is the retro-Diels-Alder reaction,
giving an abundant ring A fragment **68a** (m/e 120) which further fragments
by the usual loss of CO (**68a** → **68b**) and a less abundant ring C fragment,

Fig. 5.7

the phenylacetylene radical ion **68c** (m/e 102) (9, 30). Loss of CO from the molecular ion is also marked, leading to the phenylbenzofuran ion **68d** (m/e 194) which is associated with a considerable doubly-charged ion at $m/2e$ 97. Other features are a small peak at m/e 105 which may be ascribed

to the benzoyl ion (**68e**), possibly arising from the fragmentation of **68d**, and a considerable M-1 ion (m/e 221) whose structure is not clear. Halogen-substituted flavones fragment in the same manner as flavone itself, thus confirming the pathways described (31).

The charge distribution in the retro-Diels-Alder reaction is very sensitive to the substitution on rings A and C. Thus in flavone itself and in derivatives not oxygenated on ring C, the ring A fragment is considerably more intense than the ring C fragment, but if a methoxy group is present on ring C then the reverse is true, the ring C fragment predominating and further fragmenting by loss of a methyl radical. In some cases ions corresponding formally to the retro-Diels-Alder reaction with hydrogen transfer to the ring A fragment are observed. Thus for 4',5,7-trihydroxyflavone (**69**) the hydrogen transfer ion at m/e 153 (**69b**) is of the same intensity as the normal ring A fragment **69a** (m/e 152) (32), whereas for 5,7-dihydroxy-3',4'-dimethoxyflavone (**70**) the ion at m/e 153 (**70b**) is much more intense than that at m/e 152 (30). The origin of the transferred hydrogen atom has not been established, however, and it is significant that an appreciable hydrogen transfer ion is not observed for **71** (32).

With highly-oxygenated flavone derivatives the retro-Diels-Alder reaction may be relatively unimportant. Thus for zapotin (**72**) the two retro-Diels-Alder ions **72a** and **72b** are only of 3% the intensity of the base peak (33). However, these weak ions may still be of diagnostic value in general for they are easily distinguished as often being the only even mass-number ions in their particular region of the spectrum. The presence of a 6-methoxy group in zapotin results in the highly-favored loss of a methyl radical and formation

72a, *m/e* 180
(3%)

72b, *m/e* 162 (3%)

72, M$^{+\cdot}$, *m/e* 342 (57%)

$-\cdot CH_3$

72c, *m/e* 327 (100%)

of the conjugated oxonium ion **72c** as the base peak; this behavior is analogous to that of 6-methoxycoumarins. Further loss of CO from **72c** is only of minor importance, giving *m/e* 299 (3%), and so is loss of CO directly from the parent ion [M-CO, *m/e* 314 (3%)].

The spectra of simple isoflavones resemble those of the isomeric flavones in that the retro-Diels-Alder reaction is a very important fragmentation process, and as in flavones the charge distribution on the two fragments depends on the substitution of rings A and C. If ring C is unsubstituted, as in isoflavone (**73**) itself, then the charge resides mainly on the ring A fragment, **73a** (31) (which further fragments by a typical loss of CO), whereas the presence of a methoxy group on ring C, as in 4′,7-dimethoxyisoflavone (**74**) (34), favors the ring C fragment, **74b**. Unlike the M-CO ion of flavones, the M-CO ion of isoflavones is very weak and is not observed at all in the spectra of **74** and 5,7-dihydroxyisoflavone. As a consequence, a doubly-charged M-CO ion is not observed in the spectra of isoflavones; however,

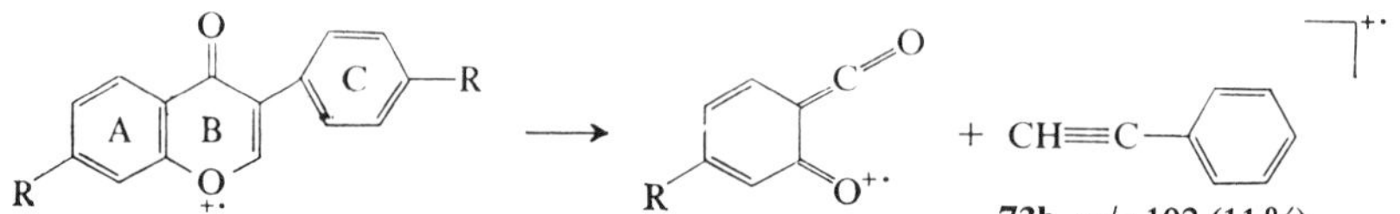

73, R = H, m/e 222 (92%)
74, R = OCH$_3$, m/e 282 (100%)

73a, m/e 120 (40%)
74a, m/e 150 (12%)

73b, m/e 102 (11%)
74b, m/e 132 (68%)

a doubly-charged parent ion is rather prominent (30). Isoflavones are also characterized by a rather intense M-1 ion which is the base peak in the spectra of isoflavone (**73**) and 7-methoxyisoflavone. It has been suggested that the lost hydrogen atom originates from the $C_{(2')}$ position as this would result in formation of a resonance-stabilized oxonium ion **73c** whose stability is reflected in the lack of further fragmentation (31).

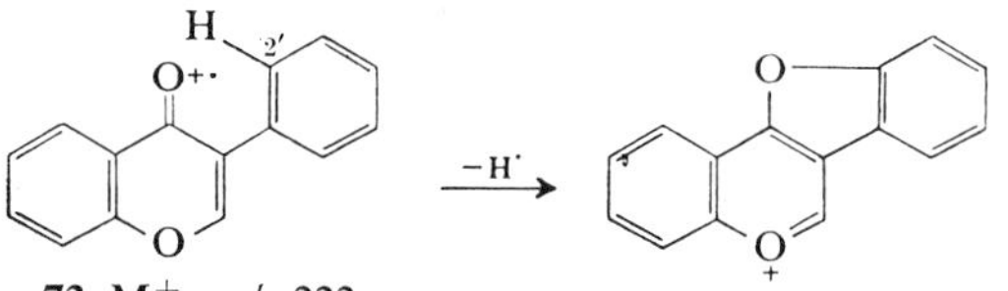

73, M$^{+\cdot}$, m/e 222

73c, m/e 221 (100%)

The retro-Diels-Alder reaction is only of minor importance in more complex isoflavone derivatives. Even the weak ions **75a** and **76a**, however, proved sufficient to indicate the ring on which the methoxyl group is situated in scandinone (**75**) and scandenone 4′-methyl ether (**76**) (35).

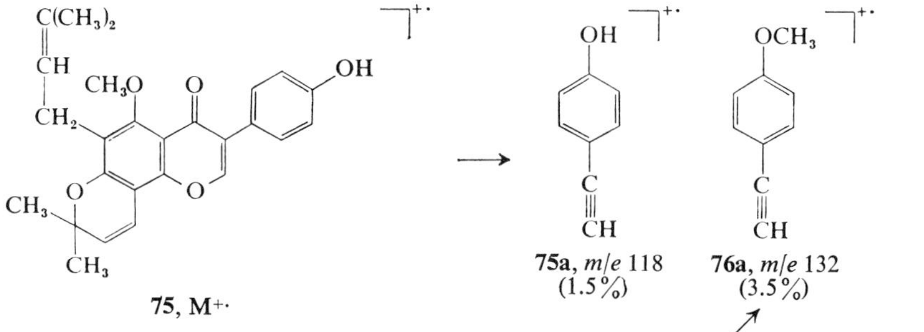

75, M$^{+\cdot}$

75a, m/e 118 (1.5%)

76a, m/e 132 (3.5%)

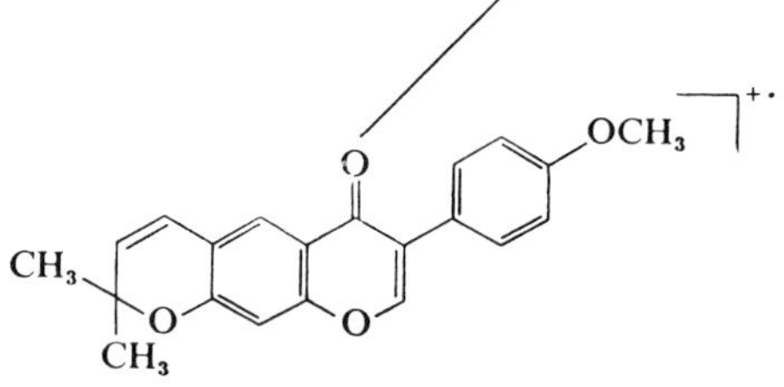

76

C.   Flavonols

Ions due to the retro-Diels-Alder reaction are either very weak or not observed at all in the mass spectra of flavonols. (In this they differ markedly from the isomeric isoflavonols, already discussed as 3-phenyl-4-hydroxycoumarins in Sec. III.C.) Hydrogen-transfer reactions are, however, important and may be most readily rationalized if the molecular ion exists in the diketo form (e.g., **77a** for **77**). They have been substantiated by deuterium labeling of the $C_{(3)}$-hydroxyl group in the case of 5,7-dimethoxyflavonol (**77**) (30). The retro-Diels-Alder reaction with hydrogen transfer to ring A is generally

77, M$^{+\cdot}$

77a

and

77a

$$\text{Ph}-\text{C}{\equiv}\text{O}^+ \xrightarrow{-\text{CO}} \text{C}_6\text{H}_5^+$$

77c, m/e 105   **77d**, m/e 77
(60%)   (54%)

77b, m/e 181 (10%)   or

78, R = H; **79**, R = CH$_3$

observed and here results in the ion **77b**. The benzoyl ion, **77c**, is an important fragment possibly formed by the concerted mechanism shown, and further fragments as expected by loss of CO and formation of the phenyl ion **77d**. Ions corresponding to **77c** and **77d** are a characteristic feature of the spectra of flavonols. More extensive substitution as in quercetin (**78**) (30) and isorhamnetin (**79**) (35) increases the stability of the molecular ion (base peak) and relatively little fragmentation is observed. A notable feature in the spectra of flavonols is the presence of a large number of doubly-charged ions of considerable intensity; for example, in the spectrum of quercetin (**78**) two important doubly-charged ions are $(\text{M-H}_2\text{O})^{2+}$ and $(\text{M-H}_2\text{O-CO})^{2+}$.

The presence of the $C_{(2')}$-hydroxyl group in 2'-hydroxyflavonol (**80**) results in the very facile loss of OH·, most probably with formation of the stable hydroxychromenyl ion **80a**. Deuterium labeling indicates that both the

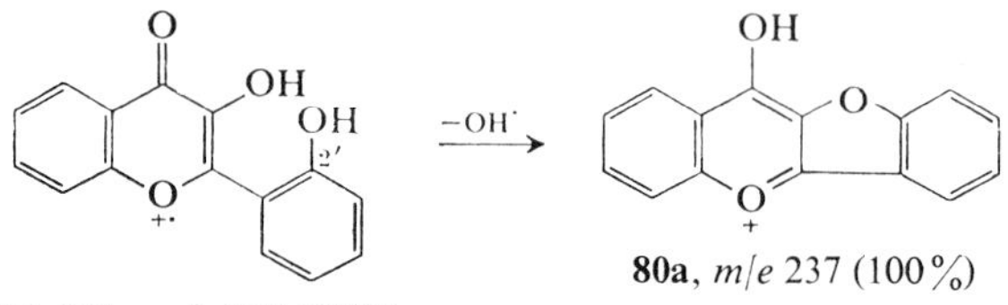

80, M$^{+·}$, m/e 254 (82%)

$-$OH· →

80a, m/e 237 (100%)

hydroxyl groups in **80** are equally involved in the loss of ·OH (36). Neither the retro-Diels-Alder reaction nor loss of CO from the molecular ion is observed in the spectra of O-alkyl-quercetagetin derivatives of general formula **81**, where the substituents R are not necessarily the same and may be H, CH$_3$, or C$_2$H$_5$ (37). Abundant molecular ions are observed, and loss of a

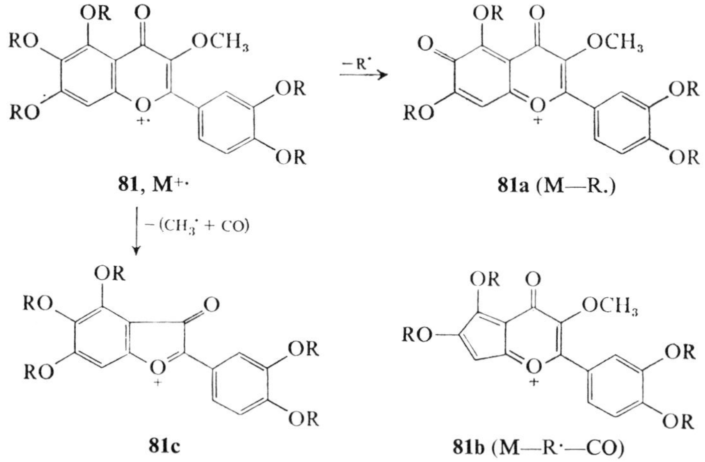

81, M$^{+·}$

$-$R· →

81a (M—R.)

$-$(CH$_3$· + CO)

81c

81b (M—R·—CO)

$C_{(6)}$-alkyl group (or H· for 6-hydroxy compounds) is highly favored as it may result in the conjugated quinonoid oxonium ion **81a** which further fragments by loss of CO (**81a → 81b**) or of water. A noteworthy feature is the one-step loss of (CH$_3$· + CO) from the molecular ion (substantiated by a metastable ion in some cases), possibly with formation of the ion **81c**.

VI. XANTHONES

The parent compound xanthone (**82**) exhibits an abundant molecular ion as the base peak of the spectrum, and the expected loss of CO results in M-CO as the most abundant fragment ion (9). The M-CO ion may be formulated as the dibenzofuran radical ion, **82a**, as its further fragmentation, by loss of CO (**82a → 82b**) followed by a more pronounced loss of a hydrogen atom (**82b → 82c**), is as observed for dibenzofuran itself (see Chap. 4, Sec. III).

$$\text{82, M}^{+\cdot},\ m/e\ 196 \quad\xrightarrow{-CO}\quad \text{82a, } m/e\ 168\ (46\%) \quad\xrightarrow{-CO}\quad C_{11}H_8^{+\cdot} \quad\xrightarrow{-H^{\cdot}}\quad C_{11}H_7^{+}$$

82, M$^{+\cdot}$, m/e 196 (100%) 82a, m/e 168 (46%) 82b, m/e 140 (7%) 82c, m/e 139 (32%)

Mass spectrometry has proved useful in the structure elucidation of a number of naturally-occurring xanthone derivatives (38–40) and in the identification of xanthones in the degradation of "humic acid" (41).

Addenda

Pirkle and Dines have further discussed the fragmentation of 2-pyrone, and conclude that the evidence excludes the furan structure for the M-CO ion. Other possible structures are discussed (42).

REFERENCES

1. W. H. Pirkle, *J. Amer. Chem. Soc.*, **87**, 3022 (1965).
2. P. Brown and M. M. Green, *J. Org. Chem.*, **32**, 1681 (1967).
3. H. Nakata and A. Tatematsu, *Tetrahedron Lett.*, **1967**, 4101.
4. H. Budzikiewicz, J. I. Brauman, and C. Djerassi, *Tetrahedron*, **21**, 1855 (1965).
5. H. Nakata, Y. Hirata, and A. Tatematsu, *Tetrahedron Lett.*, **1965**, 123.
6. H. Budzikiewicz, C. Djerassi, and D. H. Williams, *Mass Spectrometry of Organic Compounds*, Holden-Day, San Francisco, 1967, pp. 209–210.
7. P. Beak, T. H. Kinstle, and G. A. Carls, *J. Amer. Chem. Soc.*, **86**, 3833 (1964).
8. E. M. Kaiser, S. D. Work, J. F. Wolfe, and C. A. Hauser, *J. Org. Chem.*, **32**, 1483 (1967).
9. C. S. Barnes and J. L. Occolowitz, *Aust. J. Chem.*, **17**, 975 (1964).
10. J. L. Occolowitz and G. L. White, *Aust. J. Chem.*, **21**, 997 (1968).
11. R. H. Shapiro and C. Djerassi, *J. Org. Chem.*, **30**, 955 (1965).
12. A. Chatterjee, C. P. Dutta, S. Bhattacharyya, H. E. Audier, and B. C. Das, *Tetrahedron Lett.*, **1967**, 471.
13. J. F. Fisher, H. E. Nordby, A. A. Waiss, and W. L. Stanley, *Tetrahedron*, **23**, 2523 (1967).
14. J. F. Fisher and H. E. Nordby, *Tetrahedron*, **22**, 1489 (1966).
15. D. P. Chakraborty, B. K. Chowdhury, and B. C. Das, *Tetrahedron Lett.*, **1967**, 3471.
16. A. Pelter, P. Stainton, A. P. Johnson, and M. Barber, *J. Heterocycl. Chem.*, **2**, 256 (1965).
17. R. A. W. Johnstone, B. J. Millard, F. M. Dean and A. W. Hill, *J. Chem. Soc.*, *C*, **1966**, 1712.
18. A. P. Johnson, A. Pelter, and M. Barber, *Tetrahedron Lett.*, **1964**, 1267.

19. A. P. Johnson and A. Pelter, *J. Chem. Soc., C*, **1966**, 606.
20. A. Pelter and P. Stainton, *Tetrahedron Lett.*, **1964**, 1209.
21. A. Pelter and A. P. Johnson, *Tetrahedron Lett.*, **1964**, 2817.
22. A. P. Johnson, A. Pelter, and P. Stainton, *J. Chem. Soc., C*, **1966**, 192.
23. N. S. Vulf'son, V. I. Zaretskii, and V. G. Zaikin, *Dokl. Akad. Nauk SSSR*, **155**, 1104 (1964).
24. F. M. Abdel-Hay, E. A. Abu-Mustafa, B. A. H. El-Tawil, M. B. E. Fayez, C. S. Barnes, and J. L. Occolowitz, *Indian J. Chem.*, **5**, 89 (1967).
25. R. T. Aplin and C. B. Page, *J. Chem. Soc., C*, **1967**, 2593.
26. J. Aknin and D. Molho, *Bull. Soc. Chim. Fr.*, **1965**, 3025.
27. M. M. Badawi, M. B. E. Fayez, T. A. Brice, and R. I. Reed, *Chem. Ind.* (London), **1966**, 498.
28. P. H. McCabe, R. McCrindle, and R. D. H. Murray, *J. Chem. Soc., C*, **1967**, 145.
29. R. G. Cooke and J. G. Down, Unpublished results.
30. H. Audier, *Bull. Soc. Chim. Fr.*, **1966**, 2892.
31. Y. Itagaki, T. Kurokawa, S. Sasaki, C.-T. Chang, and F.-C. Chen, *Bull. Chem. Soc. Japan*, **39**, 538 (1966).
32. R. I. Reed and J. M. Wilson, *J. Chem. Soc.*, **1963**, 5949.
33. D. L. Dreyer and D. J. Bertelli, *Tetrahedron*, **23**, 4607 (1967).
34. J. H. Beynon, *Mass Spectrometry and its Applications to Organic Chemistry*, Elsevier, Amsterdam, 1960, p. 270.
35. A. Pelter, P. Stainton, and M. Barber, *J. Heterocycl. Chem.*, **2**, 262 (1965).
36. A. Pelter and P. Stainton, *J. Chem. Soc., C*, **1967**, 1933.
37. J. H. Bowie and D. W. Cameron, *Aust. J. Chem.*, **19**, 1627 (1966).
38. S. Huneck, *Tetrahedron Lett.*, **1966**, 3547.
39. B. Jackson, H. D. Locksley, and F. Scheinmann, *J. Chem. Soc., C*, **1967**, 785.
40. D. Billet, G. Massicot, C. Mercier, D. Anker, A. Matschenko, C. Mentzer, M. Chaigneau, G. Valdener, and H. Pacheco, *Bull. Soc. Chim. Fr.*, **1965**, 3006.
41. M. V. Chesire, P. A. Cranwell, C. P. Falshaw, A. J. Floyd, and R. D. Haworth, *Tetrahedron*, **23**, 1669 (1967).
42. W. H. Pirkle and M. Dines, *J. Am. Chem. Soc.*, **90**, 2318 (1968).

6 LACTONES AND CYCLIC ANHYDRIDES

I. SIMPLE LACTONES

A. β-Lactones

The simplest lactone, β-propiolactone (**1**), behaves as a typical oxetan derivative, readily fragmenting by the ring-cleavage characteristic of these compounds (see Chap. 1); indeed, fragmentation is so facile that the molecular ion (m/e 72) is not observed (1). Cleavage along line a in **1** results in charge retention on the ketene fragment **1a** which accounts for the base peak (m/e 42) of the spectrum, whereas cleavage along line b gives an

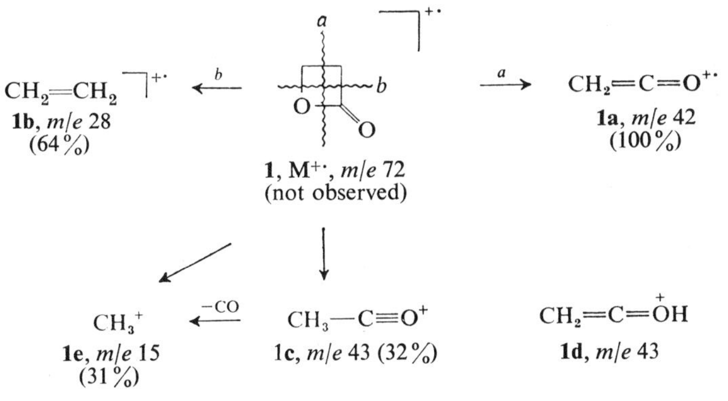

$$CH_2{=}CH_2 \rceil^{+\cdot} \xleftarrow{b}$$

1b, m/e 28
(64 %)

1, $M^{+\cdot}$, m/e 72
(not observed)

$$\xrightarrow{a} \quad CH_2{=}C{=}O^{+\cdot}$$

1a, m/e 42
(100 %)

$$CH_3^+ \xleftarrow{-CO} CH_3{-}C{\equiv}O^+ \qquad CH_2{=}C{=}\overset{+}{O}H$$

1e, m/e 15
(31 %)

1c, m/e 43 (32 %)

1d, m/e 43

abundant ion at m/e 28 due to ionized ethylene **1b** (part of m/e 28 may also be due to CO) and to a lesser extent ionized carbon dioxide (m/e 44, 5.5 %). A moderately intense ion at m/e 43 may be formulated as the acetyl ion (**1c**) or the protonated ketene ion **1d**. Its formation must involve hydrogen rearrangement. Another hydrogen-rearrangement ion is m/e 15 (CH_3^+), **1e**.

However, this probably originates both directly from the parent ion and also from the further fragmentation of m/e 43 by loss of CO.

One of the earliest applications of mass spectrometry to problems in heterocyclic chemistry was in the structure elucidation of the ketene dimer for which structures **2** and **3** were considered (2). The ketene dimer exhibits

a weak parent ion (m/e 84) and the base peak at m/e 42 is due to ionized ketene (**3a**) which strongly favors structure **3** (2). Further support for **3** is provided by the high intensity of $CH_2^{+\cdot}$ (m/e 14, 14.5%) relative to the methyl ion (CH_3^+, m/e 15, 4.9%). Structure 2 would be expected to yield an m/e 15 ion much more intense than m/e 14.

As might be expected, cleavage along line a with formation of the dimethylketene ion **4a** is an important fragmentation process for the dimethyl-ketene dimer **4** (3). Ring cleavage along line b results in loss of CO_2 and

formation of the allenic ion **4b**. An interesting feature in the spectrum of **4** is an ion at m/e 84 corresponding to sequential loss of two molecules of carbon monoxide from the molecular ion. Formation of a carbon–carbon bond must either precede or occur simultaneously with the elimination of the second molecule of CO, and a possible mechanism has been suggested (**4** → **4c**) involving the intermediacy of the cyclopropanone radical ion **4d**.

B.   Saturated γ-, δ-, and Higher Lactones

The spectra of a considerable number of simple γ- and δ-lactones (e.g., **5, 6, 7,** and **8,** R = alkyl) have been examined in detail. Loss of CO_2 from the molecular ion is an important fragmentation process for γ-butyro- and γ-valerolactones [**5,** and **6** (R = CH_3)] resulting in the ions **5b** and **6a** as the base peaks (1). These further fragment by loss of H· or CH_3·, respectively,

to give an abundant ion at m/e 41, formulated as either the allyl ion **5c** or the cyclopropyl ion **5d.** Prominent ions due to loss of CO_2 (**7b** and **8a**) are also observed for δ-valerolactone and δ-hexanolactone [**7,** and **8** (R = CH_3)] (4), but this mode of fragmentation becomes insignificant for both δ- and γ-lactones of general formulas **6** and **8** containing more than seven or eight carbon atoms (4). Loss of formaldehyde from the molecular ion of γ-butyrolactone (**5**) results in an ion of moderate intensity at m/e 56 (**5e**) (cf. the loss of formaldehyde in tetrahydrofuran, Chap. 2, Sec. I). Such a loss, however, is insignificant in δ-valerolactone (**7**).

In general the most important fragmentation process is α-cleavage with loss of the side chain to give the oxonium ion **5a** (m/e 85) for γ-lactones and the ion **7a** (m/e 99) for δ-lactones. These ions are highly characteristic and (except for a few of the lower members) correspond to the base peaks in the spectra, allowing a very easy and reliable differentiation between isomeric δ- and γ-lactones.

A further distinction between γ- and δ-lactones is the occurrence of a strong m/e 42 ion and of a pair of peaks of moderate intensity at m/e 70 and 71 which is typical of δ-lactones. The m/e 42 ion may be due to $C_3H_6^{+\cdot}$ or ketene (or both), while m/e 71 and 70 may be due to loss of CO and CHO· from the m/e 99 ion (7a), but the required high-resolution data is not available to decide the composition of these ions.

In general the molecular ion is of very low intensity (being, however, two or three times as intense for a δ-lactone as for the corresponding γ-lactone), and this together with the low intensity of the ion due to the alkyl side-chain (R^+) often makes gas-chromatographic retention time the best criterion of the molecular size of an unknown lactone, especially if higher molecular weight impurities are also present. Loss of water from the molecular ion is not observed for the smaller lactones but becomes significant if the chain length is three or more carbon atoms. In many cases the M-H_2O ion is several times the intensity of the molecular ion and thus may be useful in establishing the molecular weight in weak spectra (5). It is interesting that with increasing chain length there occur moderately intense ions formed by the loss of two water molecules (M-36). With side chains of three or more carbon atoms (9) the McLafferty rearrangement with hydrogen transfer to the lactone oxygen-atoms and cleavage β to the ring is observed to give an ion at m/e 100 or 114 (9a) which accounts for from 1 to 3% of the total ion current (5).

9, $n = 1, 2$

9a, m/e 100 ($n = 1$)
m/e 114 ($n = 2$)

11

α-Cleavage is relatively less important for 12-stearolactone (10) (5) where the ion (10a) due to loss of the six-carbon side-chain accounts for 2.4% of the total ion current whereas the M-H_2O ion accounts for 2.9%. The base peak is m/e 55 (8.3% of the total ion current). An ion of moderate intensity at m/e 168 (10b) corresponds to loss of n-heptaldehyde (cf. loss of CH_2O for γ-butyrolactone).

The ω-lactone cyclopentadecanolide (11) shows significant ions due to loss of one and two molecules of water from the parent ion (M-18 and M-36, respectively). Since such losses are observed for δ- and γ-lactones only if a

10a, m/e 197   10, M$^{+\cdot}$, m/e 282   10b, m/e 168

side chain of three or more carbon atoms is present, this indicates that ring-opening of the cyclopentadecanolide molecular ion may occur readily, thus allowing hydrogen transfer from the hydrocarbon portion to the oxygen atoms. Alternatively, the large ring size may facilitate transannular hydrogen-transfers. An unexpected ion in the spectrum of **11** is at M-60, corresponding to loss of acetic acid (or its equivalent) (5).

If two alkyl groups are substituted at $C_{(4)}$ of a γ-lactone, as is usual for α-cleavage reactions, the larger alkyl group is preferentially lost. Thus for the lactone **12** loss of the ethyl group results in **12a** as the base peak. (In the unsaturated compound **13** loss of the vinyl group is relatively unfavored and the conjugated oxonium ion **13a** (M-CH$_3$·) is the base peak (6).)

13a, m/e 111 (100%)   12, R = C_2H_5 12a, m/e 99
13, R = CH=CH$_2$ (100%)

With the lactone **14**, in which there is an isopropyl side-chain at $C_{(3)}$, only a relatively weak ion (**14a**) due to loss of C_3H_7· is observed. Such differences may provide a useful means of distinguishing between isomeric $C_{(4)}$- and

M-1, m/e 141 (3%)   14, M$^{+\cdot}$, m/e 142 (3.5%)   14a, m/e 99 (16%) $^+$CH(CH$_3$)$_2$
m/e 43
(88%)

CH$_3$—CH=C=O$^{+\cdot}$
14c, m/e 56 (100%)

CH(CH$_3$)$_2$
—CH$_3$
14b, m/e 98 (2.5%)

$C_{(3)}$-substituted γ-lactones by mass spectrometry (7). The weak M-1 ion (m/e 141) is of comparable intensity to the parent ion, another distinction from γ-lactones with an alkyl side-chain at $C_{(4)}$, in which the M-1 ion is very much weaker than the parent ion. The M-CO$_2$ ion, **14b**, is also rather weak, but abundant ions are observed at m/e 83 and 69 which probably originate from **14b** by loss of $CH_3\cdot$ or $C_2H_5\cdot$, respectively. The base peak is at m/e 56 and is probably due to the methylketene radical ion (**14c**).

II. MORE COMPLEX LACTONES

A. Lignan Lactones

The mass spectra of a number of lignans containing a 2,3-dibenzyl-γ-butyrolactone ring have been reported. A typical example is ($-$)-matairesinol (**15**) where a considerable (45%) parent ion is observed and benzylic cleavage is the dominating fragmentation process, resulting in the benzyl (or tropylium) ion **15a** as the base peak. All other fragment ions are of less than 20% relative intensity (8). Benzylic cleavage with loss of **15a** as a radical and

charge retention on the lactone moiety occurs only to a minor extent, resulting in a weak ion at m/e 221. This has been formulated as the ring-opened ion **15c** but is probably better represented as the bicyclic oxonium ion **15b** since further loss of CO_2, which may be expected to occur readily from **15c**, is not in fact observed. Cleavage of the lactone ring also occurs to only a small extent with elimination of the neutral 3-phenylpropene **15e** and formation of a weak ion at m/e 194 (**15d**). Losses of water, of carbon dioxide, or of a methyl or methoxy radical from the parent ion are not observed.

In lignans containing the 1,2,3,4-tetrahydronaphthalene ring-system the cleavage of a single benzylic bond does not result in fragmentation, so that the spectrum of α-(−)-peltatin (**16**) is dominated by the intense molecular ion and relatively little fragmentation is observed. Hydrogen-transfer reactions are important and result in the ions **16a** and **16b**, while loss of **16a** as a neutral fragment gives an ion at m/e 246 (**16c**) which further fragments

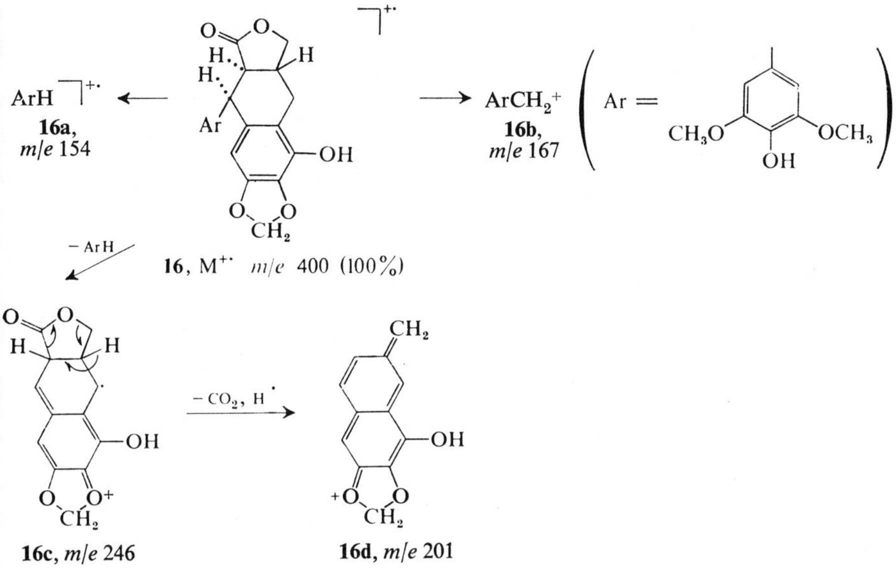

16, $M^{+\cdot}$ m/e 400 (100%)

16c, m/e 246   **16d**, m/e 201

by the concerted loss of CO_2 and H· to form the resonance-stabilized ion **16d** (m/e 201). A weak ion at m/e 355 (M-45) corresponds to the loss of CO_2 and H· directly from the parent ion (8). The prominent ion at m/e 189 in the spectrum of α-(−)-peltatin may be envisaged as arising directly from the

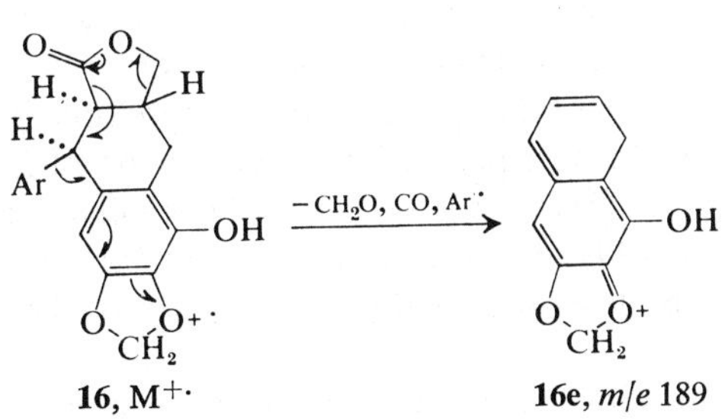

16, $M^{+\cdot}$ **16e**, m/e 189

parent ion by the concerted fragmentation of the lactone ring together with the loss of Ar· (**16** → **16e**).

The ease of loss of water from the molecular ion of a series of isomeric hydroxylignans (**17**, **18**, and their epimers) has been shown to be very markedly dependent on stereochemistry (8). Thus the spectrum of *epi*-podophyllotoxin (**17** with OH α) exhibits a much more pronounced (11 %) M-H$_2$O ion than does the epimeric (−)-podophyllotoxin (**17**). This effect is much more striking for picropodophyllin (**18**) where the lactone ring is

17

18, R = H, M$^{+\cdot}$,
 m/e 414 (4 %)
20, R = C$_2$H$_5$

−H$_2$O | (R=H)

19

18b, m/e 312 (100%)

m/e 297 (28%)   **18a**, m/e 396

fused in the *cis* configuration. In this case the molecular ion is of only 4 % relative intensity and the M-H$_2$O ion (**18a**) is over sixteen times as intense, whereas in the spectrum of the epimeric *epi*-picropodophyllin (**18** with OH α) the molecular ion is the base peak with M-H$_2$O being of 21 % relative abundance. The M-H$_2$O ion (**18a**) of picropodophyllin eliminates a neutral molecule of γ-crotonolactone (**19**) to give **18b** (m/e 312) as the base peak which then further fragments by loss of a methyl radical to give the even-electron ion m/e 297.

Picropodophyllin ethyl ether (**20**) fragments in a manner similar to **18** (8), the ion due to loss of ethanol (M-46) being of 57 % relative intensity. Very little fragmentation is observed in the spectrum of **17** (where the M-H$_2$O ion is very weak.) The most abundant fragment ion, m/e 168 (due to ArH$^{+\cdot}$ and also observed for **18**), is of only about 13 % the intensity of the molecular ion (base peak).

B. Terpene Lactones

1. *Monoterpenes*

Mass spectrometry has proved useful in the structure elucidation of two naturally-occurring δ-lactones, dihydronepetalactone and isodihydronepetalactone (**21**, stereochemistry not shown) (9). Formation of an ion at m/e 113 (**21a**) may be readily rationalized by an internal McLafferty rearrangement to give the ring-opened molecular ion **21b** which then undergoes an allylic cleavage to give the conjugated oxonium ion **21a**. The cyclopentenyl ions **21c** (m/e 81, base peak) and **21d** (m/e 67) have been accounted for by the mechanism shown, which involves initial cleavage of the carbonyl–carbon bond (**21 → 21e**).

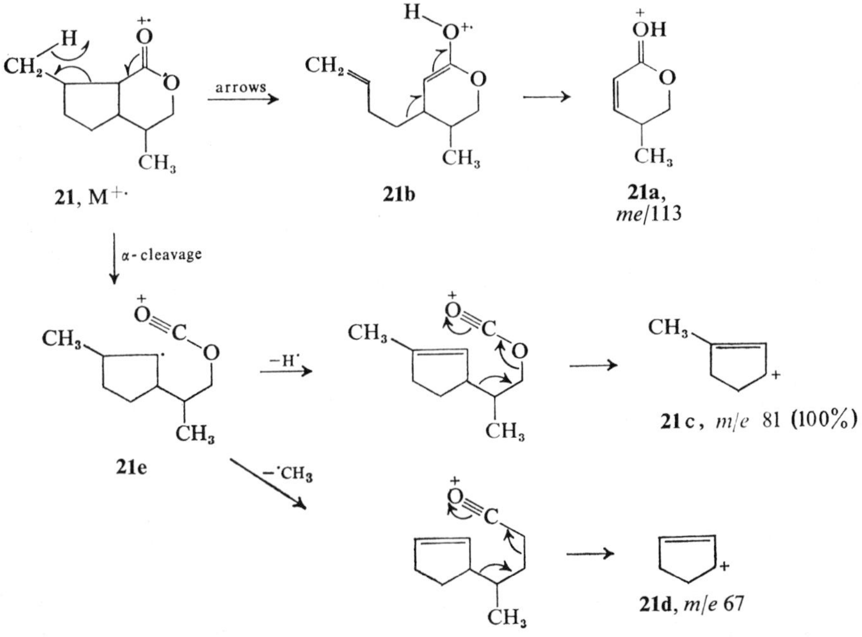

The acetoxy-lactone **22** exhibits only a very weak molecular ion but shows a ready loss of acetic acid to give the ion **22a** which further fragments both by loss of CO_2 giving an ion at m/e 92 (probably ionized toluene formed by hydrogen rearrangement) and by consecutive loss of CO and CHO· to give the phenonium ion **22b**, m/e 79 (9). The ketolactone **23** shows the expected intense acetyl ion, **23a** (m/e 43, base peak), together with an ion of almost equal intensity due to elimination of acetone from the molecular ion, probably

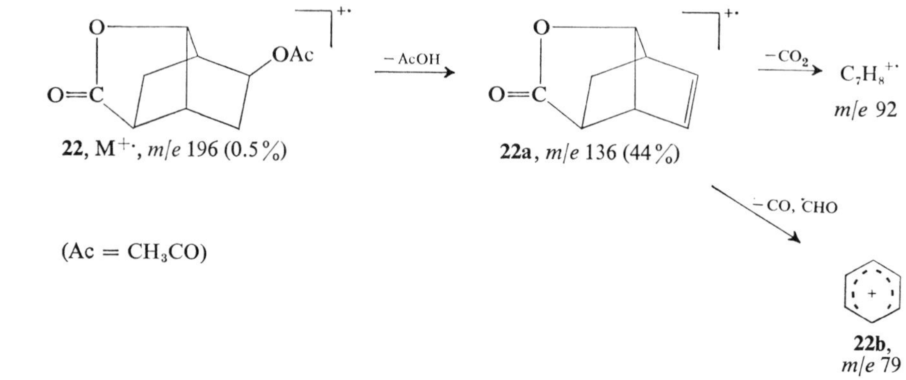

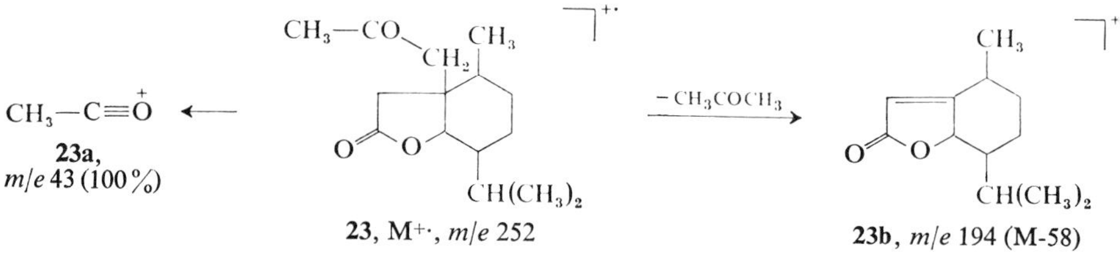

OAc
− AcOH
− CO_2
$C_7H_8^{+\cdot}$
m/e 92
22, $M^{+\cdot}$, m/e 196 (0.5%)
22a, m/e 136 (44%)
− CO, ·CHO
(Ac = CH_3CO)
22b,
m/e 79

$CH_3—\overset{+}{C}≡O$
23a,
m/e 43 (100%)
− CH_3COCH_3
23, $M^{+\cdot}$, m/e 252
23b, m/e 194 (M-58)

with formation of the unsaturated lactone **23b** (10). Fragmentation of the α-ketolactone ring of apoaplysin (**24**) occurs so readily that the molecular ion is not observed (11). The loss of the three oxygen atoms as **CO** and $\mathbf{CO_2}$

− CO_2, CO
− $\cdot CH_3$

24, $M^{+\cdot}$ (not observed)

24a, m/e 110 (50%)

24b, m/e 95 (100%)

probably occurs via a concerted cyclic mechanism with formation of the abundant trimethylcyclopentene ion **24a** which then undergoes the expected further loss of a methyl radical resulting in **24b** as the base peak of the spectrum.

2. *Sesqui- and Diterpene Lactones*

The mass spectra of a considerable number of sesquiterpene lactones have been investigated. A prominent feature in the spectra of the santonins (e.g., α-santonin, **25**) is the abundant M-73 ion which was shown by high-resolution mass-measurement to involve the elimination of $C_3H_5O_2$ including (as shown

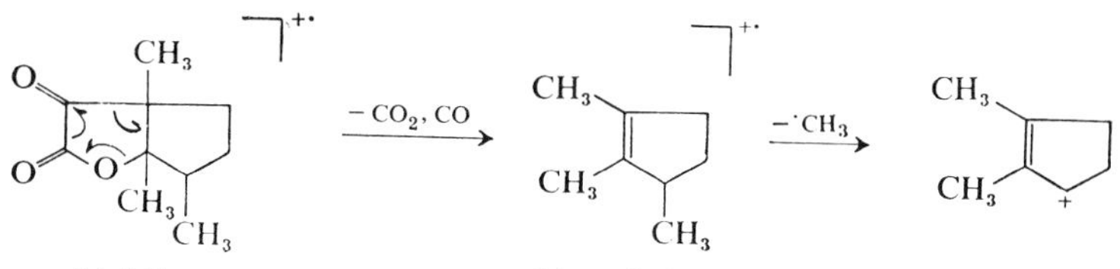

by deuterium labeling) the hydrogen atoms at $C_{(11)}$ and $C_{(13)}$ (12, 13). Hence the M-73 ion (m/e 173) corresponds to loss of the lactone ring together with an additional transferred hydrogen atom. It has been suggested that this fragmentation involves a methyl migration to give the rearranged molecular ion **25a** which may then fragment as shown to give **25b** as the M-73 ion (12).

25, M$^{+\cdot}$, m/e 246   25a   25b, m/e 173

25c, m/e 246   25d, m/e 173

It has, however, also been pointed out that postulation of a methyl migration is unnecessary and that a more plausible mechanism would be fragmentation of the unrearranged molecular ion initiated by removal of one of the π-electrons from the $C_{(4)}$-$C_{(5)}$ double bond (**25 → 25c**), resulting in the resonance-stabilized ion **25d** as M-73 (m/e 173) (14). A point in favor of the latter mechanism is that it does not involve ions such as **25a** and **25b** in which the oxygen atom is associated with the unfavored electron sextet. The exact origin of the transferred hydrogen atom has not been established by deuterium labeling (apart from showing that the $C_{(2)}$ hydrogen is not involved). It has been suggested that the transferred hydrogen originates from $C_{(8)}$. The intensity of a number of ions shows considerable dependence on stereochemistry. Thus, for example, the loss of carbon monoxide from the molecular ion (which most probably involves the $C_{(3)}$ carbonyl group) is considerably more intense for C- and D-santonin (lactone ring *cis*-fused) than for α- and β-santonin (lactone ring *trans*-fused). The M-73 ion is also observed in the spectra of tetrahydrosantonins (e.g., α-tetrahydrosantonin, **26**) although here a mechanism different from **25 → 25d** would be involved as there is no $C_{(4)}$-$C_{(5)}$ double bond (13). Compounds where ring A is aromatic, such as α-desmotroposantonin (**27**, R = H) and its methyl ether (**27**, R = CH$_3$), show a much simpler fragmentation pattern than the santonins or tetrahydrosantonins (13). An abundant molecular ion is observed (as is the case also for **25** and **26**) and the dominant fragment is M-73 (base peak) which has been formulated as the substituted naphthonium ion **27a**.

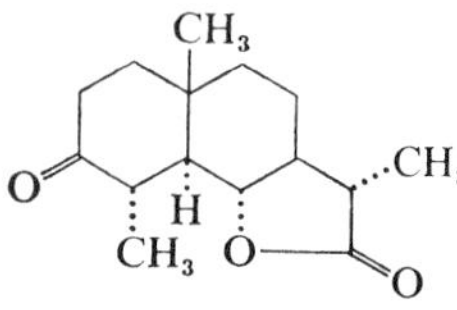

26 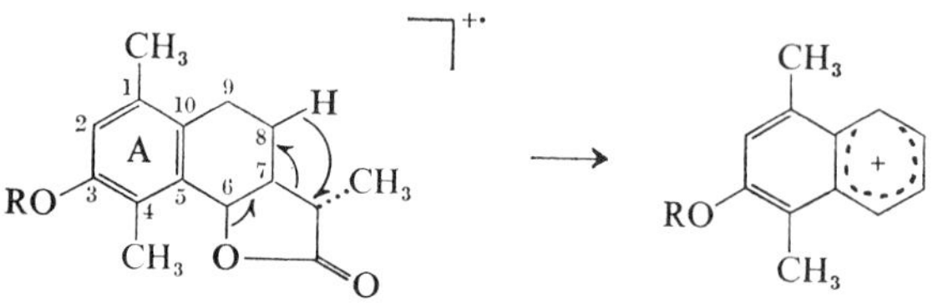   27, M$^{+\cdot}$ 27a, M-73 (100%)

The formation of **27a** may be rationalized by transfer of a $C_{(8)}$ hydrogen atom in the fragmentation **27 → 27a**.

The spectra of a number of sesquiterpene lactones of the guaianolide type (e.g., **28**, **29**, **30**) have been reported (12). 6-Deoxygeigerin (**28**) exhibits a fragmentation similar to **25 → 25d** for santonin, resulting in an M-73 ion, but cleavages α to the hydroxyl functions and subsequent hydrogen-transfer reactions become important fragmentation processes upon introduction of hydroxyl groups. Thus the m/e 151 ion in the spectrum of geigerin (**29**) (shifted to m/e 153 in 2,2,11-trideuterogeigerin) may be rationalized by α-cleavage of the $C_{(6)}$—$C_{(7)}$ bond followed by hydrogen transfer (probably

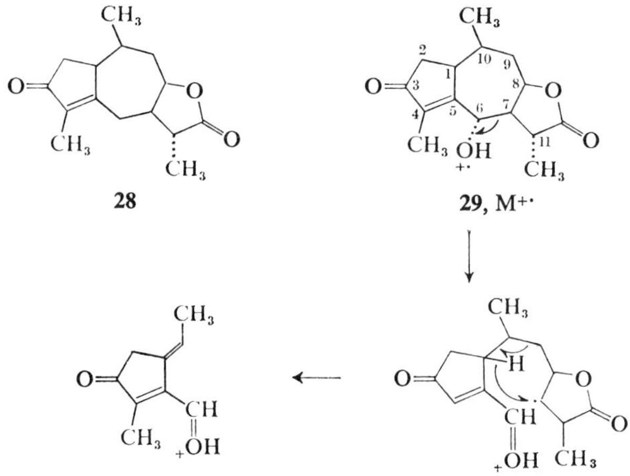

28   29, M$^{+\cdot}$

29a, m/e 151

from the allylic $C_{(1)}$ position via a cyclic six-membered transition state, as shown) and fragmentation to give the conjugated oxonium ion **29a**. Similarly, the particularly favored loss of the lactone ring together with the transfer of a hydrogen atom from either $C_{(11)}$ or the $C_{(11)}$ methyl group of isophoto-santonic lactone (30) may be regarded as being initiated by cleavage α to the hydroxyl group and as resulting in the ion 30a.

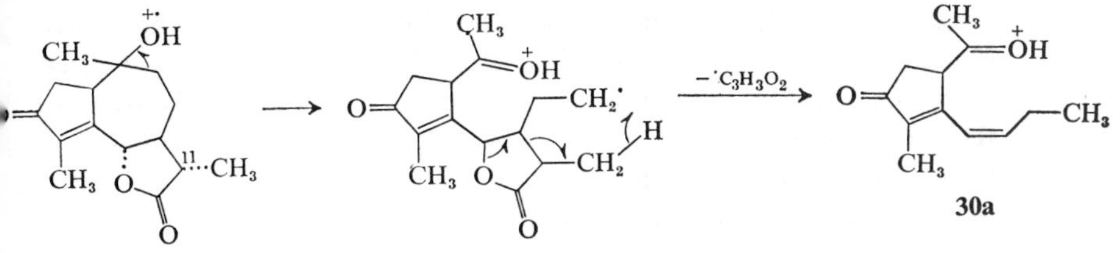

30, M+·

30a

The base peak of the spectrum of the complex epoxydilactone, picrotoxinin (31) is m/e 95, and its composition has been shown by high-resolution mass-measurement to be C_6H_7O (27). The formation of this ion may be rationalized by ionization of the $C_{(15)}$ lactone carbonyl-group (31 → 31a) followed by

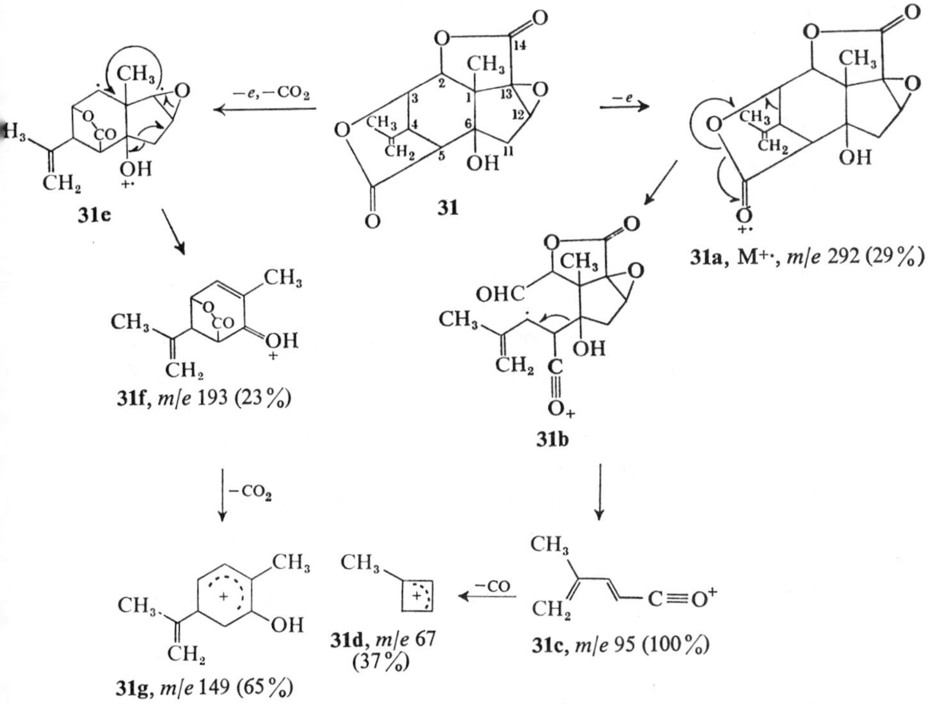

31c

31f, m/e 193 (23%)

31, m/e 292 (29%) (31a, M+·)

31b

31g, m/e 149 (65%)

31d, m/e 67 (37%)

31c, m/e 95 (100%)

cleavage of two bonds to give the ring-opened aldehydic molecular ion 31b (a reaction analogous mechanistically to the elimination of formaldehyde from the molecular ion of γ-butyrolactone). 31b may then readily fragment to give the conjugated acyl ion **31c** as the base peak. The further fragmentation of **31c** by loss of carbon monoxide (**31c** → **31d**) is unexceptional. The M-CO$_2$ ion (**31e**) is insignificant, but a fairly intense ion is observed at m/e 193 (**31f**, $C_{11}H_{13}O_3$ by mass measurement) which corresponds to the loss of CO$_2$ and the propenoyl radical, probably as a concerted process although the appropriate metastable ion is not observed. The m/e 193 ion shows a pronounced loss of carbon dioxide with formation of the stable substituted benzonium ion **31g**. Ions due to loss of a methyl group (M-15) or of water (M-18) are relatively weak (3 and 7% of the base peak, respectively) compared with the ions arising from the other facile fragmentations of the molecular ion. The related diepoxylactone, tutin (**32**), also exhibits a very abundant m/e 95 ion which may arise by the same mechanism as that postulated for picrotoxinin (**31** → **31c**) because the six-membered ring is the same in both compounds

32 (R = H), m/e 294 (0.3%)
33 (R = OH)

32b

32a, m/e 141 (100%)

(27). The base peak is m/e 141 (**32a**) which may be due to α-cleavage of the $C_{(5)}$—$C_{(6)}$ bond accompanied by hydrogen transfer from the six-membered ring to $C_{(1)}$. Deuteration of both the hydroxyl groups of **32** results in a shift from m/e 141 to m/e 142 in **32a**, showing that the transferred hydrogen atom does not originate from the $C_{(2)}$ hydroxyl group (15). Transfer of the $C_{(3)}$ hydrogen atom seems likely as it would be aided by the rearrangement of the lactone ring as shown in **32b**. The second-most abundant fragment is m/e 125 which may arise from the M-15 ion (**32c**). Several mechanisms may be envisaged, one of which involving rearrangement of the epoxide rings to give the conjugated oxonium ion **32d** is shown. Some other fragmentations of tutin have already been discussed in Chap. 1. A comparison of the mass spectrum of tutin with that of the insect metabolite mellitoxin strongly

32 $\rceil^{+\cdot}$ $\xrightarrow{-\cdot CH_3}$

32c

32d, m/e 125 (73.5%)

supported the hydroxytutin structure **33** assigned to this compound (15). The bromine-containing doublet at m/e 329/331 (M-57) in the spectrum of bromomellitoxinone (**34**) has been shown by observation of the appropriate metastable ion to arise directly from the molecular ion, and may be rationalized in terms of cleavage of the $C_{(1)}$—$C_{(2)}$ and $C_{(3)}$—$C_{(4)}$ bonds (**34** → **34a**) and fragmentation with resultant loss of $C_2HO_2^{\cdot}$ (although this loss has not been confirmed by high-resolution mass-measurement), possibly with a concomitant hydrogen migration from the $C_{(1)}$-methyl group to $C_{(1)}$ to give the ion **34b** (m/e 329/331). This ion may then further fragment with hydrogen transfer as shown and formation of the resonance-stabilized ion **34c** as the base peak.

34, M$^{+\cdot}$, m/e 386/388

34a

34b, m/e 329/331 (M-57)

34c, m/e 165 (100%)

The spectra of a considerable number of diterpene lactones have been reported. Marrubiin (**35**) exhibits an abundant molecular ion and an intense ion at m/e 109 (**35b**) whose formation may be rationalized in terms of cleavage of the $C_{(9)}$—$C_{(10)}$ bond on ionization (**35** → **35a**) followed by fragmentation of the lactone ring (possibly concerted as shown by the arrows marked a in **35a**) (16, 17). Alternatively **35a** may undergo hydrogen transfer followed by cleavage of the $C_{(6)}$—$C_{(7)}$ bond (arrows marked b in **35a**) to give an ion of moderate intensity at m/e 165 (**35c**). It has been suggested that **35b** and **35c** may originate from various fragment ions, e.g., M-H$_2$O, as well as directly from the molecular ion. Ions of the general type **35b** and **35c** are typical of many carbodicyclic diterpenes but in this case it is also possible that m/e 165 is due to **35d**, formed from **35e** by hydrogen transfer and cleavage of the $C_{(6)}$—$C_{(7)}$ bond. (A similar fragmentation accounts for the base peak in the spectrum of grindelane; see Chap. 2, Sec. III.B.) These two possibilities

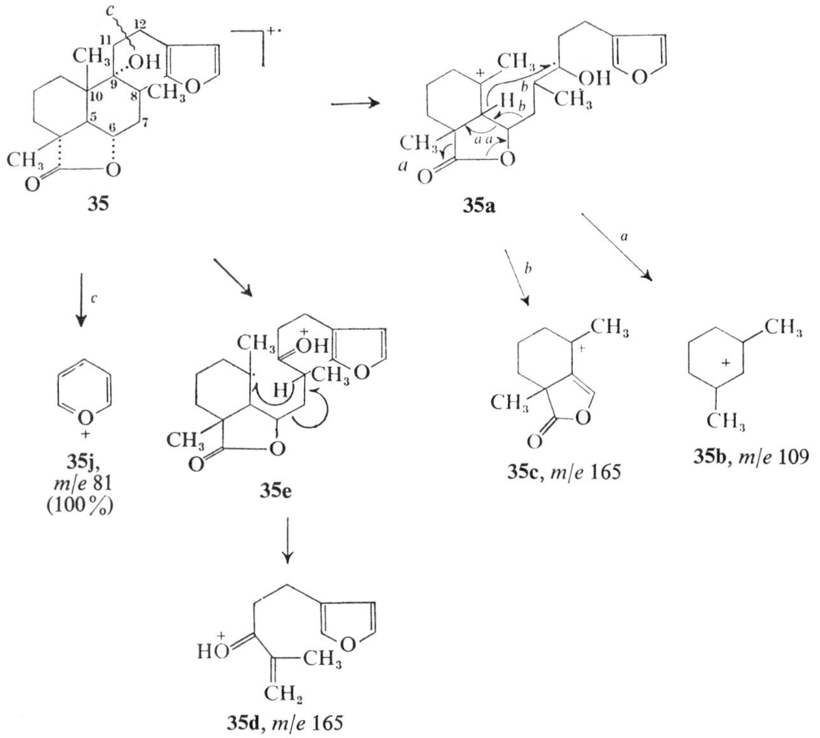

could be readily distinguished by deuterium labeling of the hydroxyl group. Weak ions are observed at M-44 and M-46, corresponding to loss of carbon dioxide and formic acid. Two prominent ions at m/e 135 (**35i**) and 152 (**35g**) may be explained by assuming that the molecular ion undergoes hydrogen rearrangement to the $C_{(5)}$-$C_{(6)}$-unsaturated acid **35f** followed by the retro-Diels-Alder reaction to give the fragments **35g** (m/e 152) and **35h**, the latter eliminating the carboxyl radical to give **35i**. However, the composition of these ions has not been established by high-resolution measurements and it is also possible that m/e 152 is due to elimination of CO from **35h**. The base peak is the pyrilium ion **35j** (m/e 81), typical of compounds containing the 3-alkylfuran group (see Chap. 4, Sec. I.B).

Mass spectrometry has been shown to be applicable to the identification and structure elucidation of gibberellins as their methyl esters (e.g., **36** and

37). Useful structural correlations can be made on the basis of the relative intensities of the M-44 and M-46 ions, corresponding to elimination of CO_2 and of HCOOH from the lactone ring. Compounds containing the $C_{(2)}$-$C_{(3)}$

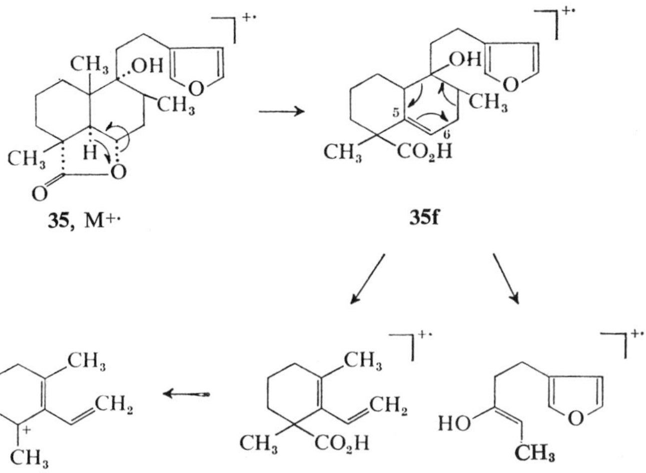

35, M$^{+\cdot}$ 35f

35i, m/e 135   35h, m/e 180   35g, m/e 152

double-bond (**36**) exhibit intense M-44 but very weak M-46 ions whereas compounds with the $C_{(2)}$ hydroxyl group (**37**) show much stronger M-46 but no M-44 ions. If the $C_{(2)}$ hydroxyl group is absent or if both the $C_{(2)}$

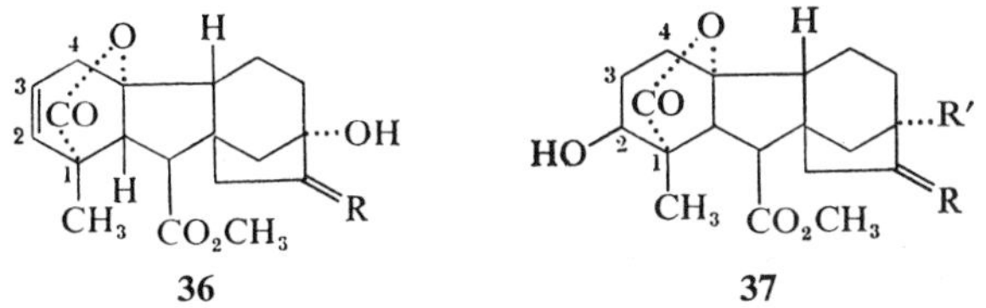

36   37

hydroxyl and a $C_{(3)}$-$C_{(4)}$ double-bond are present, both M-44 and M-46 ions are observed in appreciable intensity (18). These observations can provide useful information as to the presence of a hydroxyl group at $C_{(2)}$.

III. UNSATURATED LACTONES

A. Unsaturated γ-Lactones

Introduction of a double bond into the γ-lactone ring between $C_{(3)}$ and $C_{(4)}$, as in γ-crotonolactone (**38**) and β-angelicalactone (**39**) (Figs. 6.1 and 6.2),

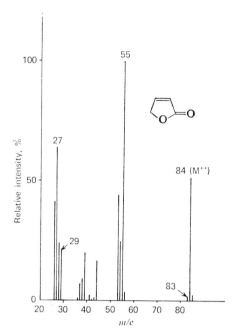

Fig. 6.1

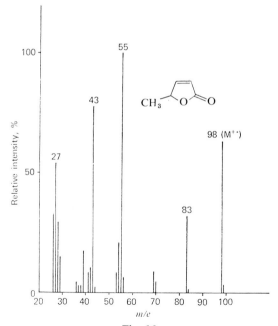

Fig. 6.2

results in parent ions of much greater intensity than in the spectra of the corresponding saturated lactones (1). The intense hydrogen-rearrangement ion at m/e 55 (base peak for both **38** and **39**; only possible composition C_3H_3O) has been shown by ^{18}O-labeling (1) to incorporate the carbonyl oxygen atom and may be formulated as the propenoyl ion (**40**). This ion is possibly formed by initial cleavage of the $O_{(1)}$—$C_{(2)}$ bond, i.e., cleavage α to the carbonyl group (**38, 39 → 38a, 39a**), followed by cleavage of the $C_{(4)}$—$C_{(5)}$ bond with simultaneous hydrogen migration from $C_{(5)}$ to $C_{(4)}$. Hydrogen migration from $C_{(5)}$ to the carbonyl oxygen has been considered unlikely on the grounds that it would give rise to the unfavored diradical ion **40a**. This is not a valid objection, however, as hydrogen migration via a six-membered transition-state (as shown) would result in the stable, aromatic hydroxycyclopropenyl

38, R = H, m/e 84
39, R = CH_3, m/e 98

38a
39a

$$CH_2\!=\!CH\!-\!C\!\equiv\!O^+$$
40, m/e 55 (100%)

$$R\!-\!C\!\equiv\!O^+$$
38b, m/e 29 (21%)
39b, m/e 43 (78%)

OH
40b, m/e 55

$$\dot{C}H\!=\!CH\!-\!\dot{C}\!=\!\overset{+}{O}H$$
40a, m/e 55

ion (**40b**) as an alternative formulation for m/e 55. Loss of $C_3H_3O\cdot$ as the neutral fragment is also observed, resulting in the formyl ion (**38b**) and a more intense acetyl ion (**39b**) in the spectra of **38** and **39**, respectively, but the metastable ions that would show whether **38b** and **39b** do in fact arise directly from the corresponding parent ions have not been reported. Cleavage α to the lactone oxygen and resultant loss of a methyl radical (**39 → 39c**) is an important process for β-angelicalactone (**39**) which has a methyl group

39, m/e 98

39c, m/e 83

at $C_{(5)}$, whereas γ-crotonolactone (**38**) shows only a rather weak M-1 ion (m/e 83). α-Cleavage may become increasingly important with larger substituents at $C_{(5)}$. Thus in the spectrum of the steroidal derivative **41** it accounts for the conjugated oxonium ion **41a** as the base peak of the spectrum (19). Charge retention on the hydrocarbon fragment (**41b**) is also observed to occur to a large extent.

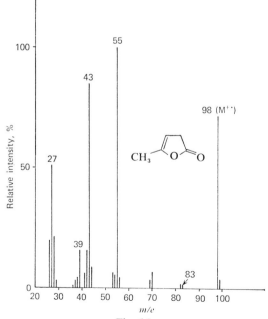

41, $M^{+\cdot}$, m/e 386 (60%)

41b, m/e 289
(87%)

41a, m/e 97 (100%)

The spectrum of the β, γ-unsaturated α-angelicalactone (**42**) (Fig. 6.3) is (except for the very much weaker m/e 83 ion) very similar to that of the α, β-unsaturated isomer, **39** (1). The formation of the m/e 55 (**42b**) and m/e 43 (**42c**) ions may be rationalized by electronic rearrangement of the molecular ion (**42d**) to the acetylcyclopropanone **42a**, from which the propenoyl (**42b**)

Fig. 6.3

and acetyl (**42c**) ions may be readily derived. In this case, however, ^{18}O-labeling to confirm the presence of the carbonyl oxygen in **42b** has not been carried out. The isomeric lactones **39** and **42** may be readily distinguished by the very weak m/e 83 (M-CH$_3$) ion in the latter case where the methyl group is attached to a double bond and α-cleavage is much less favored.

In contrast to simple saturated γ-lactones, the unsaturated lactones **38**, **39**, and **42** show relatively little loss of carbon dioxide from the parent ion. Thus γ-crotonolactone (**38**) does not show a significant ion at m/e 40 (M-44, **38c**) although an ion at m/e 39 is presumably the cyclopropenyl ion (**38d**) arising by the ready loss of a hydrogen atom from **38c**.

Mass spectrometry has proved useful in the structure investigation of a number of compounds containing the unsaturated γ-lactone ring as part of the structure. The lactone of 2,6,6-trimethyl-2-hydroxycyclohexylideneacetic acid (**43**) has been identified as a constituent of black tea aroma, and high-resolution data on the major ions has been reported (20). An appreciable parent ion (25%) is observed and the base peak is due to the loss of ·C$_5$H$_9$. This process may possibly involve cleavage along the dotted line shown in **43** with hydrogen transfer to the lactone ring to give an ion such as **43a** which may rearrange (with hydrogen migration) to a more stable structure such as the substituted pyrilium ion **43b**. Loss of carbon monoxide from the parent ion (**43** → **43c**) is of some importance; the resultant ion **43c** may then further lose a methyl group to give the oxonium ion **43d**. α-Cleavage with loss of a methyl group from the parent ion (**43** → **43e**) occurs to a small extent. A considerable portion of the m/e 165 (M-CH$_3$) ion may also be due to loss of one of the allylic *gem*-dimethyl groups.

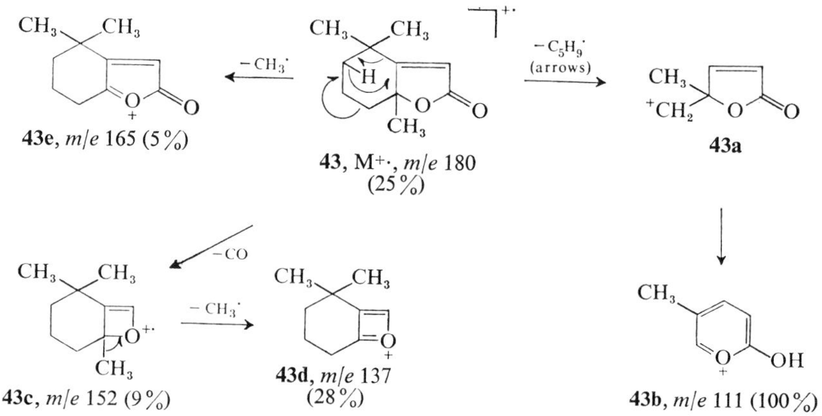

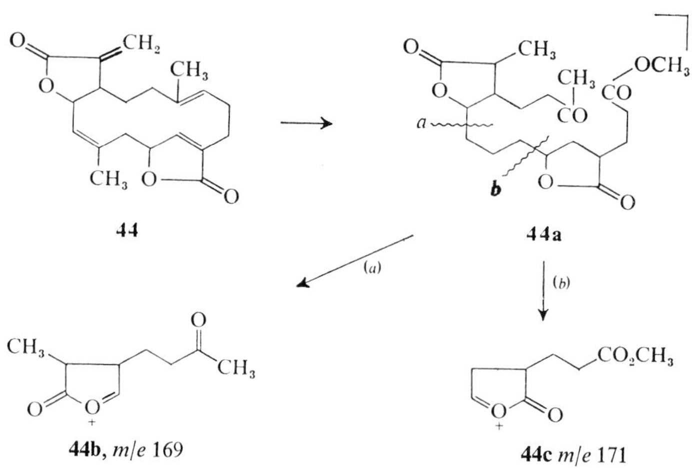

The mass spectrum of the macrocyclic diterpene, ovatodiolide (**44**), gave little useful structural information (21). A much more clear-cut and useful fragmentation could be obtained, however, by conversion (with rupture of the carbocyclic ring) of **44** to a "linear" derivative, the saturated dilactone ester **44a**. The characteristic α-cleavage of saturated γ-lactones accounts for the ions m/e 169 (**44b**, cleavage *a*) and 171 (**44c**, cleavage *b*) which are of comparable intensity and the two most important peaks in the spectrum of **44a**. Of the various possible structures for ovatodiolide consistent with the isoprene rule, the mass spectrometric data confirms structure **44** as the only structure that would give a derivative (**44a**) which could be expected to show intense ions at m/e 169 and 171 (**44b** and **44c**).

The spectra of a number of lichen constituents, derivatives of pulvic acid, have been studied (22, 23) and of these the spectrum of pulvic acid lactone (**45**) is shown in Fig. 6.4 (22). The major fragmentation pathway for **45** is symmetric cleavage of both the lactone rings which may be rationalized by initial ionization of one of the carbonyl groups followed by the concerted mechanism shown. The resulting ion (**45a**, m/e 145, base peak) then undergoes two consecutive losses of carbon monoxide (**45a** → **45b** → **45c**) to give the m/e 89 ion (formulated as the benzocyclopropenyl ion **45c**). This shows the expected loss of acetylene to give the ion at m/e 63 (**45d**). Loss of carbon dioxide from the parent ion is not observed, and loss of carbon monoxide only occurs to a small extent (**45** → **45e**). It is interesting, however, to note that **45e** (M-CO) readily loses a hydrogen atom (most probably one of the *ortho* hydrogens of the phenyl ring A in **45e**, as only aromatic hydrogen atoms are available) to give the well-stabilized oxonium ion **45f**. Of particular interest is the M-C$_2$O$_2$ ion at m/e 234 which arises in part from the loss of CO from **45e** (M-CO), but a metastable ion at m/e 188.8 indicates that m/e 234 is also due to concerted loss of C$_2$O$_2$ from the parent ion. This concerted loss has been rationalized by a skeletal rearrangement involving homolysis of both the oxygen–carbonyl bonds and rotation about the central carbon–carbon bond in **45g** followed by recyclization (**45g** → **45h**) to give the rearranged parent ion **45i** which can now lose C$_2$O$_2$ as a unit (22). The formulation of m/e 234 (M-C$_2$O$_2$) as the diphenylcyclobutene-1,2-dione ion (**45j**)

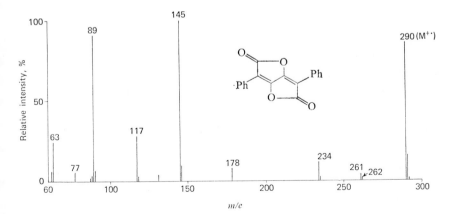

Fig. 6.4

45a, m/e 145 (100%)

$-$ CO

$O{=}C{=}\overset{+}{C}{-}Ph$

45b, m/e 117

$-$ CO

$C_7H_5{}^+$

45c, m/e 89

$-$ C$_2$H$_2$

$C_5H_3{}^+$

45d, m/e 63

45, m/e 290

$-$ CO

45e, m/e 262

$-$ H$\cdot$

45f, m/e 261

46, R$=$OCH$_3$

47, R $=$ NHCH(CH$_2$Ph)CO$_2$C

45g

45h

$Ph{-}C{\equiv}C{-}Ph \rceil^{+\cdot}$ $\overset{-\ C_2O_2}{\longleftarrow}$ **45j**, m/e 234 $\overset{-\ C_2O_2}{\longleftarrow}$ **45i**, m/e 290

45k, m/e 178

is consistent with a further concerted loss of C$_2$O$_2$ (substantiated by the appropriate metastable ion) to give the diphenylacetylene ion (**45k**, m/e 178).

Vulpinic acid (**46**) shows a prominent parent ion which undergoes a facile elimination of methanol to give an ion at m/e 290 (base peak) whose further fragmentation is virtually identical to that of **45** (22). Similarly the spectrum of rhizocarpic acid (**47**), apart from a prominent parent ion, is essentially a combination of the spectra of **45** and phenylalanine methyl ester (22). A metastable ion establishes that the formation of **45** from the parent ion of **47** is indeed an electron-impact-induced process. The spectra of the above and other compounds studied indicate that mass spectrometry may prove a useful tool in structure elucidation studies in this field.

B. Unsaturated δ-Lactones

No systematic study of the fragmentation of simple unsaturated δ-lactones is yet available in the literature. However, the spectra of a series of kawa lactones (e.g., **48**, **49**, and **50**) have been examined in some detail (24). Kawain (**48**) and 5,6-dihydroyangonin (**49**), which both have a styryl substituent at $C_{(6)}$, exhibit very intense parent ions. The elimination of cinnamaldehyde via the retro-Diels-Alder reaction is an important fragmentation process for **48**, giving m/e 98 as the most intense fragment ion, whereas for **49** the loss of *p*-methoxycinnamaldehyde is less pronounced. The m/e 98 ion (**48a**, **49a**) may be formulated as either a cyclobutenone or a conjugated ketene, and the presence of a prominent m/e 68 ion in both spectra indicates that it further fragments by loss of formaldehyde from the methoxyl group. Both compounds **48** and **49** exhibit considerable cinnamoyl ions (**48b** and **49b**, respectively); however, only **49** shows a peak due to the cinnamaldehyde radical ion (**49c**) itself. Cleavage of the $C_{(6)}$-$C_{(7)}$ and $C_{(7)}$-$C_{(8)}$ bonds with charge retention on, and hydrogen transfer to, the aromatic

48, $M^{+\cdot}$, m/e 230 (100%), R = H
49, $M^{+\cdot}$, m/e 260 (95%), R = OCH$_3$

48a, m/e 98 (68%)
49a, m/e 98 (29%)

$\xrightarrow{-CH_2O}$ m/e 68

49c, m/e 162 (R = OCH$_3$

48b, m/e 131
49b, m/e 161

moiety is also an important fragmentation process, especially for **49** where the *p*-methoxyl group stabilizes the styrene (**49d**) and tropylium (**49e**) ions so formed. Although no deuterium-labeling studies have been performed it seems likely that the transferred hydrogen originates (at least to a considerable extent) from $C_{(5)}$ of the lactone ring. The typical α-cleavage reaction of lactones is inhibited as it requires the breakage of a vinylic ($C_{(6)}$-$C_{(7)}$) bond so that the spectrum of **49** shows no ion at m/e 127 (**49f**) attributable to cleavage α to, and charge retention on, the lactone ring. As with simple unsaturated γ-lactones, loss of CO_2 from the parent ion is of very minor importance. However, **48** exhibits a prominent M-CO ion which is much reduced in intensity in the spectrum of **49**.

48, $M^{+\cdot}$ R = H

49, $M^{+\cdot}$ R = OCH_3

48d, m/e 104 (31 %)

49d, m/e 134 (100 %)

49f, m/e 127

48e, m/e 91 (36 %)

49e, m/e 121 (78 %)

Compounds with a saturated $C_{(7)}$—$C_{(8)}$ bond, such as 7,8-dihydro-methysticin (**50**), also exhibit very intense parent ions, but they differ from compounds with a double bond in this position in that loss of a dihydro-cinnamaldehyde from the parent ion is not observed (contrast **49** → **49a**). As might be expected, the base peak in the spectrum of **50** is due to the tropylium ion **50a** (24). Cleavage α to the lactone ring to give the oxonium ion **50b** is not a dominant fragmentation process for **50** but is much more important in 7,8-dihydrokawain (**51**) where the phenyl ring does not have oxygen substituents. The considerable m/e 161 ion in the spectrum of **50** (corresponding to m/e 117 in the spectrum of **51**) may be assigned the vinyl-tropylium structure **50c**. The mechanism of its formation is not clear, however, and formally requires cleavage of the $C_{(5)}$—$C_{(6)}$ and $C_{(6)}$—$O_{(1)}$ bonds together with loss or transfer of a hydrogen atom.

Cleavage α to the lactone oxygen atom is a very important fragmentation process for steroid derivatives with an unsaturated δ-lactone ring, such as dihydrodesoxy-jaborosalactone-A (**52**) where it accounts for m/e 125 (**52a**) as the base peak of the spectrum (25). (This ion may also be formulated as the tautomeric hydroxypyrilium ion **52b**.) High-resolution mass-measurement establishes the composition ($C_7H_9O_2$) of the m/e 125 ion.

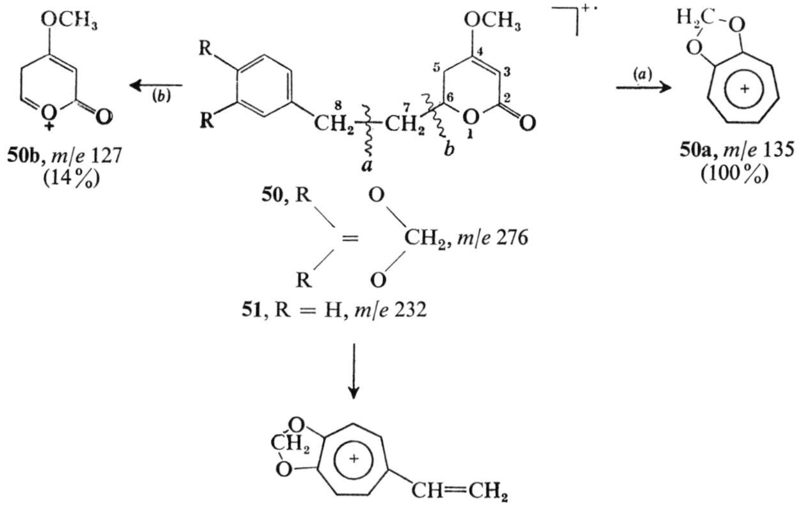

OCH₃

R

R

CH_2 CH_2
8   7

OCH₃

5   4
6   3
2
O   O
1
a   b

(a)

H₂C–O
O

(+)

50a, m/e 135
(100%)

50b, m/e 127
(14%)

(b)

50, R   O
\   \
/ = \ CH_2, m/e 276
R   O

51, R = H, m/e 232

O
CH₂
O

(+)

CH=CH₂

50c, m/e 161 (26%)

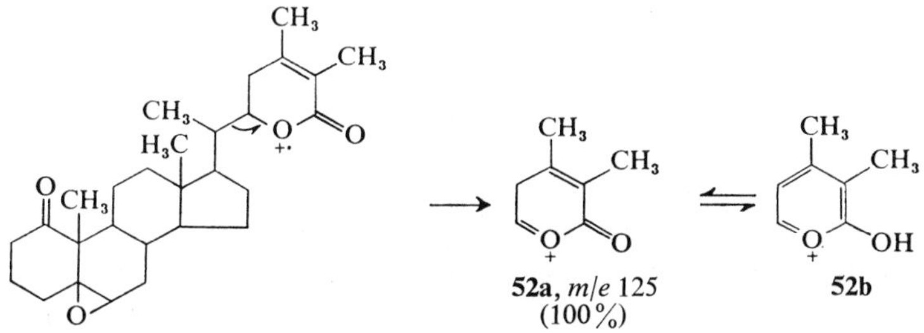

CH₃

CH₃

CH₃
H₃C

O CH₃

O

O
+·
O

52, M⁺·

CH₃

CH₃

O   O
+

52a, m/e 125
(100%)

CH₃

CH₃

O   OH
+

52b

The δ-lactol of 3,4,4-trimethyl-5-oxo-cis-2-hexenoic acid (53) does not exhibit a molecular ion but does undergo ready elimination of water to give 53a which then undergoes the retro-Diels-Alder reaction with elimination of ketene and formation of the conjugated ketene 53b (26).

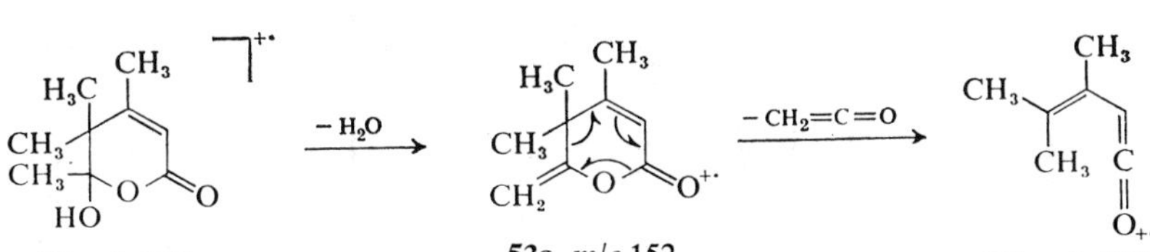

H₃C CH₃

CH₃
CH₃
O   O
HO

53, m/e 170

– H₂O

H₃C CH₃

CH₃

CH₂
O   O⁺·

53a, m/e 152

– CH₂=C=O

CH₃

CH₃

CH₃   C
O⁺·

53b, m/e 110

The fragmentation of phthalide (54) (Fig. 6.5) resembles that of γ-crotono-lactone (38) in that loss of CHO· from the parent ion with hydrogen re-arrangement is the dominant fragmentation process, giving the benzoyl ion

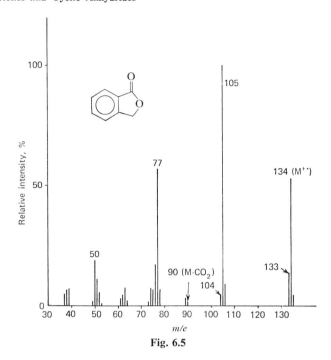

Fig. 6.5

(54a) (27). Loss of a benzylic hydrogen atom is also observed and a metastable ion establishes that loss of CO (with hydrogen rearrangement) from the M-1 ion **(54b)** is an alternative route to the benzoyl ion **(54a)**. This ion then

further fragments by the usual loss of CO to give the phenyl ion **(54c)**. Loss of formaldehyde from the parent ion occurs only to a small extent as is evidenced by the weak m/e 104 ion **(54d)**. Similarly loss of CO_2 is also of very minor importance. Coumaran-2,3-dione **(55)**, which here may be regarded as an α-ketolactone, exhibits a weak parent ion and a simple fragmentation pattern involving the consecutive loss of three molecules of carbon monoxide followed by loss of a hydrogen atom to give m/e 63 as the stable end product (28). Loss of CO_2 from the parent ion of **55** is not observed.

55, $M^{+\cdot}$, m/e 148

$\xrightarrow{-CO}$ m/e 120 (100%) $\xrightarrow{-CO}$ m/e 92

$\xrightarrow{-CO}$ m/e 64 $\xrightarrow{-H\cdot}$ m/e 63

Dihydrocoumarin (**56**) shows an abundant parent ion (base peak), and a loss of 28 mass units gives the most intense fragment ion at m/e 120 (29). Although no high-resolution data appear to be available, it is most probable that this initial loss is of carbon monoxide rather than of ethylene as the further fragmentation of the M-28 ion resembles that of dihydrobenzofuran

56, $M^{+\cdot}$, m/e 148 (100%)

$\xrightarrow{-CO}$

56a, m/e 120 (85%)

$\xrightarrow{-H\cdot}$

m/e 119 (22%)

$\downarrow -CH_2CO$

$\downarrow -CO$

56c, m/e 106 (12%)

$\xrightarrow{-CO}$ $C_6H_6^{+\cdot}$

m/e 78 (40%)

56b, m/e 91 (56%)

(**56a**), i.e., loss of a hydrogen atom followed by loss of CO to give the tropylium ion (**56b**, m/e 91). Another fragmentation pathway, of lesser importance, is the retro-Diels-Alder reaction with loss of ketene to give **56c** which then may lose carbon monoxide to give the benzene radical ion (or some isomer thereof) at m/e 78. The presence of a phenyl substituent at $C_{(4)}$, as in 4-phenyl-3,4-dihydrocoumarin (**57**, R = H), facilitates the retro-Diels-Alder reaction, but only a weak ion is observed at m/e 182 (**57a**, R = H) as very ready loss of a hydrogen atom occurs to give m/e 181 as the base peak (30). The m/e 181 ion has been shown as **57b** but is probably better formulated as the xanthenyl ion (**57c**, R = H). A xanthenyl ion (**57c**, R = Ph) is the favored

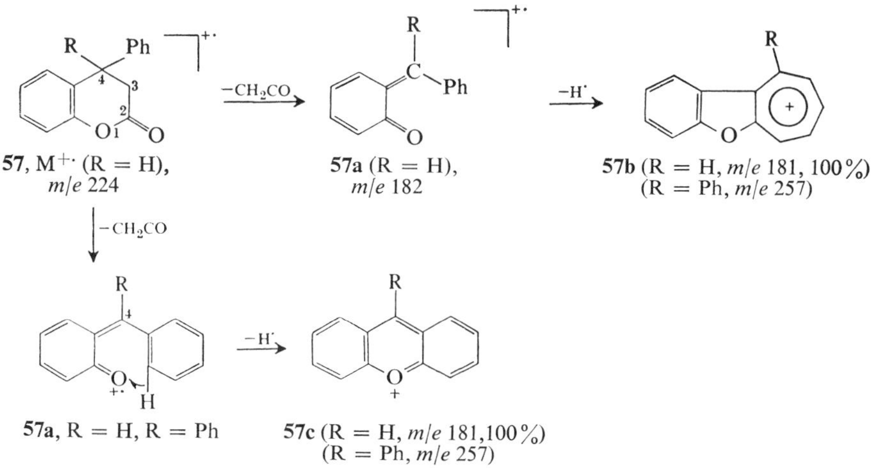

57, M$^{+\cdot}$ (R = H),
m/e 224

57a (R = H),
m/e 182

57b (R = H, m/e 181, 100%)
(R = Ph, m/e 257)

57a, R = H, R = Ph

57c (R = H, m/e 181,100%)
(R = Ph, m/e 257)

structure for the intense m/e 257 peak in the spectrum of 4,4-diphenyl-3,4-dihydrocoumarin (**57**, R = Ph) as the structure proposed (**57b**, R = Ph) would in this case require a phenyl migration (31).

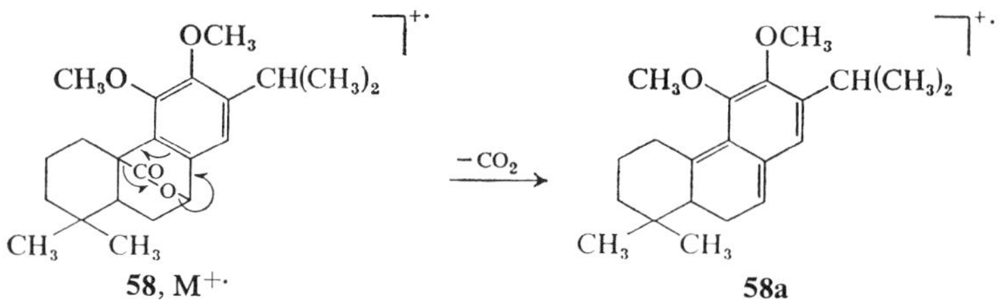

58, M$^{+\cdot}$

58a

The diterpene lactone carnosol (**58**) shows an intense ion at M-44-15 corresponding to loss of carbon dioxide via the retro-Diels-Alder reaction (**58 → 58a**) followed by loss of a methyl group (32).

IV. MACROCYCLIC LACTONES

Mass spectrometry has played a useful role in the structure elucidation of a number of macrocyclic ketolactone derivatives isolated from fungi. An example is the antibiotic monorden (**59**) whose structure was determined from various spectroscopic measurements on monorden itself and on its diacetate (33). The mass spectrum, which established the correct molecular formula, shows as the main feature a doublet at m/e 184,186 assigned to the chlorine-containing fragment **59a**. This may be formed by cleavage of the two bonds β to the benzene ring, an example of the "ortho effect" typical

59, M$^{+\cdot}$ **59a, m/e 184, 186**

of suitably *ortho*-substituted aromatic esters, and confirms the other spectroscopic evidence leading to structure **59**. Curvularin (**60**), in which the lactone carbonyl group is not in conjugation with the benzene ring, exhibits a prominent molecular ion, and the major fragmentation pathway may be rationalized in terms of cleavage α to the conjugated keto group and hydrogen transfer to the lactone oxygen atom (possibly as shown) (34). This gives the ion **60a** as the base peak which then further fragments, as expected, by loss of carbon monoxide (**60a → 60b**). A prominent ion at m/e 150 has been ascribed to loss of a hydroxyl radical from **60b** (metastable ion not observed), but this

60, M$^{+\cdot}$, m/e 292 (51%) **60a, m/e 195 (100%)**

60c, m/e 150 (53%) **60b, m/e 167 (74%)**

ion can also be formulated as **60c** which could arise directly from the parent ion by cleavage of the two bonds β to the benzene ring (cf. **59 → 59a**). A second mode of fragmentation for **60** leads to an ion at m/e 205 which may be formulated as **60d**. This has been of use in the structure elucidation of a dehydrocurvularin whose spectrum resembles that of curvularin itself but differs in that an abundant ion occurs at m/e 203 (possibly **60e**) whereas m/e 205 is virtually absent. The mass-spectroscopic observations and other evidence indicate that the double bond in the dehydrocurvularin is in conjugation with the ketone group.

Mass spectrometry has also been used to confirm the structure of the fungal metabolite zearalenone and some of its derivatives (35).

60, M$^{+\cdot}$ 60d, m/e 205 (28%)   60e, m/e 203

The structure elucidation of the macrolide antibiotics (e.g., **61**) has been greatly aided by mass spectrometry, structural information being obtained largely from various degradation products rather than from the intact molecule. One of the major problems in the structure elucidation of macrolide antibiotics (and of many polyhydroxy compounds of high molecular weight) is establishing the correct molecular formula. It has been shown that the trimethylsilyl ethers of macrolide antibiotics are in general thermally stable and sufficiently volatile to lend themselves well to mass spectrometry. As an example the case of filipin (**61**), which had been assigned the formula $C_{37}H_{62}O_{12}$, may be considered. Per-trimethylsilylation of filipin (**61**) and its

61

decahydro-derivative gives ethers for which the lightest molecular ions (i.e., those due to molecules lacking heavy isotopes) appear at m/e 1302 and 1312, respectively, corresponding to nonatrimethylsilyl ethers of composition $C_{62}H_{130}O_{11}Si_9$ and $C_{62}H_{140}O_{11}Si_9$ (36). The composition of the m/e 1312 ion of decahydrofilipin has been confirmed by high-resolution mass-measurement using the $[C_{28}F_{51}]^+$ ion at m/e 1304.9186 in the spectrum of Fluorolube as the reference, and hence the correct molecular formula for filipin (**61**) may be deduced as $C_{35}H_{58}O_{11}$. Of the more complex macrolide antibiotics in whose structure elucidation mass spectrometry has played a major role, pimaricin (**62**) and lucensomycin (**63**) are illustrative examples (37–39).

Mass spectrometry may also be applied to the structure elucidation of the "macrotetrolides," a class of *Streptomyces* metabolites of which nonactin (**64**) is the simplest member. The two major fragmentation processes for nonactin (**64**), as for other members of the series, involve a McLafferty

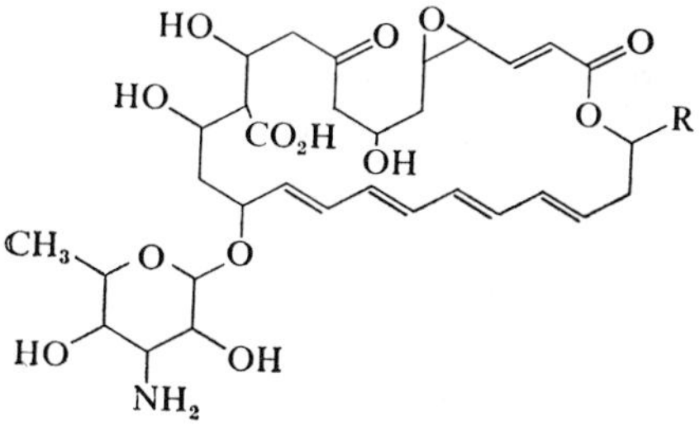

62, R = CH$_3$
63, R = n-C$_4$H$_9$

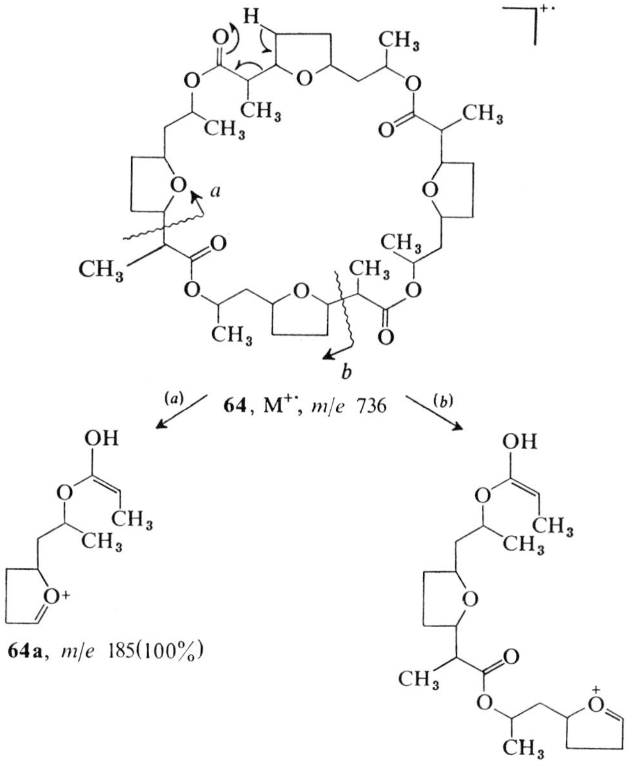

64, M$^{+\cdot}$, m/e 736

64a, m/e 185(100%)

64b, m/e 369 (74%)

rearrangement and cleavage α to the heterocyclic oxygen atom of the tetra-hydrofuran rings to give the two prominent ions **64a** (m/e 185, base peak) and **64b** (m/e 369). The naturally-occurring compounds of this series are built up from only three different hydroxy acids which differ from each other in having a methyl, ethyl, or isopropyl substituent at the carbon atom bearing the hydroxyl group, so it may readily be established which of these hydroxy acids are present in an unknown "macrotetrolide," by the presence (or absence) of the ion **64a** (m/e 185) and the higher homologs at m/e 199 or 213, while if some other homologous hydroxy acid were present it would be readily recognized by an additional fragment ion in this series of peaks. Furthermore it is possible to use the ions corresponding to **64b** in the spectrum of nonactin (**64**) in a sequence analysis of the hydroxy acids. Thus for a macrotetrolide built up of two molecules each from two different hydroxy acids (A and B), the two structures **65** and **66** may be readily distinguished, as the structure **65** would give only one ion analogous to **64b** while for **66** three fragments would be observed in this mass region (40, 41).

$$
\begin{array}{cc}
\text{A——B} & \text{A——B} \\
| \quad\;\; | & | \quad\;\; | \\
\text{B——A} & \text{A——B} \\
\mathbf{65} & \mathbf{66}
\end{array}
$$

V.   CYCLIC ANHYDRIDES

A.   Simple Anhydrides

Little information is available on the mass spectra of anhydrides of simple alkanedioic acids. No molecular ion is observed in the spectra of succinic and glutaric anhydrides; abundant ions are produced by loss of carbon dioxide from the molecular ions but no CO loss is observed in either case (42, 43). Similarly, hexahydrophthalic anhydride (**67**) does not give a molecular ion peak, nor a peak due to loss of either CO or CO_2, but a peak is observed due to loss of both groups to give the cyclohexene radical ion **67a** (42).

67, $M^{+\cdot}$, m/e 154
(not observed)

67a, m/e 82

B.   Unsaturated Anhydrides

The spectra of 1,2,5,6-tetrahydrophthalic anhydride (**68**) and a deuterated derivative show loss of CO from the molecular ion (for which again no peak is observed) but not loss of CO_2. The mechanism put forward involves initial cleavage of the anhydride CO—O bond to give the ring-opened molecular ion **68a**, and it has been suggested that homoallylic overlap, stabilizing the positive charge in **68b**, provides the driving force for the loss of CO rather than of CO_2 (42). The ion **68b** undergoes hydrogen rearrangement to give **68c** which loses a carboxy radical to give the phenonium ion **68d** at m/e 79. This ion accounts for 28.5% of the total ion current. The retro-Diels-Alder fragmentation is of little importance.

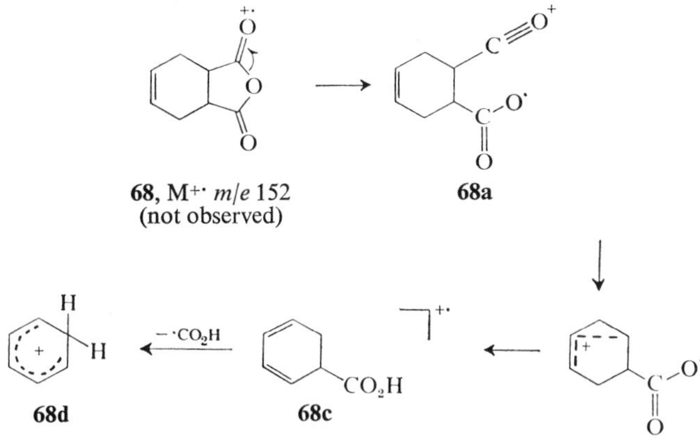

68, M+· m/e 152
(not observed)

68a

68d   − ·CO_2H **68c**   **68b**

The spectra of the bridged tetrahydrophthalic anhydrides **69** and **70** show molecular ions of considerable intensity, with marked loss of the bridge carbonyl group to give **69a** and **70a**. The latter ions achieve aromatization by further loss of CO and CO_2 to give the substituted benzenes **69b** and **70b** (44).

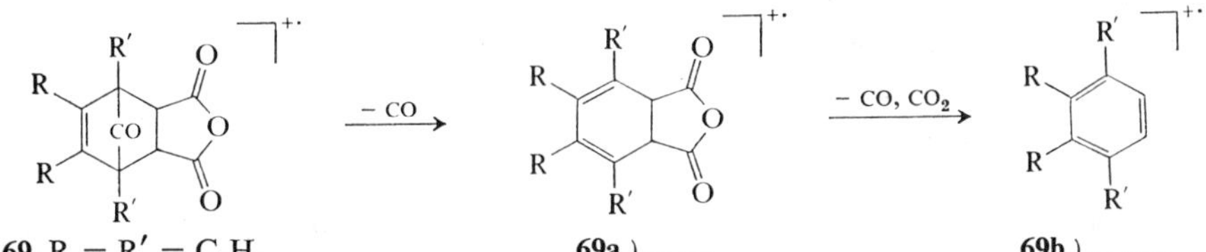

69, R = R′ = C_6H_5
70, R = C_6H_5, R′ = CH_3

− CO →

69a,
70a, } M-28

− CO, CO_2 →

69b,
70b, } M-100

Loss of the bridging group in the methylene-bridged anhydride **71** does not occur. The retro-Diels-Alder process becomes important and the ionized cyclopentadiene **71a** accounts for 36% of the total ionization (42). The

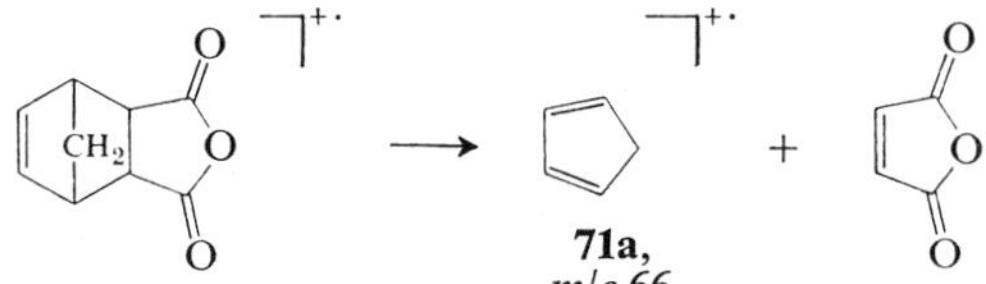

71, M+·

71a,
m/e 66

1,2-dihydrophthalic anhydride derivatives **72** show intense molecular ions which aromatize either by loss of a hydrogen molecule (**72 → 72a**) or by loss of C_2O_3 to give the base peaks in the spectra at M-72 (**72b**) (44).

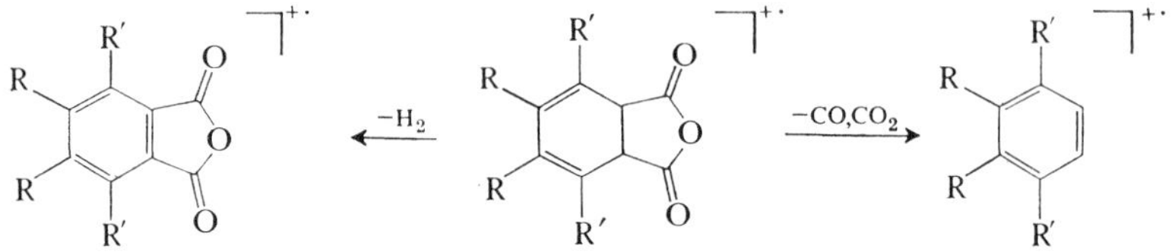

72a	**72**, $M^{+\cdot}$, R = H, R$'$ = C_6H_5 or R = C_6H_5, R$'$ = CH_3 or R = R$'$ = C_6H_5	**72b**, M-72 (100%)

The mass spectrum of maleic anhydride (**73**) shows a strong molecular ion with insignificant loss of CO. Loss of carbon dioxide occurs readily (**73 → 73a**), however, and the base peak is due to acetylene, **73b**. An inter-

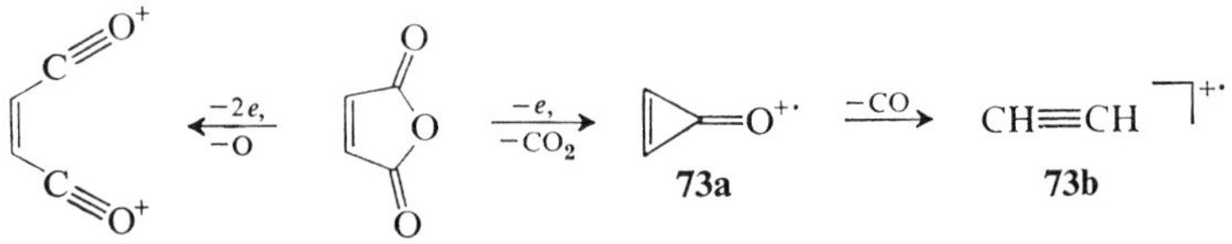

esting feature is the weak ion (1%) due to loss of an oxygen atom from the parent ion and the associated doubly-charged ion **73c** of seven times the intensity, resulting from the stabilization of each of the positive charges on one of the oxygen atoms (45).

Phthalic anhydride (**74**) fragments by initial expulsion of carbon dioxide followed by loss of carbon monoxide to give ionized benzyne (46–49). A

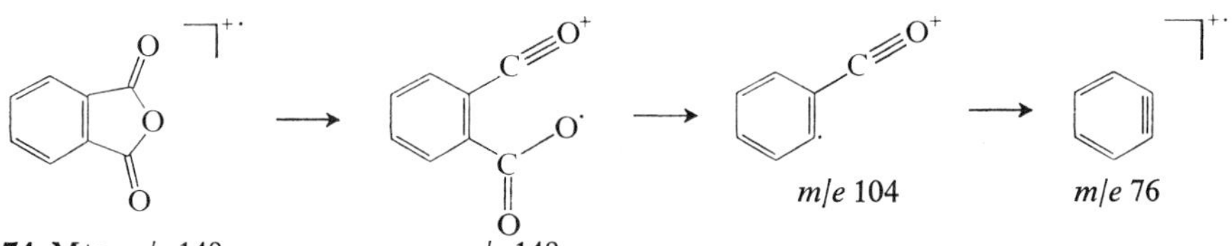

similar fragmentation pattern is observed with tetrachloro- and tetrabromo-phthalic anhydrides and with the pyridine and pyrazine compounds (**75** and **76**) (49, 50).

75   76

Addenda

A detailed study of variously methylated δ-lactones (Section I-B) has appeared (51). Two groups have discussed the spectra of tetronic acid derivatives (which are unsaturated hydroxy γ-lactones, Section III-A) (52, 53), while the spectra of the bakkenolides, a group of bitter principles containing a methylene γ-lactone have been discussed (54).

REFERENCES

1. L. Friedman and F. A. Long, *J. Amer. Chem. Soc.*, **75**, 2832 (1953).
2. F. A. Long and L. Friedman, *J. Amer. Chem. Soc.*, **75**, 2837 (1953).
3. N. J. Turro, D. C. Neckers, P. A. Leermakers, D. Seldner, and P. D'Angelo, *J. Amer. Chem. Soc.*, **87**, 4097 (1965).
4. E. Honkanen, T. Moisio, and P. Karvonen, *Acta Chem. Scand.*, **19**, 370 (1965).
5. W. H. McFadden, E. A. Day, and M. J. Diamond, *Anal. Chem.*, **37**, 89 (1965).
6. D. Felix, A. Melera, J. Seibl, and E. Sz. Kováts, *Helv. Chim. Acta*, **46**, 1513 (1963).
7. K. Biemann, *Mass Spectrometry*, McGraw-Hill, New York, 1962, pp. 127–128.
8. A. M. Duffield, *J. Heterocycl. Chem.*, **4**, 16 (1967).
9. T. Sakan, S. Isoe, S. B. Hyeon, R. Katsumura, T. Maeda, J. Wolinsky, D. Dickerson, M. Slabaugh, and D. Nelson, *Tetrahedron Lett.*, **1965**, 4097.
10. Y. Ohta, T. Sakai, and Y. Hirose, *Tetrahedron Lett.*, **1966**, 6365.
11. S. Yamamura and Y. Hirata, *Tetrahedron*, **19**, 1485 (1963).
12. D. G. B. Boocock and E. S. Waight, *Chem. Commun.*, **1966**, 90.
13. N. Wasada, T. Tsuchiya, E. Yoshii, and E. Watanabe, *Tetrahedron*, **23**, 4623 (1967).
14. P. Brown and C. Djerassi, *Angew. Chem., Int. Ed.*, **6**, 477 (1967).
15. R. Hodges, E. P. White, and J. S. Shannon, *Tetrahedron Lett.*, **1964**, 371.
16. R. I. Reed and W. K. Reid, *J. Chem. Soc.*, **1963**, 5933.
17. R. I. Reed, in *Mass Spectrometry of Organic Ions* (F. W. McLafferty, Ed.), Academic Press, New York, 1963, Chap. 13.
18. N. S. Wulfson, V. I. Zaretskii, I. B. Papernaja, E. P. Serebryakov, and V. F. Kucherov, *Tetrahedron Lett.*, **1965**, 4209.
19. W. H. Lunn, J. T. Edward, and S. Meyerson, *Can. J. Chem.*, **44**, 279 (1966).
20. J. Bricout, R. Viani, F. Müggler-Chavan, J. P. Marion, D. Reymond, and R. H. Egli, *Helv. Chim. Acta*, **50**, 1517 (1967).
21. H. Immer, J. Polonsky, R. Toubiana, and Ho Dac An, *Tetrahedron*, **21**, 2117 (1965).
22. R. M. Letcher and S. Eggers, *Tetrahedron Lett.*, **1967**, 3541.
23. R. M. Letcher, *Org. Mass Spectrosc.*, **1**, 805 (1968).
24. M. Pailer, G. Schaden, and R. Hänsel, *Monatsh. Chem.*, **96**, 1842 (1965).
25. R. Tschesche, H. Schwang, H.-W. Fehlhaber, and G. Snatzke, *Tetrahedron*, **22**, 1129 (1966).

26. P. J. Chapman, G. Meerman, I. C. Gunsalus, R. Srinivasan, and K. L. Rinehart, *J. Amer. Chem. Soc.*, **88**, 618 (1966).
27. J. Baldas, Unpublished results.
28. J. H. Bowie and J. W. W. Morgan, *Aust. J. Chem.*, **20**, 117 (1967).
29. C. S. Barnes and J. L. Occolowitz, *Aust. J. Chem.*, **17**, 975 (1964).
30. U. K. Pandit and I. P. Dirk, *Tetrahedron Lett.*, **1963**, 891.
31. W. H. Starnes, *J. Amer. Chem. Soc.*, **86**, 5603 (1964).
32. H. E. Audier, S. Bory, G. Defaye, M. Fétizon and G. Moreau, *Bull. Soc. Chim. Fr.*, **1966**, 3181.
33. F. McCapra, A. I. Scott, P. Delmotte, J. Delmotte-Plaquée, and N. S. Bhacca, *Tetrahedron Lett.*, **1964**, 869.
34. H. D. Munro, O. C. Musgrave, and R. Templeton, *J. Chem. Soc.*, C, **1967**, 947.
35. W. H. Urry, H. L. Wehrmeister, E. B. Hodge, and P. H. Hidy, *Tetrahedron Lett.*, **1966**, 3109.
36. B. T. Golding, R. W. Rickards, and M. Barber, *Tetrahedron Lett.*, **1964**, 2615.
37. B. T. Golding, R. W. Rickards, W. E. Meyer, J. B. Patrick, and M. Barber, *Tetrahedron Lett.*, **1966**, 3551.
38. G. Gaudiano, P. Bravo, and A. Quilico, *Tetrahedron Lett.*, **1966**, 3559.
39. G. Gaudiano, P. Bravo, A. Quilico, B. T. Golding, and R. W. Rickards, *Tetrahedron Lett.*, **1966**, 3567.
40. H. Gerlach, R. Hütter, W. Keller-Schierlein, J. Seibl, and H. Zähner, *Helv. Chim. Acta*, **50**, 1782 (1967).
41. J. Seibl, *Z. Anal. Chem.*, **221**, 342 (1966).
42. S. J. Weininger, V. T. Mai, and E. R. Thornton, *J. Amer. Chem. Soc.*, **86**, 3732 (1964).
43. H. Budzikiewicz, C. Djerassi, and D. H. Williams, *Mass Spectrometry of Organic Compounds*, Holden-Day, San Francisco, 1967, p. 222.
44. H. Prinzbach, R. Kitzing, E. Druckrey, and H. Achenbach, *Tetrahedron Lett.*, **1966**, 4265.
45. J. H. Beynon, *Mass Spectrometry and its Applications to Organic Chemistry*, Elsevier, Amsterdam, 1960, pp. 283–374.
46. Ref. 45, p. 374.
47. F. W. McLafferty and R. S. Gholke, *Anal. Chem.*, **31**, 2076 (1959).
48. E. K. Fields and S. Meyerson, *Chem. Commun.*, **1965**, 474.
49. M. P. Cava, M. J. Mitchell, D. C. De Jongh, and R. Y. Van Fossen, *Tetrahedron Lett.*, **1966**, 2947.
50. R. F. C. Brown, W. D. Crow, and R. K. Solly, *Chem. Ind.* (London), **1966**, 343.
51. B. J. Millard, *Org. Mass Spec.*, **1**, 279 (1968).
52. L. J. Haynes, A. Kirkien-Konasiewicz, A. G. Loudon and A. Maccoll, *Org. Mass Spec.*, **1**, 743 (1968).
53. J. A. Ballantine, R. G. Fenwick and V. Ferrito, *Org. Mass Spec.*, **1**, 761 (1968).
54. K. Shirahata, T. Kato, Y. Kitahara and N. Abe, *Tetrahedron*, **25**, 4671 (1969).

7 COMPOUNDS WITH TWO OR MORE OXYGEN ATOMS IN THE SAME RING

I. DERIVATIVES OF 1,3-DIOXOLAN

A. Simple Dioxolans

The mass spectrum of the parent compound (**1**) has as its base peak (Fig. 7.1) the M-1 species (**1a**) (1, 2). The origin of the hydrogen atom lost has not been confirmed by deuterium labeling, but consideration of the spectra of alkyl-1,3-dioxolans leaves little doubt that the $C_{(2)}$ methylene group is involved. Thus 2-methyldioxolan (**2**) loses $CH_3\cdot$ as the major fragmentation to give **1a** (2), while a series of 2,2-dialkyldioxolans (**3**) lose the radicals $R\cdot$ and $R'\cdot$ in approximately the order of amounts that might be expected from the relative radical stabilities (3).

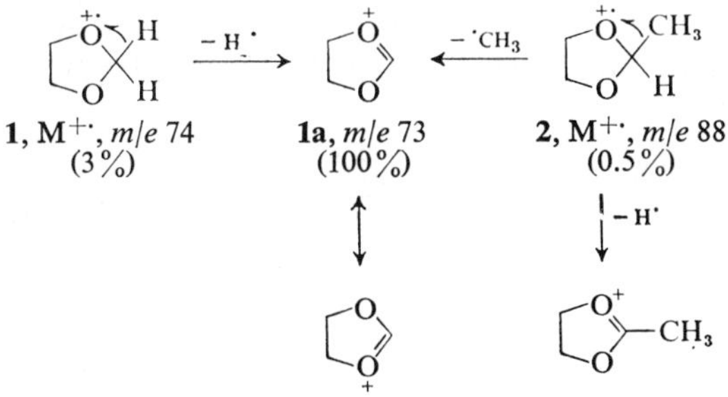

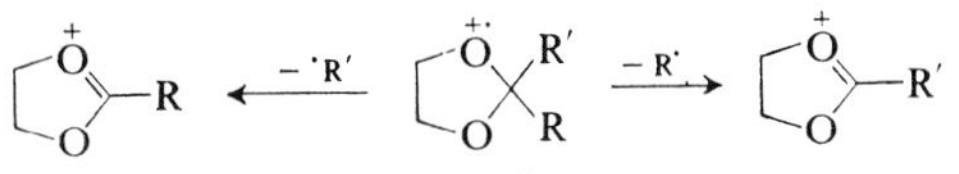

1, $M^{+\cdot}$, m/e 74
(3%)

1a, m/e 73
(100%)

2, $M^{+\cdot}$, m/e 88
(0.5%)

3, $M^{+\cdot}$

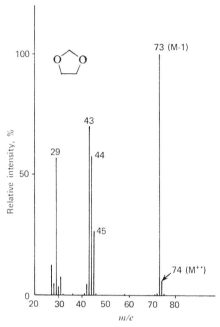

Fig. 7.1

B. Ethylene-Ketals of Cyclic Ketones

The ability of the ethylene-ketal grouping to direct fragmentation in a highly specific manner is of value in the structure determination of naturally-occurring cyclic ketones (4–8). The initial fragmentation occurs, as is to be expected, at a bond α to the oxygen function and is well demonstrated in the mass spectrum of cyclohexanone ethylene-ketal (**4**) (4).

The highly specific fragmentation of the ethylene-ketals of complex naturally-occurring ketones is demonstrated by the mass spectrum of 5α-androstan-3-one ethylene-ketal (5) (8). Only two significant fragment ions occur (*m/e* 99 and 125), and the scheme shown, supported by the spectrum of the 7-deutero-derivative of 5, has been advanced to explain their occurrence.

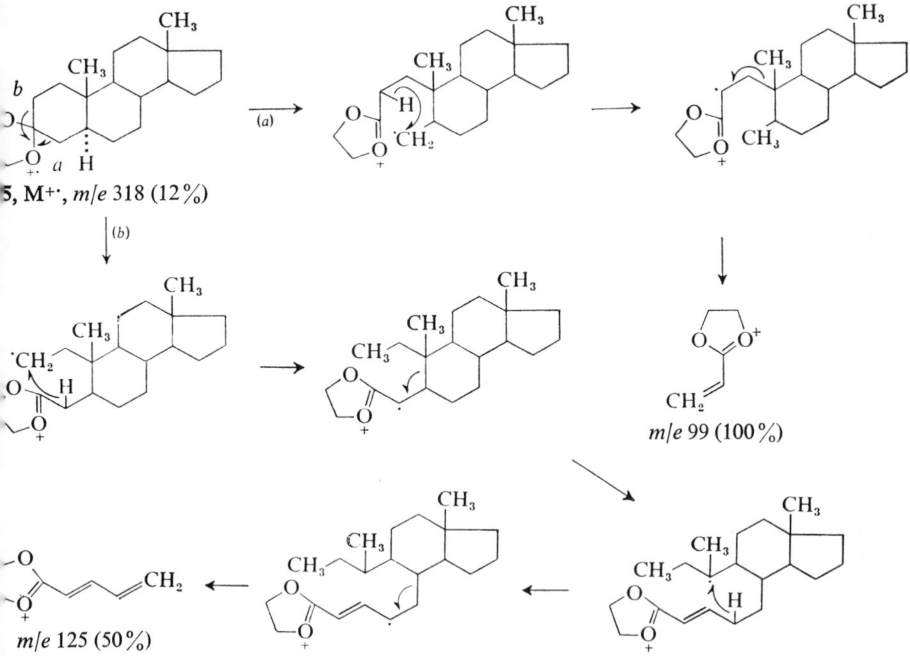

5, M$^{+\cdot}$, *m/e* 318 (12%)

m/e 99 (100%)

m/e 125 (50%)

C. Cyclic Carbonates of 1,2-Diols

Like the corresponding cyclic sulfites (9), the cyclic carbonates of 1,2-diols show fragmentations which are associated with molecular rearrangement (9). In the parent compound (6), the rearrangement involves only a hydrogen transfer, but highly alkylated derivatives have spectra best explained on the basis of alkyl shifts. A typical example is provided in the spectrum of the tetramethyl-compound 7.

The simple scheme outlined for the tetramethyl-compound 7 does not explain the formation of a strong ion $C_3H_7O^+$ at *m/e* 59. The possibility that this resulted from initial loss of carbon dioxide, *without rearrangement*, to the epoxide 8, followed by the known cleavage of this compound to *m/e* 59, was disproved by running the spectrum of 7 at low electron-voltage. This did not affect the *m/e* 100 (M-CO$_2$) peak, but the one at *m/e* 59 rapidly

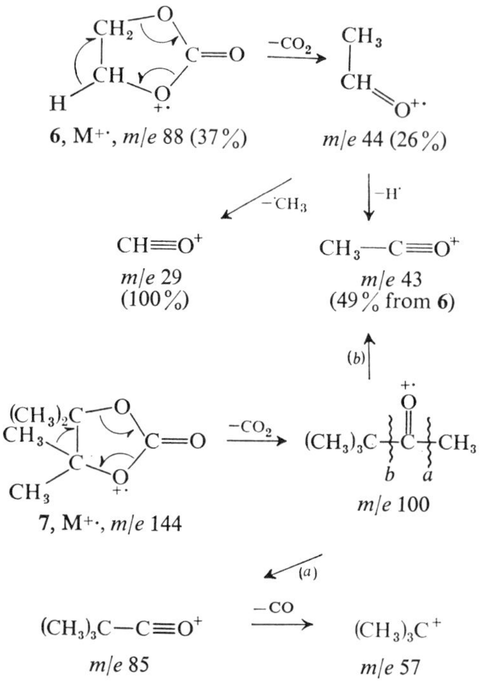

6, M⁺·, m/e 88 (37%) → (−CO₂) → m/e 44 (26%)

CH≡O⁺
m/e 29
(100%)

CH₃—C≡O⁺
m/e 43
(49% from 6)

(b)

7, M⁺·, m/e 144 → (−CO₂) → (CH₃)₃C—C(b)—C(a)—CH₃
m/e 100

(a)

(CH₃)₃C—C≡O⁺
m/e 85
→ (−CO) →
(CH₃)₃C⁺
m/e 57

increased in intensity, eventually becoming the base peak of the spectrum, showing that M-CO₂ cannot be its precursor. The scheme shown has been suggested for the genesis of m/e 59.

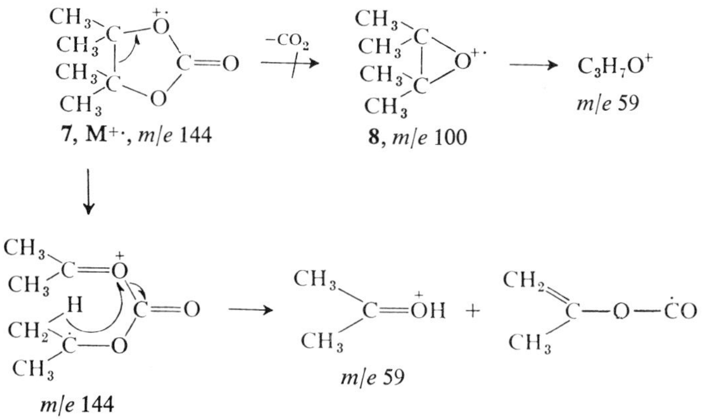

7, M⁺·, m/e 144 → (−CO₂, ✗) →
8, m/e 100 →
C₃H₇O⁺
m/e 59

m/e 144 →
(CH₃)₂C=OH⁺ + CH₂=C(CH₃)—O—ĊO
m/e 59

On the other hand it seems that aryl substitution initiates decomposition via an epoxide intermediate. Thus the spectrum of *meso*-hydrobenzoin carbonate (**9**) below M-CO$_2$ (m/e 196) is essentially identical with that of stilbene oxide (**10**), and moreover every metastable peak position, shape, and relative intensity in the spectrum of the oxide is reproduced in that of the cyclic carbonate.

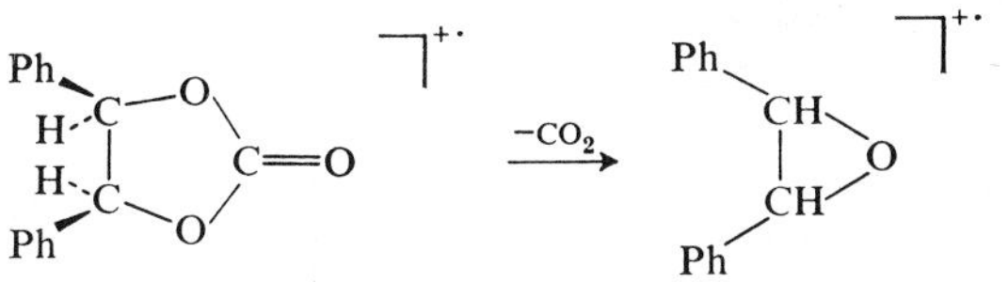

9, M$^{+\cdot}$, m/e 240 $\xrightarrow{-CO_2}$ **10**, m/e 196 (M-CO$_2$)

The mass spectrum of *o*-phenylene carbonate (**11**) shows marked loss of carbon dioxide, presumably leading to the benzene oxide **11a** (**10**).

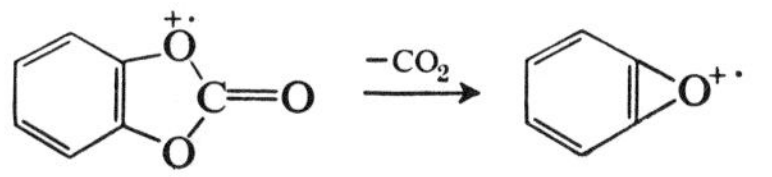

11, M$^{+\cdot}$, m/e 136 $\xrightarrow{-CO_2}$ **11a**, m/e 92

II. 1,3-DIOXANS

1,3-Dioxan (**12**) shows an intense M-1 ion (11). Although the required deuterium labeling has not been carried out, it seems highly likely that the

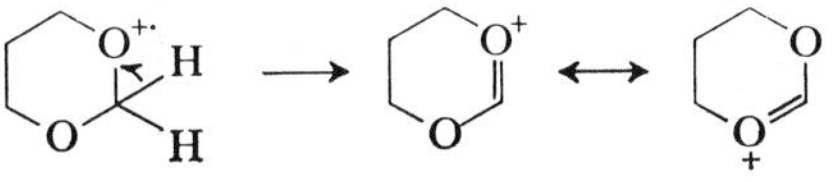

12, M$^{+\cdot}$, m/e 88 (10%) $\rightarrow$ **12a**, m/e 87 (100%) $\longleftrightarrow$

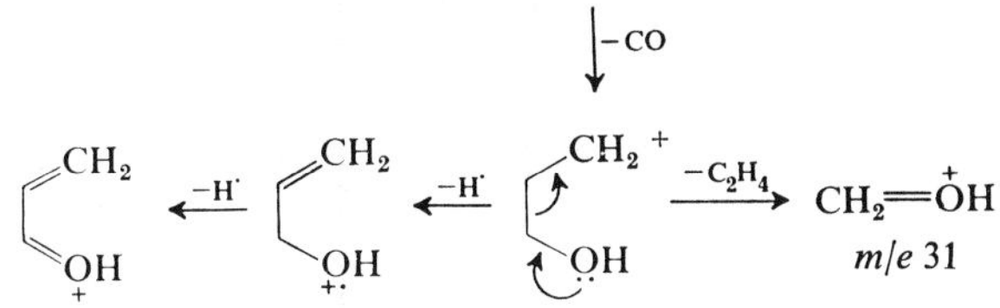

$\downarrow$-CO

m/e 57 (13.5%) $\xleftarrow{-H\cdot}$ m/e 58 (16%) $\xleftarrow{-H\cdot}$ m/e 59 (13%) $\xrightarrow{-C_2H_4}$ CH$_2$=$\overset{+}{O}$H m/e 31

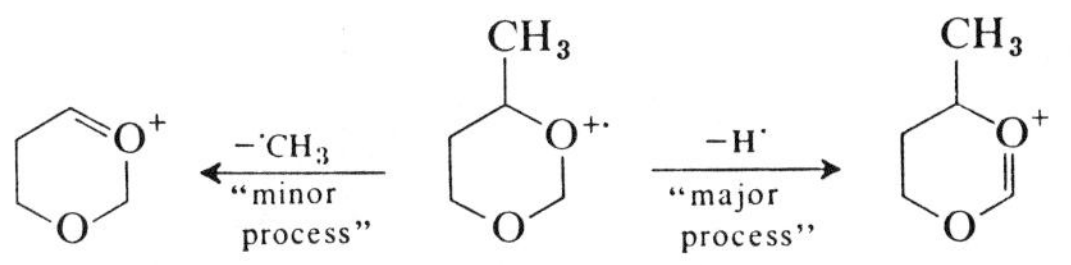

m/e 87 $\xleftarrow[\text{"minor process"}]{-\cdot CH_3}$ **13**, M$^{+\cdot}$, m/e 102 $\xrightarrow[\text{"major process"}]{-H\cdot}$ m/e 101

atom lost originates from the 2-position, the driving force being the formation of the well-stabilized oxonium ion **12a**. This conclusion is supported by the observation that the M-·CH$_3$ peak in 4-methyl-1,3-dioxan (**13**) is weaker than the M-1 peak. The accompanying scheme explains the further decompositions of these 1,3-dioxans.

III. 1,4-DIOXANS

The mass spectrum of 1,4-dioxan (**14**) (1, 11) is readily rationalized as involving initial fission α to an oxygen atom, followed by losses of the elements of formaldehyde and ethylene. In contrast to that in the 1,3-dioxans, the M-1 ion is relatively weak. Loss of a substituent by α-cleavage becomes

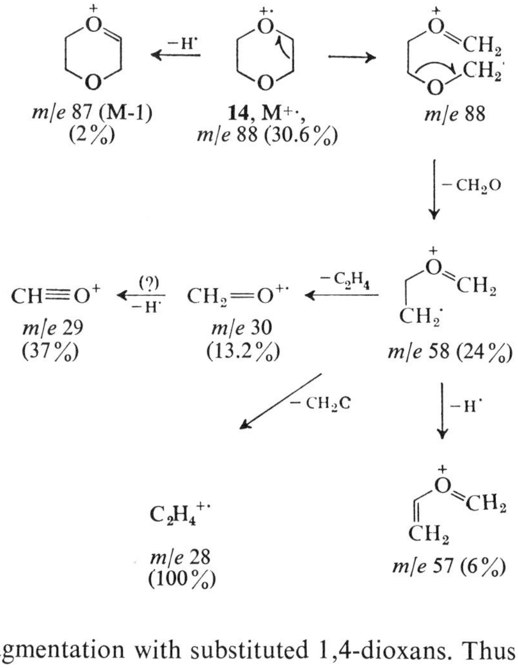

m/e 87 (M-1) (2%)   **14, M$^{+\cdot}$,** *m/e* 88 (30.6%)   *m/e* 88

$$\big| -CH_2O$$

$CH\!\equiv\!O^+ \xleftarrow[-H^{\cdot}]{(?)} CH_2\!=\!O^{+\cdot} \xleftarrow{-C_2H_4}$

m/e 29 (37%)   *m/e* 30 (13.2%)   *m/e* 58 (24%)

$- CH_2C$ $\big| -H^{\cdot}$

C$_2$H$_4^{+\cdot}$

m/e 28 (100%)   *m/e* 57 (6%)

a dominant fragmentation with substituted 1,4-dioxans. Thus the base peak of the spectrum of *n*-octyl-1,4-dioxan (**15**) is the ion of *m/e* 87 (12).

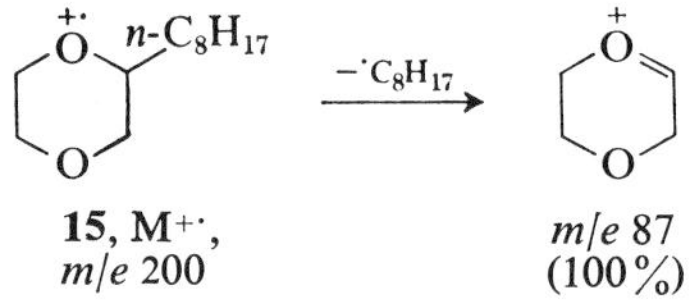

15, M$^{+\cdot}$, *m/e* 200 $\xrightarrow{-\,^{\cdot}C_8H_{17}}$ *m/e* 87 (100%)

IV. 1,3,5-TRIOXAN

Like 1,3-dioxan (and for the same reason), 1,3,5-trioxan (16) has a spectrum in which the M-1 ion is much more intense than the molecular ion. The base peak of the spectrum is the ion $[CH_3O]^+$ (m/e 31), and the scheme explains most of the fragmentation observed (11).

16, $M^{+\cdot}$, m/e 90
(1.3%)

m/e 89
(29%)

m/e 90

$-\cdot CHO$
(arrows)

m/e 61 (38%)

$-CH_2O$

$\overset{+}{HO}=CH_2$
m/e 31
(100%)

$-C_2H_4O_2$
(?)

(?) $-\cdot CH_2OH$

$CH_2=O^{+\cdot}$
m/e 30 (71%)

$-H\cdot$

$CH\equiv O^+$
m/e 29 (89%)

V. DERIVATIVES OF 1,4-DIOXIN

The only derivative of this ring system to be examined mass-spectrometrically appears to be the dibenzo-derivative **17** (13). The spectrum is of interest in showing a significant M-O peak. Loss of an oxygen atom from an oxygen heterocycle is unusual, and here the driving force is presumably the stability of the dibenzofuran radical cation produced. Also of interest is the concerted loss of two molecules of carbon monoxide from the molecular ion, as evidenced by the appropriate metastable ion. Here the driving force for the rearrangement is presumably a combination of the stabilities of the neutral fragments and the ion produced (the latter probably the naphthalene radical cation).

m/e 128

$-2CO$

$M^{+\cdot}$, m/e 184 (100%)

$-O$

m/e 168 (2.1%)

Addenda

Two papers have appeared dealing with fragmentations of the cyclic peroxide sym-tetroxane (14, 15), while further work has appeared on the mass spectrum of dioxan (Section III) (16).

REFERENCES

1. American Petroleum Institute, Research Project 44. Spectrum 87.
2. J. Collin, *Bull. Soc. Chim. Belges*, **69**, 585 (1960).
3. J. T. B. Marshall and D. H. Williams, *Tetrahedron*, **23**, 321 (1967).
4. H. Audier, M. Fétizon, J.-C. Gramain, J. Schalbar, and B. Waegel, *Bull. Soc. Chim. Fr.*, **1964**, 1880.
5. H. Audier, M. Fétizon, and J.-C. Gramain, *Bull. Soc. Chim. Fr.*, **1965**, 3088.
6. H. Audier, A. Diara, M. de J. Durazo, M. Fétizon, P. Foy, and W. Vetter, *Bull. Soc. Chim. Fr.*, **1963**, 2827.
7. Z. Pelah, D. H. Williams, and C. Djerassi, *J. Amer. Chem. Soc.*, **86**, 3722 (1964).
8. G. v. Mutzenbecher, Z. Pelah, D. H. Williams, H. Budzikiewicz, and C. Djerassi, *Steroids*, **2**, 475 (1963).
9. P. Brown and C. Djerassi, *Tetrahedron*, **24**, 2949 (1968).
10. G. Kersaint and C. Mentzer, *Bull. Soc. Chim. Fr.*, **1964**, 3271.
11. J. H. Beynon, *Mass Spectrometry and its Applications to Organic Chemistry*, Elsevier, Amsterdam, 1960, pp. 368–369.
12. D. Elad and R. D. Youssefyeh, *J. Org. Chem.*, **29**, 2031 (1964).
13. J. M. Desmarchelier and R. B. Johns, Unpublished results.
14. T. Ledaal, *Tetrahedron Lett.*, 3661 (1969).
15. M. Bertrand, S. Fliszár and Y. Rousseau, *J. Org. Chem.*, **33**, 1931 (1968).
16. G. Condé-Caprace and J. E. Collin, *Org. Mass Spec.*, **2**, 1277 (1969).

8 HETEROCYCLIC SULFUR COMPOUNDS

I. COMPOUNDS WITH ONE SULFUR ATOM PER RING

A. Ethylene Sulfide and Derivatives

Although the spectra of the parent compound and several of its alkyl derivatives have been published (1–3), there have been no detailed analyses of fragmentation in this class comparable with those described for the epoxides. No information about decomposition has been recorded from studies of metastable ions and no deuteration studies have been carried out. In the substituted ethylene sulfides studied thus far, none of the alkyl groups is of sufficient length to allow determination of the importance of the β- and γ-cleavages observed with the epoxides. Thus it is apparent that this is a field for much further research, and it is also clear that many of the conclusions we draw are tentative.

The mass spectrum of the parent compound (1) (Fig. 8.1) shows that loss of a hydrogen atom from the molecular ion (α-cleavage) is an important process leading to the abundant M-1 species 1a. The base peak at m/e 45 must have composition CHS^+, and its formation is best rationalized, as shown, by prior isomerization of the cyclic molecular ion to the thioacetaldehyde radical cation 1b. The loss of ·SH from the molecular ion to give

$$CH_2\!-\!CH_2 \longrightarrow CH_3\!-\!CH\!=\!S^{+\cdot}$$

1, $M^{+\cdot}$, m/e 60   1b, m/e 60

$-\cdot CH_3$

$C_2H_3^+$ 1d, m/e 27

$CH_2\!-\!CH$ $CH\!\equiv\!S^+$

1a, m/e 59   1c, m/e 45

225

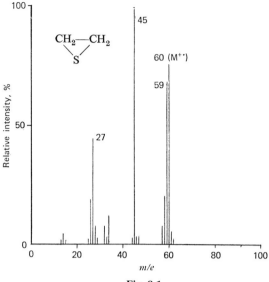

Fig. 8.1

$C_2H_3^+$ is also a prominent process. Indeed, loss of ·SH from the molecular ion is always observed with ethylene sulfides, as will be seen below.

The mass spectrum of propylene sulfide **2** (Fig. 8.2) (2) is of interest since loss of ·CH$_3$ by α-cleavage is relatively less important than loss of H· from the parent compound. This difference may be due to a greater tendency of propylene sulfide to isomerize upon impact, and most of the ions may derive from thiopropionaldehyde (**2a**) and thioacetone (**2b**) radical cations rather than from the unisomerized parent.

The relative importances of the ions of m/e 45 and 59 will depend partly upon the relative amounts of **2a** and **2b** produced. The greater stability of

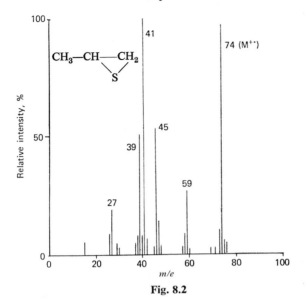

Fig. 8.2

the $C_2H_5\cdot$ radical will probably favor the m/e 45 species **1c**, other things being equal. The base peak of this spectrum is the stable allyl cation produced by loss of $\cdot SH$ from **2**.

CH$_3$—CH—CH$_2$ → CH$_2$—CH$_2$—CH$_2$$^+$ $\xrightarrow{-SH}$ CH$_2$=CH—CH$_2$$^+$

2 m/e 41

The mass spectra of *cis*- and *trans*-2-butenesulfide (**3**) (3) (Fig. 8.3) are essentially identical and are again easily rationalized if we assume that impact-induced isomerization is the initial process. It is interesting that

m/e 73

CH$_3$—C—CH$_2$—CH$_3$

m/e 59

3 — migration CH$_3$

m/e 45

CH$_3$ CH—CH=S$^{+\cdot}$

m/e 43

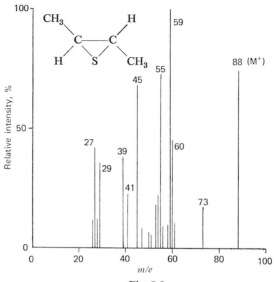

Fig. 8.3

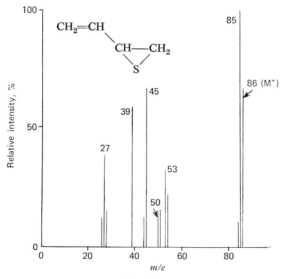

Fig. 8.4

butadiene sulfide (**4**) apparently has much less tendency to isomerize. In its spectrum (Fig. 8.4) (3) the M-1 ion is the base peak. This is presumably the result of α-cleavage leading to a well-delocalized species **4a**. The absence

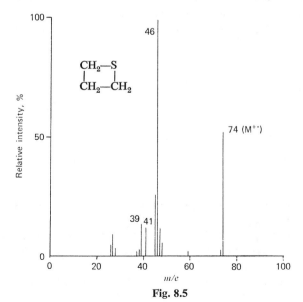

4   **4a**   **4b**

$$CH_2{=}CH{-}CH{=}\overset{+}{S}{-}CH_2^{\cdot}$$

4c

of an M-15 ion suggests that **4b** does not contribute to the molecular ion. Possibly the stability of the α-fission product **4c** inhibits further fragmentation.

B. Trimethylene Sulfides

The mass spectrum of the parent compound (**5**) (Fig. 8.5) (4, 5) is very simple and can easily be rationalized on the basis of a favorable four-center fragmentation with charge retention on the sulfur-containing fragment. (It is interesting to compare this result with that obtained for oxetan (Chap. 1, Sec. II) where similar fragmentation results in charge retention on the hydrocarbon fragment.) The relative weakness of the M-1 ion is striking compared

Fig. 8.5

with the three- and five-membered ring analogs. Presumably other fragmentation modes are suppressed as a result of the facility of the above

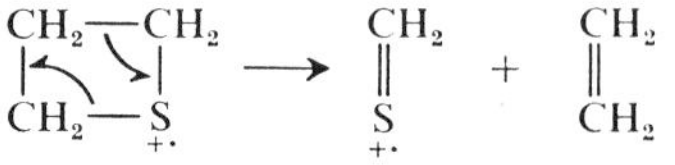

fragmentation. A similar one occurs with 3-methyltrimethylene sulfide **6** (6). In this case the M-CH_3· ion (**6a**) (7.5%) is probably stabilized by transannular charge interaction with the sulfur atom.

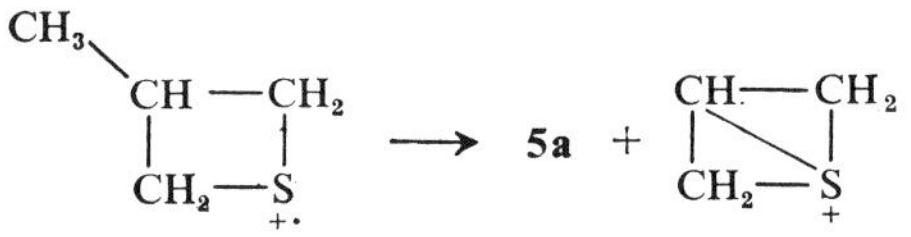

Unfortunately no mass spectra of 2-alkyl derivatives of **5** have been described, so it is impossible to judge whether α-cleavage would be of any importance in such compounds. A partial spectrum of 2,2-di(trifluoromethyl)-4-methoxytrimethylene sulfide (**7**) (7) shows fragmentation of the type already described although there is charge retention on both species produced (**7a** and **7b**). As the full spectrum has not been published it is

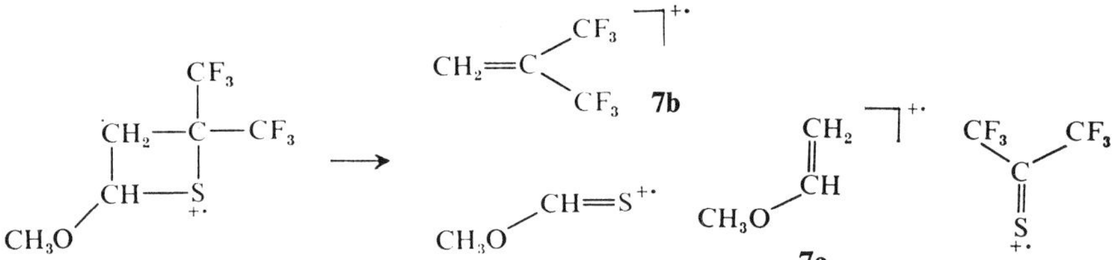

impossible to tell whether either possible α-fission occurs, but apparently neither is an important process. It is also interesting to note that the alternative four-center fragmentation to give **7c** and **7d** does not occur despite the fact that the ground-state analogs of these two ions are stable species and are, in fact, the precursors of **7**.

The elusive unsaturated compound thiete (**8**) has recently been prepared

(8) and its behavior is in sharp contrast to that of **5**. The base peak of the spectrum is the M-1 species which is formulated as the stable cation **8a**, an isoster of the cyclopentadienyl cation. The other fragment ions are shown here and are easily rationalized on the basis of the intact molecule's structure. The occurrence of a peak at m/e 58 is interesting since it represents the rarely-observed loss of methylene from the molecular ion.

No simple sulfones of this series appear to have been examined mass-spectrometrically. It would be interesting to see if the rearrangements to sulfinate esters observed with larger unsaturated and aromatic sulfones occur in this more strained system or whether elision of sulfur dioxide predominates. The complex sulfone **9** is reported (9) to lose SO and CO upon electron impact, suggesting that rearrangement to **9a** occurs, but the full spectrum has not been published. It is interesting to note that this rearrangement also occurs when the compound is heated at 300° *in vacuo*, an interesting parallel between thermal and electron-impact reactions.

C. Tetrahydrothiophens

The mass spectrum of tetrahydrothiophen (**10**) has appeared in the API index (10) and appearance potentials for the principal ions have been determined (4). Rationalization of the spectrum (Fig. 8.6) has been greatly aided by a study of the 2,2,5,5-tetradeutero compound (11). Thus the M-1 species is made up of contributions from **10a** (65%) and **10b** (35%). The rationale for the main features of the fragmentation is shown. It is seen that

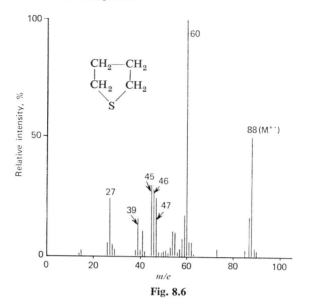

Fig. 8.6

the base peak of the spectrum ($C_2H_4S^{+\cdot}$, m/e 60) is derived by loss of $C_{(2)}$ + $C_{(3)}$ and $C_{(3)}$ + $C_{(4)}$ as ethylene. The latter loss is best explained as proceeding via the α-fission product 10c. The usual ions at m/e 45, 46, and 47 result from loss of hydrocarbon fragments from 10c.

The mass spectra of a number of alkylated tetrahydrothiophens have appeared (12–17). Their nature depends upon the position of the alkyl group.

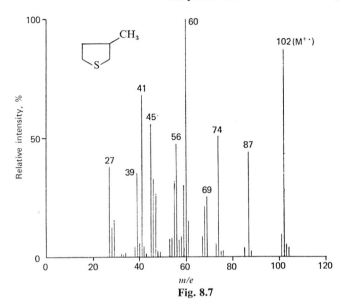

Fig. 8.7

Thus the 3-methyl-compound (**11**) (12) has a spectrum (Fig. 8.7) readily rationalized in the same way as that of the parent compound.

When the alkyl substituent is in the 2-position, α-fission is greatly facilitated. The spectrum of the 2-methyl-derivative (**12**) is shown in Fig. 8.8. It is dominated by the stable ion **10a**, and the amount of fragmentation is much diminished (13). It will be noticed, for example, that no loss of C_2H_4 to give **12b** occurs.

The mass spectra of *cis*- and *trans*-octahydrobenzo[*b*]thiophens **13** (Fig. 8.9) (18, 19) are very similar to one another and are dominated by losses of

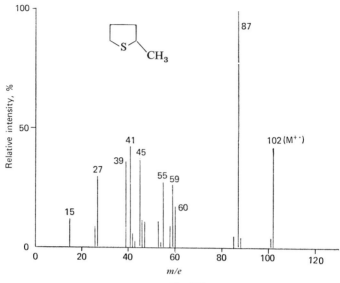

Fig. 8.8

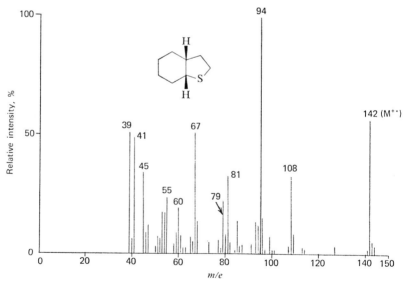

Fig. 8.9

$CH_3S \cdot$ and H_2S. In the absence of deuterium labeling it is futile to speculate on the structures of the ions produced.

$$m/e \ 108 \ (33\%) \quad \xleftarrow{-H_2S} \qquad \xrightarrow{-CH_3S \cdot} \quad m/e \ 95 \ (100\%)$$

13

D. Tetrahydrothiophen Sulfones (Sulfolanes) and Related Compounds

The mass spectrum of sulfolane (**14**) and its 2,2,5,5-d_4 analog have been published and discussed (20). The principal fragmentations involve extrusion of SO_2 and $SO_2 + \cdot CH_3$ from the molecular ion (to give m/e 56 and 41, respectively) and formation of ethylene (m/e 28). The mechanism shown is consistent with the deuterium labeling, although considerable scrambling of hydrogen atoms seems to occur before the loss of $C_{(2)}/C_{(5)}$ as a methyl group. The labeling studies show clearly that the $C_2H_4^{+\cdot}$ ion does not come

$$\xleftarrow{} \qquad \left[\quad\right]^{+\cdot} \xrightarrow{-SO_2} \qquad \overset{a}{\nrightarrow} \quad \left[\quad\right]^{+\cdot}$$

$$\text{SO}_2^+ \qquad \text{SO}_2 \qquad$$

$$\textbf{14}, \text{M}^{+\cdot}, m/e \qquad\qquad m/e \ 56$$
$$120 \ (35\%) \qquad\qquad (82\%)$$

$$- (SO_2 + C_2H_4) \searrow (a)$$

$$\xrightarrow{-(CH_3 \cdot + SO_2)} \quad CH_2{=}CH{-}CH_2^+ \qquad \left[CH_2{=}CH_2\right]^{+\cdot}$$
$$\text{SO}_2^+ \text{CH}_3 \qquad\qquad m/e \ 41 \ (100\%) \qquad m/e \ 28 \ (50\%)$$

from $C_{(3)} + C_{(4)}$ of the tetrahydrothiophen ring, and that the ion m/e 56 does not have the cyclobutane structure.

The mass spectra of a series of 3-monosubstituted and 3,4-disubstituted sulfolanes have also been reported, and in all cases formation of the most stable substituted ethylene is the dominant fragmentation. Thus the 3-hydroxy-4-amino compound (**15**) gives the ions **15a** and **15b** in the proportions shown.

$$\left[H_2N \diagdown\!\!\!\diagdown\, OH\right]^{+\cdot} \qquad \longrightarrow \qquad \left[\begin{matrix} H_2N \diagdown \\ CH \\ \| \\ CH_2 \end{matrix}\right]^{+\cdot} \qquad + \qquad \left[\begin{matrix} OH \\ CH \\ \| \\ CH_2 \end{matrix}\right]^{+\cdot}$$

$$\text{SO}_2$$

$$\textbf{15} \qquad\qquad\qquad \textbf{15a}, m/e \ 43 \qquad\qquad \textbf{15b}, m/e \ 44$$
$$\qquad\qquad\qquad (100\%) \qquad\qquad\qquad (1.9\%)$$

Introduction of unsaturation into the sulfolane molecule profoundly modifies the observed fragmentation. Thus the base peak of the spectrum of 2-sulfolene (**16**) occurs at *m/e* 89 (loss of ·CHO) and the rearrangement shown explains its formation.

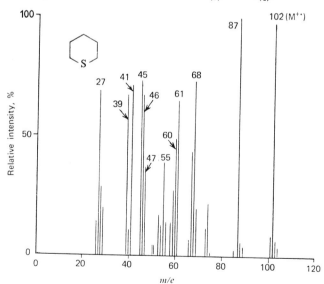

E.   Tetrahydrothiapyrans

The mass spectrum of the parent compound (**17**) is shown in Fig. 8.10 (21). The fragmentation is quite dissimilar to that of the oxygen analog, and deuterium labeling (22) has shown that the base peak (M-CH$_3$·, *m/e* 87) involves loss of an α-methylene, together with a transferred β-hydrogen atom. A rationale to explain the results observed with deuterium labeling is shown. The ion at *m/e* 74 (M-C$_2$H$_4$) appears to originate largely from C$_{(3)}$ and C$_{(4)}$ as shown, whereas the M-HS· and M-H$_2$S ions appear to result predominantly from H abstractions from C$_{(3)}$ and C$_{(4)}$.

Fig. 8.10

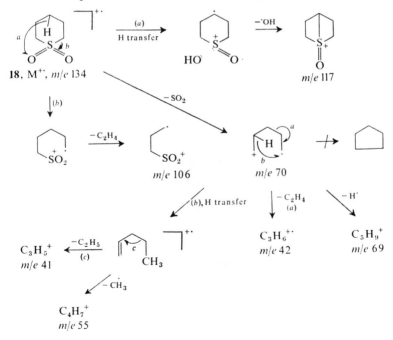

The mass spectrum of tetrahydrothiapyran 1,1-dioxide (**18**) has also been examined (22) and with the aid of the spectra of its 2,2,6,6- and 3,3,5,5-tetradeutero-derivatives the rationale shown for its fragmentation may be presented. The spectra of the 2-, 3-, and 4-methyltetrahydrothiapyrans **19**,

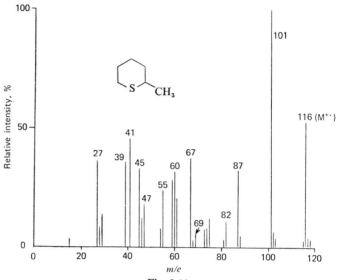

Fig. 8.11

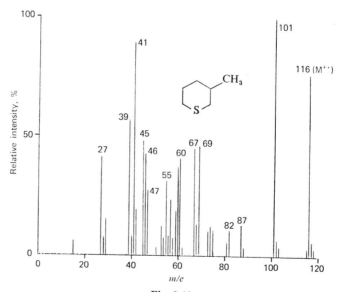

Fig. 8.12

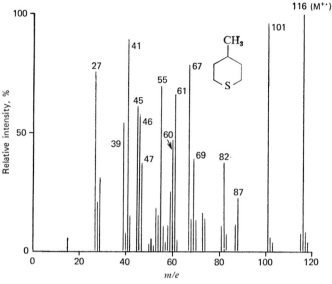

Fig. 8.13

20, and **21** are shown in Figs. 8.11, 8.12, and 8.13 (23–25). M-·CH₃ is an important ion in all three and is the base peak of the spectrum in the 2- and 3-isomers. These results demonstrate easy formation of the bridged ions **20a** and **21a**. Otherwise the spectra are quite similar, and in many details resemble

the parent compound. In fact, the M-·CH₃ ions may in part originate from ring-methylene groups. Deuterium labeling would be required to elucidate the latter point.

Cis- and *trans*-2-thiadecalin (**22**) (26, 27), like other stereoisomeric pairs in this series, show very similar complex spectra which would be resolved only by deuterium labeling combined with a study of metastable ions. The spectra are dominated by losses of alkyl radicals from the molecular ion,

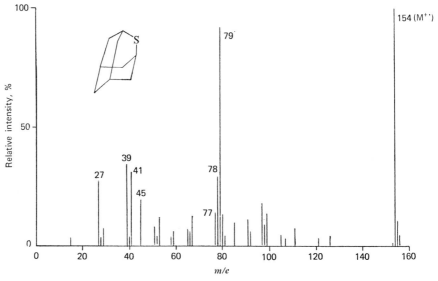

Fig. 8.14

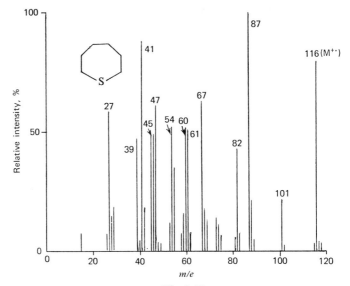

Fig. 8.15

together with losses of H_2S. Thia-adamantane (23) can be regarded as a tetrahydrothiapyran derivative, but upon electron impact its tricyclic ring structure shows a behavior (28) having no resemblance to that of the simpler systems. Its only decomposition mode (Fig. 8.14) is loss of $C_3H_7S\cdot$ from the molecular ion to give the cation 23a, a process requiring the transfer of two hydrogen atoms to the neutral fragment.

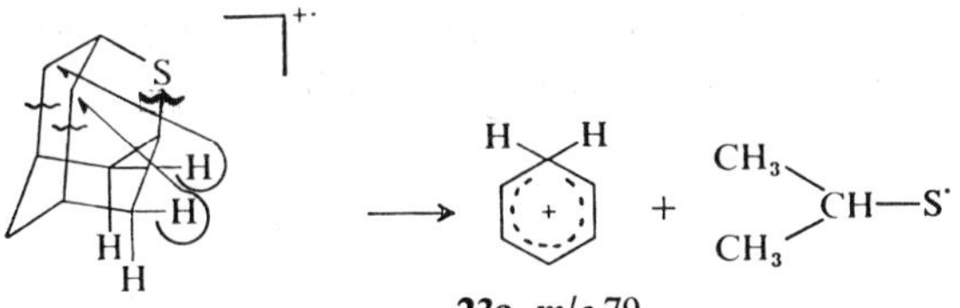

23   23a, m/e 79

F.   Hexamethylene Sulfide

Hexamethylene sulfide (24) (29) has a spectrum (Fig. 8.15) which is almost identical below m/e 87 with that of tetrahydrothiapyran (17). It therefore probably fragments in a similar fashion except that the molecular ion loses an ethyl group rather than a methyl group as the initial fragmentation.

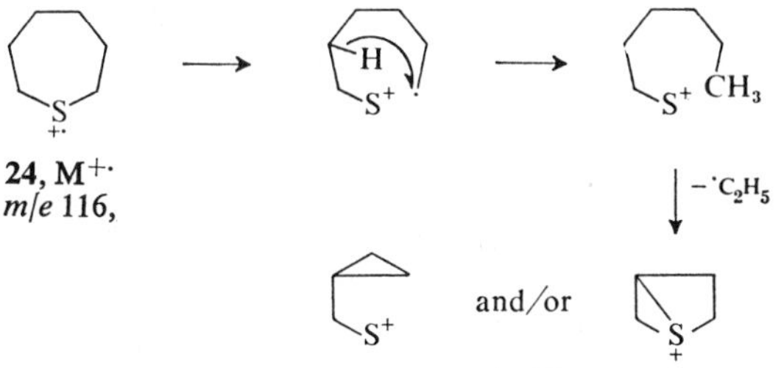

24, $M^{+\cdot}$
m/e 116,

and/or

m/e 87

Some features of the mass spectrum of the 2-hydroxy derivative of 24 (30) have been reported. Interestingly, the compound does not lose $\cdot OH$ but rather $\cdot CH_2OH$ and H_2O. These changes can probably be represented as shown.

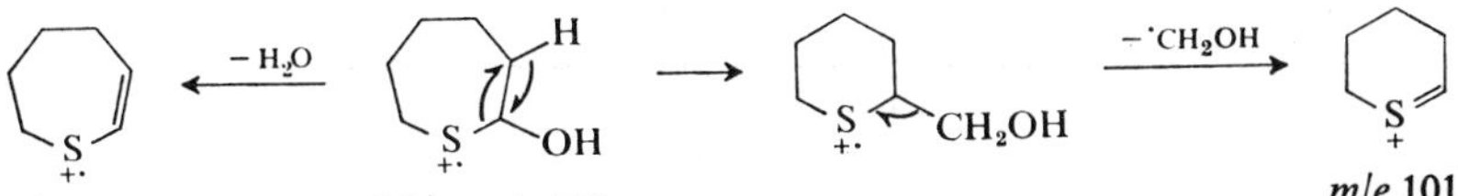

m/e 114 $M^{+\cdot}$, m/e 132 m/e 101

G.   Derivatives of Thiapyran and Thiepin

The mass spectra of 4-thiapyrone (**25**) and its 2,6-dideutero-derivative have been reported (31).

The major pathway, illustrated here, involves a retro-Diels-Alder process with loss of acetylene as the neutral product to give the ion **25a** which subsequently decomposes by loss of carbon monoxide to **25b**. A less important process is extrusion of carbon monoxide to give a species which may be formulated as the thiophen radical cation (**25c**).

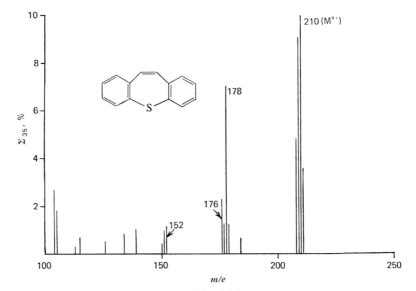

The interesting observation has been made that the loss of carbon monoxide from the molecular ion becomes the major fragmentation of **25** at lower ionizing potentials.

The mass spectrum of the dibenzo-derivative of 4*H*-thiapyran (thioxanthene, **26**) has been briefly reported (32). The base peak arises from loss

Fig. 8.16

of one hydrogen atom from the molecular ion. The thioxanthenyl cation (26a) has also been observed as a stable species in solution chemistry (32).

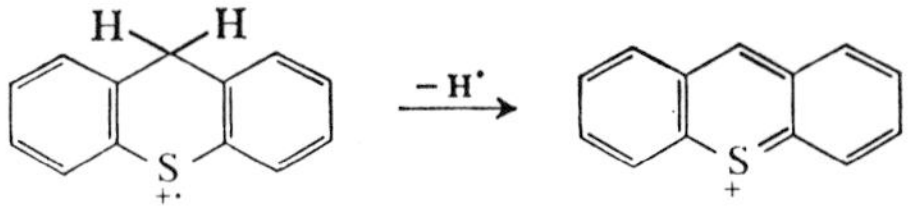

26, M$^{+\cdot}$, m/e 198   26a, m/e 197 (100%)

The mass spectrum of dibenzo[b,f]thiepin (27) has been reported and is shown in Fig. 8.16 (33). Extrusion of sulfur to give a species plausibly formulated as the phenanthrene radical cation (27a) is a prominent feature of the spectrum, and the rationale shown has been suggested.

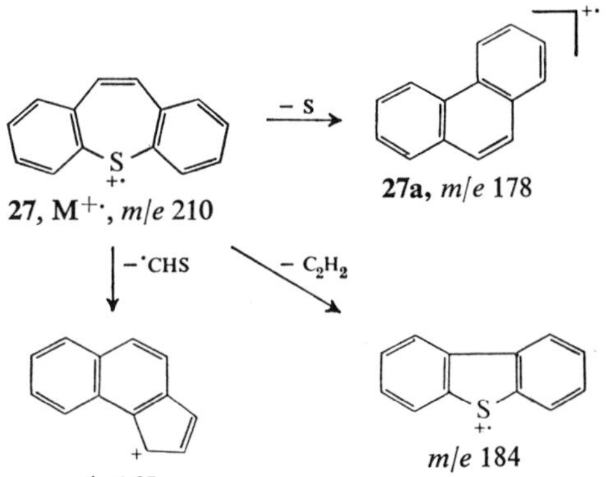

27, M$^{+\cdot}$, m/e 210   27a, m/e 178

m/e 165 m/e 184

The mass spectrum below m/e 160 is dominated by weak hydrocarbon ions and would clearly be of no diagnostic value for compounds of this type.

H. Thiophen and Derivatives

1. Thiophen

The mass spectrum of thiophen (28), which has been extensively studied (34–36), is shown in Fig. 8.17. Its general form resembles that of furan (Chap. 4, Sec. I.A) although the parent ion is relatively stronger. A rationale for the main fragmentation is shown. Mass-spectral studies of 2-deutero- and

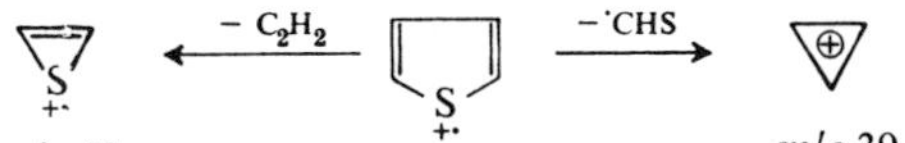

m/e 58   28, M$^{+\cdot}$,
                  m/e 84 m/e 39

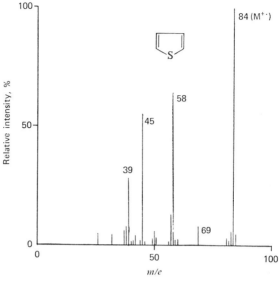

Fig. 8.17

2,3- and 2,5-dideutero-thiophens have shown them to lose C_2H_2, C_2HD, and C_2D_2 in a statistical manner, indicating that rapid randomization of hydrogen atoms precedes fragmentation of the molecular ion (37). The mechanism of this process is as yet uncertain but may resemble those postulated for the photochemical conversion of 2-phenylthiophen to its 3-substituted isomer (38). One possible mechanism, which makes use of the $3d$ orbitals of the sulfur atom, is shown.

The synchronous movement of C and H atoms required for the impact-induced analog of such a scheme could possibly be demonstrated by ^{13}C—^2H double-labeling, but experimental difficulties in such a project will be formidable.

As with benzene and furan, finer details of the structure of the thiophen molecular ion remain uncertain. While it is convenient to use the formulation **28**, open chain (**29**) or carbocyclic (**30**) structures are also possibilities.

2. Alkylthiophens

The mass spectra of 2- and 3-alkylthiophens (31) have been well studied, and in all cases the base peak is the ion $C_5H_5S^+$, m/e 97, resulting from fission of the bond in the alkyl group between the carbon atoms in positions α and β relative to the ring.

31
(R = H or alkyl)　31a, m/e 97　31b　31c, m/e 97

The close resemblance to the fragmentation of toluene and other alkyl-benzenes is immediately apparent, and the thiopyrylium ion structure 31c has been suggested for the species m/e 97 (39). Occurrence of the thiafulvene ions (31a or 31b) as intermediates is likely, but the virtual identity of fragmentation of the m/e 97 species from 2- and 3-alkylthiophens supports the ultimate symmetrical structure. The thiopyrylium ion is a stable species in ground state chemistry (40) in agreement with molecular orbital calculations (41). It is interesting to observe that the ion 31c is much more intense than the tropylium ion 32a in the mass spectrum of 2-benzylthiophen (32) (42).

32a, m/e 91
(10%)　32　31c, m/e 97
(100%)

3-α-Methylbenzylthiophen fragments in a similar manner, but unfortunately the relative intensities of the methyl-homologs of 32a and 31c were not reported (43).

When the α-carbon atom of an alkylthiophen is fully substituted, the simple cleavage and rearrangement described above cannot occur and hydrogen migrations become important. Thus the spectrum of 2-t-butylthio-phen (33) (Fig. 8.18) shows the expected strong M-15 ion (33a) together with the moderately strong ion at m/e 85 (33b), corresponding to loss of an isobutenyl radical resulting from a double hydrogen migration (44).

Simple dialkylthiophens (45) show fission of an α,β-bond in a manner analogous to that discussed above, giving (presumably) substituted thio-pyrylium ions. As expected, the stability of the neutral fragment controls the major mode of fragmentation. This is seen in the preferred loss of $\cdot C_2H_5$ rather than H· in the mass spectrum of 2-methyl-5-n-propylthiophen (34), and the preferred loss of an ethyl radical in the spectrum of 2-ethyl-5-n-propylthiophen (35). 2,3-Dimethylthiophen (36) and its derivatives have M-15

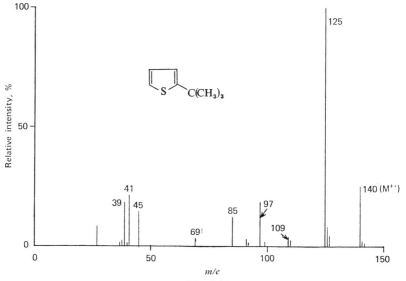

Fig. 8.18

ions as base peaks (45), probably on account of the process **36** → **36a**. Similar processes have been observed with dialkylindoles (Chap. 10, Sec. VII).

Minor differences in the mass spectra of alkylated thiophens can be of great value in determining positions and types of substitution. For further information the reader is directed to the valuable compilation of Kinney and Cook dealing with this matter (45).

3. Arylthiophens

2-Phenylthiophen (**37**) shows (Fig. 8.19) little intense fragmentation (42), and the scheme shown has been postulated to rationalize the decompositions observed.

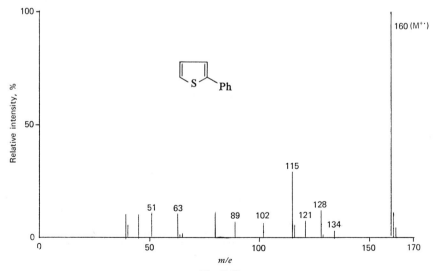

The loss of sulfur from the molecular ion to give **37a**, m/e 128, has been contrasted with the absence of direct losses of the hetero-atom from comparable furans and pyrroles. The fragmentations of 2,4- and 2,5-diphenylthiophens proceed mainly by similar pathways, and the original publication should be consulted for fuller details. The diphenylthiophen **38** (42) loses H_2S rather than S from the molecular ion, and this has been ascribed to formation of the hydrocarbon radical cation **38a**. It is interesting to note that

Fig. 8.19

38 gives an ion (4% intensity) at m/e 191 corresponding to the loss of ·CHS from the molecular ion. This must necessitate rearrangement of **38** subsequent to electron impact, probably by migration of phenyl from $C_{(2)}$ to $C_{(3)}$. This

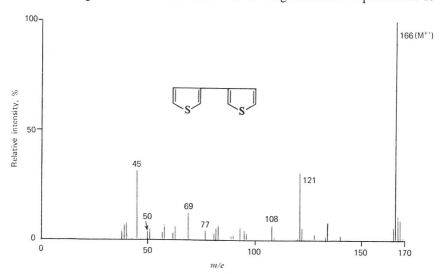

may occur by a mechanism similar to that put forward above for H/D scrambling in thiophen itself. As mentioned before, 2-phenylthiophen rearranges photolytically to 3-phenylthiophen.

Tetraphenylthiophen (**39**) has a mass spectrum (46) displaying an intense molecular ion with very little fragmentation. The only ions with intensities of 5% or more are m/e 77 ($C_6H_5^+$, 5.0%) and 121 ($C_6H_5-C\equiv S^+$, 6.5%).

4. Di- and Polythienyls

The mass spectra of a number of members of this class of compound are recorded (47–51), but only that of 3,3'-dithienyl (**40**) has been subjected to detailed analysis including high-resolution studies (42). Its spectrum is shown in Fig. 8.20 and a rationale for its fragmentation is presented. It

Fig. 8.20

seems likely that the species **40b** is the phenyl cation since it shows the characteristic loss of C_2H_2 to give the species of m/e 51. The formation of

40a, m/e 121   **40**, M$^{+\cdot}$, m/e 166

m/e 134 m/e 108

$C_6H_5^+$ $\xrightarrow{-C_2H_2}$ $C_4H_3^+$

40b, m/e 77   **40c**, m/e 51

41   **42**

40b following loss of CS from **40a** clearly requires considerable bond reorganization.

Apart from small differences in relative ion intensities, the spectra of 2,2′-dithienyl (**41**) and 2,3′-dithienyl (**42**) are similar to that of the 2,2′-isomer (**47**) and it is quite possible that all three fragment via a common isomerized molecular ion.

The 2,2′-dithienyl system occurs as a structural unit in a number of natural products, and partial mass spectra have been reported for some such compounds. Thus the chloride **43** shows the fragmentations illustrated (52), but in the absence of the full spectrum the importance of fragmentations involving the ring system cannot be judged.

43, M$^{+\cdot}$, m/e 268

$\diagup$ $-\cdot$OH $\diagdown$ $-\cdot CH_2Cl$

m/e 251 m/e 219

α-Terthienyl (**44**), which has been obtained as a natural product from the flowers of the Indian marigold *Tagetes erecta* L. (53), should be easily recognized as a component of natural-product mixtures from its intense molecular ion (m/e 248) (50). It shows only minor fragmentation, and in the absence of high-resolution and clear metastable ion data conclusions regarding fragmentation paths are necessarily tentative. Losses of S, H_2S, and ·CHS from the parent ion, however, can probably be represented as shown.

44, M⁺·, m/e 248   **44b**, m/e 214   **45**

m/e 171

44a, m/e 216

Of course sulfur could equally well be lost from a terminal thiophen ring in the formation of **44a**. The structure postulated for **44b** is by analogy to that postulated (**38a**) for the M-H_2S ion from 2,5-diphenylthiophen.

The mass spectrum of 2,2′:4′,2″-terthienyl (**45**) (51) is generally similar to that of its isomer **44** but has a rather more intense ion at m/e 203 (M-·CHS).

5.   Dithienylethylenes and [18]Annulene Trisulfide

Dithienylethylenes bear a vinylogous relationship to the corresponding dithienyls, and the spectrum of trans-2,2′-dithienylethylene (**46**) (54) is shown in Fig. 8.21. The compound resembles the simple compounds already considered in showing losses of S, CS, ·CHS, and C_2H_2 from the molecular ion. Loss of ·SH does not occur with the dithienyls, and its formulation here is difficult without knowledge of the origin of the transferred hydrogen atom. The sequence shown rationalizes most of the observations. The structure **46a** has been postulated for the M-1 ion in 5- and 7-methylbenzo[b]thiophen.

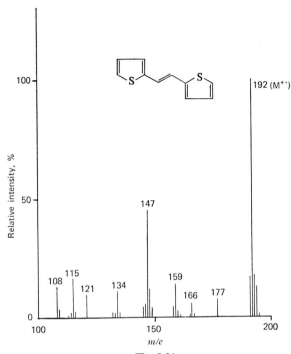

Fig. 8.21

The M-(C_2H_2 + S) ion has been formulated as the benzo[*b*]thiophen radical cation **46b** because it appears to undergo the further decompositions described for that ion (Sec. I.H.11.a).

46, M⁺·, *m/e* 192

$-C_2H_2$

$-$˙CHS

46a, *m/e* 147

m/e 166

$-$S, cyclization

46b, *m/e* 134

$-C_2H_2$

m/e 108

[18]Annulene trisulfide (**47**) (Fig. 8.22) (54) shows only minor decomposition upon electron impact, which is not surprising in view of its aromatic stability and the necessity for breaking two bonds prior to elision of fragments.

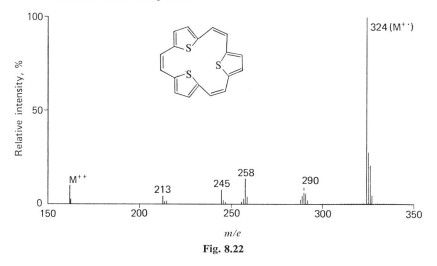

Fig. 8.22

Losses of ·SH and ·CHS units show that rearrangements precede fragmentation. These probably involve 2,3-migrations within the thiophen rings (migrations of this type have already been postulated several times in this chapter). Details of the origins of the transferred hydrogen atoms in the ·SH groups lost must await deuteration studies, although the success of a deuteration experiment in a system of this kind is highly problematical in view of the likelihood of H/D randomization.

47

6. Carbonyl Derivatives of Thiophen

a. THIENYL ALDEHYDES AND KETONES. Mass spectra of a reasonably large range of compounds in these classes have been determined and analyzed (42, 55). The spectrum of 2-formylthiophen (**48**) (Fig. 8.23) is typical of that

$$\overset{O}{\underset{\|}{}}$$

of an aromatic aldehyde, showing ready fission of the C—H bond (α-fission) to produce the stable acylium cation **48a**. Subsequent fragmentation of this species is easily rationalized as shown. As is to be expected, the acylium ion

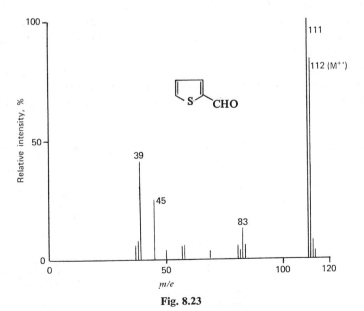

48a is a prominent feature of the fragmentation of 2-thienyl-ketones (42, 55) and is the base peak of the spectra of 2-acetyl- and 2-benzoylthiophens (**49**) although the alternative charge retention on the nonheterocyclic acylium ions is also observed. The greater relative abundance of **48a** compared with Ph—CO$^+$ in the mass spectrum of the benzoyl compound probably reflects

Fig. 8.23

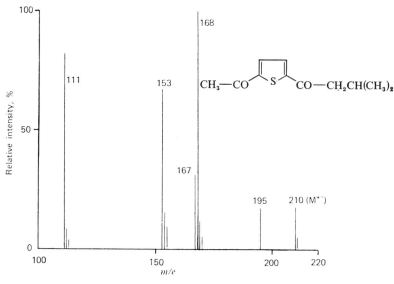

Fig. 8.24

the efficient stabilization of **48a** by resonance involving canonical forms such as **48c**, with tricovalent positively-charged sulfur.

The thienyl-ketones **49** both show fragment ions resulting from loss of CO from the molecular ion with concomitant rearrangement of the group R (42, 56). This is not a general process with aryl-ketones (57, 58), and it has therefore been suggested that the process may be initiated by migration to the electron-deficient hetero-atom.

When larger aliphatic side-chains are present in thienyl-ketones, different decomposition modes appear. Thus 2-acetyl-5-isovalerylthiophen (**50**) (Fig. 8.24) undergoes initial fragmentation via a McLafferty rearrangement

(50 → 50a) followed by a simple acyl fission (50a → 50b) (55). The peak at
m/e 154 is more intense than expected for a simple isotope peak from m/e
153, and it is suggested that the alternative hydrogen migration (50 → 50c)

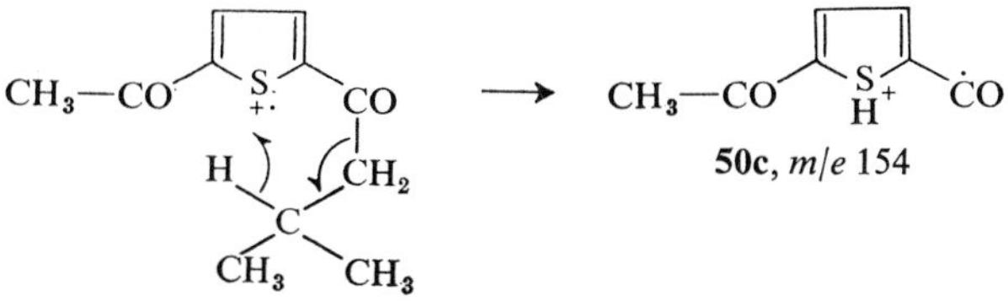

50, M⁺·, m/e 210

50c, m/e 154

is responsible for the formation of this species. A deuterium-labeling study
would be of considerable interest in this compound; in particular, replacement
of the tertiary hydrogen in the aliphatic side-chain of 50 would shed light
upon the specificity of the rearrangement mechanism.

The diketone 2-thenoylacetone (51/52) loses ketene as a major process
following electron impact. This may be formulated as either a four-center
elimination from the enol 51 (51 → 51a) or as a McLafferty rearrangement
of the β-diketone 52 (52 → 51a) (55).

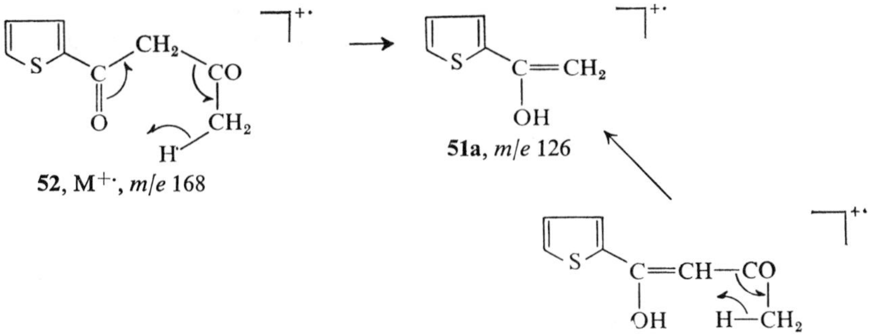

52, M⁺·, m/e 168

51a, m/e 126

51, M⁺·, m/e 168

b. CARBOXYTHIOPHENS.   2- and 3-Carboxythiophens (53 and 54) have
identical mass spectra (Fig. 8.25) (42) and, in addition to the expected sequential losses of ·OH and CO (e.g., 53 → 48a → 48b), show a rearrangement

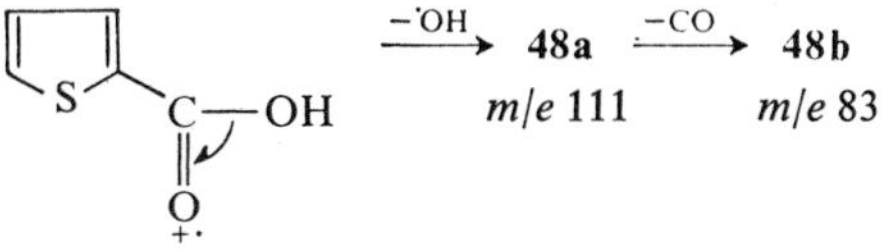

53 M⁺·, m/e 128

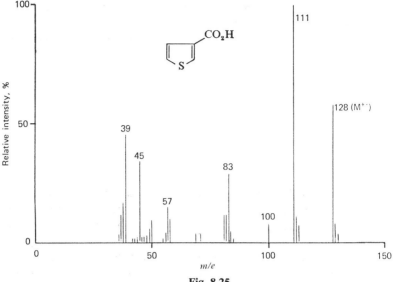

Fig. 8.25

ion resulting from expulsion of CO and migration of ·OH to the ring. Arylcarboxylic acids do not undergo this process, and again it has been speculated that specific migration to the hetero-atom may be involved. If this is so it is striking that the acids **53** and **54** both undergo the rearrangement to the same extent, and it has been suggested that in this case at least the molecular ion may no longer have a cyclic structure.

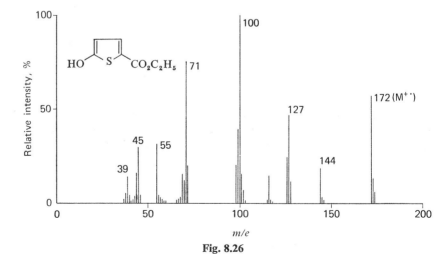

Fig. 8.26

No simple esters of the thiophen-monocarboxylic acids appear to have been examined mass-spectrometrically, but it is highly probable that they would fragment by a similar sequence to that described above for the free acids. The importance of alkoxyl-group migration compared with the hydroxyl-group migration observed with the free acids would be of some interest.

Ethyl 5-hydroxythiophen-2-carboxylate (55) (Fig. 8.26) shows the expected loss of an ethoxy group to produce 55a, and also loss of the ester group both with (55 → 55b) and without hydrogen transfer (55 → 55c). A full rationale is put forward in the original communication (59).

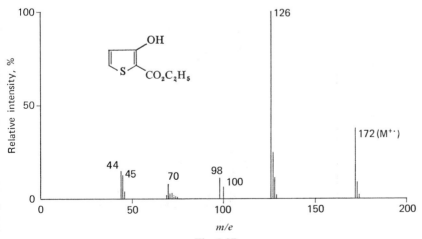

Marked modification of the fragmentation pattern is observed with ethyl 3-hydroxythiophen-2-carboxylate (56) (Fig. 8.27) wherein the presence of a pronounced "ortho effect" results in the loss of ethanol becoming the

Fig. 8.27

dominant decomposition mode (59). The species **56a** decomposes by sequential loss of two molecules of carbon monoxide, and the entire decomposition

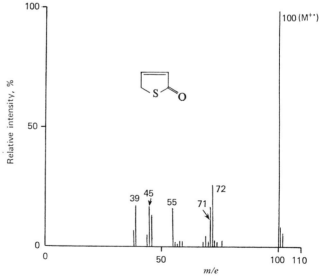

sequence parallels that observed with the isosteric ethyl *o*-hydroxybenzoate (60).

7. Hydroxythiophens

Compounds formally represented as 2-hydroxythiophens exist mainly in the keto form as thiolactones (**57** ⇋ **58**) (61–63). The simple mass spectrum of **58** is shown in Fig. 8.28, and is rationalized as shown (59).

The mass-spectra of a number of substituted thiolactones are explicable in terms of similar fragmentations, and the original paper should be consulted for fuller details (59).

Fig. 8.28

Thiophen-2-thiol (**59**) appears to lose ·CHS rather than CS as the major fragmentation (64), which may suggest that it fragments as the thiol.

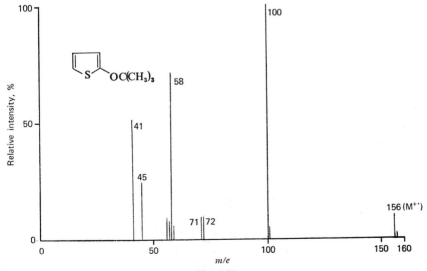

The lack of clear information about metastable ions does not disallow the possibility that the weak ion m/e 72 is an intermediate in the sequence. It is, however, difficult to see why an apparently identical ion of m/e 72 should be more prominent in the spectrum of **58** if this compound and "thiophen-2-thiol" have similar structures.

The mass spectrum of the benzoate of 2-hydroxythiophen (**60**) has been recorded (42, 65). The base peak of the spectrum is the benzoyl cation (**60a**) and few ions in the spectrum appear to originate from the thiophen portion of the molecule. Thus the ion **60b** has an intensity of only 6%.

The t-butyl ethers of 2- and 3-hydroxythiophens (**61** and **62**) have been subjected to electron impact (66). Their spectra are very similar. That of **61** is shown in Fig. 8.29, and it will be seen that the main decomposition is by

Fig. 8.29

loss of isobutene, followed by fragmentation of the resulting "hydroxy-thiophen."

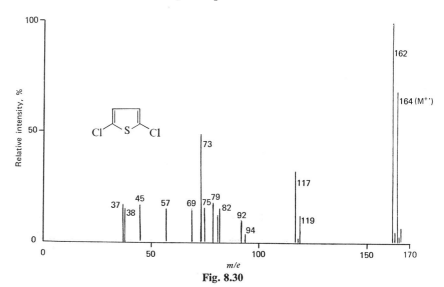

2,5-Di-*t*-butoxythiophen fragments in the same way by successive losses of two molecules of isobutene, and the spectra of several substituted *t*-butoxythiophens have also been so rationalized (66).

8. Halothiophens

A series of mono- and polyhalothiophens have been subjected to mass spectrometry (42). Their spectra are simple, and it is not possible to distinguish isomeric compounds by the technique. The spectrum of 2,5-dichlorothiophen (63) (Fig. 8.30) is fairly typical of most of the spectra encountered in this

class of compound. Features to be noted are the successive losses of chlorine atoms, the loss of the C_2HCl unit from the molecular ion, and the sequential loss of Cl· and CS, which may be represented as shown.

Fig. 8.30

It seems likely that the species (*m/e* 82) resulting from loss of both chlorine atoms may have an open chain structure such as $HC{\equiv}C{-}\overset{+\cdot}{S}{-}C{\equiv}CH$.

9.   Substituted Thiophens of Complex Structure

A number of polyfunctional thiophens, and thiophens substituted with a single complex group, have been examined mass-spectrometrically, and two interesting examples will be discussed.

a.   ETHYL 2-AMINO-4-HYDROXY-3-THIOPHENCARBOXYLATE.   This compound (**64**) has a spectrum (Fig. 8.31) which does not show the expected loss of ethanol by an "ortho effect" (**42**). Instead, the compound undergoes complex fragmentation by the route shown in Fig. 8.31. It is suggested that this anomaly is due to the presence of two *ortho* substituents, and it points to the care which is necessary in extending simple fragmentation schemes to explain the spectra of complex molecules.

b.   1,1,1-TRIFLUORO-4-(2-THIENYL)-4-MERCAPTOBUT-3-EN-2-ONE.   Although this compound (**65**) is formally related to the 1,3-diketone **51**, it is not surprising in view of the number of hetero-atoms in the molecule that its fragmentation is very different, and the rationale shown has been presented (**67**).

Of considerable interest in this fragmentation is the cyclization postulated in the conversion **65** → **65a**. Cyclization processes as a consequence of electron impact have been observed in several other cases, such as the formation

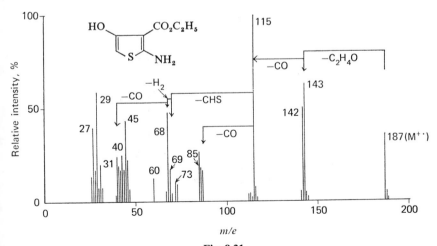

m/e

Fig. 8.31

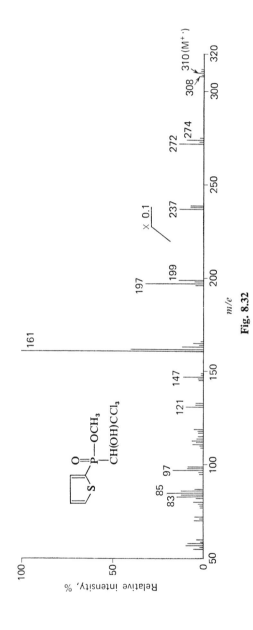

Fig. 8.32

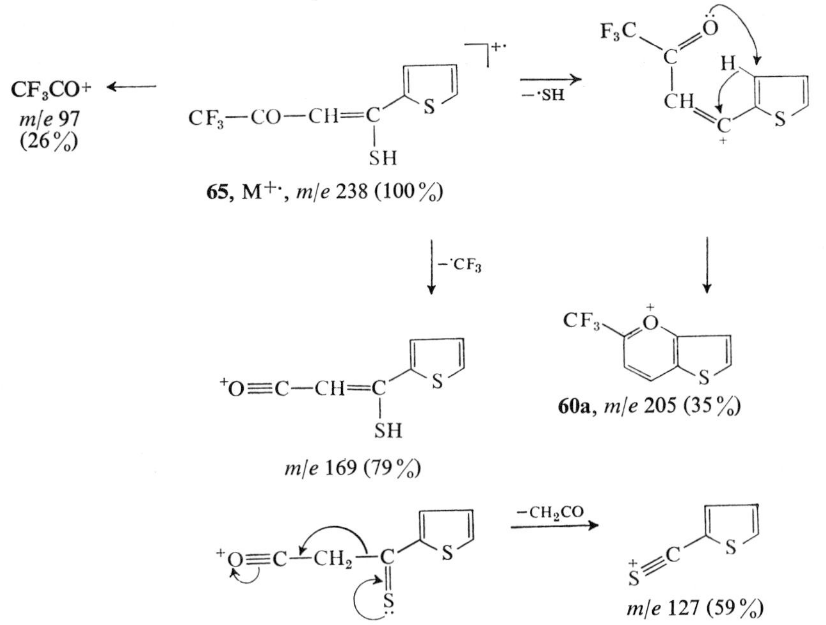

65, $M^{+\cdot}$, m/e 238 (100%)

CF_3CO^+
m/e 97
(26%)

CF_3—CO—CH=C
|
SH

$-\cdot SH$

$-\cdot CF_3$

^+O≡C—CH=C
|
SH

m/e 169 (79%)

60a, m/e 205 (35%)

$-CH_2CO$

m/e 127 (59%)

of the M-1 peak from benzalacetophenone (**66**), and their occurrence is the subject of considerable study (68, 69).

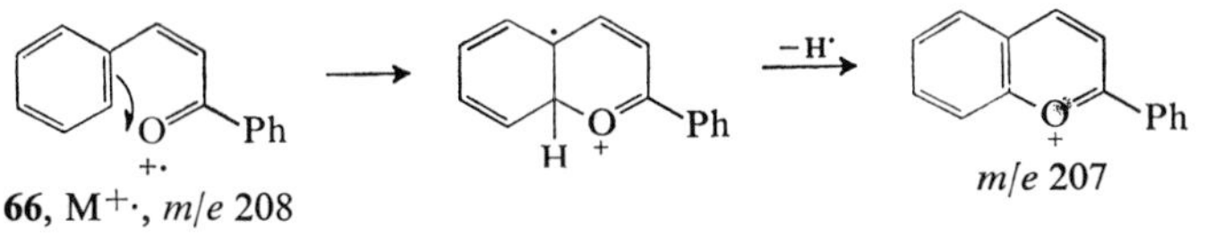

66, $M^{+\cdot}$, m/e 208

$-H^{\cdot}$

m/e 207

10.   Thienylalkylphosphonates

The mass spectrum of the phosphonate **67** is shown in Fig. 8.32. It will be seen that the base peak results from fission of the P—CH(OH) bond (70). Subsequent fragmentation of the resulting species is minimal. A rationale of the main features of the spectrum is shown. Of considerable interest is

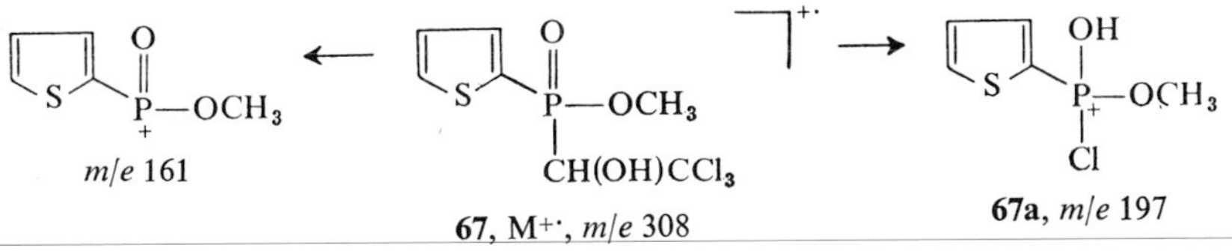

m/e 161

67, $M^{+\cdot}$, m/e 308

67a, m/e 197

the presence of the ion of m/e 197, the genesis of which requires a four-center chlorine migration together with hydrogen transfer.

The spectrum of the *n*-butoxy-analog (**68**) has also been determined and is complicated by the presence of a McLafferty rearrangement (**68a** → **68b**).

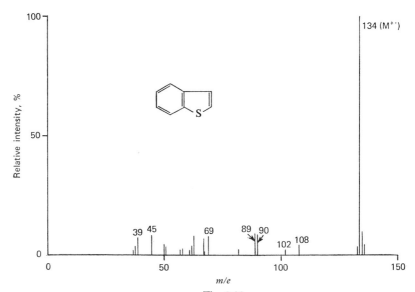

In this series fragmentation involving the thiophen ring seems to be quite minimal.

11.   Condensed Thiophens

a.   BENZO[*b*]THIOPHEN.   The mass spectrum of benzo[*b*]thiophen (**69**) has been reported on a number of occasions (71, 72) and is shown in Fig. 8.33. The rationale shown has been suggested (72). Most of the decompositions are fairly similar to those observed with benzo[*b*]furan.

A recent publication by Williams and his co-workers describes the mass spectrum of 2,3-dideuterobenzo[*b*]thiophen (73). It shows that C_2H_2, C_2HD, and C_2D_2 are expelled approximately statistically from the molecular ion,

Fig. 8.33

implying complete randomization of all hydrogen atoms in **69** before fragmentation. Further elucidation of the scrambling mechanism must await
labeling using carbon as well as hydrogen isotopes.

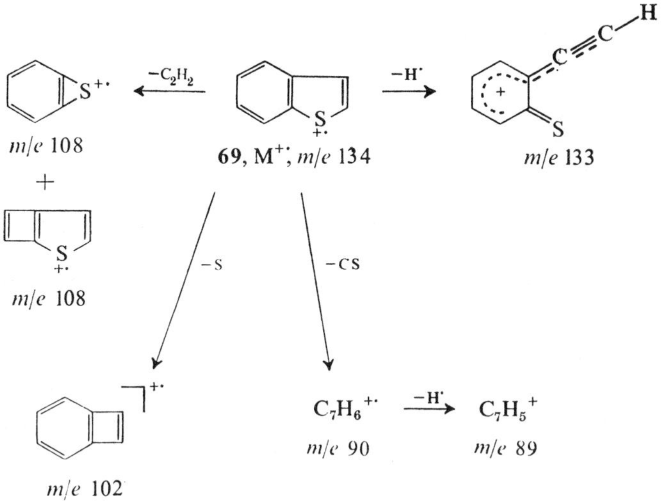

m/e 108

$+$

m/e 108

69, M$^+$, m/e 134

m/e 133

m/e 102

$C_7H_6^{+\cdot}$ $\xrightarrow{-H^{\cdot}}$ $C_7H_5^+$

m/e 90 $\qquad$ m/e 89

b. ALKYLBENZO[*b*]THIOPHENS. Those alkylbenzo[*b*] thiophens which have
been examined mass-spectrometrically behave similarly to the simple alkylthiophens, and the results have been rationalized in terms of expansion of
either the 5- or the 6-membered ring (72). Benzo[*b*]thiophens in which the

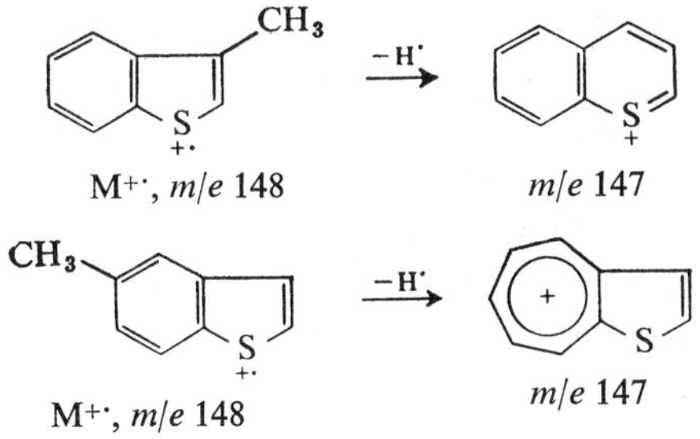

M$^{+\cdot}$, m/e 148 $\qquad$ m/e 147

M$^+$, m/e 148 $\qquad$ m/e 147

substituent group is a longer alkyl chain undergo the expected β-fission as the
major decomposition process. As with the alkylthiophens, minor differences
in spectra may be of diagnostic value.

When more than one alkyl group is present in a benzo[*b*]thiophen, the
preferred fission is, as usual, the one leading to the most stable neutral

fragment. Thus with 2-*n*-propyl-7-ethylbenzo[*b*]thiophen (**70**) the M-·C$_2$H$_5$ ion (**70a**) is much more intense than the M-·CH$_3$ ion (**70b**) (74).

70b, *m/e* 189

$-$·CH$_3$

70, M$^{+·}$, *m/e* 204

$\xrightarrow{-\cdot C_2H_5}$

70a, *m/e* 175

The mass spectrum of 2,3-dihydrobenzo[*b*]thiophen (**71**) is simple. The base peak is the M-1 ion. Labeling with deuterium is essential to elucidate its structure, but it may well be a mixture of **71a** and **71b**. The main fragmentation observed is loss of CS from the M-1 ion to give C$_7$H$_7$$^+$ (31 %) (75).

71, M$^{+·}$, *m/e* 136   **71a** *m/e* 135   **71b**

c.   PHENYLBENZO[*b*]THIOPHENS. The mass spectra of 2- and 3-phenylbenzo[*b*]thiophens (**72, 73**) are very similar, but not identical, suggesting that the randomization of position that occurs in simple thiophens may not be complete with these benzo[*b*]thiophens (72). This is reasonable if the mechanism postulated (Sec. I.H.1) is correct since the fused benzo-ring will prevent formation of the required intermediates.

The fragmentations of **72** and **73** are basically similar to that described for the parent compound and will not be described in detail. It is interesting that both compounds lose ·CHS from the molecular ion to give a species represented as **73a**. This result requires a migration of the phenyl group in **72** prior to fragmentation.

72, M$^{+·}$, *m/e* 210   **73,** M$^{+·}$, *m/e* 210   **73a**

d.   BENZO[c]THIOPHENS.   The mass spectrum of the parent compound has not been reported, but that of the 1,3-diphenyl derivative (74) is reported in the API index (76) and is shown in Fig. 8.34. It will be seen that fragmentation is not very marked, and that, like 2,5-diphenylthiophen (38), the molecule undergoes impact-induced loss of H₂S, which may perhaps be represented as involving the formation of 74a.

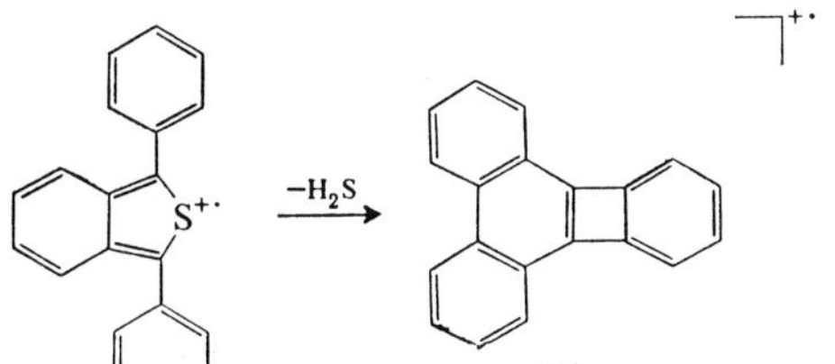

As with most polycyclic compounds, ionization of 74 produces a spectrum containing many doubly-charged ions, and the triply-charged species corresponding to $M^{3+\cdot}$ (m/e 93.3) can also be seen.

Some features of the mass spectrum of 3,3′-diphenyl-1,1′-bibenzo[c]thienyl (75) have been reported (77). Apparently the major fragmentation involves loss of both sulfur atoms, perhaps because of the formation of 9,9′-dianthryl (75a).

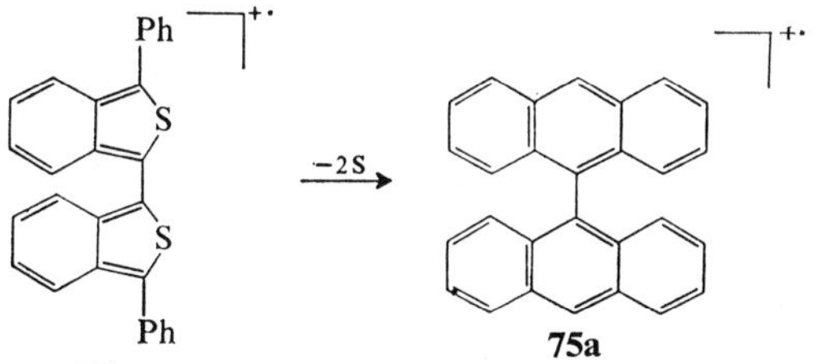

12.   Dibenzothiophens

Although the spectra of a number of compounds in this class have been reported, there are few detailed analyses of fragmentation.

Dibenzothiophen (76) has recently been the subject of a high-resolution mass-spectrometric study (78), and it will be seen that its spectrum (Fig. 8.35) is dominated by an intense molecular ion, as is expected for an aromatic

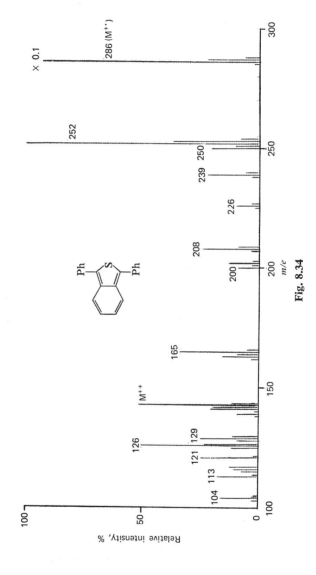

Fig. 8.34

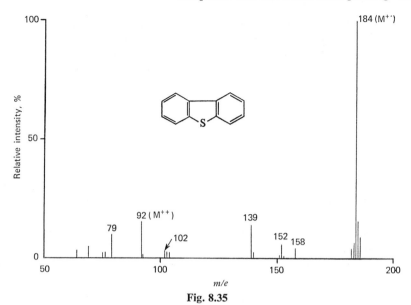

Fig. 8.35

molecule. The M-·CHS ion is, interestingly, produced by the three sequences M→(M-·H)→(M-·CHS), M→(M-CS) → (M-·CHS), and M → (M-·CHS). The decompositions shown have been suggested to account for these pathways.

3-Methyldibenzothiophen (77), as expected, shows an intense M-1 ion (79), and this has been rationalized in terms of formation of 77a. Further fragment ions are weak, the only significant one being at m/e 165 (9%), probably corresponding to loss of sulfur from the M-1 species.

It is interesting that 1-methyldibenzothiophen (**78**) is reported to give an identical fragmentation pattern to that of the di-ene-di-yne **79** (80).

$$CH_3—CH{=}CH—C{\equiv}C \quad C{\equiv}C—CH{=}CH_2$$

79

78

This is reminiscent of the observation that 1,3-hexadiene-5-yne and benzene have identical spectra (81). Although it was implied that rearrangement of **79** to **78** had occurred upon impact, it is possible that the reverse isomerization occurs, or even that the molecular ions of both **78** and **79** are completely noncyclic. In view of recent interest in the structures of molecular ions of aromatic compounds, this observation is worthy of further study.

The mass spectra of 1,2,3,4-tetrahydro-dibenzothiophen (**80**, R = H) (82) and its 7-methyl derivative (**80**, R = CH_3) (83) have been reported, and in each case the base peak of the spectrum corresponds to the product from a retro-Diels-Alder reaction involving loss of ethylene.

80

$$\xrightarrow{-C_2H_4}$$

13. Naphthothiophens

The mass spectrum of 3-phenylnaphtho[2,1-*b*]thiophen (**81**) has been reported (84), and is interesting in that it shows a very intense M-2 ion, presumably the cyclic species (**81b**) formed via **81a** as shown.

81, M+·, m/e 260

cyclization
−H·

81a, m/e 259
(73%)

−H··

81b, m/e 258

The mass spectra of a number of derivatives of 2*H*-naphtho[1,8-*bc*]thiophen (**82**, R = H) have been described (85). The spectra are dominated by loss of the substituent in the 2-position to give a species formulated as the naphtho[1,8-*bc*]thienylium cation **82a**.

82, M+·   82a, *m/e* 171

82c, *m/e* 127   82b, *m/e* 139

The species **82a** seems reasonably stable (salts of cations possessing the ring system of **82a** have been prepared) and shows only minor fragmentation by losses of S and CS to give species that have been formulated as **82b** and **82c**, respectively.

The mass spectra of a number of oxygenated derivatives possessing the naphtho[1,8-*bc*]thiophen skeleton have been described, and the original paper should be consulted for full details.

14.   Thienothiophens

Although thieno[2,3-*b*]thiophen (**83**) and thieno[3,2-*b*]thiophen (**84**) are well known, their mass spectra do not appear to have been reported. The [2,3-*c*] isomer (**85**) has the spectrum shown in Fig. 8.36 (86).

The spectrum is a fairly simple one. The relatively intense M-1 ion (22%) may result from ring-opening of the type already postulated for the isosteric benzo[*b*]thiophen, leading to the well-stabilized ion **85a**. The only other significant fragment ions result from losses of CS and ·CHS from the molecular ion.

83   84   85, M+·, *m/e* 140   85a, *m/e* 139

The mass spectra of the mono- and dibenzo derivatives of **84** have been recorded. Both show intense molecular ions, with fragmentation similar

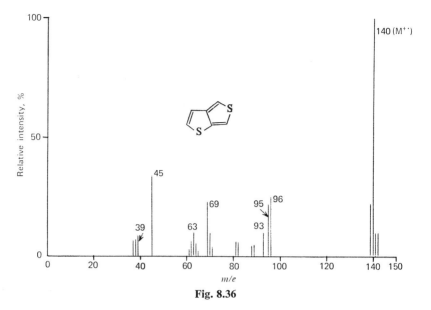

Fig. 8.36

to benzo[*b*]thiophen and dibenzothiophen, respectively. The original spectra should be consulted for more details (87, 88).

15.   *4,9-Dithiabicyclo[5,3,0]deca-1(10),2,7-triene*

This compound (**86**) has been obtained by the base-catalyzed rearrangement of 1,6-dithiacyclodeca-3,8-diyne (**87**) (89).

The tabulated mass spectrum of **86** was not discussed by the original authors, but a rationale for some of the observed features is suggested in the scheme.

16.   Sulfoxides and Sulfones of Thiophen Derivatives

Thiophen itself does not form a stable sulfone, but sulfones and sulfoxides derived from certain substituted and condensed thiophens are well known. Molecular rearrangements dominate the mass spectra of such compounds, and it is convenient to consider the spectra of the thiophen sulfoxides and sulfones separately from their parent thiahydrocarbons.

Simple dialkyl sulfones and sulfoxides show straightforward fragmentation without abundant skeletal rearrangement (90). Introduction of unsaturated or aryl groups appears to modify the fragmentation paths and induce skeletal rearrangements. Thus dibenzyl sulfoxide (88) shows relatively intense ions corresponding to $(M-SO)^{+\cdot}$ and $(M-H_2SO)^{+\cdot}$ (90, 91).

$$Ph—CH_2—SO—CH_2—Ph$$
88

89

Direct attachment of the aryl (or heteroaryl) group to the sulfur function further modifies the fragmentation. Thus 3,3'-dithienyl sulfone (89) has as base peak the ion $C_4H_3S_2O^+$ (m/e 131) corresponding to loss of the fragment $C_4H_3SO\cdot$ from the molecular ion. These findings have been rationalized in terms of an impact-induced rearrangement to the sulfinate ester (89a), followed by fragmentation.

Dibenzothiophen 5,5-dioxide (90) undergoes a similar rearrangement, which initiates losses of CO units (90).

Such a rearrangement is also observed in the mass spectrum of benzo[b]-thiophen 1,1-dioxide (72).

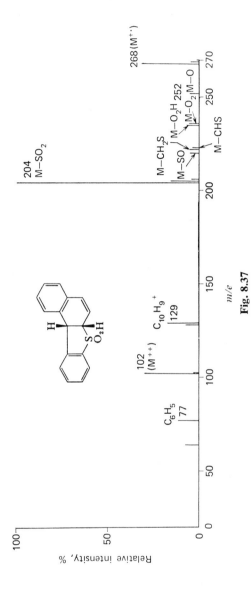

Fig. 8.37

A brief report has appeared dealing with the mass spectrum of 6a,11b-dihydronaphtho[2,1-b]benzo[d]thiophen 7,7-dioxide (91) obtained with a sample reservoir at 315°, at which temperature pyrolysis is probably signifi-cant. Ions corresponding to losses of SO_2, H_2, O, H_2O, H_2O_2, and O_2 were observed, and the scheme shown was put forward (92).

91, M⁺·

92

This result is interesting since thermolysis of 91 at 190° in the condensed phase gives 92 as the major product (93). The mass spectrum of 92 shows rearrangements of the type described for dibenzothiophen dioxide (90) (losses of CO). The spectrum of 91 measured with direct insertion into the source (MS9) at 120° gives a spectrum (Fig. 8.37) showing significant differences to that obtained at the higher temperature (94). In particular the ions corresponding to loss of H_2, H_2O, and H_2O_2 from the parent ion are absent. No ions explicable in terms of an —SO_2— → —SO—O— rearrangement are present, but the M-·CHS and M-CH₂S species certainly require skeletal rearrangements for their genesis.

The mass spectrum of dibenzothiophen sulfoxide (93) is very different from that of the corresponding sulfone (78) and is dominated by loss of oxygen to give the dibenzothiophen radical cation (m/e 184) which decomposes in the way already discussed.

A rearrangement to the cyclic sulfenate (93a) is, however, suggested as initiating a minor decomposition pathway to account for the presence of ions corresponding to M-CO (m/e 172), M-·CHO (m/e 171), and M-S (m/e 168). The rationale shown has been suggested.

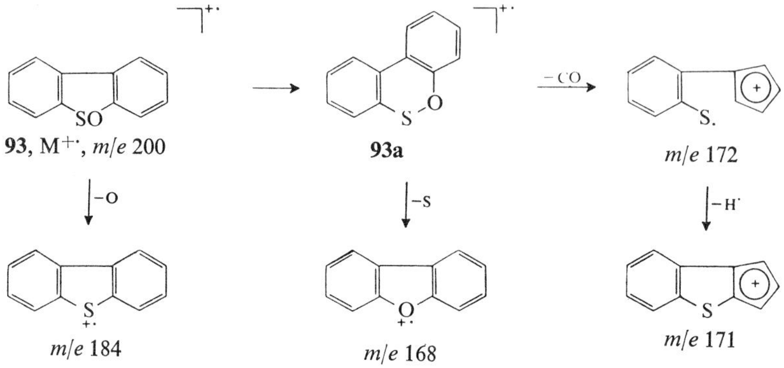

93, M$^{+\cdot}$, m/e 200   **93a** m/e 172

m/e 184 m/e 168 m/e 171

II. COMPOUNDS CONTAINING MORE THAN ONE SULFUR ATOM PER RING

A. 1,2-Disulfides

The mass spectrum of 1,2-dithiacyclohexane (**94**) has been reported, and the major process is loss of the elements of HS$_2$ to give the ion C$_4$H$_7$$^+$ (**95**). No significant M-C$_2$H$_4$ ion occurs, although in 1,2-dithiacyclopentane (**95**) loss of C$_2$H$_4$ is apparently a prominent process (**96**). In corresponding carboxy-

C$_4$H$_7$$^+$ **94, M$^{+\cdot}$,**   **95, M$^{+\cdot}$,** m/e 78
m/e 55 m/e 120 m/e 106

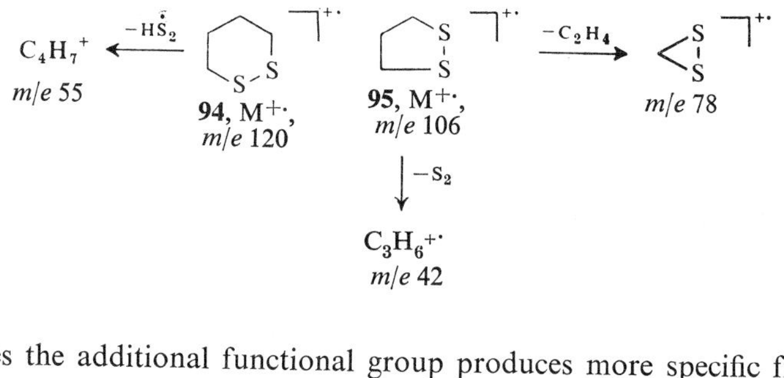

C$_3$H$_6$$^{+\cdot}$
m/e 42

disulfides the additional functional group produces more specific fragmentation (**95**). The spectra of 3-carboxy-1,2-dithiacyclopentane (**96**) and 3- and 4-carboxy-1,2-dithiacyclohexanes have been reported and analyzed. The 3-substituted compounds show relatively strong M-·CO$_2$H ions as a result of the formation of well-stabilized sulfonium ions (e.g., **96a**). The scheme shown rationalizes the decomposition of 3-carboxy-1,2-dithiacyclohexane.

The 4-carboxy-compound (97), as expected, shows a much weaker M-·CO₂H ion, but its spectrum is qualitatively similar to that of 96.

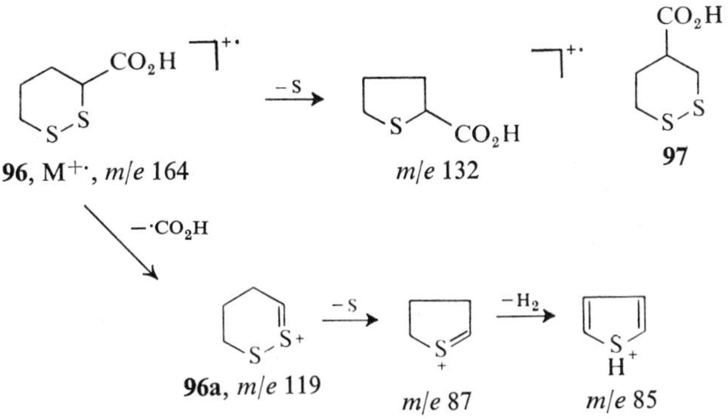

96, M⁺·, m/e 164   m/e 132   97

96a, m/e 119   m/e 87   m/e 85

Mass spectrometry has been of assistance in determining the structure of the bis-cyclic disulfide alkaloid gerrardine (98) (97). The alkaloid shows no parent ion, and the base peak is the α-cleavage ion at m/e 204 (98a). Insufficient data are presented to determine whether the complementary ion

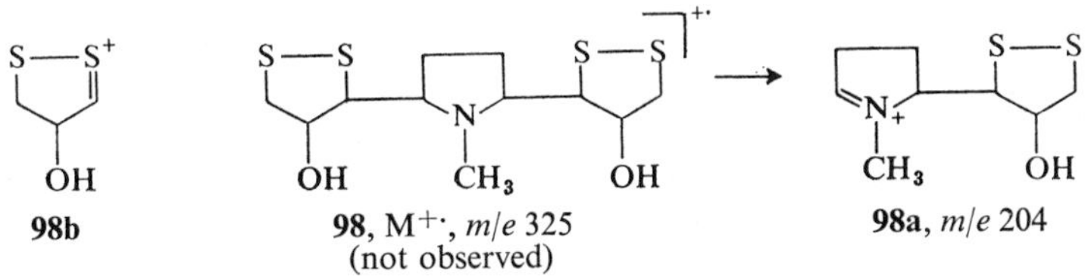

98b   98, M⁺·, m/e 325   98a, m/e 204
(not observed)

(98b) is formed in appreciable amount. Its presence would indicate charge localization on sulfur as well as on nitrogen in the molecular ion.

B.   1,3-Disulfides (Thioacetals and Thioketals)

The mass spectra of a number of ethylene-thioketals have been examined to investigate the ability of this grouping to direct specific fragmentations in complex natural products. In general, molecular ions are stronger and significant fragment ions are weaker than those observed with the corresponding ethylene-ketals. These features are exemplified in the fragmentation of 5α-androstan-3,11-dione 3-ethylene-thioketal (99) (98). The proposed hydrogen shifts have been confirmed by examination of the spectra of the 6,6-d₂- and 2,2,4,4,9,12-d₆-derivatives of 99.

The mass spectra of a number of ethylene-thioacetals derived from fully acetylated carbohydrates have been described, and unlike the corresponding diethyl-thioacetals show no molecular ion peak (99). In fact, most fragment

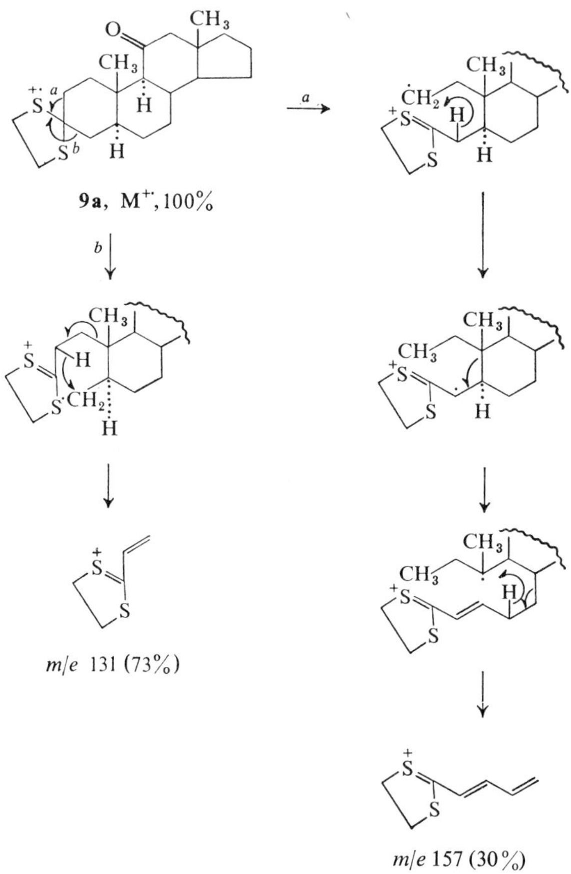

9a, M$^+$, 100%

m/e 131 (73%)

m/e 157 (30%)

ion peaks are of very low intensity as is seen with the spectrum of D-glucose ethylene-thioacetal pentaacetate (**100**). The fragmentation can be explained in terms of successive losses of acetic acid and/or ketene, and the scheme shown is proposed. The base peak of the spectrum in most cases is the well-delocalized species at *m/e* 105 which can be represented as **100a**, and the acetyl ion (*m/e* 43) also carries a considerable portion of the ion current.

Analysis of the impact-induced fragmentations of a number of bis-ethylene- and -trimethylene-thioacetals of malondialdehyde have been reported (100).

$$
\begin{array}{c}
\text{S} \diagdown \diagup \text{S} \\
\text{C} \\
\text{H} \\
\text{H—C—OAc} \\
\text{AcO—C—H} \\
\text{H—C—OAc} \\
\text{H—C—OAc} \\
\text{CH}_2\text{OAc}
\end{array}^{+\cdot}
\quad \xrightarrow{-\text{AcOH}} \quad
\begin{array}{c}
\text{S} \diagdown \diagup \text{S} \\
\text{C} \\
\text{OH} \\
\text{AcO—C—H} \\
\text{H—C—OAc} \\
\text{H—C—OAc} \\
\text{CH}_2\text{OAc}
\end{array}^{+\cdot}
\quad \xrightarrow{-\text{AcOH}}
$$

100, M⁺· (not observed),
m/e 664

$$
\begin{array}{c}
\text{S} \diagdown \diagup \text{S} \\
\text{C} \\
\text{CH} \\
\text{AcO—C} \\
\text{CH} \\
\text{H—C—OAc} \\
\text{CH}_2\text{OAc}
\end{array}^{+\cdot}
\quad \xrightarrow{-\text{CH}_2\text{CO}} \quad
\begin{array}{c}
\text{S} \diagdown \diagup \text{S} \\
\text{C} \\
\text{CH} \\
\text{C=O} \\
\text{CH}_2 \\
\text{H—C—OAc} \\
\text{CH}_2\text{OAc}
\end{array}^{+\cdot}
\quad \xrightarrow[\text{CH}_2\text{CO}]{-\text{AcOH,}} \quad
\begin{array}{c}
\text{S} \diagdown \diagup \text{S} \\
\text{C} \\
\text{CH} \\
\text{C=O} \\
\text{CH} \\
\text{CH} \\
\text{CH}_2\text{OH}
\end{array}^{+\cdot}
$$

$$
\begin{array}{c}
\text{S} \diagdown \diagup \text{S}^+ \\
\end{array}
$$

100a, *m/e* 105

The fragmentation of the parent trimethylene compound (**101**) may be taken as typical; three routes can be discerned and are shown in the scheme. The same basic fragmentation pathways occur with the five-membered ring analog **102**.

A series of spiro compounds (**103**, **104**, and **105**) has been prepared from the interaction of 2-cyanoimino-1,3-dithiacyclopentane and -1,3-dithiacyclohexane with ethane- and propane-dithiols (**101**). The mass spectra of the three compounds show easily-recognizable parent ions, but in the absence of high-resolution and metastable-ion data, analysis of the fragmentation observed is not possible. A strong ion in the spectra of all three compounds occurs at *m/e* 76 and is presumably the species $CS_2^{+\cdot}$. Its formation clearly requires considerable bond cleavage. Related to these spiro compounds is the trimer of thiocyclohexanone (**106**) (102). The spectrum of the compound has not been fully analyzed, but the occurrence of a peak at *m/e* 114 ($C_6H_{10}S$) presumably indicates depolymerization to the parent thione. Simple fragmentations lead to the species at *m/e* 146 ($C_6H_{10}S_2$) and the dimeric ion at *m/e* 228 ($C_{12}H_{20}S_2$). The base peak of the spectrum is the hydrocarbon ion $C_6H_9^+$ (*m/e* 81).

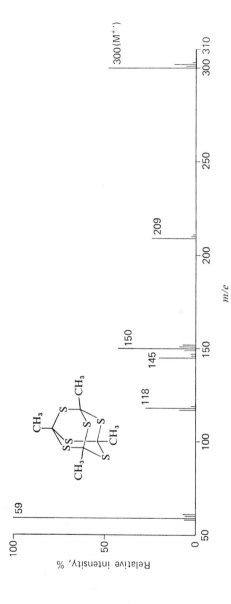

Fig. 8.38

(a)

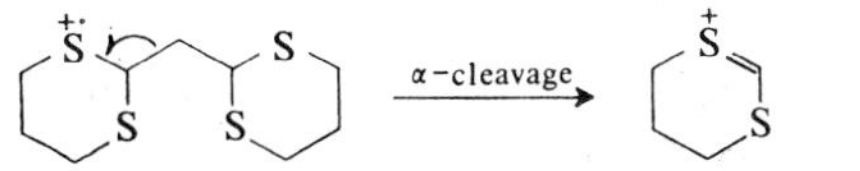

101, M$^{+\cdot}$, m/e 252 α–cleavage m/e 119 (80%)

(b)

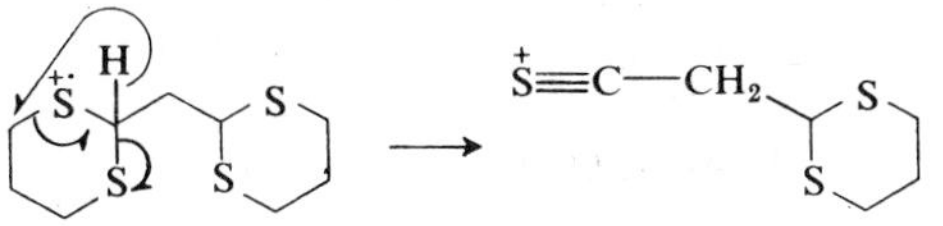

m/e 177 (15%)

(c)

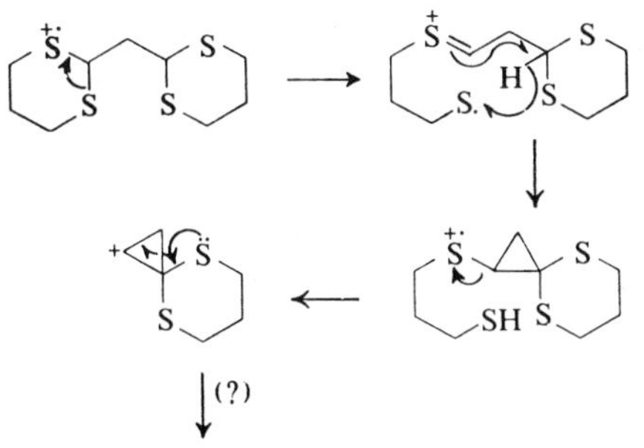

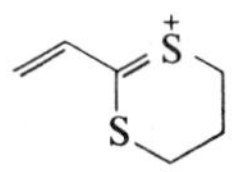

m/e 145 (100%) 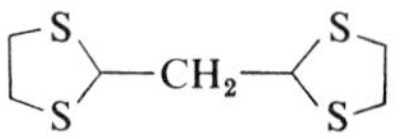   **102**

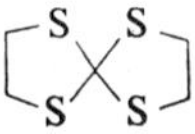

103 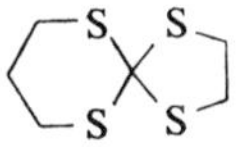   **104** 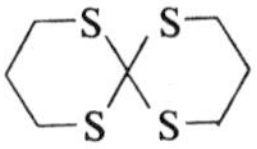   **105**

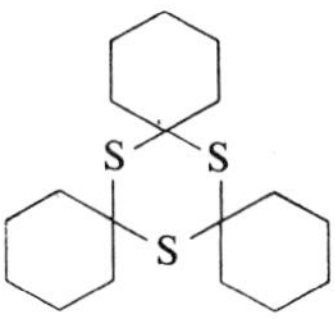

106

C. Polythia-adamantanes

The mass spectrum of the tetramethylhexathia-adamantane **107** has been reported and is shown in Fig. 8.38 (103). Although no metastable-ion information is available, a satisfactory rationale for the spectrum can be advanced. Of considerable interest is the presence of skeletal rearrangements which results from elimination of S$_2$ and S fragments.

It is probable that a key driving force in this fragmentation is the formation of the stable, well-delocalized ions containing the structural unit [S⋯C⋯S]$^+$.

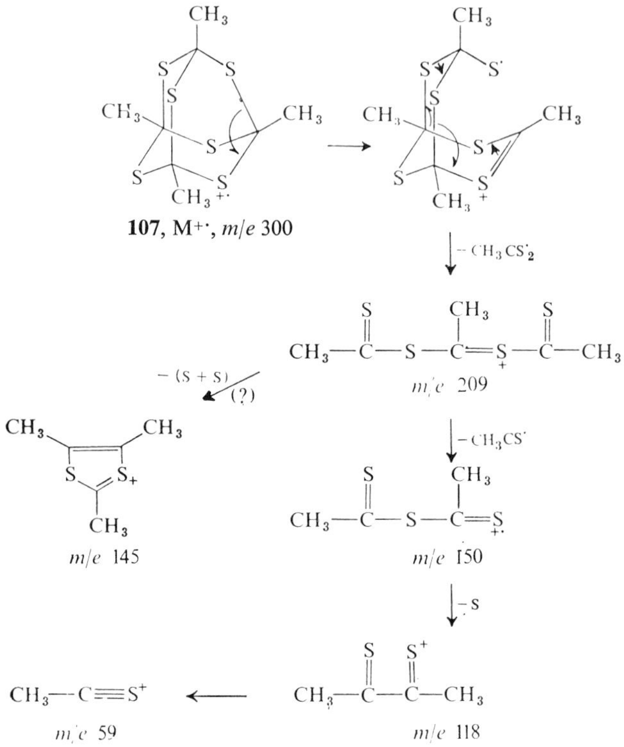

107, M$^{+\cdot}$, m/e 300

$-CH_3CS_2^\cdot$

$$CH_3-\overset{\overset{\displaystyle S}{\|}}{C}-S-\overset{\overset{\displaystyle CH_3}{|}}{C}=\overset{+}{S}-\overset{\overset{\displaystyle S}{\|}}{C}-CH_3$$

m/e 209

$-(S+S)$ (?)

CH_3, CH_3

m/e 145

$-CH_3CS^\cdot$

$$CH_3-\overset{\overset{\displaystyle S}{\|}}{C}-S-\overset{\overset{\displaystyle CH_3}{|}}{C}=\overset{+\cdot}{S}$$

m/e 150

$-S^\cdot$

$$CH_3-\overset{\overset{\displaystyle S}{\|}}{C}-\overset{\overset{\displaystyle S^+}{\|}}{C}-CH_3$$

m/e 118

$$CH_3-C\equiv S^+$$

m/e 59

A similar fragmentation occurs with the analogous tetra(chloromethyl) compound.

D.   Derivatives of Thianthren

The fragmentation of thianthren (**108**) resembles that of dibenzothiophen (**76**) (78). Thus it eliminates CS and then H·, and also ·CHS in a concerted process.

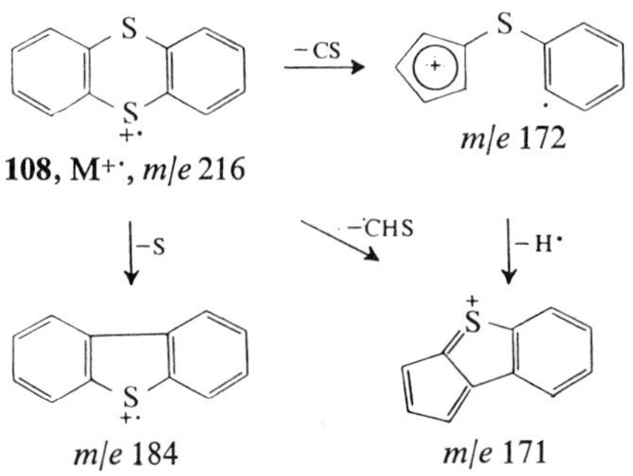

108, M$^{+\cdot}$, m/e 216

$-CS$

m/e 172

$-S$

$-CHS$

$-H\cdot$

m/e 184

m/e 171

The mass spectra of the mono- (**109**) and di- (**110**) sulfoxides of thianthren and of its monosulfone (**111**) have also been examined. With **109** the major fragmentation involves loss of SO to give dibenzothiophen, but with all three compounds there are also observed rearrangements analogous to those already described for dibenzothiophen sulfoxide and sulfone. It will be noticed in the scheme that mass spectrometry clearly distinguishes the isomeric disulfoxide (**110**) and sulfone (**111**).

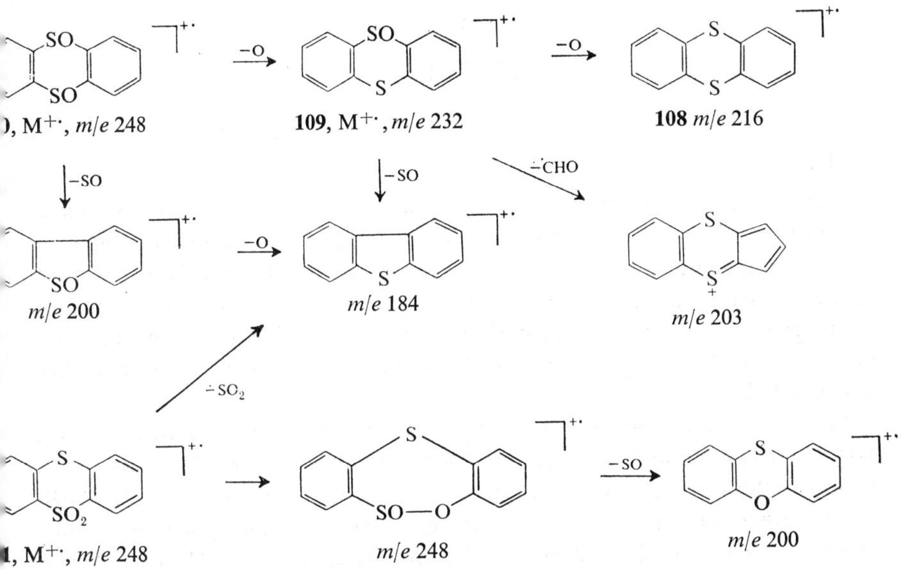

E. Large-Ring Polysulfur Heterocycles

The characteristic odor of the edible mushroom *Shiitake* has been shown to be due to the presence of the interesting pentasulfide lenthionine (**112**). Mass spectrometry was of very considerable value in the elucidation of the structure of this compound. The postulated formula is the only one which satisfactorily accommodates the series of ions observed corresponding to $(CH_2)_2S_n$ ($n = 3, 4, 5$) and $(CH_2)S_m$ ($m = 1, 2, 3, 4, 5$) as well as the species S_2, S_3, and $\cdot CHS$ (104).

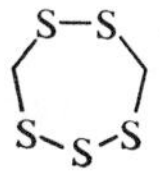

112

The mass spectrum of 1,6-dithiacyclodeca-3,8-diyne (113) has been briefly reported but not analyzed (89). The results may perhaps be accommodated by the fragmentation scheme shown.

m/e 167   113, M$^{+\cdot}$, m/e 168 m/e 153

m/e 135 m/e 122

Addenda

Two papers deal briefly with the fragmentations of derivatives of thiete 1,1-dioxide (Section I-B) (105, 106). A detailed comparative investigation of the fragmentations of tetrahydrothiophen, tetrahydrothiapyran, and hexamethylene sulfide has appeared (Section I-C, -E, and -F) (107), and the fragmentations of the oxide and dioxide of tetrahydrothiapyran and some ketonic derivatives together with 4-thiapyran 4,4-dioxide have been further discussed (108).

More work has appeared on hydrogen randomization in thiophen (Section I-H-1) (109), and analysis of the fragmentation of 1-(2-thienyl) hexane-1-^{13}C supports the ring expanded formulation for m/e 97. (Section 1-H-2) (110). Other areas of thiophen chemistry examined include phenylthiophens (Section 1-H-3) (111), 1-(2-thienyl)alkylalkanones (Section (1-H-6a) (112), dibenzothiophen and naphthothiophen (Section 1-H-12 and 13) (113), and the structure of the M-SO$_2$ ion in the spectrum of tetraphenylthiophen 1,1-dioxide (Section 1-H-16) (114).

Several further papers have appeared dealing with 1,3-dithiolanes and 1,3-dithianes (Section II-B) (115–117), and the spectrum of 1,4-dithiane has been discussed (118).

REFERENCES

1. E. J. Gallegos and R. W. Kiser, *J. Phys. Chem.*, **65**, 1177 (1961).
2. B. G. Hobrock and R. W. Kiser, *J. Phys. Chem.*, **66**, 1551 (1962).
3. K. S. Sidhu, E. M. Lown, O. P. Strausz, and H. E. Gunning, *J. Amer. Chem. Soc.*, **88**, 254 (1966).
4. E. J. Gallegos and R. W. Kiser, *J. Phys. Chem.*, **66**, 136 (1962).
5. American Petroleum Institute Research Project 44, Catalog of Mass Spectral Data. Spectrum 514.
6. Ref. 5. Spectrum 1412.

7. W. J. Middleton, *J. Org. Chem.*, **30**, 1395 (1965).
8. D. C. Dittmer, K. Takahashi, and F. A. Davis, *Tetrahedron Lett.*, **1967**, 4061.
9. R. W. Hoffmann and W. Sieber, *Justus Liebigs Ann. Chem.*, **703**, 96 (1967).
10. Ref. 5. Spectrum 516.
11. A. M. Duffield, H. Budzikiewicz, and C. Djerassi, *J. Amer. Chem. Soc.*, **87**, 2920 (1965).
12. Ref. 5. Spectrum 571.
13. Ref. 5. Spectrum 570.
14. Ref. 5. Spectrum 572.
15. Ref. 5. Spectrum 573.
16. K. I. Zimina, A. A. Polyakova. R. A. Khmel'nitskii, and R. D. Obolentsev, *Zh. Obshch. Khim.*, **30**, 1264 (1960).
17. M. Pailer, W. Oesterreicher, and E. Simonitsch, *Monatsh. Chem.*, **96**, 1391 (1965).
18. Ref. 5. Spectrum 1048.
19. Ref. 5. Spectrum 1049.
20. D. S. Weinberg, C. Stafford, and M. W. Scoggins, *Tetrahedron*, **24**, 5409 (1968).
21. Ref. 5. Spectrum 916.
22. Q. N. Porter and G. H. Upstill, Unpublished observations.
23. Ref. 5. Spectrum 566.
24. Ref. 5. Spectrum 567.
25. Ref. 5. Spectrum 568.
26. Ref. 5. Spectrum 1050.
27. Ref. 5. Spectrum 1051.
28. Ref. 5. Spectrum 940.
29. Ref. 5. Spectrum 1227.
30. J. M. Cox and L. N. Owen, *J. Chem. Soc.*, *C*, **1967**, 1130.
31. J. Bonham, E. McLeister, and P. Beak, *J. Org. Chem.*, **32**, 639 (1967).
32. E. K. Fields and S. Meyerson, *J. Org. Chem.*, **30**, 937 (1965).
33. H. Seidl and K. Biemann, *J. Heterocycl. Chem.*, **4**, 209 (1967).
34. Ref. 5. Spectrum 158.
35. H. Budzikiewicz, C. Djerassi, and D. H. Williams, *Interpretation of Mass Spectra of Organic Compounds*, Holden-Day, San Francisco, 1964, pp. 231–235.
36. H. Budzikiewicz, C. Djerassi, and D. H. Williams, *Mass Spectrometry of Organic Compounds*, Holden-Day, San Francisco, 1967, pp. 625–630.
37. D. H. Williams, R. G. Cooks, J. Ronayne, and S. W. Tam, *Tetrahedron Lett.*, **1968**, 1777.
38. H. Wynberg, R. M. Kellogg, H. van Driel, and G. E. Berkhius, *J. Amer. Chem. Soc.*, **89**, 3501 (1967).
39. V. Hanuš and V. Čermák, *Collect. Czech. Chem. Commun.*, **24**, 1602 (1959).
40. R. Pettit, *Tetrahedron Lett.*, **No. 23**, 11 (1960).
41. J. Koutecký, *Collect. Czech. Chem. Commun.*, **24**, 1608 (1959).
42. J. H. Bowie, R. G. Cooks, S.-O. Lawesson, and C. Nolde, *J. Chem. Soc.*, *B.*, **1967**, 616.
43. E. C. Kooyman and J. B. H. Kroon, *Rec. Trav. Chim. Pays-Bas*, **82**, 464 (1963).
44. Ref. 5. Spectrum 506.
45. I. W. Kinney and G. L. Cook, *Anal. Chem.*, **24**, 1391 (1952).
46. Ref. 5. Spectrum 1408.
47. Ref. 5. Spectrum 1537.
48. Ref. 5. Spectrum 1538.
49. Ref. 5. Spectrum 1539.

50. Ref. 5. Spectrum 1540.
51. Ref. 5. Spectrum 1541.
52. R. E. Atkinson, R. F. Curtis, and G. T. Phillips, *J. Chem. Soc., C.*, **1966**, 1101.
53. F. Bohlmann and P. Herbst, *Chem. Ber.*, **95**, 2945 (1962).
54. G. M. Badger, J. H. Bowie, J. A. Elix, G. E. Lewis, and U. P. Singh, *Aust. J. Chem.*, **20**, 2669 (1967).
55. T. Nishiwaki, *Tetrahedron*, **23**, 2979 (1967).
56. J. H. Bowie, R. G. Cooks, S.-O. Lawesson, P. Jakobsen, and G. Schroll, *Chem. Commun.*, **1966**, 539.
57. V. H. Dibeler, R. M. Reese, and F. L. Mohler, *J. Chem. Phys.*, **26**, 304 (1957).
58. S. Meyerson and P. N. Meyerson, *J. Amer. Chem. Soc.*, **79**, 1058 (1957).
59. R. Grigg, H. J. Jakobsen, S.-O. Lawesson, M. V. Sargent, G. Schroll, and D. H. Williams, *J. Chem. Soc., B.*, **1966**, 331.
60. F. W. McLafferty and R. S. Gohlke, *Anal. Chem.*, **31**, 2076 (1959).
61. S. Gronowitz and R. A. Hoffmann, *Ark. Kemi*, **15**, 499 (1960).
62. A.-B. Hörnfeldt and S. Gronowitz, *Ark. Kemi*, **21**, 239 (1963).
63. H. J. Jakobsen, E. H. Larsen, and S.-O. Lawesson, *Tetrahedron*, **19**, 1867 (1963).
64. Ref. 5. Spectrum 162.
65. Q. N. Porter, Unpublished results.
66. H. J. Jakobsen, G. Schroll, S.-O. Lawesson, D. A. Lightner, A. M. Duffield, and C. Djerassi, *Ark. Kemi*, **26**, 199 (1967).
67. S. H. H. Chaston, S. E. Livingstone, T. N. Lockyer, V. A. Pickles, and J. S. Shannon, *Aust. J. Chem.*, **18**, 673 (1965).
68. J. Ronayne, D. H. Williams, and J. H. Bowie, *J. Amer. Chem. Soc.*, **88**, 4980 (1966).
69. J. Baldas, K. A. Maurer, Q. N. Porter, and C. C. R. Ramsay, Unpublished observations.
70. H. Budzikiewicz and Z. Pelah, *Monatsh. Chem.*, **96**, 1739 (1965).
71. Ref. 5. Spectrum 917.
72. Q. N. Porter, *Aust. J. Chem.*, **20**, 103 (1967).
73. R. G. Cooks, I. Howe, S. W. Tam, and D. H. Williams, *J. Amer. Chem. Soc.*, **90**, 4064 (1968).
74. Ref. 5. Spectrum 1857.
75. Q. N. Porter and K. A. Maurer, Unpublished observations.
76. Ref. 5. Spectrum 1501.
77. F. G. Bordwell and T. W. Cutshall, *J. Org. Chem.*, **29**, 2019 (1964).
78. J. Heiss, K.-P. Zeller, and B. Zeeh, *Tetrahedron*, **24**, 3255 (1968).
79. Ref. 5. Spectrum 1306.
80. R. E. Atkinson and R. F. Curtis, *Tetrahedron Lett.*, **1965**, 297.
81. J. Momigny, L. Brakier, and L. d'Or, *Bull. Class. Sci. Acad. Roy. Belg.*, **48**, 1002 (1962).
82. Ref. 5. Spectrum 1375.
83. Ref. 5. Spectrum 1307.
84. Ref. 5. Spectrum 1378.
85. D. G. Hawthorne and Q. N. Porter, *Aust. J. Chem.*, **21**, 171 (1968).
86. F. C. James, Personal communication.
87. Ref. 5. Spectrum 1496.
88. Ref. 5. Spectrum 1308.
89. G. Eglinton, I. A. Lardy, R. A. Raphael, and G. A. Sim, *J. Chem. Soc.*, **1964**, 1154.
90. J. H. Bowie, D. H. Williams, S.-O. Lawesson. J. Ø. Madsen, C. Nolde, and G. Schroll, *Tetrahedron*, **22**, 3515 (1966).

91. J. Ø. Madsen, C. Nolde, S.-O. Lawesson, G. Schroll, J. H. Bowie, and D. H. Williams, *Tetrahedron Lett.*, **1965,** 4377.
92. W. F. Taylor, J. M. Kelliher, and T. J. Wallace, *Chem. Ind.* (London), **1968,** 651.
93. W. Davies, B. C. Ennis and Q. N. Porter, *Aust. J. Chem.*, **21,** 1571 (1968).
94. Q. N. Porter and B. C. Ennis, Unpublished observations.
95. J. H. Bowie, S.-O. Lawesson, J. Ø. Madsen, C. Nolde, G. Schroll, and D. H. Williams, *J. Chem. Soc., B*, **1966,** 946.
96. T. J. Wallace, *J. Amer. Chem. Soc.*, **86,** 2018 (1964).
97. W. G. Wright and F. L. Warren, *J. Chem. Soc., C*, **1967,** 284.
98. G. v. Mutzenbecher, Z. Pelah, D. H. Williams, H. Budzikiewicz, and C. Djerassi, *Steroids*, **2,** 475 (1963).
99. D. C. De Jongh, *J. Amer. Chem. Soc.*, **86,** 4027 (1964).
100. R. H. Shapiro, T. E. McEntee, and D. L. Coffen, *Tetrahedron*, **24,** 2809 (1968).
101. J. J. D'Amico and R. H. Campbell, *J. Org. Chem.*, **32,** 2567 (1967).
102. C. Djerassi and B. Tursch, *J. Org. Chem.*, **27,** 1041 (1962).
103. K. Olsson, H. Baeckström, and R. Engwall, *Ark. Kemi*, **26,** 219 (1967).
104. K. Morita and S. Kobayashi, *Tetrahedron Lett.*, **1966,** 573.
105. D. C. Dittmer and J. M. Balquist, *J. Org. Chem.*, **33,** 1364 (1968).
106. L. A. Paquette, M. Rosen and J. Stucki, *J. Org. Chem.*, **33,** 3020 (1968).
107. R. Smakman and Th. J. De Boer in *Advances in Mass Spectrometry* Vol. 4, Institute of Petroleum, London, 1968, p. 357.
108. A. A. Kutz and S. J. Weininger, *J. Org. Chem.*, **33,** 4070 (1968).
109. S. Meyerson and E. K. Fields, *Org. Mass Spec.*, **2,** 241 (1969).
110. N. G. Foster and R. W. Higgins, *Org. Mass Spec.*, **1,** 191 (1968).
111. S. Meyerson and E. K. Fields, *Org. Mass Spec.*, **1,** 263 (1968).
112. N. G. Foster and R. W. Higgins, *Org. Mass Spec.*, **2,** 1005 (1969).
113. S. Meyerson and E. K. Fields, *J. Org. Chem.*, **33,** 847 (1968).
114. M. M. Bursey, T. A. Elwood and P. F. Rogerson, *Tetrahedron*, **25,** 605 (1969).
115. R. S. Shapiro, T. E. McEntee and D. L. Coffen in *Advances in Mass Spectrometry* Vol. 4, Institute of Petroleum, London, 1968, p. 231.
116. D. L. Coffen, K. C. Bank and P. E. Garrett, *J. Org. Chem.*, **34,** 605 (1969).
117. J. H. Bowie and P. Y. White, *Org. Mass Spec.*, **2,** 611 (1969).
118. G. Condé-Caprace and J. E. Collin, *Org. Mass Spec.*, **2,** 1277 (1969).

9 COMPOUNDS WITH OXYGEN AND SULFUR ATOMS IN THE SAME RING

The number of homoannular O,S heterocycles examined mass-spectro-metrically is relatively small, and in only a few cases has a full high-resolution study been carried out. Most compounds described are cyclic esters—sulfinates, sulfonates, sulfites, and sulfates.

I. CYCLIC SULFINATE ESTERS

The mass spectrum of 1,2-benzoxathian 2-oxide (**1**) has been described in part (1). The base peak of the spectrum results from elimination of SO from the molecular ion to give a species best represented as the 2,3-dihydro-benzofuran radical cation. It is interesting to note that the impact-induced decomposition of 2,3-dihydrobenzo[b]thiophen 1,1-dioxide (**2**) has been rationalized in terms of initial isomerization to **1** (2), and it would be worth-while to carry out a detailed comparison of the mass spectra of this dioxide and **1**.

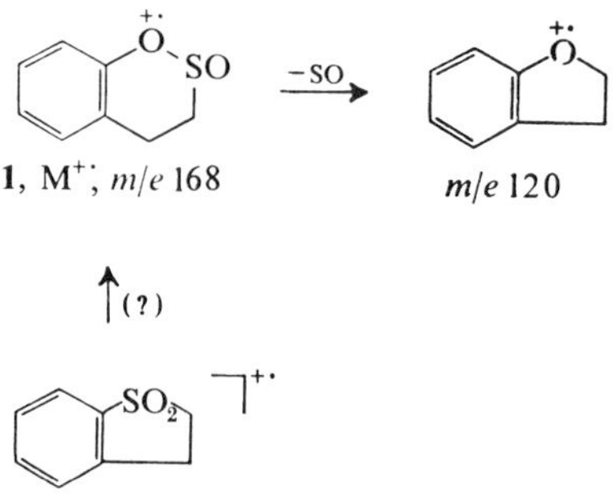

1, M⁺; *m/e* 168   *m/e* 120

2, M⁺·, *m/e* 168

II. CYCLIC SULFONATE ESTERS

Several cyclic sulfonates have been subjected to electron impact, but no detailed studies of their fragmentation have been carried out so only very limited conclusions about fragmentation can be drawn. Extrusion of sulfur dioxide is often the major decomposition and has been commented upon in the mass spectra of 6- and 8-chloro-1,2-benzoxathian 2,2-dioxides (3) (1).

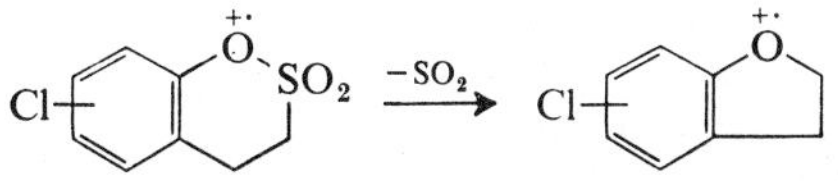

3, M+·, m/e 218   m/e 154

The production of the stable cyclic allylic radical **4a** appears to modify the fragmentation of camphene sultone (**4**) (3, 4), and the fragmentation may be represented as shown.

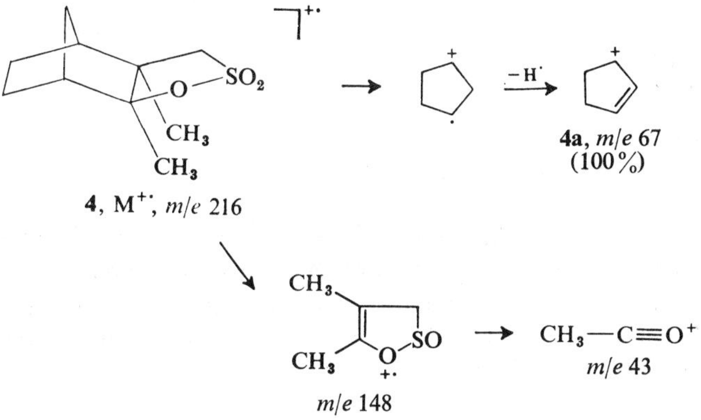

4, M⁺·, m/e 216

4a, m/e 67 (100%)

CH₃—C≡O⁺
m/e 43

m/e 148

o-Sulphobenzoic anhydride (**5**) is related to these cyclic sulfonates and its decomposition has been shown to involve three competing pathways (5).

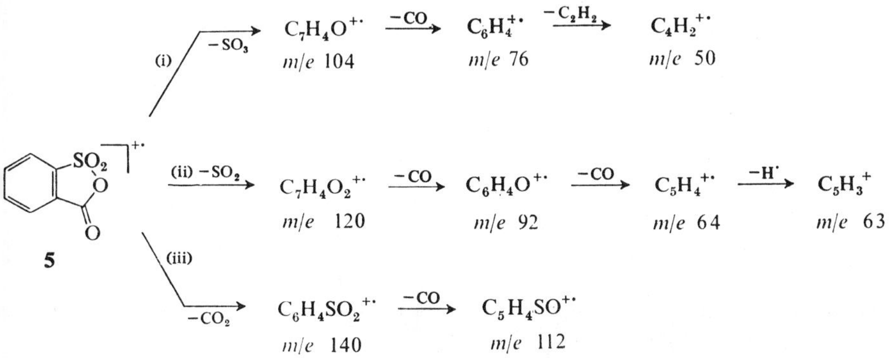

(i) $\xrightarrow{-SO_3}$ $C_7H_4O^{+\cdot}$ m/e 104 $\xrightarrow{-CO}$ $C_6H_4^{+\cdot}$ m/e 76 $\xrightarrow{-C_2H_2}$ $C_4H_2^{+\cdot}$ m/e 50

(ii) $-SO_2$ $\longrightarrow$ $C_7H_4O_2^{+\cdot}$ m/e 120 $\xrightarrow{-CO}$ $C_6H_4O^{+\cdot}$ m/e 92 $\xrightarrow{-CO}$ $C_5H_4^{+\cdot}$ m/e 64 $\xrightarrow{-H\cdot}$ $C_5H_3^{+}$ m/e 63

(iii) $\xrightarrow{-CO_2}$ $C_6H_4SO_2^{+\cdot}$ m/e 140 $\xrightarrow{-CO}$ $C_5H_4SO^{+\cdot}$ m/e 112

5

At 70 eV the relative contributions of the pathways are (i) 68.7%, (ii) 30.9%, and (iii) 0.40%. The position is different at 9.5 eV as at this low ionizing-potential loss of SO_2 predominates, and it was predicted that loss of SO_2 would also be the predominant fragmentation mode in a pyrolytic reaction. This was in fact found to be so, but the other low molecular-weight product at 690° was carbon dioxide, not carbon monoxide as observed after electron impact. The rationales for the two processes may be put forward as shown here.

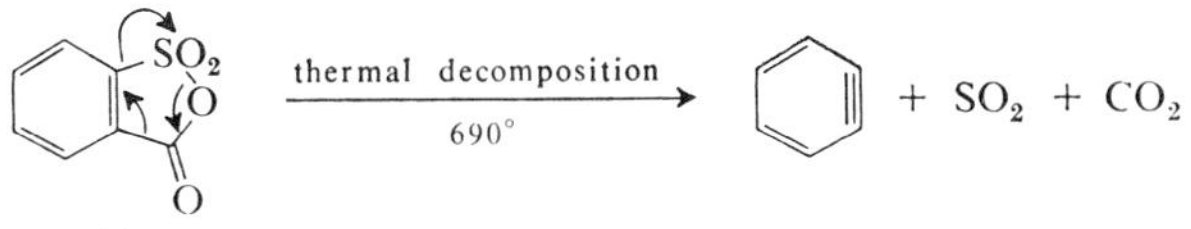

5

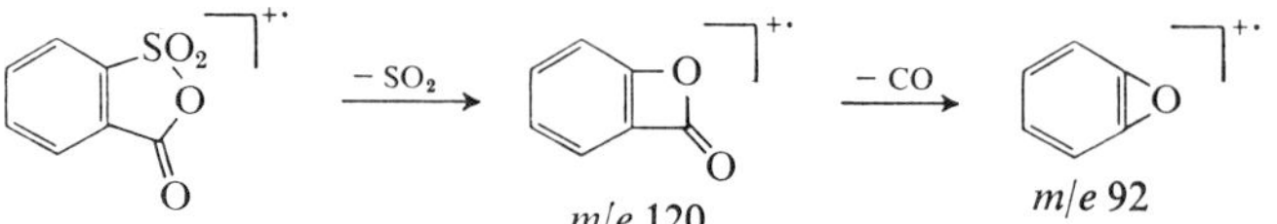

5, M+·
(9.5 eV)

III. CYCLIC SULFITE ESTERS

The mass spectrum of o-phenylene sulfite (**6**) has been recorded and analyzed and the impact-induced fragmentation has been compared with thermal breakdown of the same compound (**6**). At 70 eV loss of sulfur monoxide is followed by successive losses of two molecules of carbon monoxide, and a possible rationale is shown.

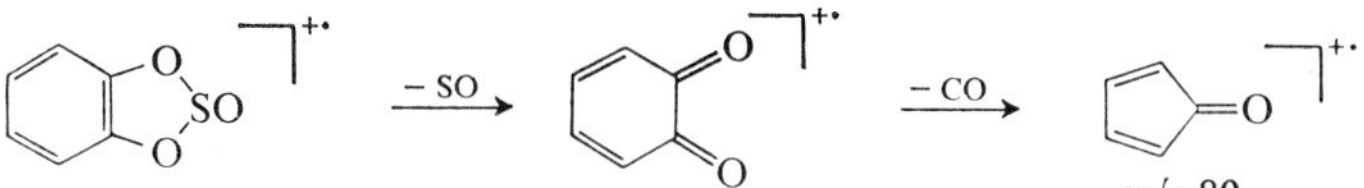

6, M+·,
m/e 156 (100%)

m/e 108 (40%)

m/e 80
(98%)

7

$C_4H_4^{+·}$
m/e 52
(64%)

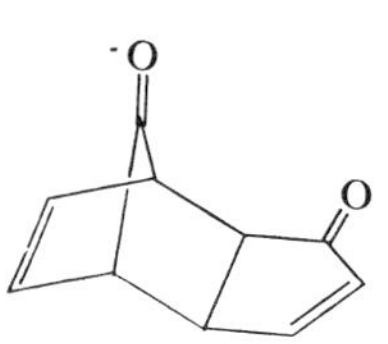

It is apparent that thermal breakdown of **6** closely parallels the impact-induced process since pyrolysis at 800° over a Nichrome coil produces

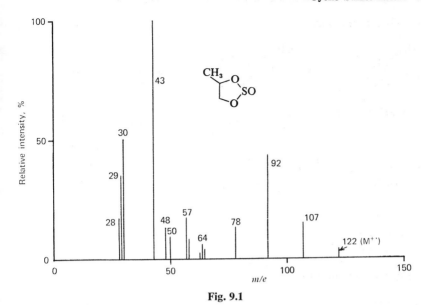

Fig. 9.1

1,8-diketo-4,7-methano-3a,4,7,7a-tetrahydroindene (**7**), the dimer of cyclo-pentadienone.

The cyclic sulfites of a series of 1,2-diols have been fully analyzed mass-spectrometrically, making use of deuterium-labeling and high-resolution techniques (7). Many of the features of their decomposition upon electron impact are exemplified in the spectrum of propane-1,2-diol sulfite (**8**) (Fig. 9.1). Two major decomposition pathways can be seen. They are initiated (i) by α-cleavage of the C—C bond in the ring and (ii) by cyclic elimination of SO_2 or $\cdot SO_2H$ with concurrent hydrogen migration.

With the cyclic sulfite of 2,3-dimethylbutane-2,3-diol (**9**) a hydrogen atom is not available for transfer in the cyclic elimination of SO_2, and methyl migration occurs to give ionized 3,3-dimethylbutan-2-one (pinacolone) (**9a**) which decomposes in its characteristic way (7). Differences in metastable peak characteristics render tetramethylethylene oxide (**9b**) an unlikely formal representation for the $M-SO_2$ peak, but in certain phenyl-substituted sulfite esters (e.g., **10**) the intermediacy of an ethylene oxide is more likely. The reader is referred to the original article for further details.

Spectra of *dl*- and *meso*-hydrobenzoin sulfites (**10**) were also reported by Pritchard and Funke (8). Their observations were rather different to those outlined above and may be explicable in terms of initial thermal decompositions of the compounds since they were apparently introduced via a heated inlet system.

(i) α-fission pathway

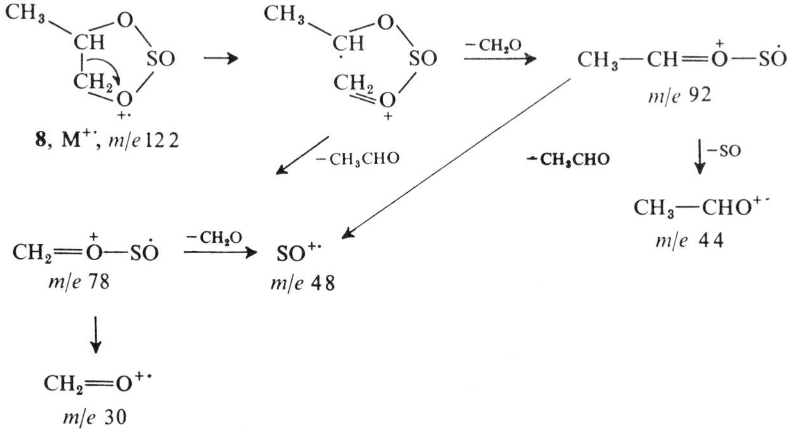

(ii) cyclic eliminations

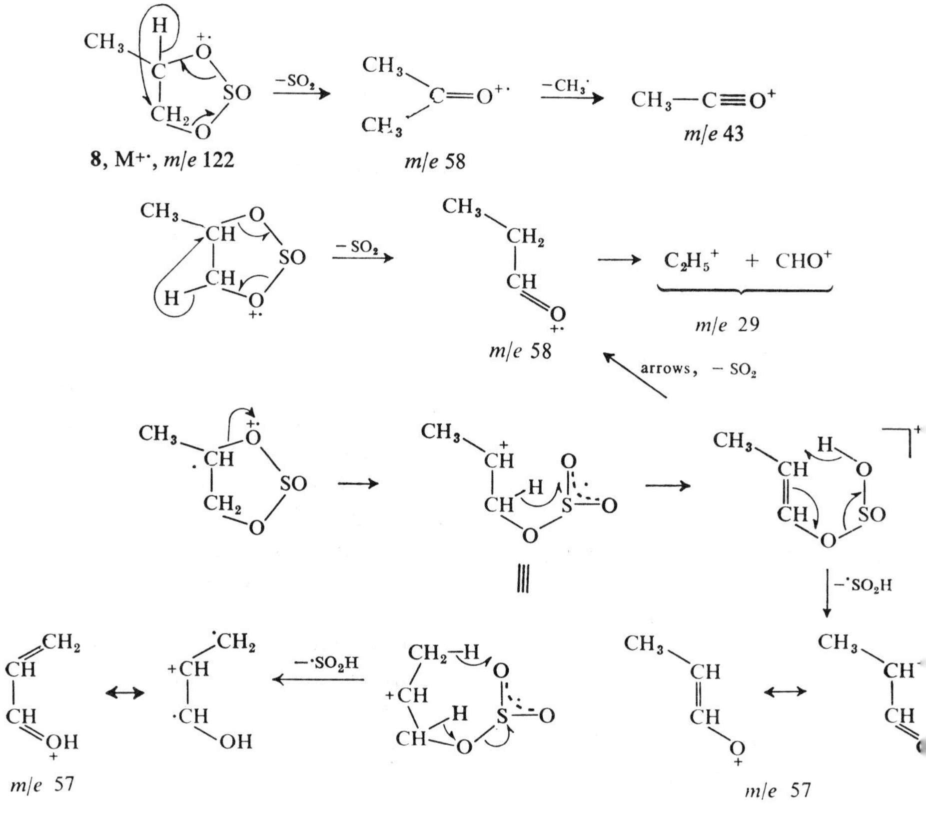

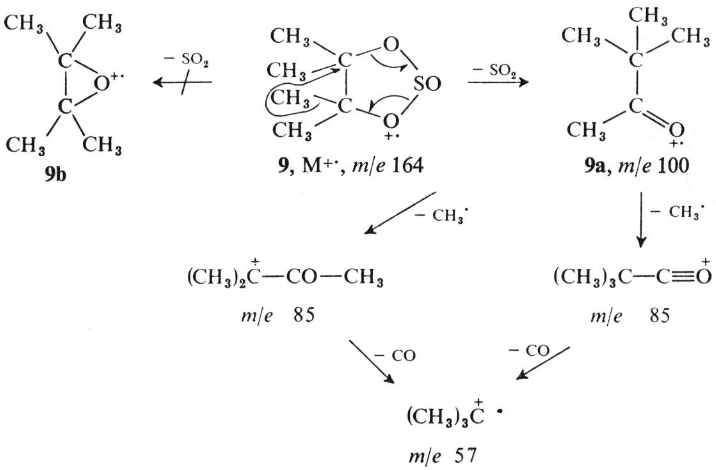

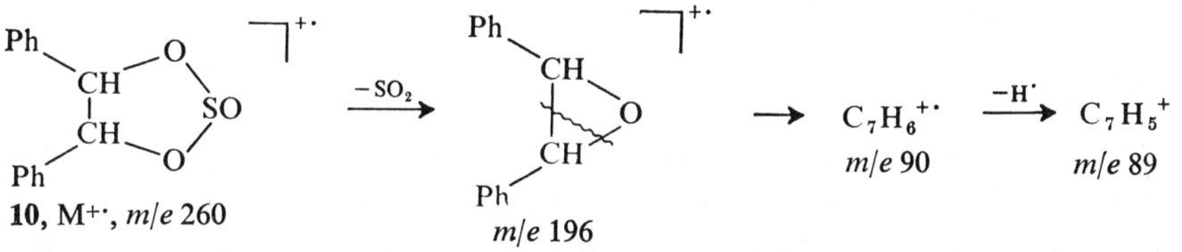

9b   9, M$^{+\cdot}$, m/e 164   9a, m/e 100

$(CH_3)_2\overset{+}{C}-CO-CH_3$

m/e 85

$(CH_3)_3C-C\equiv\overset{+}{O}$

m/e 85

$(CH_3)_3\overset{+}{C}\cdot$

m/e 57

10, M$^{+\cdot}$, m/e 260 m/e 196 $C_7H_6^{+\cdot}$ m/e 90 $C_7H_5^{+}$ m/e 89

IV. CYCLIC SULFATES

Very few investigations of the mass spectra of cyclic sulfates have been reported, and those that have appeared only contain partial spectra. The mass spectra of a number of bis-sulfates resulting from the interaction of polyhydroxyanthraquinones and oleum show successive losses of SO_2 and CO (9), and a rationale for the decomposition of bis(1,9-4,10)-sulfuryldioxy-anthraquinone (**11**) is shown.

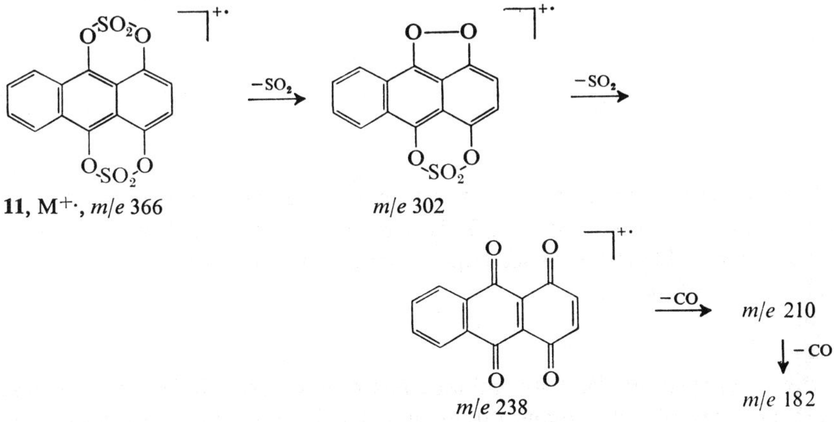

11, M$^{+\cdot}$, m/e 366 m/e 302

m/e 238 $-CO$ m/e 210 $-CO$ m/e 182

V. PHENOTHIOXIN

After electron impact this compound (12) decomposes by separate pathways involving each hetero atom, but as will be seen, the sulfur atom is lost directly from the molecular ion as S rather than CS or ·CHS, while the oxygen atom is lost as CO. A rationale for the spectrum is shown (10).

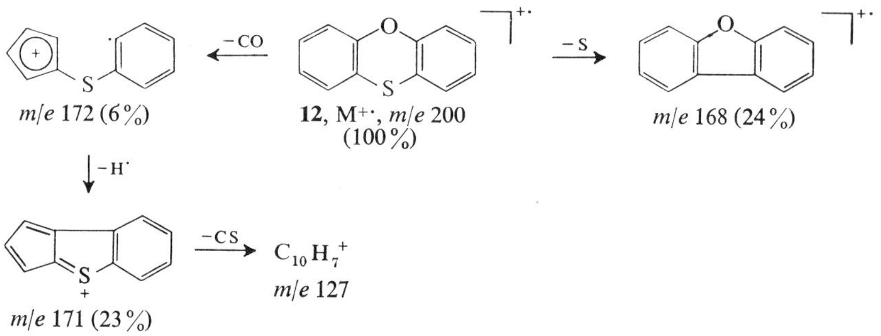

m/e 172 (6%)   12, M$^{+\cdot}$, m/e 200 m/e 168 (24%)
                            (100%)

m/e 171 (23%) $C_{10}H_7^+$
                         m/e 127

Phenothioxin sulfoxide (13) decomposes mainly by elimination of SO to give the dibenzofuran radical cation, the base peak of the spectrum, and by loss of O to give benzothioxin. These facile fragmentations apparently repress formation and fragmentation of the intermediate sulfenate ester (13a).

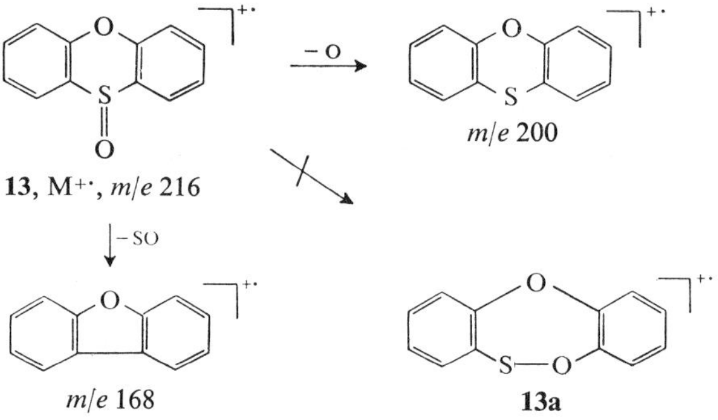

13, M$^{+\cdot}$, m/e 216 m/e 200

m/e 168   13a

Phenothioxin sulfone (14) has a markedly different spectrum from that of the sulfoxide (13). Loss of SO$_2$ is a much less important process than loss of SO from 13, and most decompositions are initiated from the rearranged molecular ion (14a).

Addenda

The mass spectrum of 1,4-oxathiane has been described. Fragmentations resemble those of 1,4-dithiane rather than 1,4-dioxane (11). The spectra of

14, M⁺·, m/e 232   **14a**   m/e 184

m/e 168   m/e 108   C₆H₄SO⁺· m/e 124   m/e 96

steroidal compounds in which ketone groups have been converted into ethylene hemithioketals have been discussed (12), as have simpler compounds of this kind (13).

REFERENCES

1. E. N. Givens and L. A. Hamilton, *J. Org. Chem.*, **32**, 2857 (1967).
2. Q. N. Porter, *Aust. J. Chem.*, **20**, 103 (1967).
3. J. Wolinsky, D. R. Dimmel, and T. W. Gibson, *J. Org. Chem.*, **32**, 2087 (1967).
4. D. R. Dimmel and J. Wolinsky, *J. Org. Chem.*, **32**, 2735 (1967).
5. S. Meyerson and E. K. Fields, *Chem. Commun.*, **1966**, 275.
6. D. C. De Jongh, R. Y. Van Fossen, and C. F. Bourgeois, *Tetrahedron Lett.*, **1967**, 271.
7. P. Brown and C. Djerassi, *Tetrahedron*, **24**, 2949 (1968).
8. J. G. Pritchard and P. T. Funke, *J. Heterocycl. Chem.*, **3**, 209 (1966).
9. J. Winkler and W. Jenny, *Helv. Chim. Acta*, **48**, 119 (1965).
10. J. Heiss, K.-P. Zeller, and B. Zeeh, *Tetrahedron*, **24**, 3255 (1968).
11. G. Condé-Caprace and J. E. Collin, *Org. Mass Spec.*, **2**, 1277 (1969).
12. C. Fenselau, L. Milewich and C. H. Robinson, *J. Org. Chem.*, **34**, 1374 (1969).
13. D. J. Paslo, *J. Het. Chem.*, **6**, 175 (1969).

10 COMPOUNDS WITH ONE NITROGEN ATOM IN A THREE-, FOUR-, OR FIVE-MEMBERED RING

I. ETHYLENEIMINE DERIVATIVES

The mass spectra of the parent heterocycle (**1**) and its 1- and 2-methyl derivatives are shown in Figs. 10.1, 10.2, and 10.3 (1–4). The great intensity of the M-CH$_3\cdot$ ion in **2** compared with that in **3** shows that this series, unlike simple amines has a marked tendency to undergo C—N cleavage. Deuterium-labeling experiments with **1** show that the M-1 ion results largely from elimination of the imino proton (5) in agreement with the fragmentation of the methyl homologs. The results are best explained in terms of ring-opening to form the well-stabilized ion **1a** and its homolog **3a**. High-resolution measurements (5) show that the m/e 28 species is CH$_2$N$^+$ rather than C$_2$H$_4^{+\cdot}$ as suggested earlier (1).

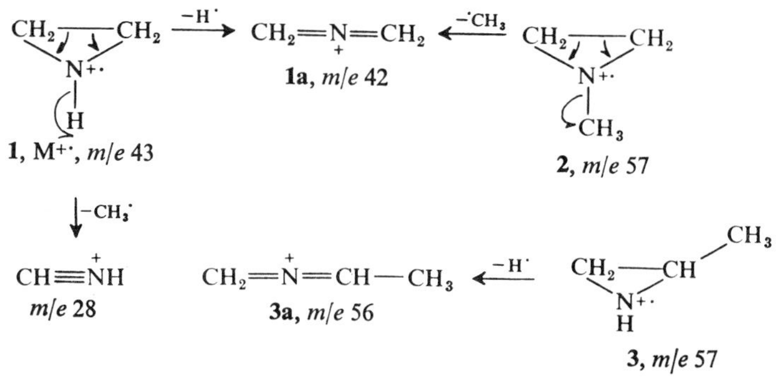

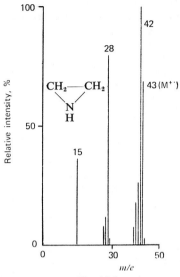

Fig. 10.1

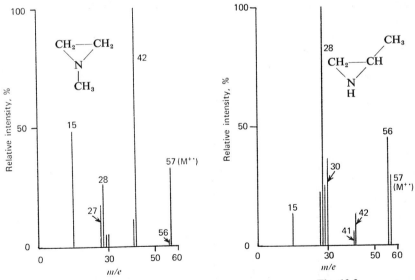

Fig. 10.2

Fig. 10.3

The mass spectra of a number of N,N'-methylene-bisethyleneimines have been described (6). The parent compound (4) shows predominant fragmentation resulting from $N—CH_2$ fission. Similar fragmentation is observed with the phenyl derivative [5 → 5a].

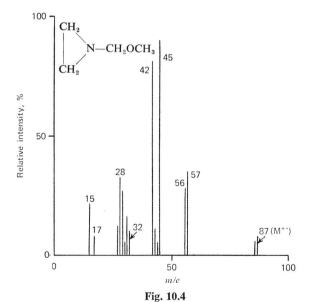

Fragment ion intensities in the mass spectrum of N-methoxymethylethyleneimine (6) (Fig. 10.4) are interesting (7). It will be noticed that charge retention on the oxygen fragment $CH_3—O^+=CH_2$ is much more marked than on the corresponding nitrogenous fragment $\rangle N^+=CH_2$, and again the C—N cleavage fragment is an intense ion in the spectrum.

Fig. 10.4

Several fluorinated ethyleneimines obtained by thermal decomposition of difluorodiazirine and difluoroaminofluorodiazirine have been examined mass-spectrometrically (8). The perfluorobisaziridine 7, rather surprisingly,

$$\begin{array}{c} CH_2 \\ | \quad\ \ \diagdown \\ \ \quad\ \ N\!\!-\!\!\{CH_2\}\!\!-\!\!OCH_3 \\ | \quad\ \ \diagup \\ CH_2 \\ \quad\ \ a \quad\ \ b \end{array}\Bigg]^{+\cdot} \xrightarrow{a} \quad 1a \quad + \quad CH_2\!\!=\!\!\overset{+}{O}\!\!-\!\!CH_3$$

6, M$^{+\cdot}$, m/e 87 $\qquad\qquad\qquad$ m/e 42 $\qquad$ m/e 45

$-CH_2O$, b, H transfer $\qquad\qquad b$

$$\begin{array}{c} CH_2 \\ | \quad\ \ \diagdown \\ \quad\ \ \overset{+\cdot}{N}\!\!-\!\!CH_3 \\ | \quad\ \ \diagup \\ CH_2 \end{array}$$

m/e 57

$$\begin{array}{c} CH_2 \\ | \quad\ \ \diagdown \\ \quad\ \ \overset{+}{N}\!\!=\!\!CH_2 \\ | \quad\ \ \diagup \\ CH_2 \end{array}\Bigg/ CH_3\!\!-\!\!CH\!\!=\!\!\overset{+}{N}\!\!=\!\!CH_2 \ + \ CH_3O^+$$

m/e 56 $\qquad\qquad\qquad\qquad\qquad\qquad m/e$ 31

shows some similarity to its simple analogs in that its initial fragmentation from the (unobserved) parent ion is loss of ·CF$_3$. The base peak of the spectrum is C$_2$F$_4^{+\cdot}$. A partial rationale of the spectrum is shown.

$$\begin{array}{c} CF_2 \qquad\qquad\ CF_2 \\ | \quad \diagdown \qquad\qquad \diagup \ | \\ \ \ \ \ N\!\!-\!\!N \\ | \quad \diagup \qquad\qquad \diagdown \ | \\ CF_2 \qquad\qquad\ CF_2 \end{array}\Bigg]^{+\cdot} \xrightarrow{-\cdot CF_3} \begin{array}{c} CF_2 \\ | \quad \diagdown \\ \quad\ N\!\!-\!\!\overset{+}{N}\!\!\equiv\!\!CF \\ | \quad \diagup \\ CF_2 \end{array}$$

7, M$^{+\cdot}$, (not observed) $\qquad\qquad\qquad\qquad\qquad\qquad m/e$ 159

$-C_2F_4$

$-C_2F_4N_2$

C$_2$F$_4^{+\cdot}$
m/e 100

$$\begin{array}{c} CF_2 \\ | \quad \diagdown \\ \quad\ N\!\!-\!\!N^{+\cdot} \\ | \quad \diagup \\ CF_2 \end{array}$$

m/e 128

II. AZETIDINE DERIVATIVES

The mass spectrum of the parent compound (8) is shown in Fig. 10.5 (9). Again no high-resolution or deuterium-labeling work has been done on this compound so that definite fragmentation assignments are impossible. Gallegos and Kiser (9) have suggested from appearance-potential measurements that the base peak (m/e 28) has the composition CH$_2$N, but it seems likely that the simple cyclic cleavage observed with trimethylene sulfide (Chap. 8, Sec. I.B) also occurs, and that the species m/e 28 is (at least in part) C$_2$H$_4^{+\cdot}$.

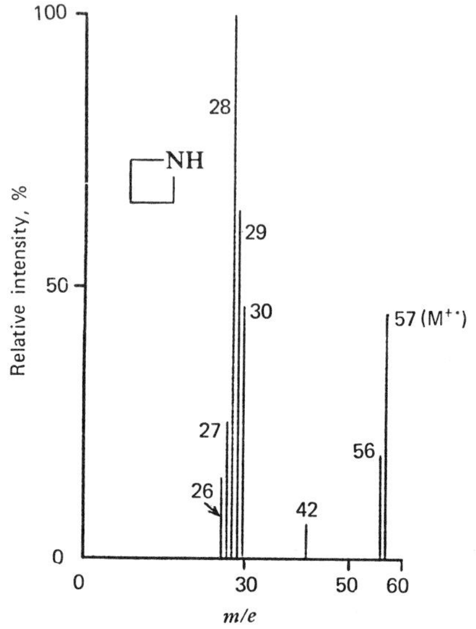

Fig. 10.5

Clearly, fragmentation with hydrogen transfer must accompany any simple cleavage of this kind to explain the intense ions at *m/e* 30 and 27. The M-1 species is much less intense than that observed with ethyleneimine, and it

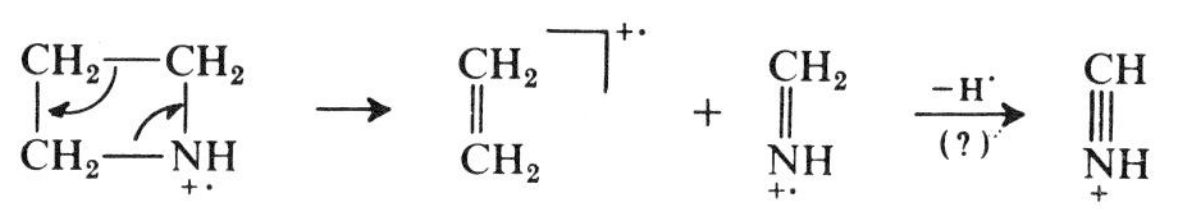

8, $M^{+\cdot}$, *m/e* 57 *m/e* 28 *m/e* 29 *m/e* 28

would be interesting to know whether it originates from N—H or α-C—H cleavage. Such information will only come with a labeling study.

The azetidine **9** shows some interesting features in its fragmentation (10), and it is clear that the initial C—N fission mainly involves the bond to the ethyleneimine ring.

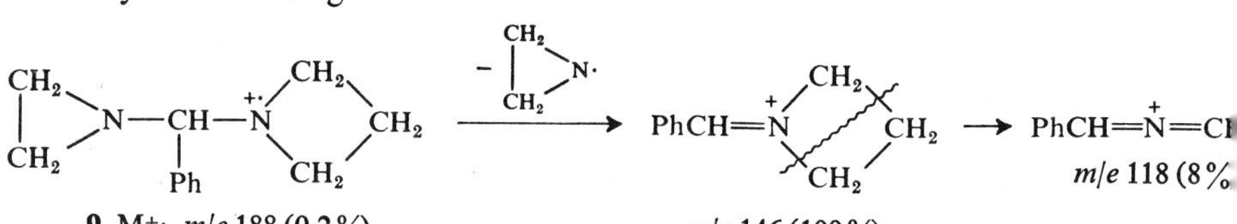

9, $M^{+\cdot}$, *m/e* 188 (0.2%) *m/e* 146 (100%)

The mass spectra of a series of 1-alkyl-2-phenyl-3-aroylazetidines have been reported and analyzed in detail (11). An interesting observation is the possibility of differentiation between the *cis* and *trans*-2-phenyl-3-aroyl compounds by the absence or presence of an M-·OH peak in the spectrum.

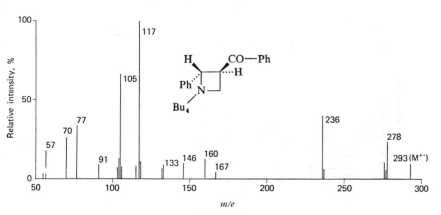

Apart from the ·OH loss, the pairs of isomers show very similar fragmentation. The 1-*t*-butyl-2-phenyl-3-benzoyl compound (**10**) shows the main fragmentation pathways illustrated in its spectrum (Fig. 10.6). As is to be expected, the benzoyl cation (*m/e* 105) is a prominent feature, and a rearrangement ion (*m/e* 167) is observed as a result of phenyl migration.

$$Ph—\overset{+}{C}H—Ph$$

m/e 167

The mass spectra of several β-lactams have been reported (12). They undergo simple ring cleavage upon electron impact, and the main fragmentations

Fig. 10.6

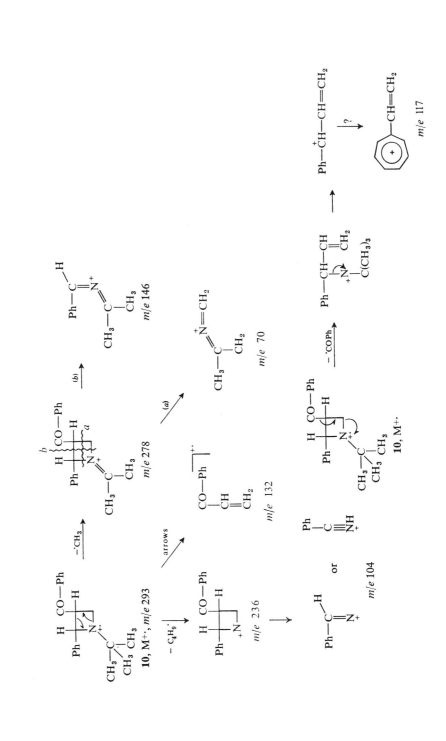

observed are given in the rationale shown for the spectrum of the 1,4-diphenyl-3-methyl compound (**11**).

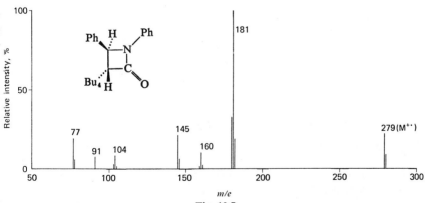

The relative intensities of the ions produced by the competing fissions a and b depend upon the stereochemistry at $C_{(3)}$. A large group at this position in a *cis* relationship to the $C_{(4)}$ phenyl facilitates fragmentation a as a consequence of the release of steric compression. The effect is clearly seen in the mass spectra of the two 3-*t*-butyl isomers (Figs. 10.7 and 10.8).

Similar fission apparently occurs very readily when 1,3,3-triphenyl-4-morpholino-azetidinone (**12**) is subjected to electron impact, since the compound is reported not to show a parent ion (**13**). It is interesting to note that all four fission products are observed in the mass spectrum in this case. By contrast the azetidinone **13** shows ions corresponding only

Fig. 10.7

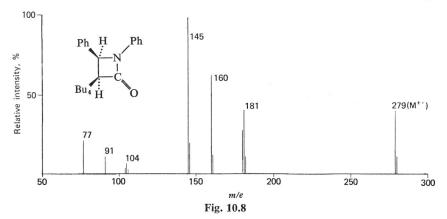

Fig. 10.8

to the anil and ketene fragments, so no definite pattern of charge retention emerges (14).

III.   PYRROLIDINE DERIVATIVES

A.   Simple Pyrrolidines

The mass spectrum of pyrrolidine (14) is shown in Fig. 10.9 (9). The mass spectra of deuterated pyrrolidines clearly show that the M-1 ion results from loss of a hydrogen atom from position 2, and thus the resultant ion is best represented as the iminium cation 14a. Labeling experiments also

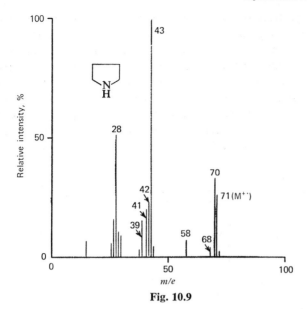

Fig. 10.9

show that the weaker M-3 species is best represented as a mixture of **14b** and **14c** (15). The base peak of the spectrum (*m/e* 43) results from loss of

14a, *m/e* 70   **14**, M⁺·, *m/e* 71   **14b**, *m/e* 68   **14c**, *m/e* 68

C_2H_4 from the molecular ion, and deuterium labeling makes it clear that the carbons in positions 3 and 4 are lost. The process is probably initiated by internal α-cleavage of the 2,3-carbon–carbon bond.

14, M⁺·, *m/e* 71   *m/e* 71   *m/e* 43   *m/e* 43

Appearance-potential data has been published for the *m/e* 43 species and, even before high-resolution and labeling data became available, this work made it clear that this species was the ion $C_2H_5N^+$ rather than the isobaric $C_3H_7^+$ (9).

The labeling work has allowed satisfactory rationales to be postulated for the formation of the minor ions in the spectrum of **14**. For more detail the original paper should be consulted (15).

The mass spectrum of *N*-methylpyrrolidine (**15**) has been discussed, and like the parent compound it has an intense M-1 ion. Deuterium labeling shows that no hydrogen loss occurs from the *N*-methyl group, suggesting the formulation **15a** for this species. The preference has been explained in terms of the greater energy that would be necessary to rupture a primary C—H bond instead of a secondary one. The peak at *m/e* 57 is produced by the same internal α-cleavage that produces the *m/e* 43 species in the spectrum of the parent compound whereas the species of *m/e* 42 involves a hydrogen transfer and loss of C_3H_7· (15).

m/e 57

15, M+·, *m/e* 85

m/e 85

m/e 42

m/e 84

15a, *m/e* 84

m/e 82

As expected, the substituent in the 2-position of a pyrrolidine ring is readily lost, a feature which can be of value in structure determination. Thus the hydroxy alkaloid (+)-hygroline (**16**, R = HOH) and the related ketone hygrine (**16**, R = O) both have as base peak the ion **15a**, *m/e* 84 (16, 17).

16

17

Similarly the pyridyl substituent in nicotine (**17**) is lost to give *m/e* 84 as base peak of its spectrum (18).

The mass spectra of two steroidal amides derived from 2-carboxypyrrolidine (**18** and **19**) have been published, but little of the fragmentation appears to be initiated by the heterocyclic ring. In fact, in both cases a prominent process appears to be the loss of the radical **20** as a neutral fragment (19).

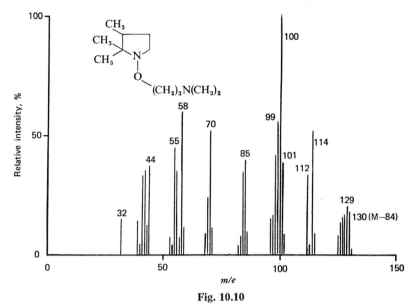

Fig. 10.10

An interesting example of how fragmentation patterns can be modified by extra functionalization is shown by the mass spectrum of **21** which is effectively a hydroxylamine derivative (**20**). The spectrum (Fig. 10.10) shows

no molecular ion, and the first-observed fragments correspond to fission of the C—O bond with and without hydrogen transfer. A rationale for the major features of the spectrum is shown.

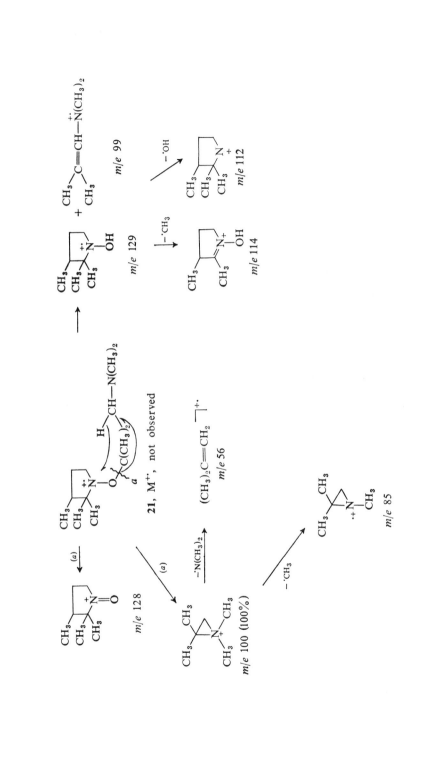

It was noted that the intensity of the m/e 99 peak relative to that at m/e 100 increased when the sample was heated in the inlet, so that the four-center hydrogen transfer postulated in the formation of m/e 99 and m/e 129 may be largely thermal in nature.

B. Pyrrolidine Enamines

The mass spectra of several enamines derived from pyrrolidine have been analyzed (21). Compounds of type **22** (R = H, CH$_3$, C$_2$H$_5$, or n-C$_5$H$_{11}$) all have base peaks corresponding to the conjugated iminium ion **22a** (m/e 110).

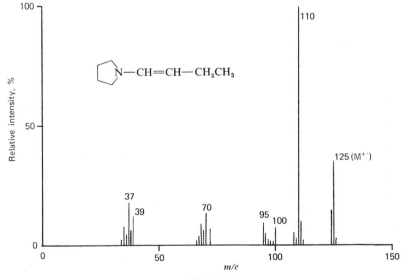

$$\underset{\textbf{22}}{\boxed{}\overset{+\cdot}{N}\!-\!CH\!=\!CH\!-\!CH_2\!-\!R} \xrightarrow{-R\cdot} \underset{\textbf{22a},\ m/e\ 110}{\boxed{}\overset{+}{N}\!=\!CH\!-\!CH\!=\!CH_2} \xrightarrow{-\cdot H_3C_4} \underset{\substack{\textbf{22b},\\ m/e\ 70}}{\boxed{}\overset{+}{NH}}$$

Little further fragmentation of **22a** occurs, the only process of any importance being loss of the N-substituent with hydrogen transfer to give the species **22b**, m/e 70.

The spectrum of an enamine may be used to demonstrate the presence of α-branching in the parent aliphatic aldehyde. Thus the derivatives from pyrrolidine and butyraldehyde (**22**, R = CH$_3$) and isobutyraldehyde (**23**) give very different mass spectra (Figs. 10.11 and 10.12, respectively). It will

Fig. 10.11

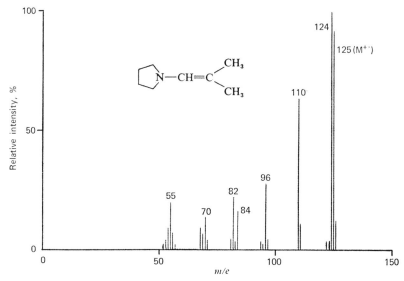

Fig. 10.12

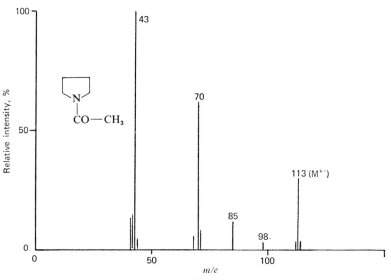

Fig. 10.13

be noticed that **23** shows a relatively weak M-·CH$_3$ peak and a strong M-1 ion, features easily explained on the basis of its structure. For further examples of analysis of mass spectra of pyrrolidine enamines of branched aldehydes, the reader is referred to the original article.

C. *N*-Acylpyrrolidines

The spectra of a number of *N*-acylpyrrolidines have been studied by means of deuterium labeling together with high-resolution mass-spectrometry, and their modes of decomposition are now well understood (22). The spectrum of *N*-acetylpyrrolidine (**24**) is shown in Fig. 10.13, and the proposed rationale for the fragmentation observed is also shown here. The ion at *m/e* 43 is composed of the isobaric species **24a** and **24b** in the proportions shown.

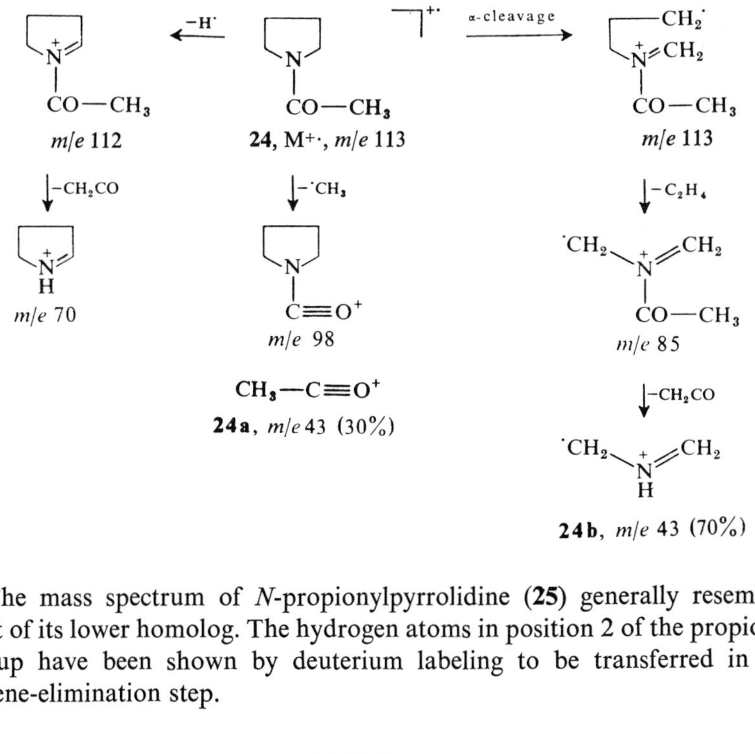

The mass spectrum of *N*-propionylpyrrolidine (**25**) generally resembles that of its lower homolog. The hydrogen atoms in position 2 of the propionyl group have been shown by deuterium labeling to be transferred in the ketene-elimination step.

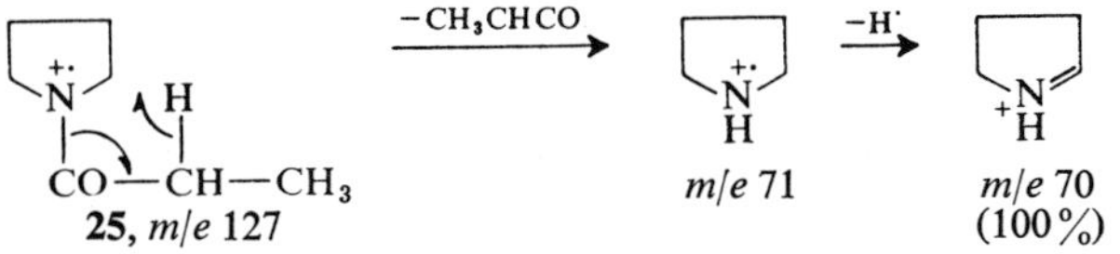

Longer alkyl chains than ethyl in the *N*-acylpyrrolidine result in the occurrence of marked McLafferty rearrangements. Thus the base peak of the mass spectrum of *N*-valerylpyrrolidine (**26**) is the ion **26a** (*m/e* 113) formed by this process. The reader is referred to the original paper for the full analysis of the rather complex mass spectrum of **26**.

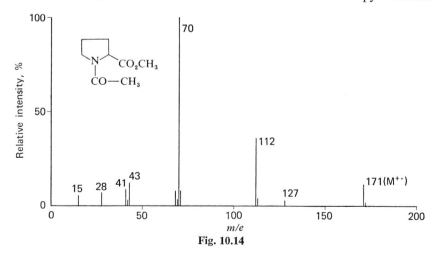

An important application of mass spectrometry is the analysis of amino acids and peptides, and it is interesting to consider the mass spectrum (Fig. 10.14) of *N*-acetyl-(±)-proline methyl ester (**27**) (23). It will be seen that the fragmentation of this compound is completely explicable on the basis of the fragmentations observed with 1- and 2-substituted pyrrolidines.

Fig. 10.14

The characteristic ion at m/e 70 would allow easy identification of proline in esterified and N-acylated amino acid mixtures. In fact esterification is not essential, and satisfactory mass spectra of N-acetylproline and N-acetyl-hydroxyproline have been reported (24).

D. N-Oxides

The mass spectra of the N-oxides (nitrones) of some Δ^1-pyrrolines have been briefly described (25). Unlike the N-oxides of aromatic heterocycles such as pyridine (Chap. 11, Sec. II.A), pyrroline-N-oxides show only weak M-16 peaks (intensities of 1 to 10% were reported for a range of alkyl-Δ^1-pyrroline N-oxides).

Full spectra have not been published, so a fragmentation rationale cannot be produced. It is reported that the base peak of most of the spectra is the ion $C_3H_5^+$ (m/e 41).

E. Pyrrolidones

The mass spectra of the simple lactam 2-pyrrolidone (**28**) (Fig. 10.15) and some of its N-alkyl derivatives have been fully analyzed with the aid of deuterium labeling and high-resolution mass spectrometry (26).

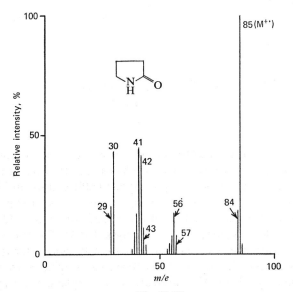

Fig. 10.15

The M-1 ion in the spectrum of **28** results from loss of a hydrogen atom from position 5 and is thus best represented as **28a**. The ion at m/e 56 is a triplet composed of the isobaric species $C_3H_4O^+$, $C_3H_6N^+$, and $C_2H_2NO^+$ in the ratio 6:2:1. The rationale shown has been suggested for these and other major ions in the spectrum.

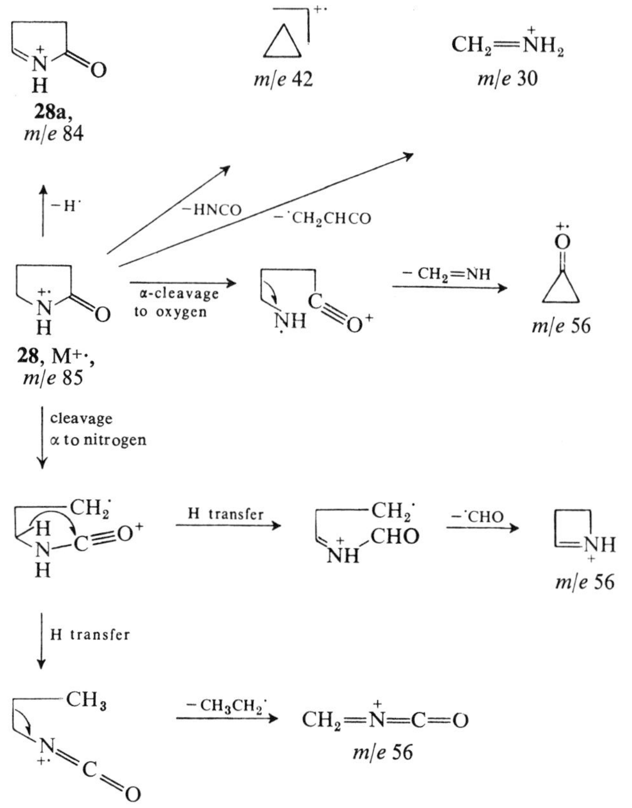

The N-methyl derivative of **28** fragments in essentially the same manner as the parent compound, but longer N-alkyl chains produce complicated fragmentation. The spectrum of the n-butyl compound (**29**) is shown in Fig. 10.16. The base peak (m/e 98) dominates the spectrum and is produced by α-cleavage of the alkyl chain. The ions of low abundance at m/e 112 and 126 correspond to the less favored β- and γ-cleavages. The M-42 species occurring at m/e 99 is a rearrangement ion and deuterium labeling established that 90% of the hydrogen transferred in its formation originates from the γ-position of the alkyl chain. The rationale shown explains the major fragmentations of this compound. The 2-imino-3,3-diphenyl-4-methylpyrrolidine

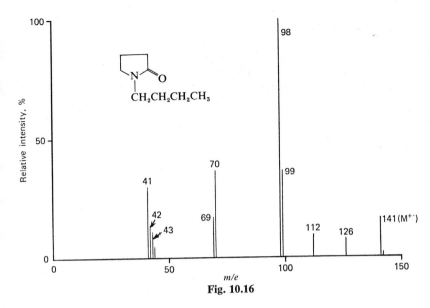

Fig. 10.16

30 is reported to show an ion at m/e 264 which from metastable ion information given appears to originate from the molecular ion (27). It may be produced by α-cleavage with γ-hydrogen transfer, as is observed with *n*-butylpyrrolidone.

F. N-Alkylsuccinimides

The mass spectra of several members of this class have been fully analyzed, and that of the *n*-butyl compound (**31**) is shown in Fig. 10.17 (26). The base peak (*m/e* 100) has a composition $C_4H_6NO_2^+$, and the labeling experiments carried out clearly show that this ion corresponds to loss of the alkyl chain from the molecular ion with transfer of two hydrogen atoms, mainly originating from the β and γ carbon atoms.

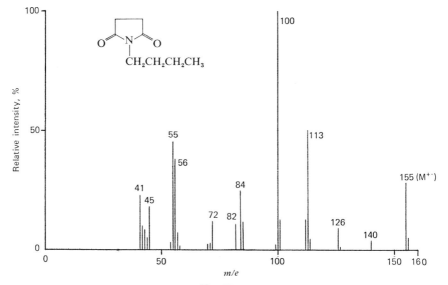

It has been pointed out that the double hydrogen transfer observed is analogous to the corresponding process which occurs in higher esters of fatty acids (28–30).

The ions at *m/e* 112, 126, and 140 correspond to α-, β-, and γ-cleavages of the alkyl chain, but the abundant ion at *m/e* 113 is formed to the extent of 90% by transfer of a γ-hydrogen atom from the side chain to (probably) the α-carbon atom.

Fig. 10.17

The compositions and origins of the less intense ions in the mass spectrum of **31** have been established, and the reader is referred to the original article for further detail.

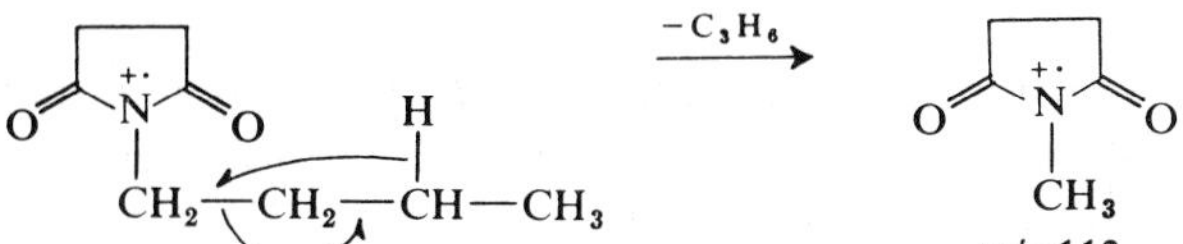

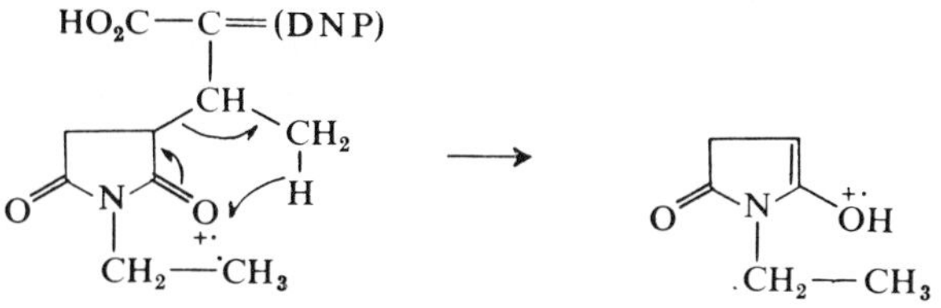

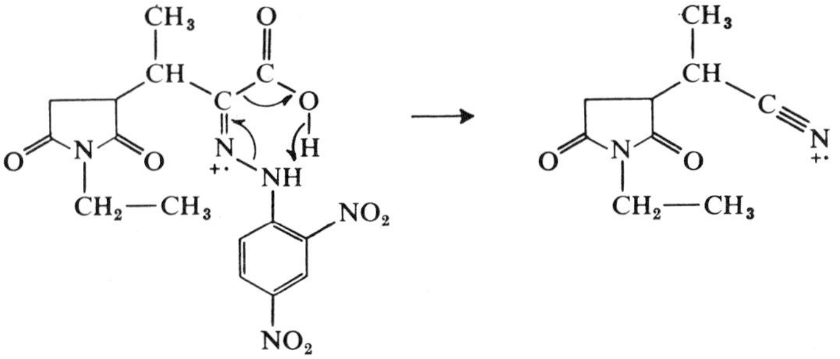

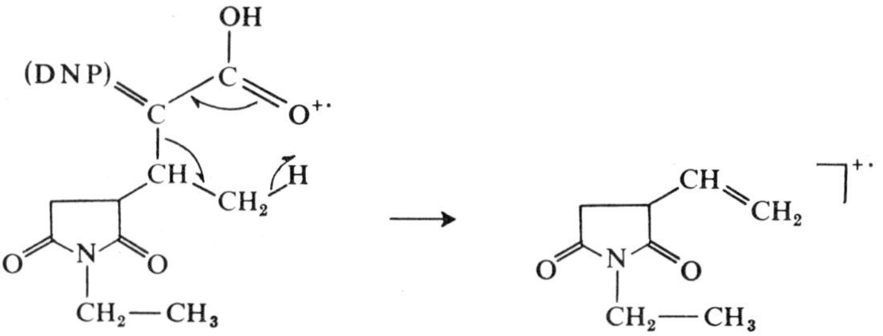

The mass spectrum of the 2,4-dinitrophenylhydrazone (**32**) of 2-oxo-3-(*N'*-ethyl-2',5'-dioxo-3'-pyrrolidyl)butyric acid has been reported (31). As is to be expected, much of the fragmentation is not characteristic of the succinimide portion of the molecule. The highest peak observed in the spectrum corresponds to M-CO_2, but a series of McLafferty rearrangements was postulated to explain the formation of a number of peaks at lower masses.

The mass spectra of two acetamido-*N*-methylsuccinimides (**33**, R = H or CH_3) have been examined as models for certain fragmentations observed with ydiginic acid (**34**, R = R′ = H) derivatives (32). Considerable charge localization apparently occurs on the secondary amide nitrogen atom since

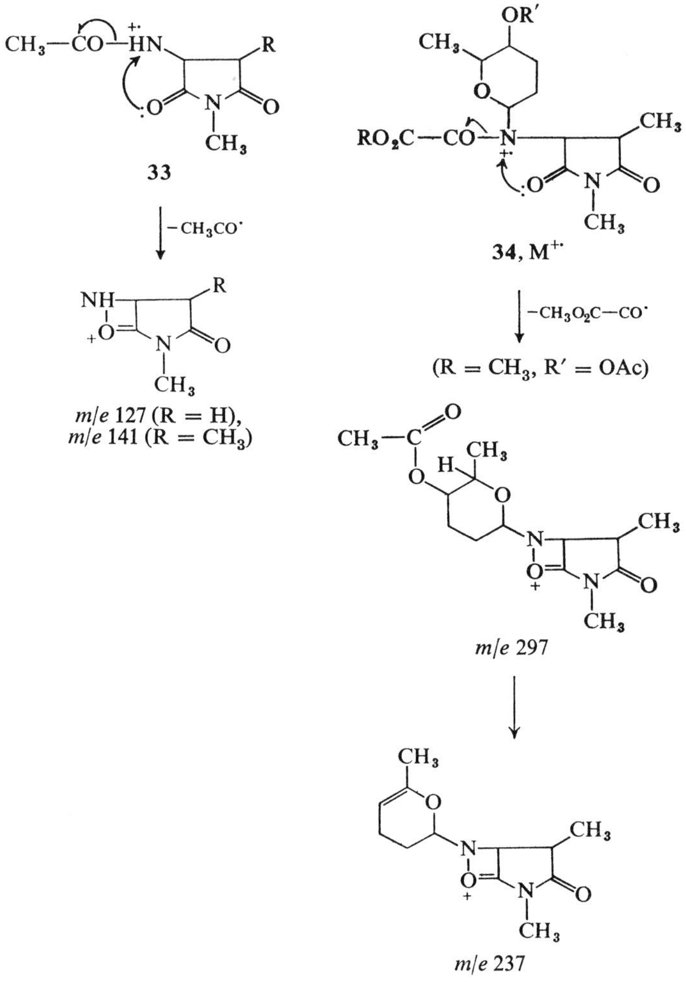
33
$-CH_3CO^{\cdot}$
m/e 127 (R = H),
m/e 141 (R = CH_3)
34, $M^{+\cdot}$
$-CH_3O_2C—CO^{\cdot}$
(R = CH_3, R′ = OAc)
m/e 297
m/e 237

the compounds show marked loss of CH_3—CO·. The occurrence of the normally unfavored CO—N fission is surprising, but the ion produced may perhaps be stabilized by interaction of the electron-deficient nitrogen atom with the neighboring oxygen atom as shown.

Fission of methyl O-acetylydiginate (**34**, R = CH_3, R' = $COCH_3$) occurs at bond *a* in a similar fashion to give the species at *m/e* 297, which further loses acetic acid to give *m/e* 237.

IV. PYRROLINE DERIVATIVES

No mass spectra of simple pyrrolines appear to have been recorded, and only fragmentary information is available about the spectrum of one complex derivative. 5-Methoxy-3,5-dimethyl-3-pyrrolin-2-one (**35**) is reported to fragment by loss of methanol as the major pathway (33).

35, M+·, *m/e* 141 → −CH₃OH → *m/e* 109

Clearly there is scope for much fundamental work on the fragmentation of the unsubstituted pyrrolines and their derivatives.

V. BICYCLIC COMPOUNDS CONTAINING NITROGEN IN A FIVE-MEMBERED RING: PYRROLIZIDINE DERIVATIVES

The pyrrolizidine ring system is of considerable interest in view of its occurrence in those natural products commonly grouped together as the pyrrolizidine alkaloids. Analyses of the spectra of these compounds are comparatively straightforward, and since the fragmentations observed may be of considerable value in structure determination, several typical examples will be discussed.

Laburnin (**36**) is typical of saturated members of this group of alkaloids, and its spectrum (Fig. 10.18) is easily rationalized on the basis of an initial α-fission (34).

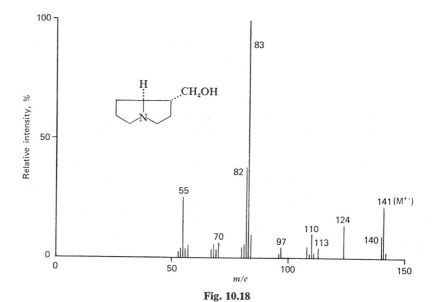

Of the two possibilities (*a* and *b*), fission *b* is the preferred one as shown by the great intensity of the species **36b** (*m/e* 83) compared with **36c** (*m/e* 113). The origin of the difference presumably lies in the greater stability of the secondary radical **36a**.

The spectrum of retronecanol (**37**) is explicable on the basis of α-fission giving the ion **37a** (34) which subsequently decomposes to the species **37b** (*m/e* 82) by the pathway shown. The same basic pathway is observed with the

Fig. 10.18

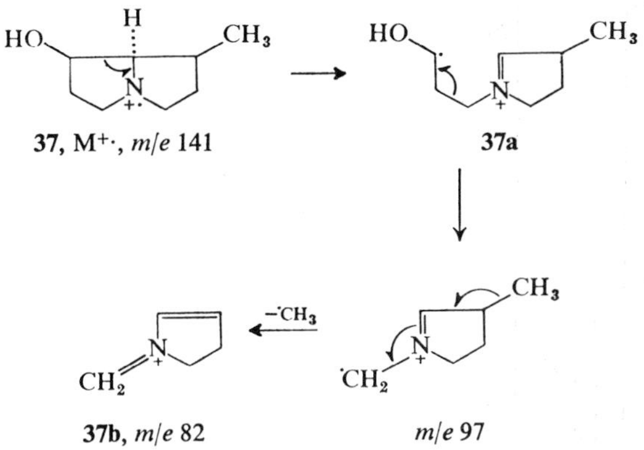

HO
H
CH₃
N
37, M+·, m/e 141
37a
CH₃
–·CH₃
CH₂
37b, m/e 82
m/e 97

unsaturated alkaloid retronecin (**38**), with α-fission occurring at the allylic bond in the saturated ring (34).

It will be noticed that in this case further decompositions of the ion **38a** (*m/e* 111) have been postulated to result in formation of pyridinium (**38b**) and methylpyridinium (**38c**) ions.

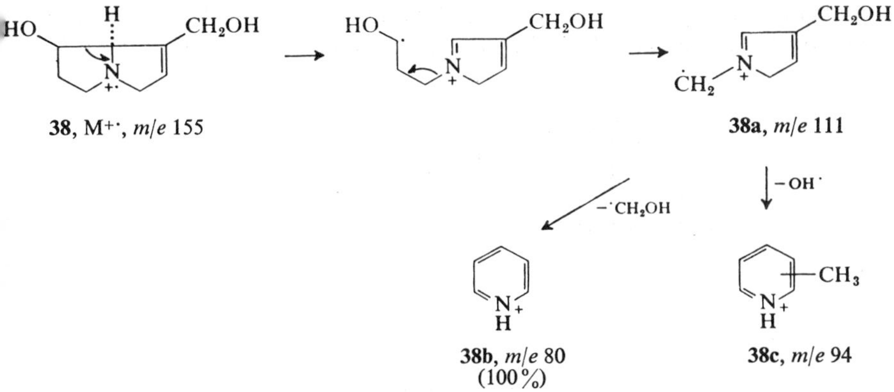

HO
H
CH₂OH
N
38, M+·, m/e 155
38a, m/e 111
–OH·
–·CH₂OH
CH₃
H
38b, m/e 80
(100%)
38c, m/e 94

The further functionalization present in the epoxypyrrolizidine alkaloid **39** produces a marked modification in the spectrum (Fig. 10.19). Decompositions leading to the species *m/e* 83 (**36b**) are repressed, and the base peak is now the ion **39a** (*m/e* 70) whose formation requires a hydrogen transfer. It may be rationalized as shown (35).

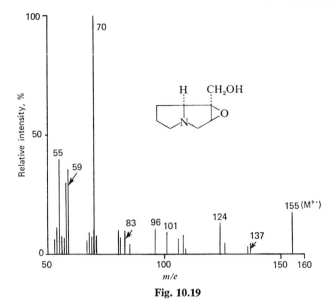

Fig. 10.19

Many naturally-occurring pyrrolizidine alkaloids contain an additional macrocyclic dilactonic ring, and mass spectrometry is a useful structural tool for studying compounds of this type (34, 36). Typical fragmentations observed are rationalized for the spectrum of monocrotalin (**40**) (Fig. 10.20).

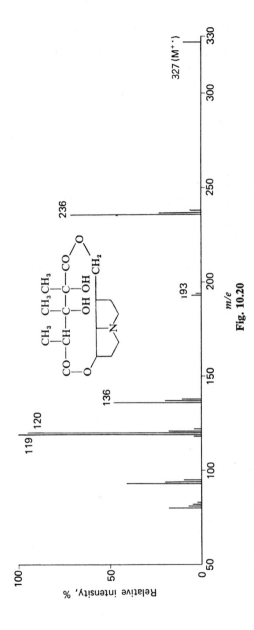

Fig. 10.20

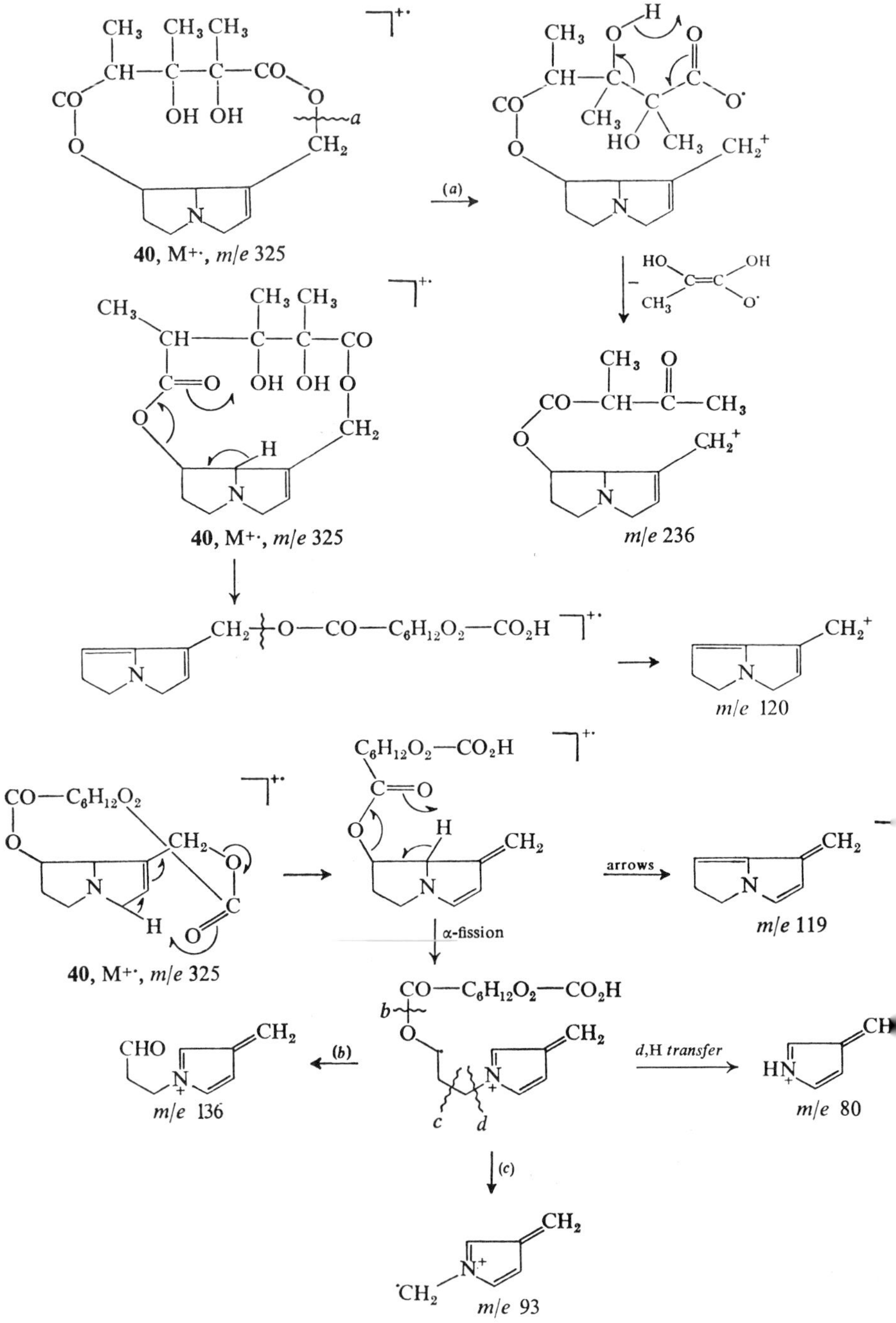
40, M+·, m/e 325
(a)
m/e 236
40, M+·, m/e 325
m/e 120
40, M+·, m/e 325
arrows
m/e 119
α-fission
(b)
m/e 136
d,H transfer
m/e 80
(c)
m/e 93

Zinc-dust distillation of the sulfur-containing alkaloid cassipourine (**41**) gives pyrrolo[1,2-*a*]pyrrolidine (**41a**) (37). The simple mass spectrum of **41a** has been rationalized as shown.

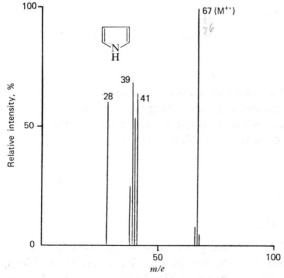

VI.   PYRROLE DERIVATIVES

A.   Pyrrole

The spectrum of pyrrole (**42**) is shown in Fig. 10.21, and the simple decomposition scheme suggested here is based upon high-resolution data for most ions in the spectrum (38).

Fig. 10.21

As with thiophen (Chap. 8) and furan (Chap. 4), it is difficult to decide upon the best representation for the parent ion of pyrrole. Budzikiewicz,

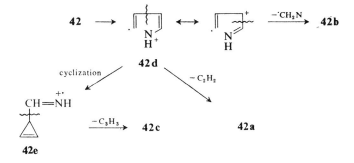

Djerassi, and Williams (39) have considered the acyclic (**42d**) and formyl-cyclopropeneimine (**42e**) representations to explain the easy formation of **42a**, **42b**, and **42c**.

No studies with isotopic substitutions have been reported for pyrrole. ^{13}C labeling should assist in establishing the various C—C fissions postulated in these schemes, while it would be interesting to observe the extent of randomization occurring with deuterated pyrrole after electron impact at various energies.

B. Alkylpyrroles

1. N-Alkylpyrroles

Since N- and C-alkylated pyrroles show marked differences in fragmentation, it is desirable to discuss the two classes separately. The spectrum of 1-methylpyrrole (**43**) is shown in Fig. 10.22 (37). It will be noted that the

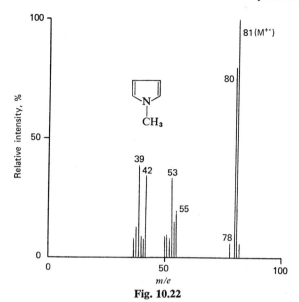

Fig. 10.22

chief feature of the spectrum is the strong M-1 ion which may be the aza-fulvene cation (**43a**) or perhaps the ring-expanded species **43b**. The spectrum is otherwise relatively straightforward.

High-resolution data on the ion of *m/e* 53 has not been published, but it most probably results from expulsion of HCN from **43a** or **43b**.

Some 5-substituted 1,2-dimethylpyrroles fragment by expulsion of the 5-substituent with (presumably) hydrogen migration from the *N*-methyl group to the 5-position (**38**).

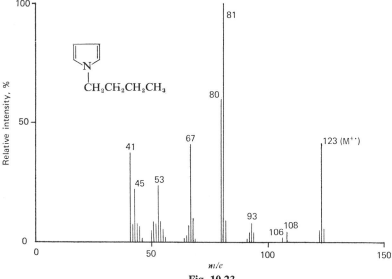

Fig. 10.23

The fragmentations of certain long-chain N-alkylpyrroles have been studied in some detail by means of labeling and high-resolution techniques (40). The spectrum of the N-butyl compound (**44**) is shown in Fig. 10.23. The α-cleavage ion **43a/b** has already been encountered, and it will be seen that β- and γ-cleavage are negligible. The base peak of the spectrum (m/e 81) was initially thought to result from transfer of the terminal methyl group to nitrogen with concomitant homolysis of the N—C bond (**44 → 44a**) or hydrogen transfer from the terminal methyl group via a six-center intermediate followed by α-cleavage (**44 → 44b → 44c**). The first suggestion was shown to be clearly untenable from the mass spectrum of N-n-(butyl-4,4,4-d_3)pyrrole whereas the second possibility accounted for only 8 % of the ion. In the spectrum of N-n-(butyl-3,3-d_2)pyrrole (45), however, the m/e 81 ion was shifted to m/e 82 to the extent of 78 %. Two mechanisms have been considered to account for this transfer: 45 → 45a and 46 → 46a.

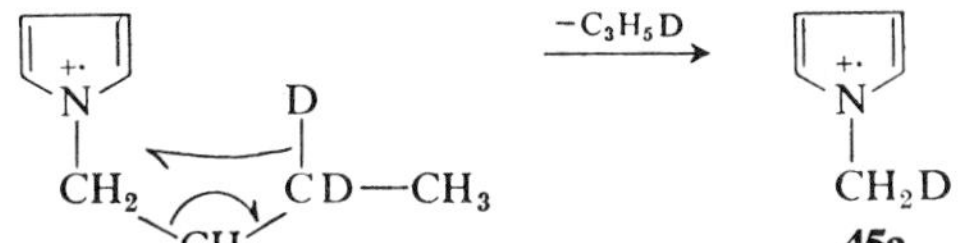

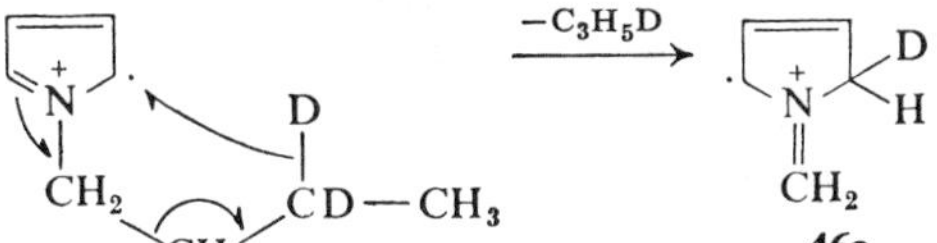

The picture is further complicated by the observation that the spectrum of 45/46 contains a peak at m/e 83 of more than 20 % of the intensity of the base peak; this ion requires the transfer of two deuterium atoms from the alkyl chain to the charged ring. To account for this observation it is necessary to assume reciprocal hydrogen transfer between the ring and the side chain, and the intermediate 46b has been postulated to explain this. Further evidence for the reciprocal transfer has been obtained from the mass spectrum of the ring-labeled compound 47, and the reader is referred to the original article for further detail. It should be noted that the mass spectrum of N-n-(butyl-2,2-d_2)pyrrole shows that 15 % of the m/e 81 peak in the spectrum of 44 originates from the 2-position of the n-butyl group, and that a series of nearly identical observations have been made for N-n-pentylpyrrole. The

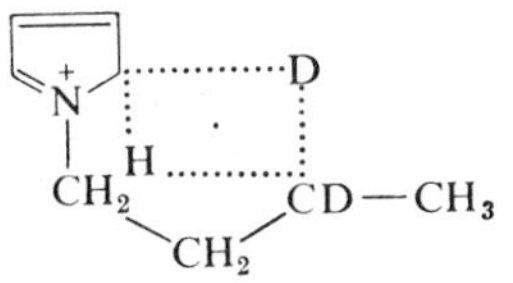

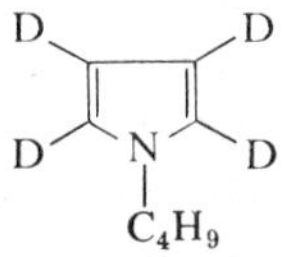

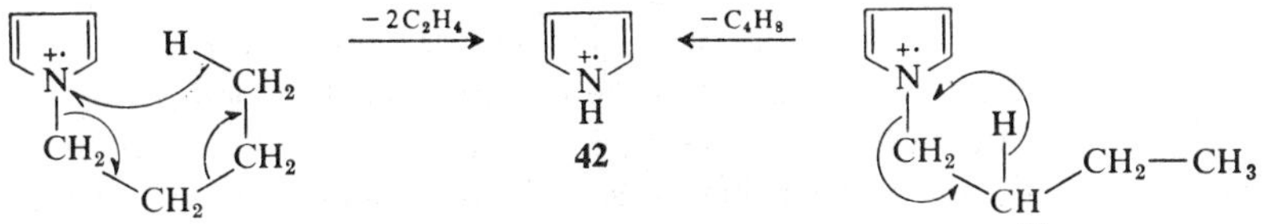

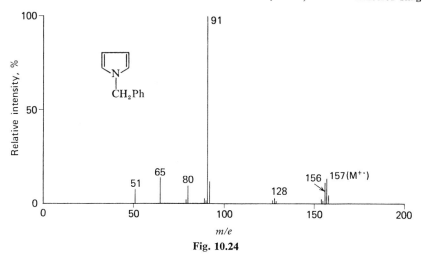

Fig. 10.24

mass spectra of the labeled compounds show that the m/e 67 ion ($C_4H_5N^{+\cdot}$) is best represented as ionized pyrrole (42) and that 40% of the hydrogen transferred originates from the terminal methyl group together with 50% from $C_{(2)}$ and a small contribution from $C_{(3)}$.

N-Benzylpyrrole (48) has a simple mass spectrum (Fig. 10.24) (41) in which it will be seen most of the ion current is carried by the tropylium ion (m/e 91). This preferred cleavage probably reflects the stability of the neutral fragment (48a) in which the unpaired electron is well-delocalized. It will be noticed that the alternative fission to give 43a/b is much reduced in importance; this probably again reflects the relatively lower stability of the neutral fragment $C_6H_5\cdot$.

2. C-Alkylpyrroles

No detailed study of long-chain C-alkylpyrroles of the type carried out for the N-alkyl compounds has yet been reported, but from the spectra available it is clear that β-cleavage is the predominant fragmentation. Thus 2-methylpyrrole (49) (38) shows (Fig. 10.25) a strong M-1 ion which has been represented as 49a or its ring-expanded isomer 43b. Similar β-cleavage of 2-ethylpyrrole (50), with loss of a methyl group, leads to the same ion(s).

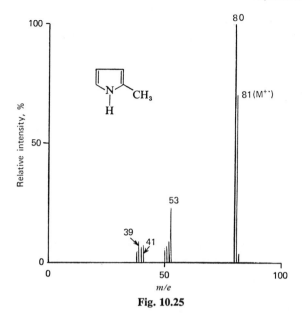

Fig. 10.25

The mass spectrum of 2,3-dimethylpyrrole (**51**) shows a moderately strong (24%) M-CH$_3$· ion as well as the expected very intense M-1 species. The

M-CH$_3$· ion, however, is relatively much weaker than that observed with analogous xylenes, perhaps indicating less tendency towards ring expansion in the heterocyclic case (38).

C. Acylpyrroles

The spectra of 2-formyl- and 2-acetylpyrroles (**52**, R = H or CH$_3$) show the expected fragmentation (loss of R·) with the intense acylium cation (**52a**) being presumably well-stabilized by resonance as shown (38). It has been suggested that formation of the relatively intense m/e 66 ion in the spectra of the 2-acylpyrroles involves a 1,2-hydrogen shift (**52b** → **52c**) to give a species in which the positive charge is capable of delocalization over five

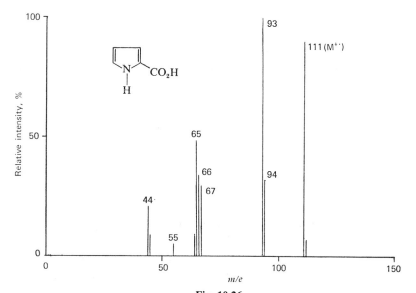

atoms (38). It is noteworthy that the M-RCO· species is relatively weaker in the spectra of 3-acylpyrroles where the necessary hydrogen shift is less favored.

D. Pyrrole-Carboxylic Acids and Esters

The mass spectrum of pyrrole-2-carboxylic acid (**53**) is shown in Fig. 10.26 (38). The ion at m/e 94 corresponds to loss of the hydroxyl group and is clearly the species **53a**, but the intense ion at m/e 93 corresponds to loss of water from the molecular ion, a process not observed with most aromatic

Fig. 10.26

carboxylic acids. The results of deuterium labeling show that the imino-hydrogen is involved in the fragmentation which can be represented as **53** → **53b**. The fragment $C_4H_3N^{+\cdot}$, derived by loss of the stable neutral

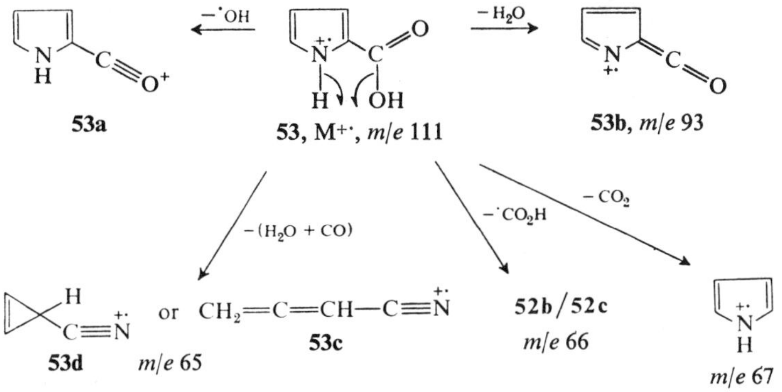

−·OH
−H₂O
53a
53, M+·, m/e 111
53b, m/e 93
−(H₂O + CO)
−·CO₂H
−CO₂
or CH₂=C=CH—C≡N
53d
m/e 65
53c
52b/52c
m/e 66
m/e 67

fragments CO and H_2O (or D_2O in the case of dideutero-**53**), may well have a structure such as **53c** or **53d**.

A very similar series of decompositions occurs with the ethyl ester of **53**, the only important additional process being loss of ethylene from the molecular ion (38).

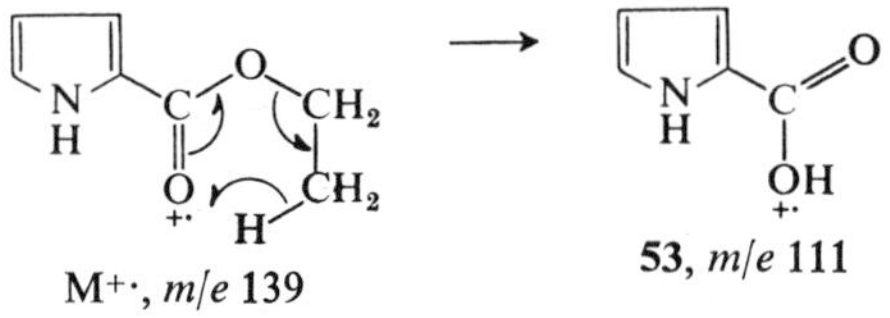

M+·, m/e 139
53, m/e 111

Since various substituted pyrrole-carboxylic esters result from the Knorr pyrrole synthesis from β-oxo-esters and are of interest in the synthesis of

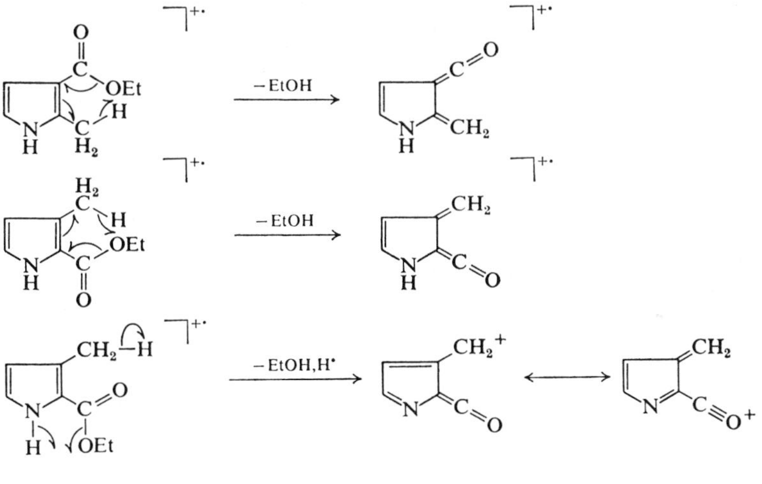

−EtOH
−EtOH
−EtOH,H·

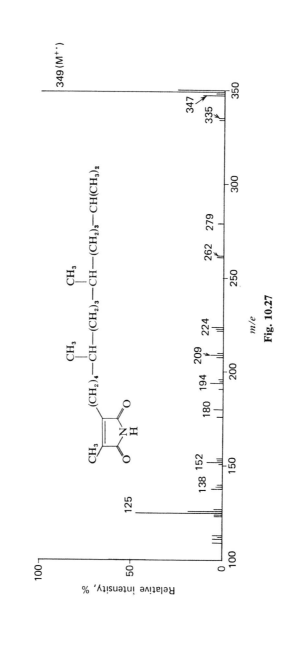

Fig. 10.27

polypyrroles, many of their spectra have been thoroughly analyzed. The fragmentations described for the simple pyrrole-2-carboxylic acids and esters can usually be observed as well as the "ortho effects" with neighboring methyl groups outlined on page 334. For further detail the reader is referred to the original article (38).

E. Pyrrolones and Pyrrolediones

Little work on either of these classes of nuclear-oxygenated pyrrole derivatives has been reported. The partial mass spectrum of 5-phenyl-3-[1-phenyl-3-carboxypropylidene]pyrrolone-(2) (**54**) has been reported (42), but in the absence of high-resolution data it is difficult to reach firm conclusions about the modes of fragmentation. The occurrence of an M-18 peak may indicate cyclization to **54a** or **54b**.

54, M$^{+\cdot}$, m/e 319   **54a**   **54b**

The mass spectrum (Fig. 10.27) of the dione 4-methyl-3-(5,9,13-trimethyl-tetradecyl)maleimide (**55**), which has been obtained by palladium-catalysed reduction of cytohemin in formic acid followed by chromic acid oxidation (43), shows a clear molecular ion together with weak ions resulting from successive fragmentations of the side chain. The strong ion at m/e 125 is best formulated as the result of a four-center hydrogen transfer to give **55a** as shown.

55

arrows

+ CH$_2$=CH–C$_{14}$H$_{29}$ $^{+\cdot}$

m/e 224

55a, m/e 125

F. Arylpyrroles

Little work on the mass spectra of simple arylpyrroles has been reported. The pentabromohydroxyphenylpyrrole **56** is an antibiotic isolated from a marine bacterium, and its mass spectrum is reported to show a sequential loss of one, two, and three bromine atoms from the molecular ion, together with loss of hydrogen cyanide (44). No fragmentation corresponding to a simple cleavage of the phenol and pyrrole portions was seen.

56

G. Polypyrrolic Compounds

Many polypyrrolic compounds have been prepared as intermediates in the synthesis of porphyrins, and the mass spectra of a number of such compounds have been analyzed as have those of the naturally occurring pigments, prodigiosin, bilirubin, and D-urobilin.

1. Dipyrroles

Three classes of dipyrroles have been examined mass-spectrometrically, namely pyrromethanes, pyrromethenes, and pyrroketones. These are dealt with separately and their fragmentations related to those of the more complex derivatives.

a. PYRROMETHANES.   The fragmentations of pyrromethanes (45) are, not unexpectedly, dependent upon the nature of the substituents present, but in general they all show intense molecular ions, allowing molecular weights (and compositions) to be determined. Exceptions are the carboxylic acids which undergo very ready decarboxylation.

The fragmentations observed can be classified into three main groups:

(i) fragmentation of side chains,
(ii) internuclear cleavage,

and

(iii) formation of tricyclic fragmentation products (when the methylene bridge is flanked by neighboring carbonyl substituents).

Some of these processes will be exemplified by fragmentations of typical pyrromethanes, but for a full discussion of the mass spectra of 15 pyrromethanes the reader is referred to the original publication.

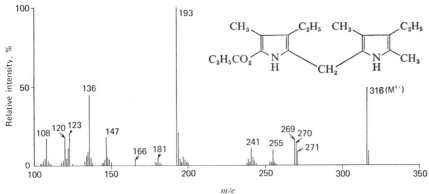

Fig. 10.28

The mass spectrum of ethyl 3′,4,5′-trimethyl-3,4′-diethylpyrromethane-5-carboxylate (**57**) (45) is shown in Fig. 10.28.

The group of peaks resulting from losses of $C_2H_5O\cdot$, C_2H_5OH, and $C_2H_5OH + H\cdot$ probably corresponds to processes already described for simple ethyl alkylpyrrolecarboxylates and may be represented as **57a**, **57b**, and **57c**.

The ions at m/e 241 and 255 correspond to M-($C_2H_5OH + C_2H_5\cdot$) and M-($C_2H_5OH + CH_3\cdot$), respectively, and it has been suggested that these species are best represented as **57e** and **57d**. These structures are the result of α-cleavage of a 3-ethyl or 3′-methyl group from the ring together with migration of a hydrogen atom from the methylene group to the position vacated. The resulting structures (**57d** and **57e**) represent stable species (methene salt cations) (45). This group of fragmentations thus includes those of type (i), namely fragmentation of side chains.

The base peak (**57f**) of the spectrum of **57** is produced by a type (ii) (internuclear cleavage) fragmentation, and involves a six-center hydrogen transfer reaction leading to **57b**. The conjugate fragment from this reaction (**57g**) is also seen in the spectrum.

Internuclear cleavages without hydrogen transfer (fission at a and b) produce the species **57i** and **57h**, respectively. It is noteworthy that **57i** is considerably more intense than **57h** (some 60% of the m/e 194 peak is due to the isotope peak from m/e 193), and this has been taken to indicate that in the simple fissions a and b inductive factors influencing the stability of the fragment ions are of paramount importance, whereas hydrogen is preferentially transferred from the ring bearing the ethoxycarbonyl substituent and the positively-charged fragment derived from this ring predominates in the fragmentation involving hydrogen transfer (**57** → **57f** + **57g**).

Further fragmentation of **57f** is limited to losses of C_2H_5OH and $\cdot CO_2Et$, and the resultant ions are probably best represented as **57j** and **57k**.

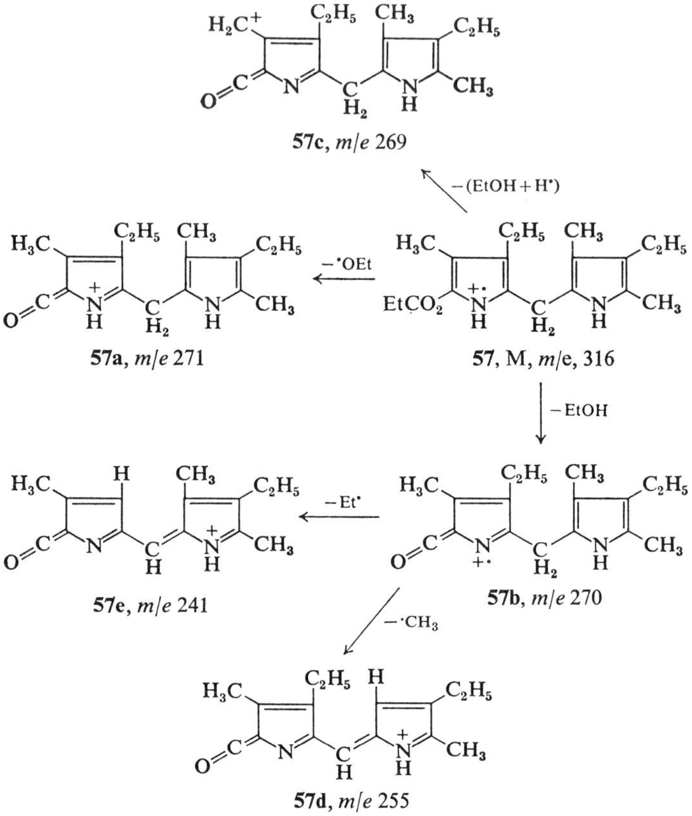

57c, m/e 269

$-(\text{EtOH} + \text{H}^\bullet)$

57a, m/e 271

$-^\bullet\text{OEt}$

57, M, m/e, 316

$-\text{EtOH}$

57e, m/e 241

$-\text{Et}^\bullet$

57b, m/e 270

$-^\bullet\text{CH}_3$

57d, m/e 255

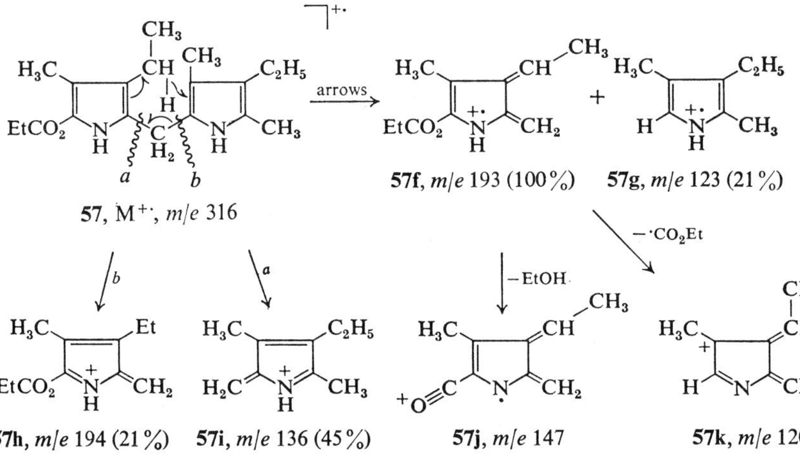

57, M$^{+\bullet}$, m/e 316

arrows

57f, m/e 193 (100%)   +   57g, m/e 123 (21%)

$-^\bullet\text{CO}_2\text{Et}$

b a

$-\text{EtOH}$

57h, m/e 194 (21%)   57i, m/e 136 (45%)   57j, m/e 147   57k, m/e 120

Process (iii), formation of a tricyclic fragment, has been invoked to explain the formation of an M-($\cdot CO_2Et$ + EtOH) ion in the 3,3'-dicarbethoxy-pyrromethane **58** (45). That the reaction is sequential (**58** → **58a** → **58b**) is shown by the appearance of a metastable ion for the loss of C_2H_5OH.

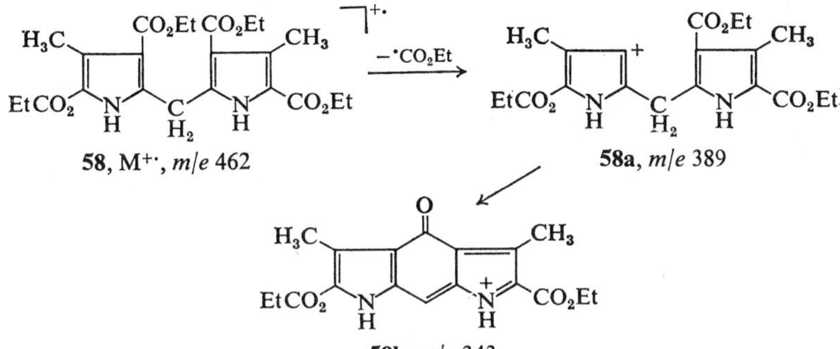

58, M$^{+\cdot}$, m/e 462

58a, m/e 389

58b, m/e 343

b. PYRROMETHENES. The reported pyrromethene spectra all display relatively intense molecular ions, reflecting the greater possibilities for charge delocalization over the fully-conjugated dipyrrolic ring system. In general fragmentations of side chains predominate over cleavage at the methene bridge (45). The fragmentation scheme for crypto-pyrromethene (**59**) shows typical cleavages observed with this group of compounds.

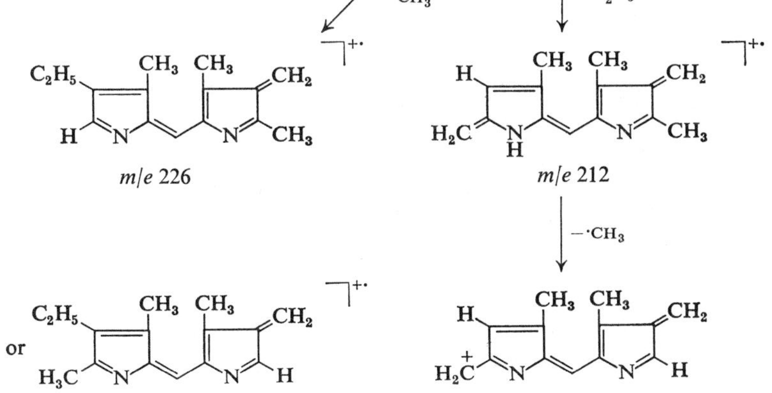

59, M$^{+\cdot}$, m/e 256

m/e 241 (100%)

m/e 226

m/e 212

or

m/e 226

m/e 197

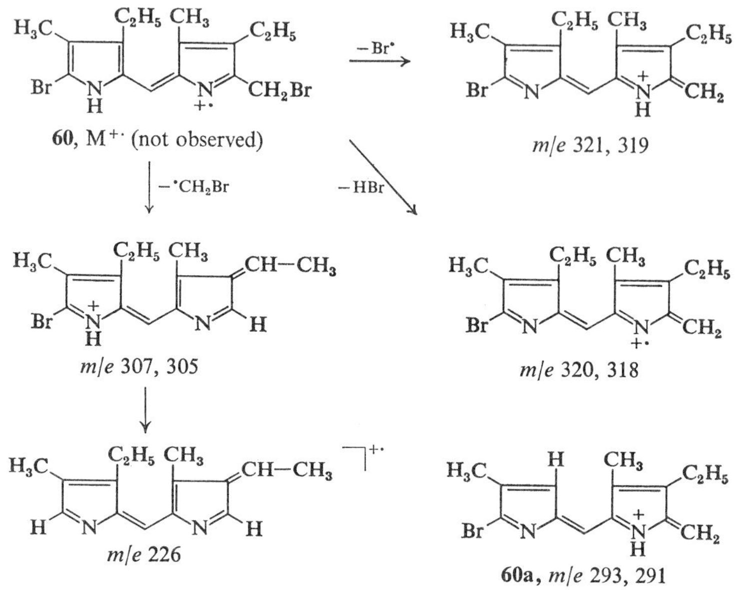

60, M+· (not observed)
−Br·
m/e 321, 319
−·CH2Br
−HBr
m/e 307, 305
m/e 320, 318
m/e 226
60a, m/e 293, 291

The spectra of brominated pyrromethenes are of some interest since these and analogous compounds have been used in the Fischer-type porphyrin synthesis. Such compounds as **60** do not show parent ions because of very ready cleavages of bromine, HBr, or $\cdot CH_2Br$, and the chief fragmentations

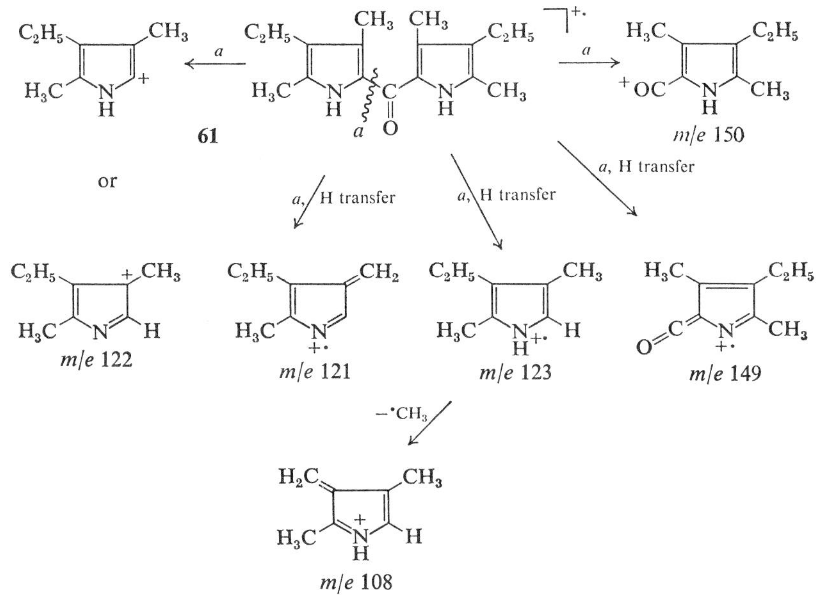

61
a
m/e 150
a, H transfer
or
m/e 122
m/e 121
m/e 123
m/e 149
−·CH3
m/e 108

have been rationalized (45) as shown. The species m/e 321/319 loses ethylene, and structures such as **60a** have been proposed for the resulting ion.

c. PYRROKETONES. The pyrroketones examined, like the pyrromethanes show ready internuclear cleavage, reflecting the facile fission of a bond α to a carbonyl group (45). The fragmentation of the symmetrical diethyltetramethyl ketone **61** is typical.

A notable feature of the fragmentations described in the scheme is the frequent occurrence of hydrogen-transfer reactions, both to and from the charged species observed in the spectrum.

2. Tri- and Tetrapyrrolic Compounds

A number of polypyrrolic compounds prepared as intermediates in synthetic studies in the porphyrin field have been examined mass-spectrometrically, as have the natural products prodigiosin, bilirubin, and D-urobilin (45). In general the cleavages observed parallel those described for the simpler dipyrroles and will not be discussed in detail here. It is interesting to note that the benzyl ethyl diester **62** undergoes cleavage to give the

fragments **62a** and **62b** but not the alternative fragments (**62c** and **62d**) bearing the ethoxycarbonyl group. This has been ascribed to the greater ability of the benzyl group to assist in the stabilization of the positive charge in the fragment ions.

Prodigiosin (**63**) has a simple mass spectrum in which the molecular ion is the base peak. This is to be expected in view of its structure which allows very considerable charge delocalization. The main fragmentation observed is β-cleavage of the alkyl side chain together with a less important loss of 15 mass units from the molecular ion (probably cleavage of the ether group as shown).

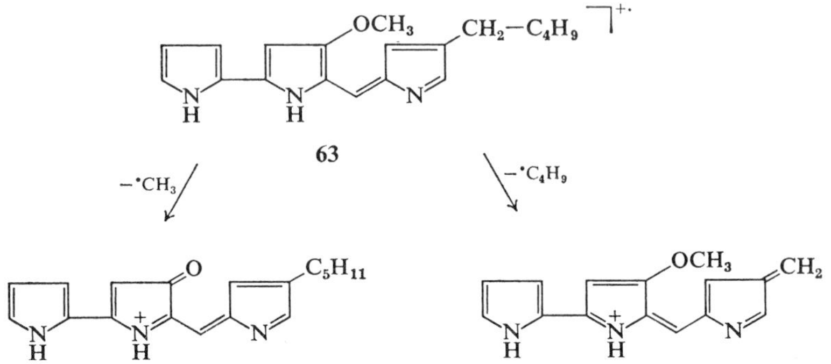

63

The mass spectrum of bilirubin (**64**) is dominated by fragmentations arising from cleavages at the central methylene group as is to be expected from our knowledge of the fission of the simple pyrromethanes.

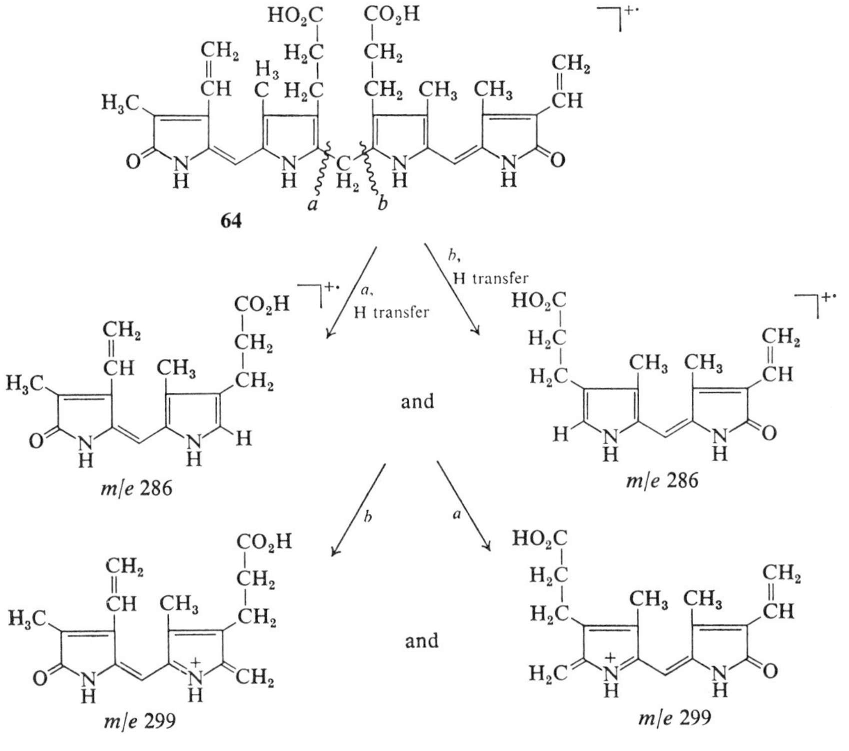

64

The mass spectrum of D-urobilin, a metabolite of bilirubin, confirms that it is a tetrahydro-derivative of its precursor, and either of its proposed structures (65, R = ·CH=CH$_2$, R' = ·C$_2$H$_5$ or vice versa) explain the formation of a strong ion at m/e 466. The origin of another strong ion at m/e 302 is less certain and clearly must involve cleavage of the strong methene linkage or extensive hydrogen rearrangement.

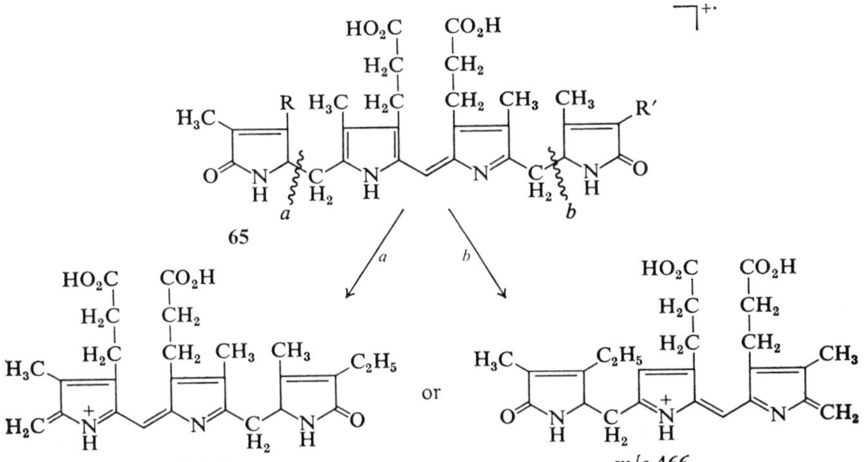

65

m/e 466

or

m/e 466

The mass spectrum of bis-[3,5-dimethylpyrryl-(2)]-cyclobutene (66) shows an intense molecular ion and (as a major fragment) an M-2CO species best represented as the acetylenic cation 66a. As the complete spectrum was not published, it is uncertain whether a C$_2$O$_4$ fragment is lost as such or as two successive CO units (46).

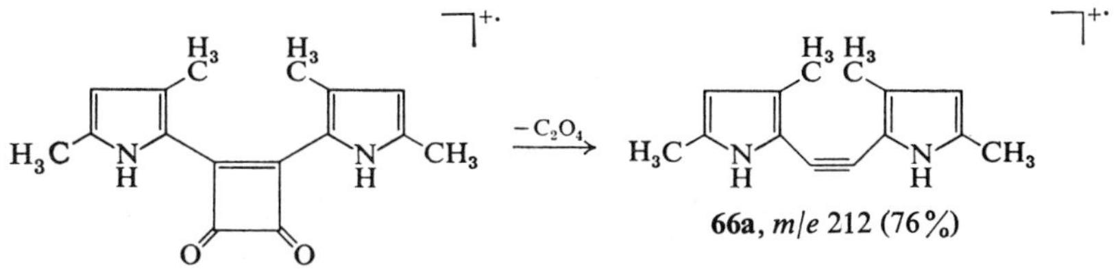

66, M$^{+·}$, m/e 268 (100%)

66a, m/e 212 (76%)

VII. INDOLE AND DERIVATIVES

The mass spectrum of indole (67) is shown in Fig. 10.29 (47). As is to be expected for an aromatic heterocycle, the molecular ion is the base peak of the spectrum and relatively little fragmentation occurs. The peaks at m/e 89

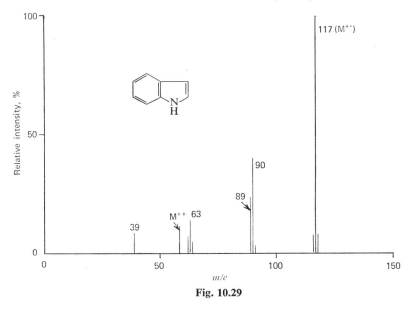

Fig. 10.29

and 90 ($C_7H_5^+$ and $C_7H_6^{+\cdot}$) appear in a wide range of benzoheterocycles and are of little structural significance. As has already been mentioned, these ions may well have the noncyclic structures **67a** and **67b** (48).

$$67, M^{+\cdot}, m/e\ 117\ (100\%) \xrightarrow{-HCN} CH_2=CH-CH_2-C\equiv C-C\equiv CH \quad \rceil^{+\cdot}$$

67b, m/e 90

$-H^\cdot$

$-(HCN+H)$

$$CH_2=CH-\overset{+}{C}H-C\equiv C-C\equiv CH$$

67a, m/e 89

Introduction of alkyl substituents into the indole system results in the expected α,β-cleavage, and thus 2-methylindole (**68**) has as its base peak the M-1 ion (Fig. 10·30) (49). As with the sulfur and oxygen analogs and the methylpyrroles, it has been suggested that this fragmentation is accompanied by ring expansion to the quinolinium cation (49). This formulation for the M-1 species allows easy rationalization of the subsequent loss of the elements of HCN to give m/e 103. If ring enlargement does occur, it may proceed by a C—C migration [pathway (i)] or C—N migration [pathway (ii)]. Marx and Djerassi (50) studied the isotope distribution in the M-(H· + HCN)

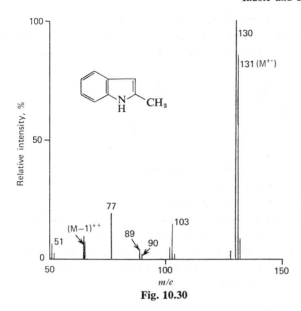

Fig. 10.30

ion (*m/e* 103) using 2-methylindole labeled in the methyl and the C$_{(2)}$ positions with ^{13}C. They concluded that pathway (i) is favored although some 14% of the species at *m/e* 103 cannot be accounted for by either pathway and possibly arises from rupture of the phenyl ring.

The mass spectra of a number of dimethylindoles have been reported (49), and like the monomethyl compounds they show intense M-1 ions which can be formulated as methylquinolinium ions. M-CH$_3$· ions also occur and are most intense when the methyl groups occupy adjacent carbon atoms, suggesting the occurrence of the hydrogen transfer reaction shown

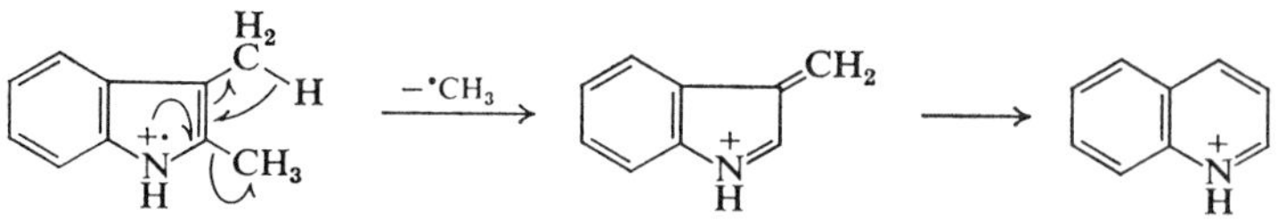

M$^{+·}$, m/e 145 (100%) m/e 130 (51%)

(49). In 2,5- and 2,6-dimethylindoles the intensities of the M-CH$_3$· ions drop to 16 and 14.6%, respectively.

The mass spectra of N-alkylindoles have not been studied in as much detail as the simple N-alkylpyrroles. Like its C-methylated analogs, 1-methylindole (69) (50) has a strong ion at M-1 which may well be a ring-expanded species. In this case C—C migration is the only possibility for the ring-expansion process and can yield either the quinolinium ion (68a) or the isoquinolinium ion (69a).

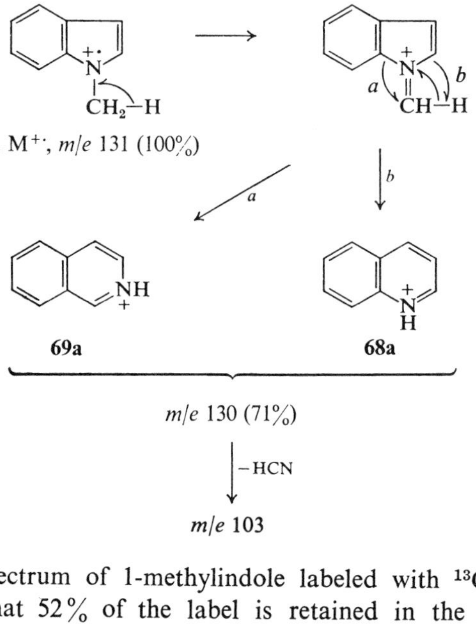

69, M$^{+·}$, m/e 131 (100%)

69a   68a

m/e 130 (71%)

$-$HCN

m/e 103

The mass spectrum of 1-methylindole labeled with ^{13}C at the methyl group shows that 52% of the label is retained in the M-(H· + HCN) fragment (m/e 103) (50). This rules out pathway b as a major contributor to the M-1 species since 68a would be labeled at C$_{(2)}$ and loss of HCN would remove all the label (except for any portion of the M-H$_2$CN ion involving loss of carbon from the phenyl ring).

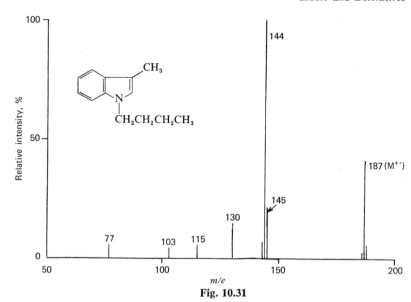

m/e

Fig. 10.31

The mass spectrum of 1-*n*-butyl-3-methylindole (**70**) (Fig. 10.31) (49) shows that here α-fission with loss of the radical $C_3H_7\cdot$ is the major fragmentation. The ion at *m/e* 145 is only partly the isotope peak from *m/e* 144, and some 50% must be the ion $C_{10}H_{11}N^{+\cdot}$ resulting from hydrogen transfer from the butyl group and elimination of C_3H_6. The results already described for *n*-butylpyrrole suggest that $C_{(2)}$ and $C_{(4)}$ of the butyl group are probably the major sources of transferred hydrogen, but deuterium-labeling experiments will be essential to clarify this point.

Cleavage of the bond situated α,β to the indole ring has been widely used in structural diagnostic work. Thus γ-indolemycenic acid (71), an acid-hydrolysis product of the antibiotic indolemycin, has been examined mass spectrometrically as its methyl ester. Ions at m/e 174 (71a) and 188 (71b) clearly result from the two alternative cleavages shown and support the proposed structure (51).

71   71b, m/e 188   71a, m/e 174

Tryptophan ethyl ester (72) shows the expected cleavage leading to the quinolinium cation (68a) as the base peak of the spectrum (52). Only minor fragmentation is initiated by charge localization on the primary amino group.

72, M$^+$, m/e 232 (5%)

68a, m/e 130 (100%)

$\overset{+}{N}H_2=CH-CO_2C_2H_5$

m/e 102 (2%)

$CH_2-CH=\overset{+}{N}H_2$

m/e 159 (8%)

The presence of tryptophan in simple peptides is easily detected by the occurrence of the intense quinolinium cation at m/e 130. Thus this ion provides the base peak of the spectrum of N-benzyloxycarbonyl-valyl-valyltryptophan methyl ester (73) (53). An intense ion at m/e 201 (73a) results from cleavage of the β,γ bond with hydrogen transfer, probably via a McLafferty rearrangement as shown.

A similar competition between α,β-fission and β,γ-fission with hydrogen transfer is shown in the mass spectrum of 3-(2-acetylaminoethyl)-6-hydroxy-5-methoxyindole (74), a metabolite of the pineal gland hormone melatonin (54).

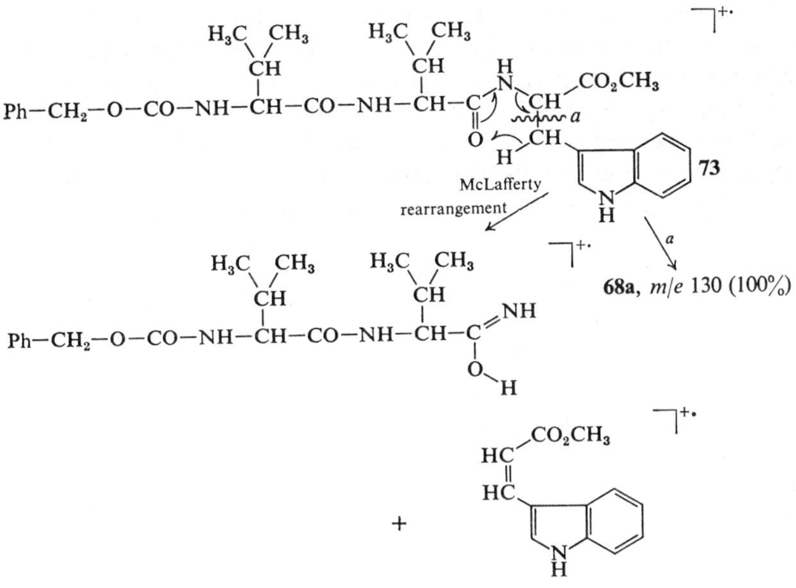

73

68a, *m/e* 130 (100%)

+

73a, *m/e* 201

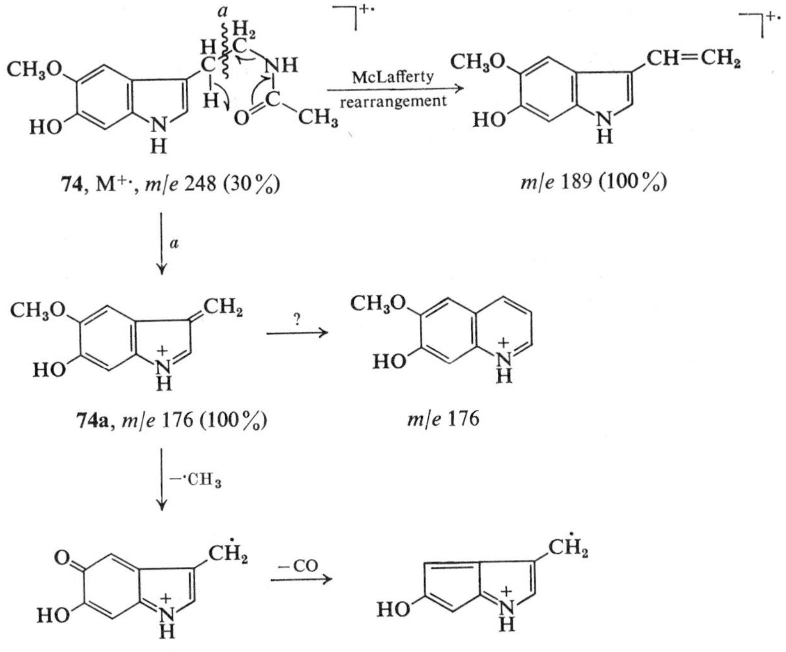

74, M$^{+\cdot}$, *m/e* 248 (30%)

m/e 189 (100%)

74a, *m/e* 176 (100%)

m/e 176

74b, *m/e* 161 (66%)

m/e 133 (17%)

It will be noticed that O-CH$_3$ fission (**74a** → **74b**) is an important process leading to the intense m/e 161 ion. This process is probably facilitated by delocalization of the positive charge and free electron in **74b** as shown.

The mass spectrum of 1,5-dimethoxy-3-(dimethylaminomethyl)indole (**75**) has been described (55). Again α,β-fission (loss of Me$_2$N·) is the dominant fragmentation, but here the next group lost is ·OCH$_3$, almost certainly that in the 1-position. This result may be readily rationalized as shown.

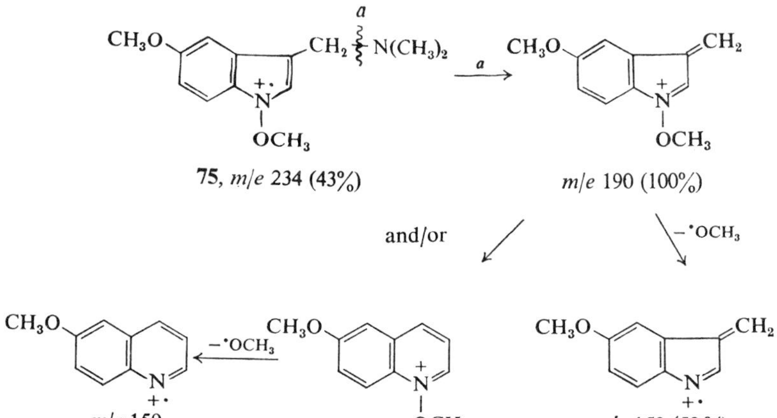

75, m/e 234 (43%)

m/e 190 (100%)

and/or

m/e 159

m/e 159 (52%)

It is interesting to note that α,β-fission (formation of a strong M-1 ion) is apparently suppressed in the 3,3′-azoindole **76** (56). The base peak occurs at m/e 159 and has been formulated as **76a**. It results from hydrogen transfer

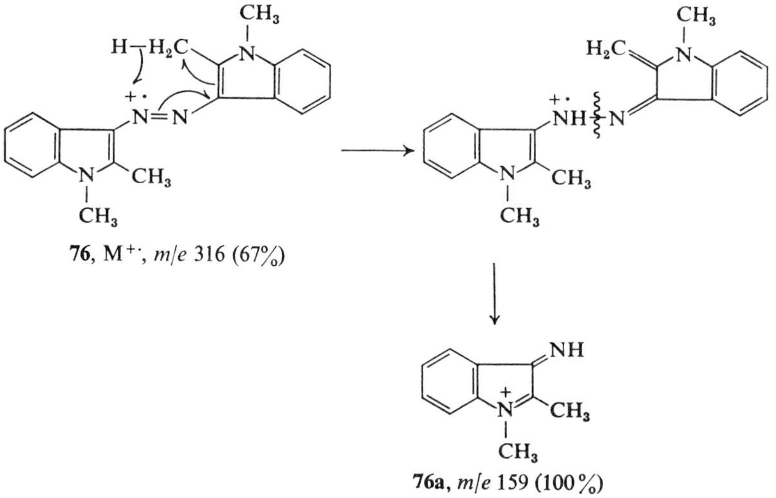

76, M$^{+\cdot}$, m/e 316 (67%)

76a, m/e 159 (100%)

from a methyl group to an azo-nitrogen atom, followed by N—N bond fission. The process is best imagined as starting from charge localization on an azo-nitrogen atom.

The mass spectrum of ethyl indole-2-carboxylate (**77**) (57) shows fragmentations analogous to those observed with pyrrole-2-carboxylates and may be rationalized as shown.

77, M$^{+\cdot}$, m/e 189 m/e 143 m/e 115   C$_7$H$_5^+$ m/e 89

Partial mass spectra of 1- and 3-acetylindoles (**78** and **79**) have been published. For 1-acetylindole the base peak of the spectrum is the M-C$_2$H$_2$O ion (58), corresponding to the usual loss of ketene from an acetylated amine. Loss of C$_2$H$_3$O is the major process for **79** (59), and the resultant ion may be formulated as **79a** or perhaps its monocyclic isomer **79b**.

78, M$^{+\cdot}$, m/e 159 m/e 117

79   79a, m/e 116   79b, m/e 116

VIII. REDUCED INDOLES

The mass spectrum of 2-carboxy-3-methoxycarbonyl-4,5,6,7-tetrahydroindole (**80**) is shown in Fig. 10.32 (60). No molecular ion is observed and the base peak of the spectrum is the ion **80a** resulting from decarboxylation. It is uncertain whether this is impact-induced or a thermal process. Further decompositions of **80a** are straightforward. Thus the ion m/e 151 results from loss of ethylene in a retro-Diels-Alder reaction whereas m/e 148 and m/e 120 are normal ester fission products as shown. The ion at m/e 164, corresponding to loss of a methyl group from **80a**, has been allocated the structure **80b**. Such a loss, resulting in a species containing an energetically unfavored sextet of electrons on oxygen, is not normally an important

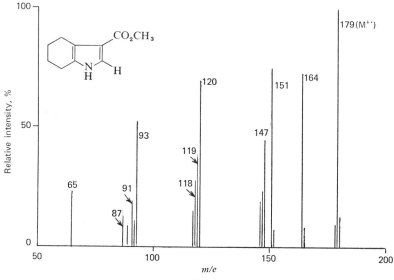

Fig. 10.32

fragmentation process with esters, and the methyl group lost may instead originate from the reduced six-membered ring as shown, leading to structure **80c** for the m/e 164 species.

The two possibilities could be clearly distinguished by labeling the ester group with CD_3 in place of CH_3. It is noteworthy that the analogous reduced benzoindoles do not show strong M-CH$_3$ ions.

4,7-Dihydrotryptophan (**81**) has been obtained by ultraviolet irradiation of aqueous solutions of L-tryptophan in the presence of sodium borohydride (61). Its mass spectrometric fragmentation is unexceptional and shows

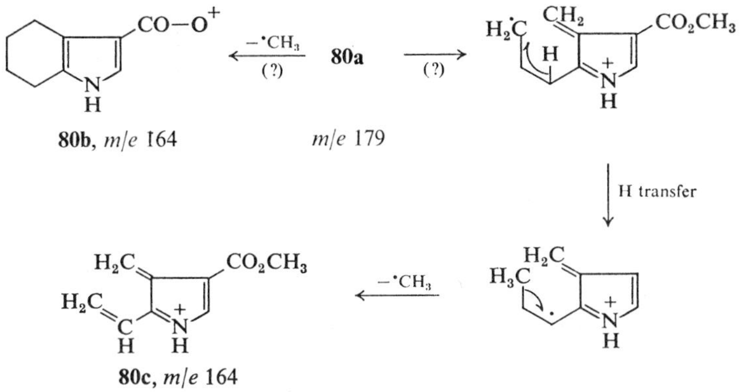

CO—O+
−•CH3
(?)
80a
(?)
CH2
H2C
CO2CH3
H
H
80b, m/e 164
m/e 179
H transfer
H2C
CO2CH3
H2C
−•CH3
H2C
H3C
H H
H
80c, m/e 164

processes which can be regarded as starting with charge localization on either nitrogen atom. Similar fragmentation occurs with the 4,5,6,7-tetrahydro-compound obtained by catalytic hydrogenation of **81**. No ion resulting from

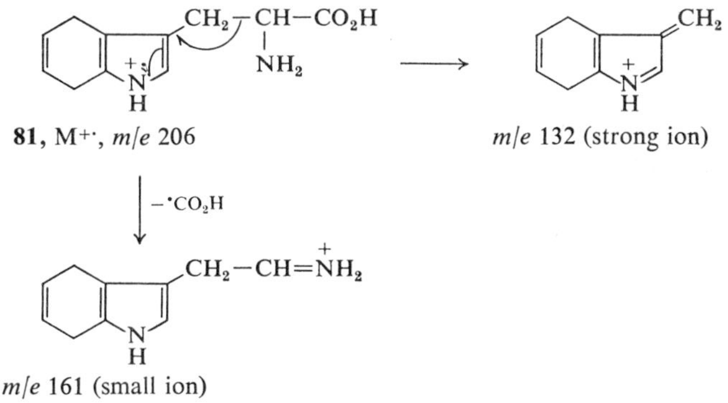

CH2—CH—CO2H
NH2
CH2
H
H
81, M+·, m/e 206
m/e 132 (strong ion)
−•CO2H
CH2—CH=NH2
H
m/e 161 (small ion)

a retro-Diels-Alder reaction is quoted in the limited data given for this compound.

IX. BENZINDOLES

A. Benz[*e*]Indoles

Attempts to determine the mass spectrum of 2-carboxybenz[*e*]indole resulted in decarboxylation and the recording of the spectrum of the parent heterocycle **82** (Fig. 10.33) (62). The fragmentation observed is analogous to that already discussed for indole itself and may be represented as shown. The corresponding 8,9-dihydro-compound also exhibits an M-CO_2 ion (**83**) rather

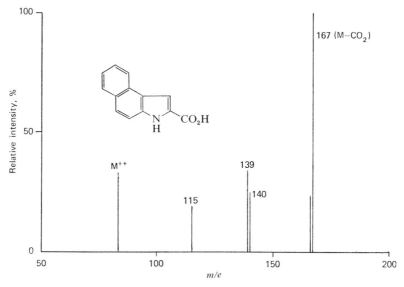

Fig. 10.33

than the parent ion, together with M-45 and M-46 ions. The fragmentation observed may be rationalized as shown. Esters derived from these acids are stable in the mass spectrometer and display intense parent ions. Their fragmentations closely parallel those already discussed for the pyrrole- and

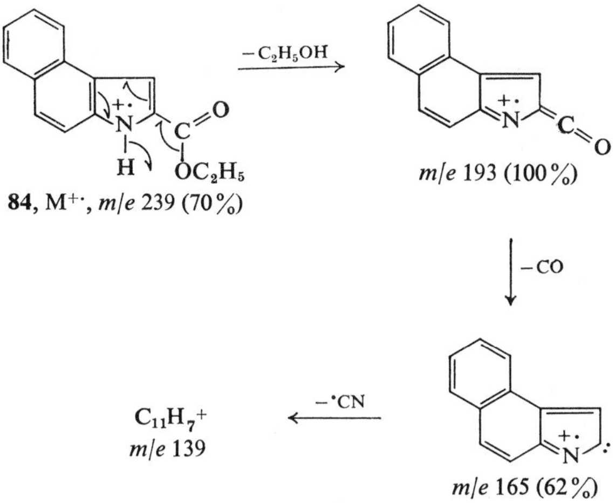

84, M$^{+\cdot}$, m/e 239 (70%)

m/e 193 (100%)

$-CO$

$C_{11}H_7^+$
m/e 139

$\xleftarrow{-\,^{\cdot}CN}$

m/e 165 (62%)

indole-2-carboxylates with M-ROH ions prominent in the spectra. The fragmentation sequence shown for the parent ethyl ester **84** is typical.

B. Carbazole

Carbazole shows little fragmentation upon electron impact (63) and the only ions of significant intensity occur at m/e 139 ($C_{11}H_7$) (13%) and 140 ($C_{11}H_8$) (11%). These clearly result from losses of HCN and H_2CN from the molecular ion, and their formation involves considerable rupture of the carbon skeleton of the molecule. A close analogy is the formation of the M-CS and M-$^{\cdot}$CHS ions (m/e 140 and 139) from dibenzothiophen, and the structures already proposed for these species (Chap. 8, Sec. I.H.12) may be applicable here.

X. TETRACYCLIC SYSTEMS CONTAINING AN INDOLE NUCLEUS

Some features of the mass spectrum of 2,3-phenylenedihydroindole, the adduct of indole and benzyne, have been described (64). The main fragmentation is loss of $CH_2N^{\cdot}$ to give the species at m/e 165, which probably is the fluorenyl cation. It is interesting to notice that simple fragmentation of the

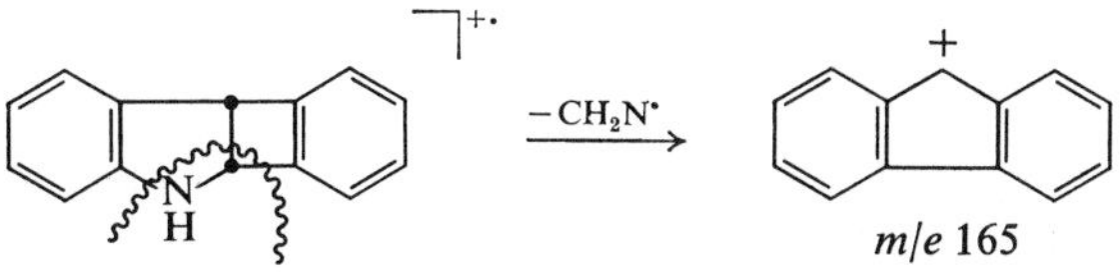

m/e 165

four-membered ring to give indole and benzyne fragments is not an important process.

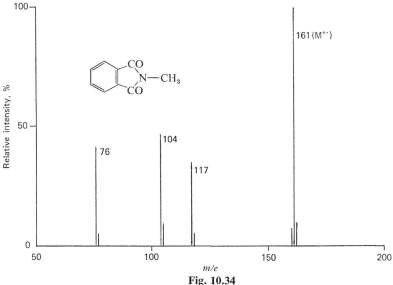

Fig. 10.34

XI. ISOINDOLE DERIVATIVES

Members of this class of compound which have been examined mass-spectrometrically are all derivatives of phthalimide and are notable for novel impact-induced rearrangements involving oxygen transfer. The spectrum of *N*-methylphthalimide (**85**) is shown in Fig. 10.34 (65).

Of considerable interest is the M-CO_2 ion at *m/e* 117, and the mechanism shown has been suggested for its formation.

Similar rearrangement with loss of CO_2 occurs with N-phenylphthalimide and 4-phthalimidobiphenyl, but not significantly with n-propylphthalimide, isopropylphthalimide, benzylphthalimide, allylphthalimide, or 2-phthalimido-biphenyl (**86**). The original publications (66, 67) should be consulted for details of fragmentation of these compounds. That of **86** is noteworthy, since it loses both ˙OH and ˙OH + CO in separate processes as evidenced by the appropriate metastable ions. The schemes shown have been postulated to explain these observations.

86, M⁺·, m/e 299

m/e 282

and

86, M⁺·

m/e 254

XII. COMPOUNDS WITH INDOLE NUCLEI FUSED TO NITROGEN-CONTAINING RINGS

The vast majority of compounds of this type are indole alkaloids and, as stated in the Introduction, it is not our intention to discuss the fragmentation of alkaloids in detail in this volume. Certain simple ring systems will, however, be considered and generalizations of their fragmentation made where possible.

A. Pyrrolo[2,3-b]indoles

Desoxybisnoreseroline (**87**) is one of the simplest indolines (tetrahydro-derivatives of this ring system) and its spectrum is shown in Fig. 10.35 (68). The peaks at m/e 144, 145, and 146 and at 130 and 131 correspond to losses of $C_{(1)}N$ and $C_{(2)}N$ fragments from the parent ion. In the absence of meta-stable ion and labeling data, conclusions as to ion structure must be tentative,

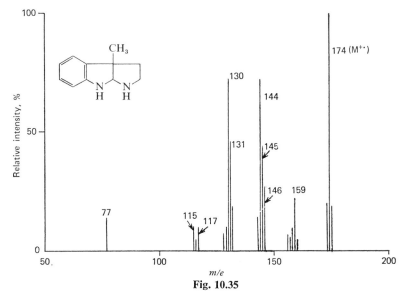

m/e
Fig. 10.35

but it seems likely that simple indole and quinolinium ions are produced and possible structures are suggested in the scheme.

The appropriate mass shifts occur in variously substituted indolines, and mass spectra can be useful in structure allocation. Several examples may be consulted in the original literature (68–70).

The "dimeric" indoline alkaloids isolated from the Calycanthaceae have also been subjected to mass-spectrometric examination and the fragmentations observed are easily explicable in terms of reasonable ion structures. Thus chimonanthine (**88**) undergoes fission of the bond connecting the two benzylic quaternary centers, followed by loss of a hydrogen atom, to give the base peak of the spectrum at m/e 172 (**71**).

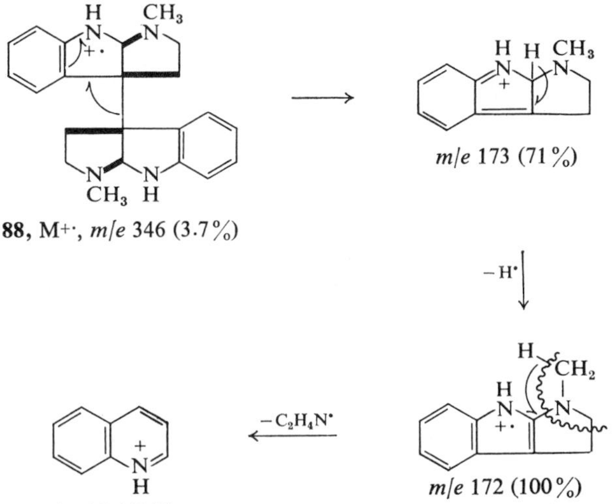

88, M$^{+\cdot}$, m/e 346 (3.7%)

m/e 173 (71%)

$-$H$^{\cdot}$

m/e 130 (49%)

$-$C$_2$H$_4$N$^{\cdot}$

m/e 172 (100%)

B.   The Sporidesmins

Sporidesmin (**89**), sporidesmin B (**90**), and sporidesmin C (**91**) have been isolated as metabolic products of the mold *Pithomyces chartarum* and are

89

90

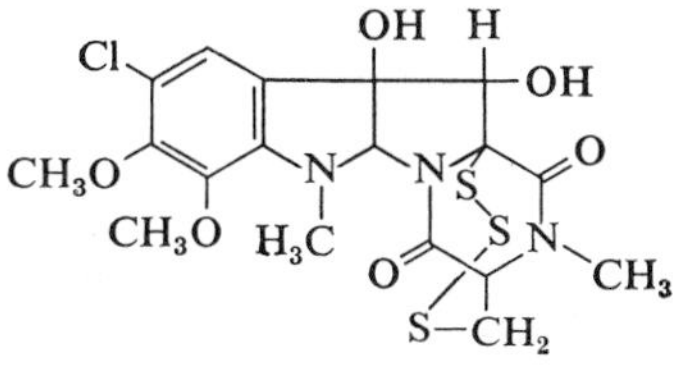

91

of considerable commercial importance since they cause liver damage and facial eczema in sheep. Their mass spectra (and those of their acetates) have been analyzed and provide much useful structural information (72).

The free metabolites are of low volatility and have low thermal stabilities. At high temperatures only the spectrum of S_8 was observed, but under optimum conditions satisfactory parent ions and spectra were obtained for **89** and **90**. The chief fragmentations of sporidesmin are loss of S_2 and fission of ring C to produce the hydroxyindole radical cation. The genesis of this fragment is complex and has been studied by means of deuterium labeling of the two hydroxy groups. The mechanisms shown have been postulated.

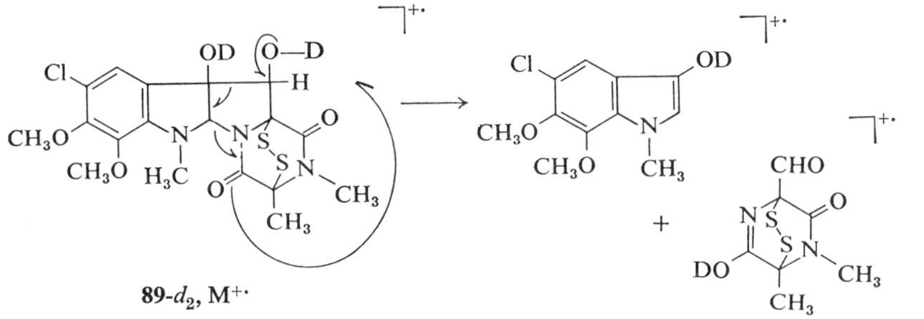

89-d_2, $M^{+\cdot}$ — *m/e* 242 (241) — *m/e* 233 (232)

and

m/e 171 (169) — *m/e* 241 (241)

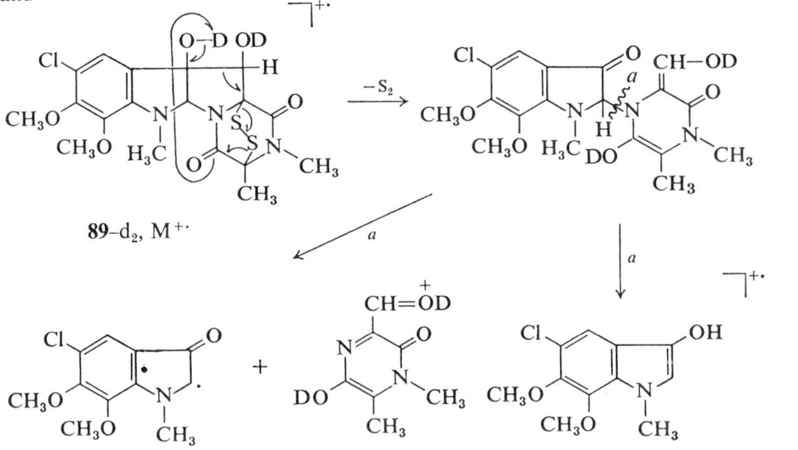

(Figures in parentheses are the masses of the corresponding *undeuterated* ions.)

Sporidesmin B shows very similar fragmentation with prominent ions corresponding to $[M\text{-}S_2]^{+\cdot}$ and *m/e* 241, together with an ion at *m/e* 153 which corresponds to the *m/e* 169 peak in the spectrum of **89**.

Sporidesmin C has been isolated and examined mass-spectrometrically only as its diacetate. This extra functionalization profoundly modifies the fragmentation, and for the rationale of this spectrum and that of sporidesmin diacetate the reader is referred to the original publication (73).

Anhydrous acid converts sporidesmin diacetate into the indole derivative **92**. Under electron impact this compound undergoes the interesting cyclic fragmentation of ring D shown (74).

92

Addenda

Some speculations have appeared on the fragmentations of simple aziridines (Section I) (75). Further work has been published on 1-acyl-aziridines (76) and fluoroaziridines (77) and two publications have appeared on aziridinones (α-lactams) (78, 79). A detailed discussion of the fragmentation of simple azetidines and azetidinones (Section II) is now available (80).

Further studies on the mass spectra of 2-alkyl-N-methyl-pyrrolidines (Section III-A) have been reported (81), and the fragmentation of a range of phthalimides and succinimides (Section III-F and XI), including N-hydroxy-compounds, have been described (82). Further work on pyrrolizidine compounds (Section V) covers the fragmentation of diols (83), and a study of simple indoles and related indolines is reported (84). High resolution and labelling studies are reported for isatin, oxindole, and 2-oxo-3-indolyl compounds (Section VII) (85). Interesting comparisons with photochemical behaviour have been made in the case of some N-alkyl-2,4-diphenylpyrroles (Section VI-B-1) (86).

REFERENCES

1. E. J. Gallegos and R. W. Kiser, *J. Phys. Chem.*, **65**, 1177 (1961).
2. American Petroleum Institute, Research Project 44. Spectrum 763.
3. Ref. 2. Spectrum 1141.
4. Ref. 2. Spectrum 1140.
5. Q. N. Porter and R. Spear, Unpublished observations.
6. R. G. Kostyanovskii and O. A. Pan'shin, *Izv. Akad. Nauk SSSR, Ser. Khim.*, **1965**, 567.

7. R. G. Kostyanovskii and O. A. Pan'shin, *Izv. Akad. Nauk SSSR, Ser. Khim.*, **1965**, 740.

8. R. A. Mitsch, E. W. Neuvar, and P. H. Ogden, *J. Heterocycl. Chem.*, **4**, 389 (1967).

9. E. J. Gallegos and R. W. Kiser, *J. Phys. Chem.*, **66**, 136 (1962).

10. O. A. Pan'shin, V. P. Nechiporenko, and R. G. Kostyanovskii, *Izv. Akad. Nauk SSSR, Ser. Khim.*, **1966**, 228.

11. J.-L. Imbach, E. Doomes, N. H. Cromwell, H. E. Baumgarten, and R. G. Parker, *J. Org. Chem.*, **32**, 3123 (1967).

12. H. E. Audier, M. Fétizon, H. B. Kagan, and J. L. Luche, *Bull. Soc. Chim. Fr.*, **1967**, 2297.

13. A. K. Bose and I. Kugajevsky, *Tetrahedron*, **23**, 957 (1967).

14. L. A. Singer and G. A. Davis, *J. Amer. Chem. Soc.*, **89**, 942 (1967).

15. A. M. Duffield, H. Budzikiewicz, D. H. Williams, and C. Djerassi, *J. Amer. Chem. Soc.*, **87**, 810 (1965).

16. J. S. Fitzgerald, *Aust. J. Chem.*, **18**, 589 (1965).

17. B. Lüning and K. Leander, *Acta Chem. Scand.*, **19**, 1607 (1965).

18. A. M. Duffield, H. Budzikiewicz, and C. Djerassi, *J. Amer. Chem. Soc.*, **87**, 2926 (1965).

19. G. R. Pettit, R. L. Smith, A. K. Das Gupta, and J. L. Occolowitz, *Can. J. Chem.*, **45**, 501 (1967).

20. N. J. Leonard, D. A. Durand, and F. Uchimaru, *J. Org. Chem.*, **32**, 3607 (1967).

21. H. J. Jakobsen, S.-O. Lawesson, J. T. B. Marshall, G. Schroll, and D. H. Williams, *J. Chem. Soc., B*, **1966**, 940.

22. A. M. Duffield and C. Djerassi, *J. Amer. Chem. Soc.*, **87**, 4554 (1965).

23. C.-O. Andersson, R. Ryhage, and E. Stenhagen, *Ark. Kemi*, **19**, 417 (1962).

24. K. Heyns and H.-F. Grützmacher, *Justus Liebigs Ann. Chem.*, **667**, 194 (1963).

25. R. Grigg and B. G. Odell, *J. Chem. Soc., B*, **1966**, 218.

26. A. M. Duffield, H. Budzikiewicz, and C. Djerassi, *J. Amer. Chem. Soc.*, **87**, 2913 (1965).

27. A. F. Casy and M. M. A. Hassan, *Tetrahedron*, **23**, 4075 (1967).

28. R. Ryhage and E. Stenhagen, *Ark. Kemi*, **14**, 483 (1959).

29. F. W. McLafferty and M. C. Hamming, *Chem. Ind.* (London), **1958**, 1366.

30. F. W. McLafferty and R. S. Gohlke, *Anal. Chem.*, **31**, 2076 (1959).

31. M. Flavin and S. Tsunakawa, *J. Biol. Chem.*, **241**, 3340 (1966).

32. K. L. Rinehart and D. B. Borders, *J. Amer. Chem. Soc.*, **85**, 4037 (1963).

33. H. C. Volger and W. Brackman, *Rec. Trav. Chim.*, **86**, 243 (1967).

34. N. Neuner-Jehle, H. Nesvadba, and G. Spiteller, *Monatsh. Chem.*, **96**, 321 (1965).

35. C. C. J. Culvenor, G. M. O'Donovan, and L. W. Smith, *Aust. J. Chem.*, **20**, 757 (1967).

36. C. K. Atal, R. K. Sharma, C. C. J. Culvenor and L. W. Smith, *Aust. J. Chem.*, **19**, 2189 (1966).

37. R. G. Cooks, F. L. Warren and D. H. Williams, *J. Chem. Soc., C*, **1967**, 286.

38. H. Budzikiewicz, C. Djerassi, A. H. Jackson, G. W. Kenner, D. J. Newman, and J. M. Wilson, *J. Chem. Soc.*, **1964**, 1949.

39. H. Budzikiewicz, C. Djerassi, and D. H. Williams, *Mass Spectrometry of Organic Compounds*, Holden-Day, San Francisco, 1967, p. 597.

40. A. M. Duffield, R. Beugelmans, H. Budzikiewicz, D. A. Lightner, D. H. Williams, and C. Djerassi, *J. Amer. Chem. Soc.*, **87**, 805 (1965).

41. Ref. 2. Spectrum 632.

42. A. Treibs, K. Jacob, and A. Dietl, *Justus Liebigs Ann. Chem.*, **702**, 112 (1967).

43. M. Grassl, U. Coy, R. Seyffert, and F. Lynen, *Biochem. Z.*, **338**, 771 (1963).

44. S. Hannessian and J. S. Kaltenbronn, *J. Amer. Chem. Soc.*, **88**, 4509 (1966).

45. A. H. Jackson, G. W. Kenner, H. Budzikiewicz, C. Djerassi, and J. M. Wilson, *Tetrahedron*, **23**, 603 (1967).
46. A. Treibs and K. Jacob, *Justus Liebigs Ann. Chem.*, **699**, 153 (1966).
47. Ref. 2. Spectrum 623.
48. J. L. Occolowitz and G. L. White, *Aust. J. Chem.*, **21**, 997 (1968).
49. J. H. Beynon and A. E. Williams, *Appl. Spectrosc.*, **13**, 101 (1959).
50. M. Marx and C. Djerassi, *J. Amer. Chem. Soc.*, **90**, 678 (1968).
51. M. S. von Wittenau and H. Els, *J. Amer. Chem. Soc.*, **85**, 3425 (1963).
52. K. Biemann, J. Seibl and F. Gapp, *J. Amer. Chem. Soc.*, **83**, 3795 (1961).
53. P. Pfaender, *Justus Liebigs Ann. Chem.*, **707**, 209 (1967).
54. D. E. Hall and A. H. Jackson, *J. Chem. Soc.*, *C*, **1967**, 1681.
55. S. R. Johns, J. A. Lamberton, and J. L. Occolowitz, *Aust. J. Chem.*, **20**, 1737 (1967).
56. A. S. Bailey and J. J. Merer, *J. Chem. Soc.*, *C*, **1966**, 1345.
57. U. K. Pandit, H. J. Hofman, and H. O. Huisman, *Tetrahedron*, **20**, 1679 (1964).
58. O. Buchardt, J. Becher, C. Lohse, and J. Møller, *Acta Chem. Scand.*, **20**, 262 (1966).
59. A. L. Jennings and J. E. Boggs, *J. Org. Chem.*, **29**, 2065 (1964).
60. U. K. Pandit and H. O. Huisman, *Rec. Trav. Chim.*, **85**, 311 (1966).
61. O. Yomemitsu, P. Cerutti, and B. Witkop, *J. Amer. Chem. Soc.*, **88**, 3941 (1966).
62. U. K. Pandit and H. O. Huisman, *Rec. Trav. Chim.*, **83**, 50 (1964).
63. K. G. Das, P. T. Funke, and A. K. Bose, *J. Amer. Chem. Soc.*, **86**, 3729 (1964).
64. M. E. Kuehne and T. Kitagawa, *J. Org. Chem.*, **29**, 1270 (1964).
65. R. A. W. Johnstone, B. J. Millard, and D. S. Millington, *Chem. Commun.*, **1966**, 600.
66. J. L. Cotter and R. A. Dine-Hart, *Chem. Commun.*, **1966**, 809.
67. J. Sharvit and A. Mandelbaum, *Israel J. Chem.*, **5**, 33 (1967).
68. E. Clayton and R. I. Reed, *Tetrahedron*, **19**, 1345 (1963).
69. B. Robinson and G. Spiteller, *Chem. Ind.* (London), **1964**, 459.
70. G. Spiteller and M. Spiteller-Friedmann, *Tetrahedron Lett.*, **1963**, 147.
71. E. Clayton, R. I. Reed, and J. M. Wilson, *Tetrahedron*, **18**, 1495 (1962).
72. J. S. Shannon, *Tetrahedron Lett.*, **1963**, 810.
73. R. Hodges and J. S. Shannon, *Aust. J. Chem.*, **19**, 1059 (1966).
74. R. Hodges, J. W. Ronaldson, J. S. Shannon, A. Taylor, and E. P. White, *J. Chem. Soc.*, **1964**, 26.
75. J. H. Beynon, R. A. Saunders, and A. E. Williams *The Mass Spectra of Organic Molecules*, Elsevier, London, 1968, p. 277.
76. R. G. Kostyanovskii, T. Z. Laloyan and I. I. Tchervin, *Izvest. Akad. Nauk. SSSR.*, **7**, 1530 (1968).
77. R. G. Kostyanovskii, Z. E. Samojlova and I. I. Tchervin, *Dokl. Akad. Nauk. SSSR.* **186**, 835 (1969).
78. H. E. Baumgarten, R. G. Parker and D. L. von Minden, *Org. Mass Spec.*, **2**, 1221 (1969).
79. I. Lengyel, D. B. Uliss, M. Mehdi Nafissi-V, and J. C. Sheehan, *Org. Mass Spec.*, **2**, 1239 (1969).
80. M. B. Jackson, T. M. Spotswood and J. H. Bowie, *Org. Mass Spec.*, **1**, 857 (1968).
81. S. Osman, C. J. Dooley and T. Foglia, *Org. Mass Spec.*, **2**, 977 (1969).
82. J. H. Bowie, M. T. W. Hearn and A. D. Ward, *Austral. J. Chem.*, **22**, 175 (1969).
83. A. J. Aasen, C. C. J. Culvenor and L. W. Smith, *J. Org. Chem.*, **34**, 4137 (1969).
84. A. H. Jackson and P. Smith, *Tetrahedron*, **24**, 2227 (1968).
85. J. A. Ballantine, R. G. Fenwick and M. Alam, *Org. Mass Spec.*, **1**, 467 (1968).
86. A. Padwa, R. Gruber, D. Pashayan, M. Bursey and L. Dusold, *Tetrahedron Lett.* 3659 (1968).

11 COMPOUNDS WITH ONE NITROGEN ATOM IN A SIX-MEMBERED RING

I. PIPERIDINES

The mass spectrum (Fig. 11.1) (1) of piperidine (1) has been fully interpreted by using mass shifts observed in the spectra of the N-d, 2,2-d_2, and 3,3-d_2 derivatives as aids to understanding the observed fragmentation (2, 3).

As shown below, the M-1 ion comes only from loss of an α-hydrogen atom. It is interesting to note that a substantial isotope effect discriminating against loss of deuterium is observed in the spectrum of the 2,2-d_2 compound.

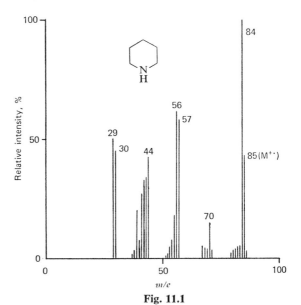

Fig. 11.1

The remaining ions can be explained by various hydrogen transfer reactions and bond fissions occurring in the internal α-cleavage ion **1a**, and the rationale

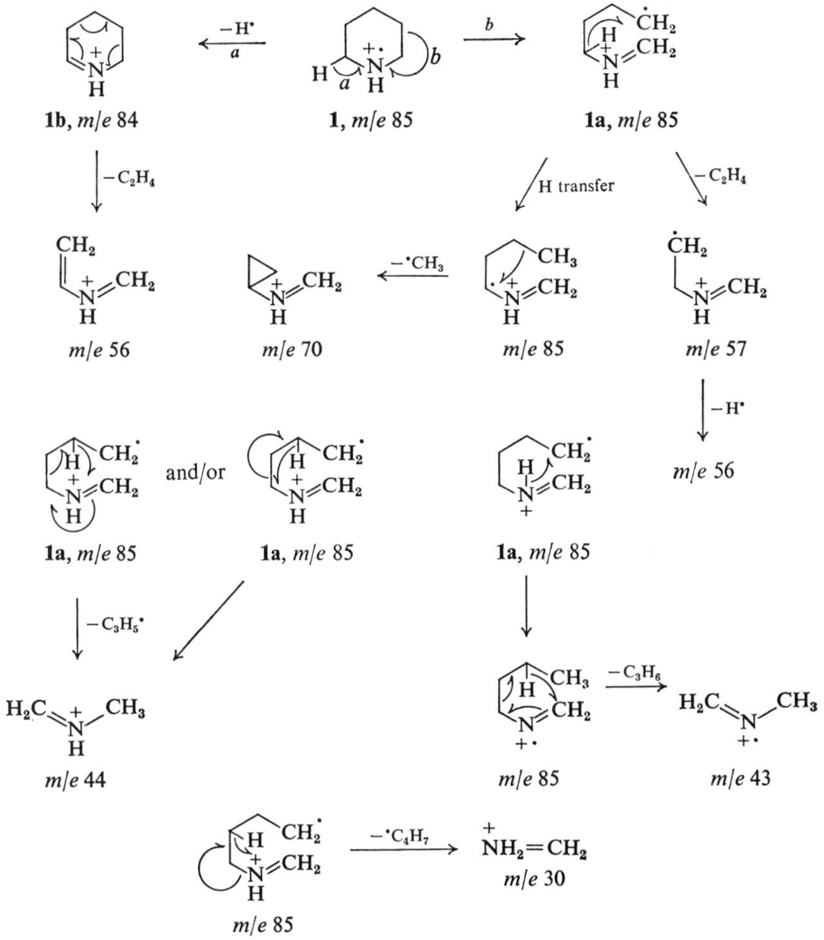

1b, m/e 84   **1**, m/e 85   **1a**, m/e 85

$-C_2H_4$ H transfer $-C_2H_4$

CH_2
m/e 56 m/e 70 m/e 85 m/e 57

$-$ ·CH₃

$-H$·

m/e 56

1a, m/e 85   and/or   **1a**, m/e 85   **1a**, m/e 85

$-C_3H_5$·

H_2C—CH_3
m/e 44

m/e 85 $\xrightarrow{-C_3H_6}$ H_2C—CH_3
m/e 43

$\xrightarrow{-·C_4H_7}$ NH_2=CH_2
m/e 30

m/e 85

shown has been put forward. The major fragmentations of *N*-methyl-piperidine follow the routes described for the parent compound. For minor differences the reader is referred to the original publications (2, 3).

A.   *C*-Alkylpiperidines

The mass spectra of 2-, 3- and 4-methylpiperidines are shown in Figs. 11.2, 11.3, and 11.4 (2). Since no deuterium labeling experiments have yet been

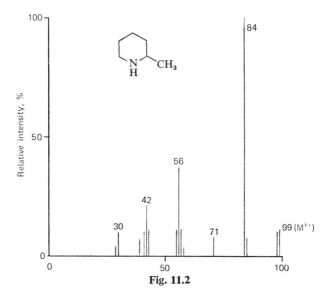

Fig. 11.2

Fig. 11.3

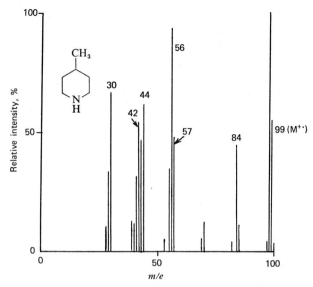

Fig. 11.4

reported for these compounds, it is clearly impossible to present complete rationales for their complex fragmentations but some points of interest emerge from comparison of the spectra. It is clear that a large proportion of the total ion current is carried by the m/e 84 species in the 2-methyl compound (**2**) and it seems fairly certain that most of this ion is the result of simple α-cleavage (**2** → **1b**). The higher stability of the methyl radical lost here (relative to the hydrogen atom lost in the analogous process for piperidine) probably reduces the importance of the internal α-cleavage ion (2a) and of the fragmentations initiated by this species.

The amount of fragmentation in the 3- and 4-methylpiperidines is clearly much greater and is presumably initiated by the species 3a and 4a. In each case the M-˙CH$_3$ ion (m/e 84) is relatively more intense than the corresponding ion in piperidine; this leads to the suspicion that they may well result (at least in part) from loss of the intact C-methyl group, in other words, that the M-˙CH$_3$ peaks may be enhanced by the presence of structures 3b and 4b.

3b   4b

Deuterium-labeling experiments to elucidate the finer details of the fragmentations of these compounds would be of considerable interest.

The 2-substituted piperidine ring is present in many alkaloids, and its presence can be detected from the appropriate α-cleavage ion. Thus coniine (2-propylpiperidine) (5) shows the expected fragmentation, as do conhydrine (6), isopelletierine (7), and sedamine (8) as well as the dipiperidine alkaloid anaferine (9) (4, 5).

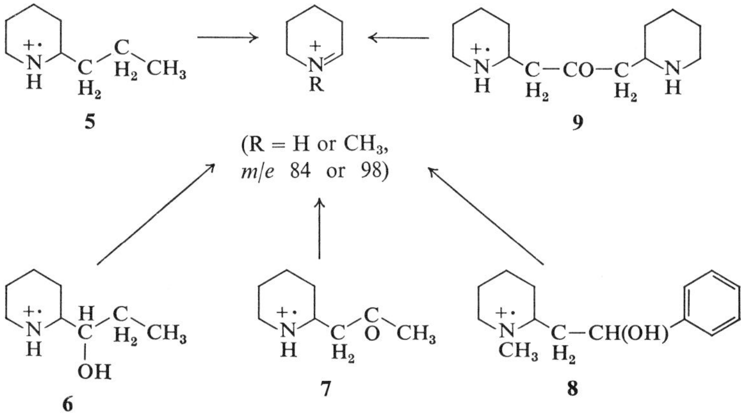

It is interesting to notice that the facile α-cleavage represses the McLafferty rearrangement in isopelletierine (7 ↛ 7a or 7b) whereas in 2,6-diphenacyl-piperidine (10) (cis-norlobelanine) the McLafferty rearrangement competes

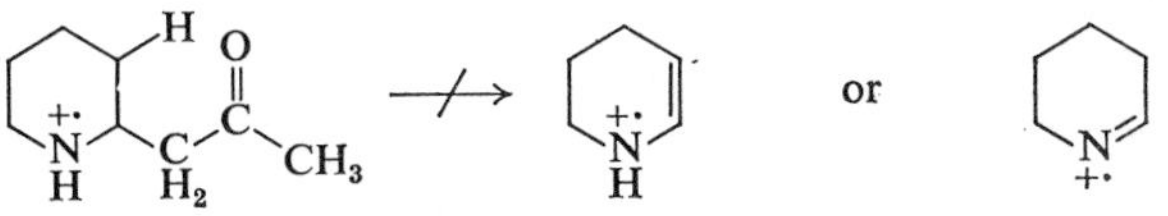

7, M$^{+\cdot}$, *m/e* 141   **7a**   *m/e* 83   **7b**

with α-fission for the removal of the first phenacyl substituent (4). It should be noted that the origin of the transferred hydrogen atom has not been demonstrated by deuterium labeling.

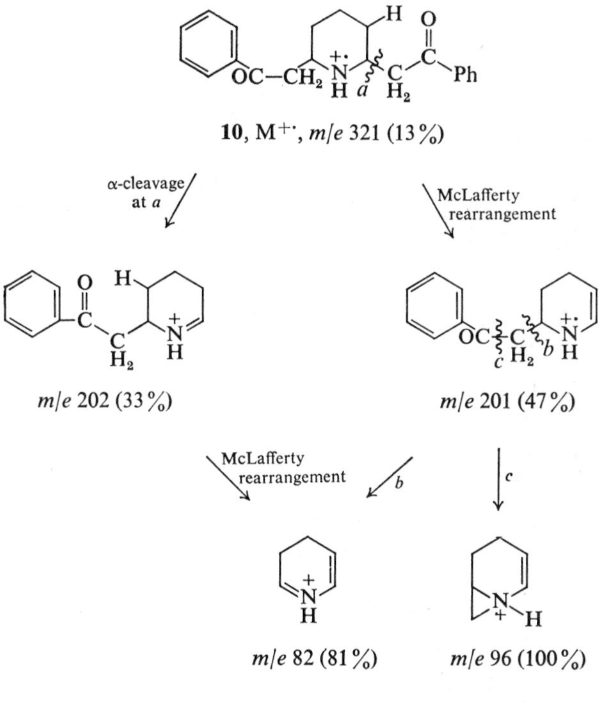

10, M$^{+\cdot}$, *m/e* 321 (13%)

α-cleavage at *a*

McLafferty rearrangement

m/e 202 (33%)   *m/e* 201 (47%)

McLafferty rearrangement   *b*   *c*

m/e 82 (81%)   *m/e* 96 (100%)

The alkaloid carpaine (**11**) is a macrocyclic lactone with two piperidine units incorporated in the ring. Mass spectrometry provided vital information in the deduction of the correct structure for this compound, and its major fragmentations are outlined (6). It will be noticed that a McLafferty rearrangement, leading to fission of the macrocyclic ring, initiates the fragmentation. The hydrogen transferred is shown arbitrarily as α, but could equally well originate from the γ-position of the piperidine ring.

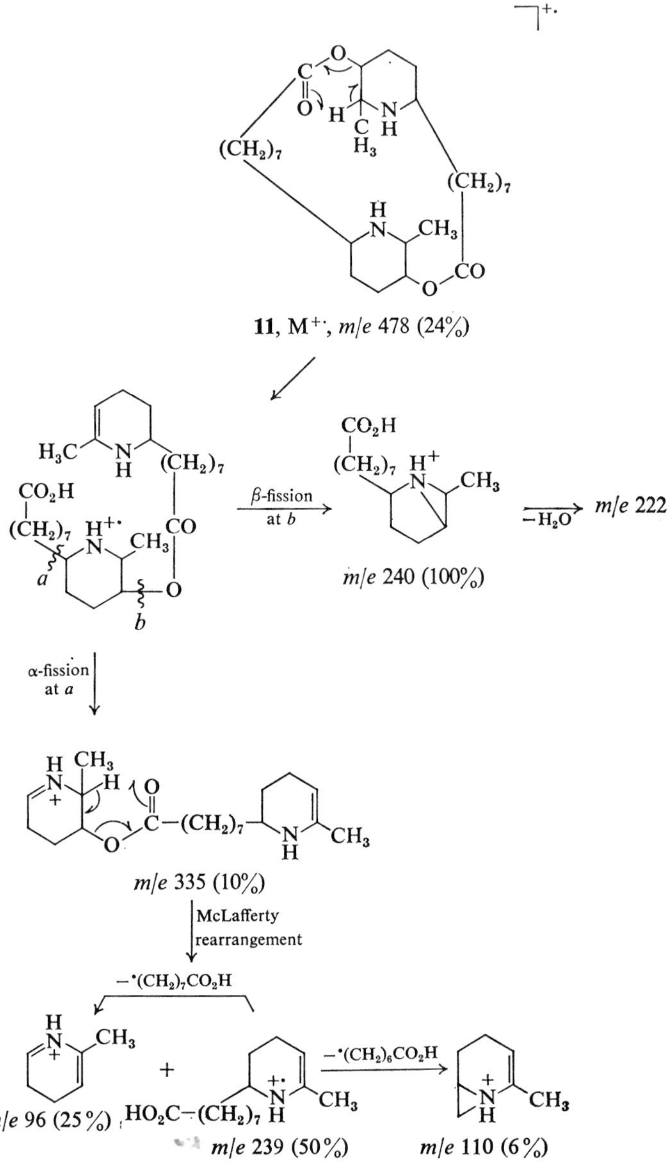

11, $M^{+\cdot}$, m/e 478 (24%)

β-fission at b

m/e 240 (100%)

$\xrightarrow{-H_2O}$ m/e 222

α-fission at a

m/e 335 (10%)

McLafferty rearrangement

$-{}^{\cdot}(CH_2)_7CO_2H$

m/e 96 (25%)   + $HO_2C-(CH_2)_7$ H

$-{}^{\cdot}(CH_2)_6CO_2H$

m/e 239 (50%)

m/e 110 (6%)

B. N-Substituted Piperidines

The mass spectrum of N-methylpiperidine has already been described. Little information is published on the mass spectra of higher homologs of this compound. It seems likely that α-cleavage of the N-alkyl group would be a major fragmentation mode. It would be of interest to compare the relative importances of this type of cleavage with the competing α-fission of the piperidine ring.

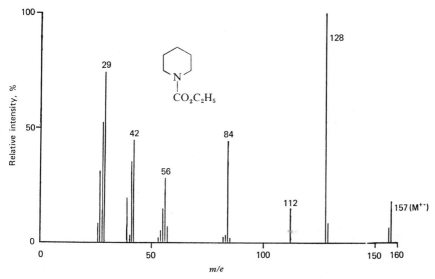

N-Acylpiperidines have not been subject to the detailed study which has been described for their pyrrolidine analogs but it seems probable that basically similar fragmentations would be observed.

The low resolution spectrum (7) of ethyl N-piperidino-carboxylate (12) is shown in Fig. 11.5. Fragmentation mainly involves fissions of the ester group, and it is only below m/e 84 that fragmentation of the ring becomes apparent. The importance of cleavage of the carbon–oxygen bond of the ester to give m/e 128 as the base peak of the spectrum is surprising; it is

Fig. 11.5

possible that the ion is cyclic (**12b**), avoiding the energetically unfavored sextet of electrons on oxygen present in **12a**.

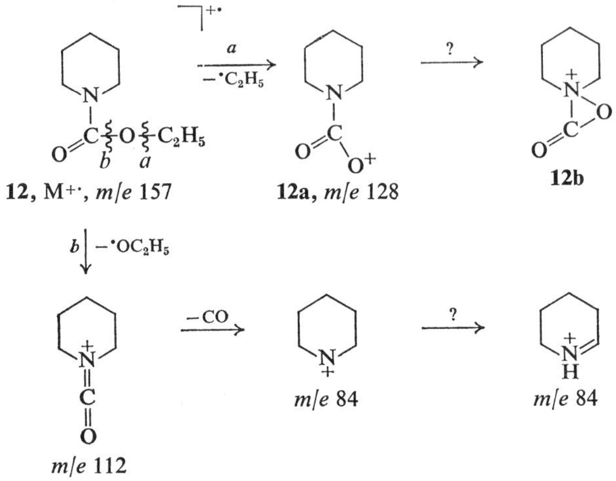

The mass spectrum of the dipiperidinothiuramdisulfide **13** has been reported in connection with a detailed study of this sulfur-containing functional group (8). Like that of other thiuramdisulfides, the spectrum is marked by skeletal rearrangements and contains ions corresponding to M-S, M-2S, and M-4S. The major features of the fragmentation are shown.

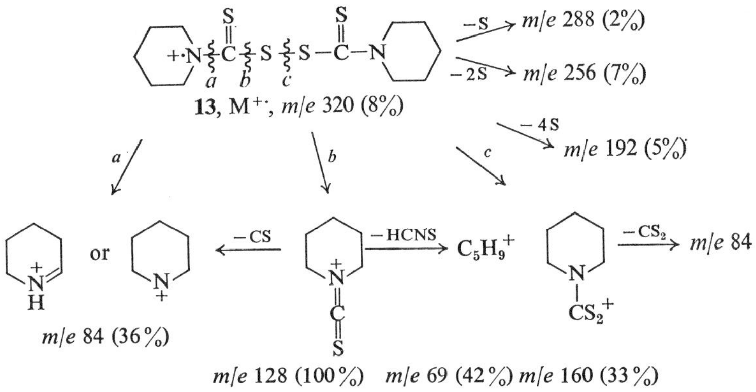

C. Piperidones and Glutarimides

The spectrum of the parent compound (**14**) has been analyzed with use of deuterium-labeling and high-resolution data (9). The fragmentation resembles that already discussed for pyrrolidone and is outlined in the scheme. A similar fragmentation pathway is observed with N-methyl-2-piperidone.

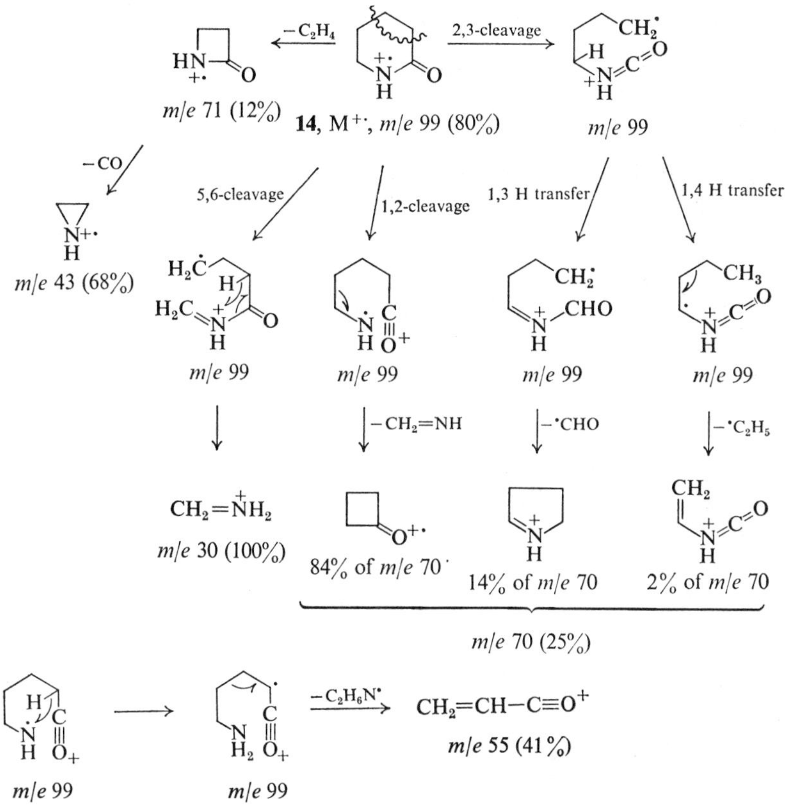

m/e 71 (12%)   14, M$^{+\cdot}$, m/e 99 (80%) m/e 99

$-C_2H_4$ 2,3-cleavage

m/e 43 (68%) $-CO$

5,6-cleavage   1,2-cleavage   1,3 H transfer   1,4 H transfer

m/e 99 m/e 99 m/e 99 m/e 99

$-CH_2=NH$ $-\cdot CHO$ $-\cdot C_2H_5$

$CH_2=\overset{+}{N}H_2$
m/e 30 (100%)

84% of m/e 70

14% of m/e 70   2% of m/e 70

m/e 70 (25%)

m/e 99 m/e 99 $-C_2H_6N^{\cdot}$ $CH_2=CH-C\equiv O^+$
m/e 55 (41%)

The mass spectrum of glutarimide has not been reported. Low-resolution spectra have been published for the glutarimide antibiotics cycloheximide (15), inactone (16), and streptimidone (17) (10). The main fragmentations are clearly recognized as allylic cleavages and McLafferty rearrangements. They do not involve fragmentation of the glutarimide ring. No doubt fragments characteristic of this part of the molecule do occur but, in view of the large number of heteroatoms present, careful high-resolution studies combined with deuterium labeling would be required for complete analysis of the spectra. The main fragmentations of 15, 16, and 17 are outlined in the scheme.

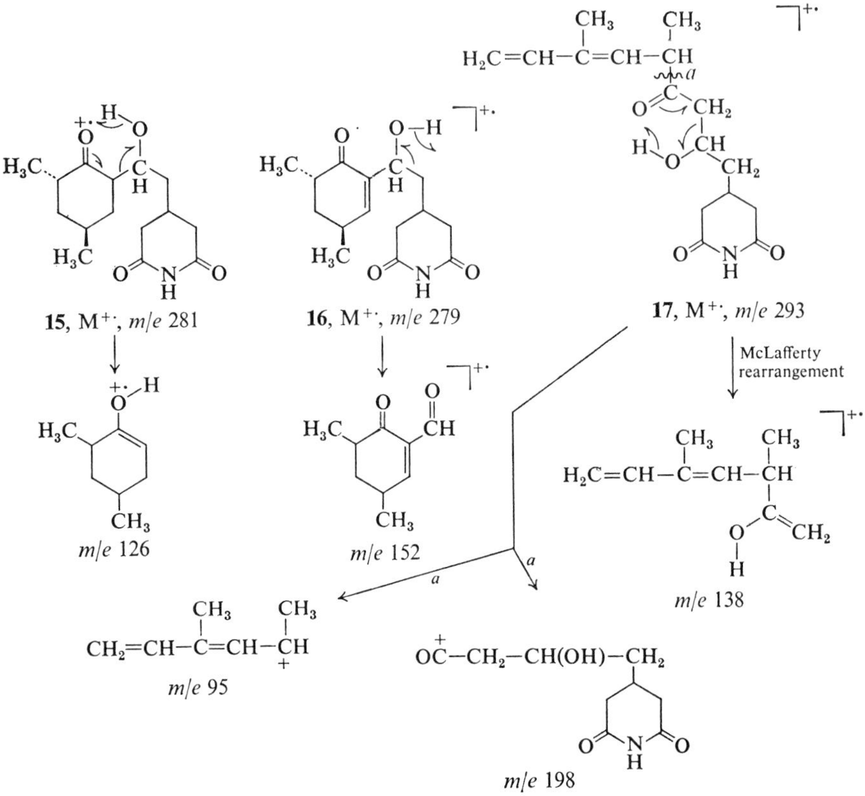

$$CH_3 \quad CH_3$$
$$H_2C=CH-C=CH-CH$$

15, M^+, m/e 281

16, M^+, m/e 279

17, M^+, m/e 293

McLafferty rearrangement

m/e 126

m/e 152

$$CH_3 \quad CH_3$$
$$H_2C=CH-C=CH-CH$$

m/e 138

$$CH_3 \quad CH_3$$
$$CH_2=CH-C=CH-CH$$

m/e 95

$$\overset{+}{O}C-CH_2-CH(OH)-CH_2$$

m/e 198

D. Tropane Alkaloids

The fragmentations of a number of members of this class of compound have been carefully studied with use of deuterium labeling to identify points of fragmentation (11). In general fragmentations are initiated by cleavages of C—C bonds next to the nitrogen atom, and a rationale for the mass spectrum of tropinone (**18**) is shown.

The mass spectrum of tropine has also been examined and a similar fragmentation sequence has been proposed. The original paper should be consulted for more detail (11).

It is clear that tropane alkaloids are a group of heterocycles whose mass-spectrometric fragmentation is easily explicable in terms of intact molecule structures, and mass spectrometry should be a powerful analytical tool for these compounds.

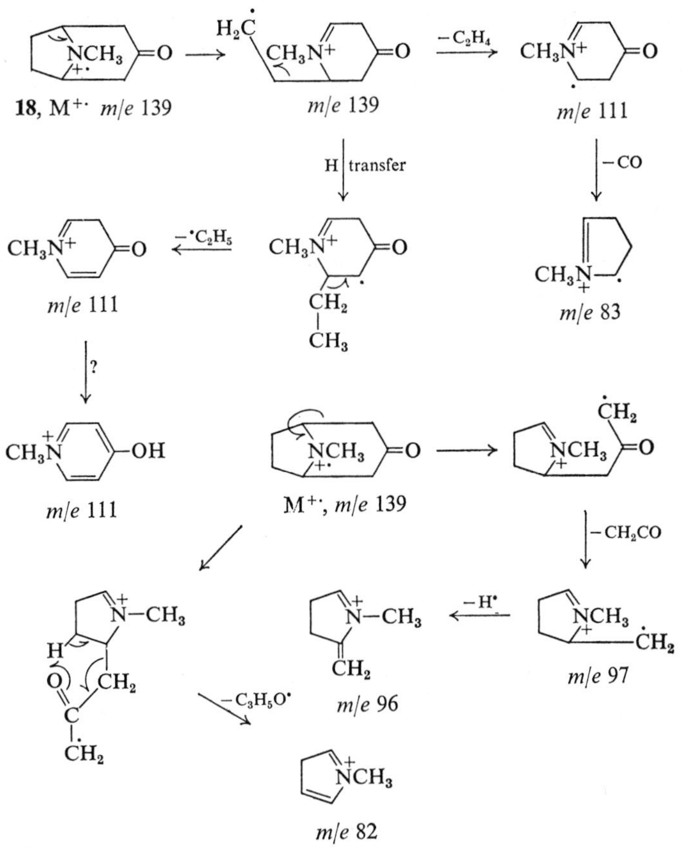
18, M+· m/e 139
m/e 139
−C2H4
m/e 111
H transfer
−CO
−·C2H5
m/e 111
m/e 83
?
m/e 111
M+·, m/e 139
−CH2CO
−H·
m/e 97
−C3H5O·
m/e 96
m/e 82

1. Ψ-Pelletierine

This homolog (**19**) of tropinone fragments similarly to that compound, with ions derived from fission of the six-membered keto ring in tropinone being, as expected, displaced upwards by 14 mass units (12). Basically similar fragmentation is observed with 9-methyl-3-oxagranatan-7-one (**20**),

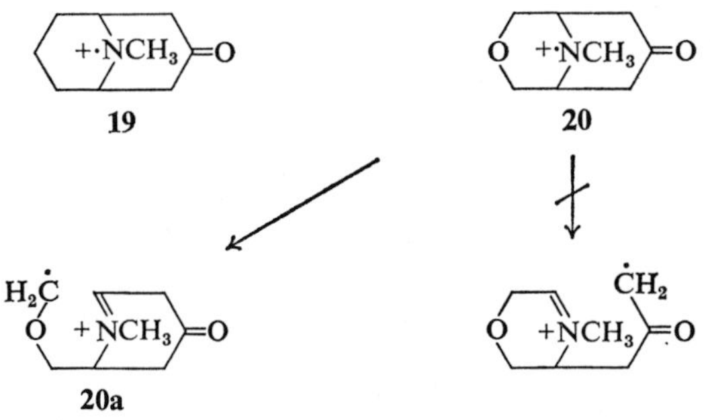
19
20
20a

but the oxygen heteroatom favors fragmentation of its own ring because of the increased stability of the initially-produced fragment **20a**. Fragmentation by the route involving fission of the keto ring is much reduced in importance.

II.   PYRIDINES

A.   Pyridine

The mass spectrum of pyridine (**21**) is shown in Fig. 11.6 (13). It will be seen to be simple and dominated by loss of the elements of HCN from the

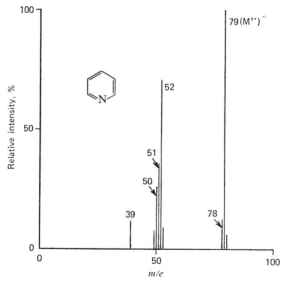

molecular ion, A minor process is loss of HCN from the M-1 ion, leading to the peak at m/e 51. Williams and Ronayne (14) have studied the fragmentation of 2,6-d_2-pyridine (**22**) and 2-d_1-pyridine (**23**). From the intensities

Fig. 11.6

of the metastable ions for the processes illustrated, these workers show that, even at ionizing potentials as low as 14 eV, statistical involvement of the α-, β-, and γ-hydrogen atoms occurs in the loss of HCN from the molecular

$$C_4H_2D_2 \quad \xleftarrow[m^* 36\cdot0]{\text{−HCN}} \quad C_5H_3D_2N \quad \xrightarrow[m^* 34\cdot7]{\text{−DCN}} \quad C_4H_3D$$

$m/e\ 54 \qquad\qquad 22,\ M^{+\cdot},\ m/e\ 81 \qquad\qquad m/e\ 53$

and

$$C_4H_4 \quad \xleftarrow[m^* 33\cdot8]{\text{−DCN}} \quad C_5H_4DN \quad \xrightarrow[m^* 35\cdot2]{\text{−HCN}} \quad C_4H_3D$$

$m/e\ 52 \qquad\qquad 23,\ M^{+\cdot},\ m/e\ 80 \qquad\qquad m/e\ 53$

$\swarrow_{-H\cdot} \qquad \searrow_{-D\cdot}$

$$C_4H_2D^+ \quad \xleftarrow[m^* 34\cdot2]{\text{−HCN}} \quad C_5H_3DN^+ \qquad C_5H_4N^+ \quad \xrightarrow[m^* 33\cdot4]{\text{−HCN}} \quad C_4H_3^+$$

$m/e\ 52 \qquad\qquad m/e\ 79 \qquad\qquad m/e\ 78 \qquad\qquad m/e\ 51$

$\underline{\qquad\qquad\qquad\qquad}\nearrow$

$-DCN,\ m^*\ 32\cdot9$

ion. This randomization is analogous to that which occurs with the hydrogen atoms of benzene (15) and thiophen (Chap. 8). The mechanism of the process is so far quite uncertain. For benzene the intervention of prismane and benz-valene structures has been suggested (14, 15) as one possibility, and with pyridine the aza-analogs (e.g., 21a and 21b) of these structures could be involved.

Such tautomerism implies the synchronous movement of carbon atoms with their attached hydrogen atoms. The alternative is a rapid equilibration of hydrogen atoms only, via a series of 1,2-shifts. The crucial experiment to differentiate these mechanisms will require labeling carbon atoms with ^{13}C. Thus, will 2,6-$^{13}C_2$-pyridine (24) lose only $H^{13}CN$ in the major fragmentation process or will it lose statistical amounts of $H^{13}CN$ and $H^{12}CN$?

24

As with benzene, yet another possibility must also be borne in mind, namely that the pyridine molecular ion may, in part at least, be represented by an open-chain structure such as **21c**, **21d**, **21e** or **21f**:

$$CH_2{=}CH{-}CH{=}CH{-}C{\equiv}N \quad \rceil^{+\cdot} \quad \textbf{21c}$$

$$CH_2{=}N{-}CH{=}CH{-}C{\equiv}CH \quad \rceil^{+\cdot} \quad \textbf{21d}$$

$$CH_2{=}CH{-}N{=}CH{-}C{\equiv}CH \quad \rceil^{+\cdot} \quad \textbf{21e}$$

$$NH{=}CH{-}CH{=}CH{-}C{\equiv}CH \quad \rceil^{+\cdot} \quad \textbf{21f}$$

B. Alkylpyridines

The spectra of the three monomethylpyridines (picolines) (**25**) are shown in Figs. 11.7, 11.8, and 11.9 (16, 17). It will be seen again that loss of HCN from the molecular ion is an important process but that a significant M-1 ion is now formed. Labeling experiments to determine the origin of the hydrogen atoms lost in the formation of the M-1 species have not been carried out, and indeed might well provide little useful information if randomization is as important in these cases as in pyridine itself. It seems likely, however, that to a large extent the loss of hydrogen is (instantaneously, at least) from the methyl groups, leading to either an azabenzyl cation (**25a**) or, if ring

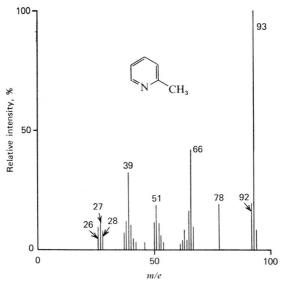

Fig. 11.7

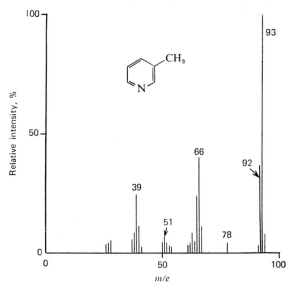

Fig. 11.8

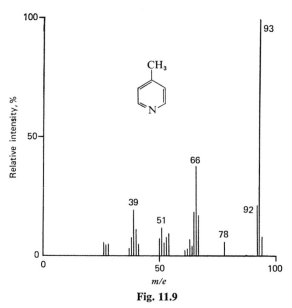

Fig. 11.9

expansion occurs (either before or after loss of hydrogen from the molecular ion), an azatropylium ion (**25b**).

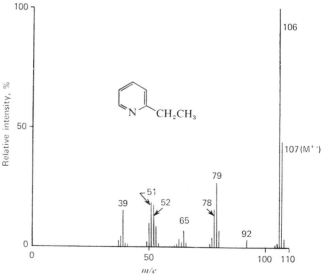

25, M$^{+\cdot}$, m/e 93   25a m/e 92

m/e 93   25b   C$_5$H$_5^+$ m/e 65

It will be noticed that the M-1 species is most intense for 3-picoline. This is probably a reflection of the high electron-density at the 3-position of the pyridine ring stabilizing **25c** relative to **25d** and **25e**.

25c   25d   25e

Similar effects are noticed in the β-cleavage of the ethylpyridines (17) (Figs. 11.10, 11.11, and 11.12). It will be noticed that M-·CH$_3$ is the base

Fig. 11.10

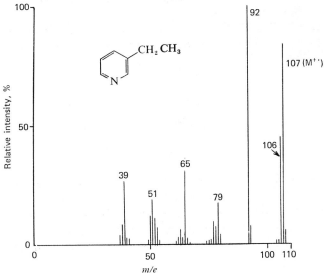

Fig. 11.11

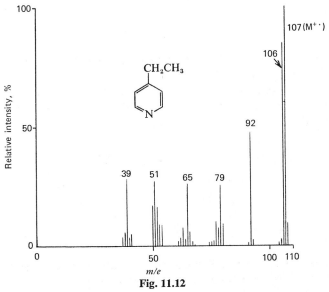

Fig. 11.12

peak of the spectrum of 3-ethylpyridine but is considerably weaker in those of the 2- and 4-ethyl-compounds.

The m/e 79 ion, resulting from expulsion of C_2H_4 from the parent, is significant in the spectra of the three ethylpyridines. It presumably results from a four-center hydrogen-transfer reaction.

26, $M^{+\cdot}$, m/e 107   **21**, m/e 79

Of some interest is the intense M-H· ion in the mass spectrum of 2-ethylpyridine (**26**) (17). Although the point has not been explicitly demonstrated by deuterium labeling, it seems likely that γ-fission is involved with stabilization of the resultant fragment (**26a**) by ring formation (18). This interpretation is supported by the observation that 2-n-propylpyridine (**27**) shows

26   **26a**   **27**

a very intense M-CH₃· ion whereas in 3- or 4-n-propylpyridines such methyl loss is a minor process only (19, 20).

The base peak of the spectrum of 2-n-propylpyridine is the ion **27a** resulting from a McLafferty rearrangement with hydrogen transfer to the nitrogen atom.

27   **27a**

The McLafferty rearrangement is of greatly reduced importance in 3-substituted pyridines containing a γ-hydrogen (e.g., **28**), but again provides the base peak in the spectrum of 4-n-propylpyridine (**29**) (20).

The reduced importance of the McLafferty rearrangement in the 3-substituted pyridine is probably due to the easy β-cleavage already described,

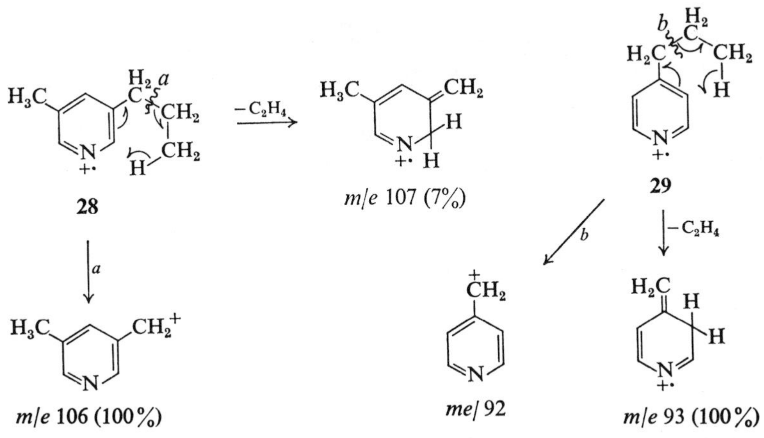

28   *m/e* 107 (7%)   29

m/e 106 (100%)   *me/* 92   *m/e* 93 (100%)

which in fact leads to the base peak of the 3-*n*-propyl-5-methyl compound **28**. The low intensity of a simple β-cleavage ion and the base peak at *m/e* 107 (McLafferty rearrangement) were used to confirm the attachment of the hydrindandione substituent at $C_{(2)}$ of the pyridine ring in compound **30** (21).

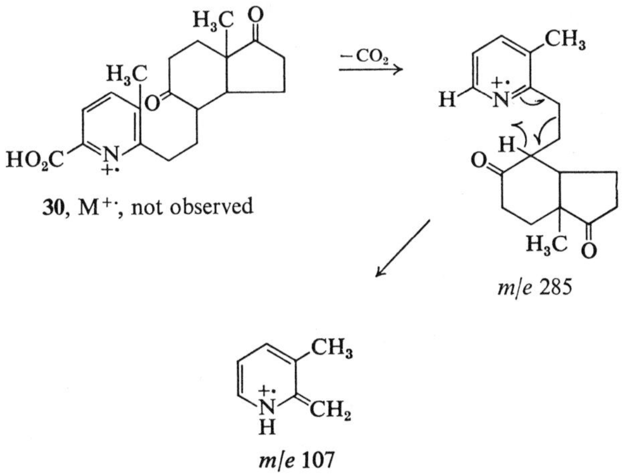

30, M$^+$; not observed

m/e 285

m/e 107

Mass spectrometry was also useful in differentiating between the two possible products (**31** and **32**) of the mixed reductive coupling of dibutyl maleate and 2-vinylpyridine (22). The product was clearly **31** since the base peak of its mass spectrum appeared at *m/e* 93 (**31a**) as expected for a McLafferty rearrangement ion from this structure [rather than *m/e* 107 (**32a**) as expected for the isomer **32**].

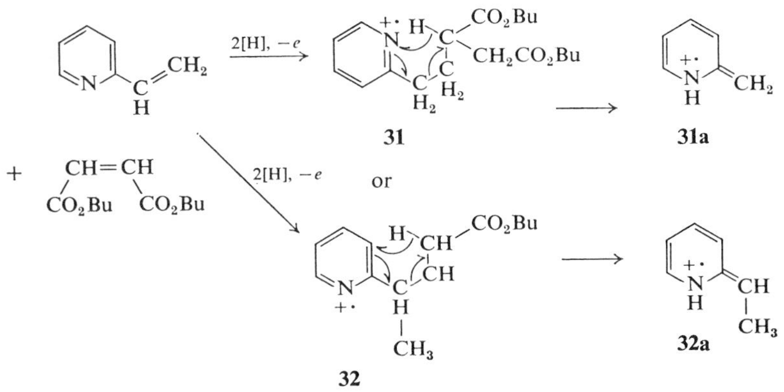

C.   Pyridines Containing Oxygen Substituents

1.   N-Oxides

The mass spectrum of pyridine-N-oxide (33) shows marked loss of oxygen, but the intensity of the M-16 peak presumably depends upon the conditions

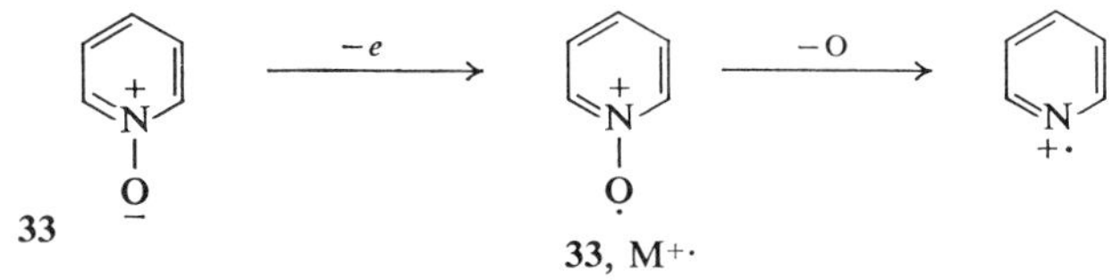

of the determination since its intensity has been reported at 87% (23) and 29% (24).

TABLE 1   INTENSITIES OF SOME IONS IN THE SPECTRA
OF SUBSTITUTED PYRIDINE N–OXIDES (24)

Substituent	Intensities				
	$M^{+\cdot}$	M-16	M-17	M-18	Base peak (m/e)
None	75	29	17	—	39
4-NO$_2$	58	10	—	—	39
2-CH$_3$	69	28	100	2	92
3-CH$_3$	100	25	11	1	109
4-CH$_3$	100	16	8	4	109
2-CH$_2$OH	7	9	35	17	79
2-CD$_2$OH	18	9	11	42	79
3-CH$_2$OH	60	100	75	34	109
4-CHO	4	79	27	—	51
2-CO$_2$H	28	0.8	0.2	—	78
3-CO$_2$H	100	72	3	—	139
3-CO$_2$C$_2$H$_5$	100	14	2	—	167

The presence of an alkyl substituent in the 2-position reduces the tendency towards oxygen loss on account of competition with an "ortho effect" and loss of ·OH. Thus 2-methylpyridine-*N*-oxide (**34**) decomposes as shown upon electron impact (23).

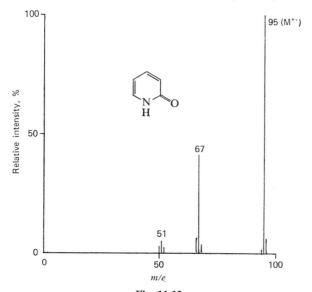

34, M⁺·, *m/e* 109

m/e 92 (100%)

The intensities of the M-16 ions for a series of substituted pyridine-*N*-oxides as well as those of the M-17 and M-18 ions, are tabulated in Table 1 (24) and it will be seen that only with 2-carboxypyridine-*N*-oxide (**35**) is the M-16 ion of negligible intensity. In this case there appears to be loss of both ·CO₂H and :O, possibly via a four-center reaction of the type shown.

35, M⁺·, *m/e* 139

78*m/e*

2. Hydroxypyridines/Pyridones

Partial spectra of 2-pyridone (**36**), 3-hydroxypyridine (**37**), and 4-pyridone (**38**) are shown in Figs. 11.13, 11.14, and 11.15, respectively (25).

Fig. 11.13

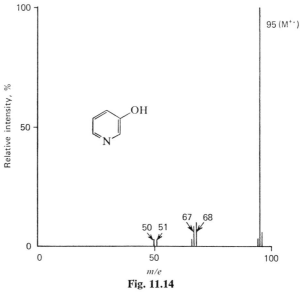

Fig. 11.14

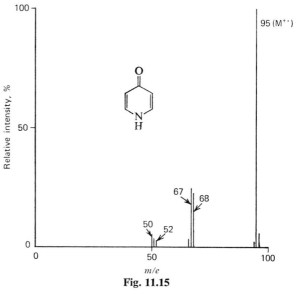

Fig. 11.15

It will be noticed that fragmentation is at a minimum with the 3-hydroxy compound, and that losses of CO and HCN lead to the ions at m/e 67 and 68, respectively, which are best represented as the pyrrole and furan radical cations. Similar behavior occurs with 4-pyridone (**38**) but 2-pyridone (**36**) only loses CO. These results are summarized in the scheme.

It is interesting to note that 4-pyridone does not lose the elements of acetylene in a retro-Diels-Alder reaction. It will be recalled that this is an important

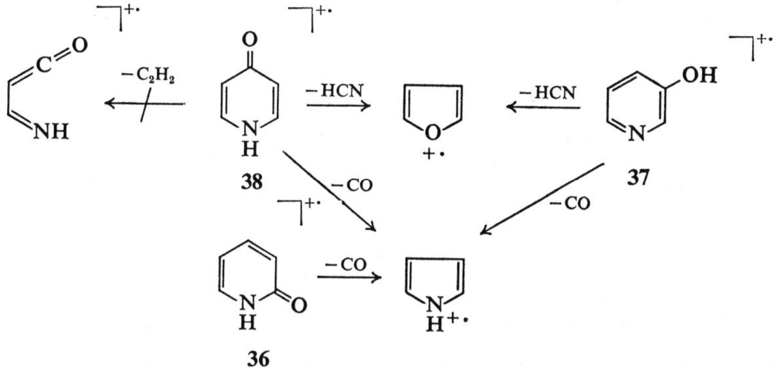

process with 4-pyrone (Chap. 5, Sec. II) and its sulfur analog (Chap. 8, Sec. I.G).

Naturally the presence of other substituents can profoundly modify the fragmentations observed with pyridones. Thus in the mass spectrum of 4-hydroxy-6-methyl-2-pyridone (**39**) the main decomposition mode is loss of $C_2HO\cdot$ from the molecular ion, formulated as **39 → 39a** (26):

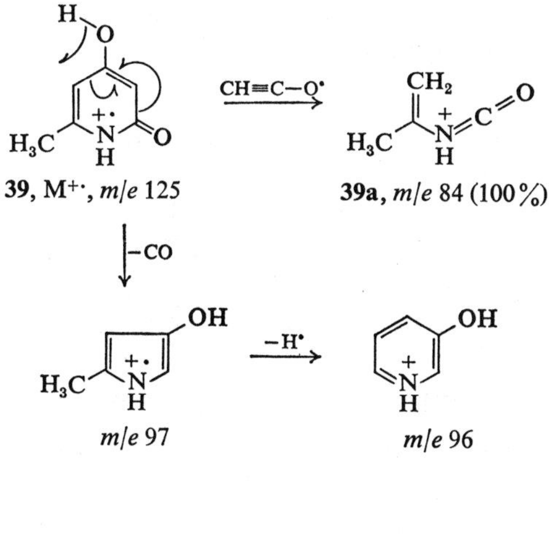

Introduction of an ethoxycarbonyl group into the 5-position of **39** further modifies the fragmentation, and the familiar "ortho effect" involving the ester and adjacent alkyl group of **40** becomes important (26). Easy cyclization

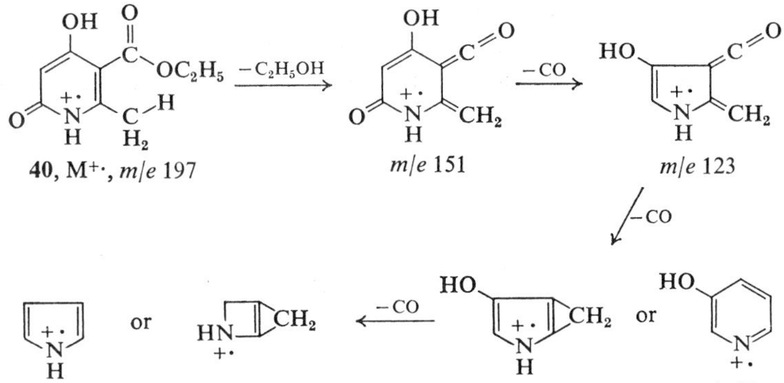

40, M$^{+\cdot}$, m/e 197 m/e 151 m/e 123

m/e 67 m/e 95 m/e 95

to form an ion best formulated as an oxazolopyridinium cation (**41a**) leads to the base peak in the spectrum of dimethyl 1-(2-pyridone)-fumarate (**41**)

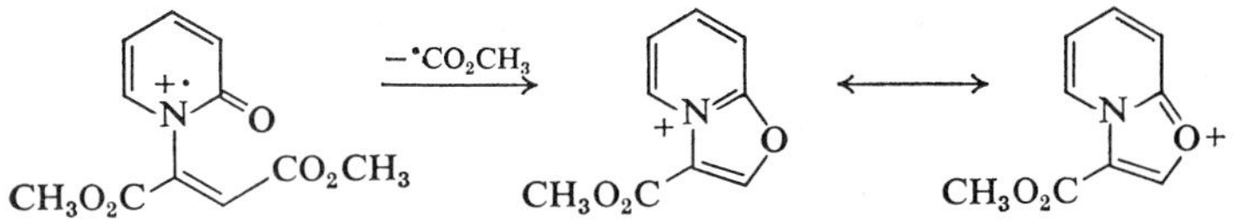

41, M$^{+\cdot}$, m/e 237 (4%)   **41a**, m/e 178 (100%)

(27). A similar process can be invoked to explain the formation of a strong M-CH$_3\cdot$ ion in the spectrum of **42** (26).

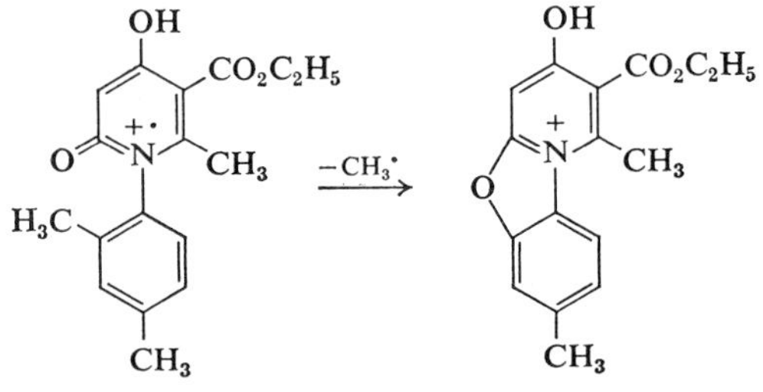

42, M$^{+\cdot}$, m/e 301 m/e 286

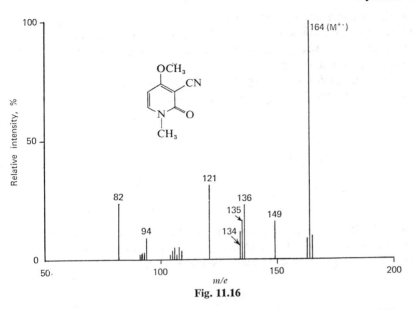

Fig. 11.16

The alkaloid ricinine (**43**) has been subjected to mass-spectrometric fragmentation and the resultant spectrum (above m/e 50) is shown in Fig. 11.16 (28). The sequence $M^{+\cdot} \rightarrow M\text{-}28 \rightarrow M\text{-}(28 + 15)$ is easily rationalized in terms of the usual loss of the ring carbonyl to give **43a**, then loss of $\cdot CH_3$ to give the well-stabilized cation **43b**. More interesting is the sequence $M^{+\cdot} \rightarrow M\text{-}15 \rightarrow m/e$ 82 (C_4H_4NO), and the rationale shown has been suggested to explain it.

That the nitrogen atom in the fragment m/e 82 is the ring nitrogen was established by biosynthetic labeling of the ricinine. Nicotinamide, which is known to be an intermediate in the biosynthesis of the alkaloid, was labelled with ^{15}N at the amide group.

2-Phenyl-5-benzoyl-6-methyl-4-pyridone (**44**) shows an intense M-1 ion (29) which is presumably the cyclized species **44a**.

Pyridoxol (**45**), an important member of the vitamin B6 group of substances, has been subjected to electron-impact fragmentation together with its isomeric isopropylidene derivatives **46** and **47** (30).

The spectra of pyridoxol and the isopropylidene derivative **46** are essentially identical down from m/e 151. This is easily rationalized in terms of the initial decompositions shown, but the remaining ions in the two spectra are much more difficult to explain and a great deal of ring expansion has been invoked to derive the rationale given.

The concerted loss of CO and ·OH in the formation of m/e 106 from m/e 151 is surprising (a metastable ion is observed for the process), but other cases have been reported in which remotely-situated portions of a molecule undergo the "concerted" loss consistent with the observation of a metastable ion.

Losses of CO and HCN from the molecular ion which are observed with the parent 3-hydroxypyridine are completely suppressed in pyridoxol by the facile fragmentations resulting from the additional functional groups.

The isomeric isopropylidene compound **47** shows marked differences from **46** in its fragmentation. Ions at m/e 150 and 152 accompany a species at m/e 151, which is probably of different structure to the m/e 151 species in the spectra of **45** and **46**, and an appreciable M-15 ion is formed. A complete rationale is difficult to derive, but some of the fragmentation may be explained as shown.

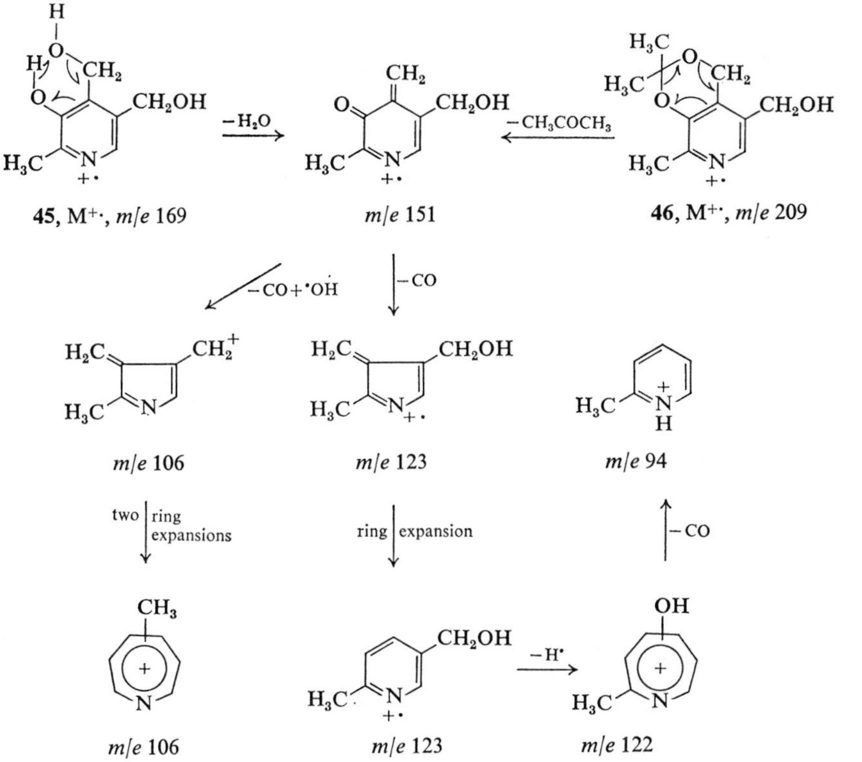

45, M+·, m/e 169
− H₂O
m/e 151
− CH₃COCH₃
46, M+·, m/e 209
− CO + ·OH
− CO
m/e 106
m/e 123
m/e 94
two ring expansions
ring expansion
− CO
m/e 106
m/e 123
− H·
m/e 122

47, M+·, m/e 209
− CH₃COCH₃
m/e 151
− CO
m/e 123
− H·
m/e 150
− CO
m/e 122
− ·CH₃
m/e 194
m/e 194
− CH₂CO
m/e 152

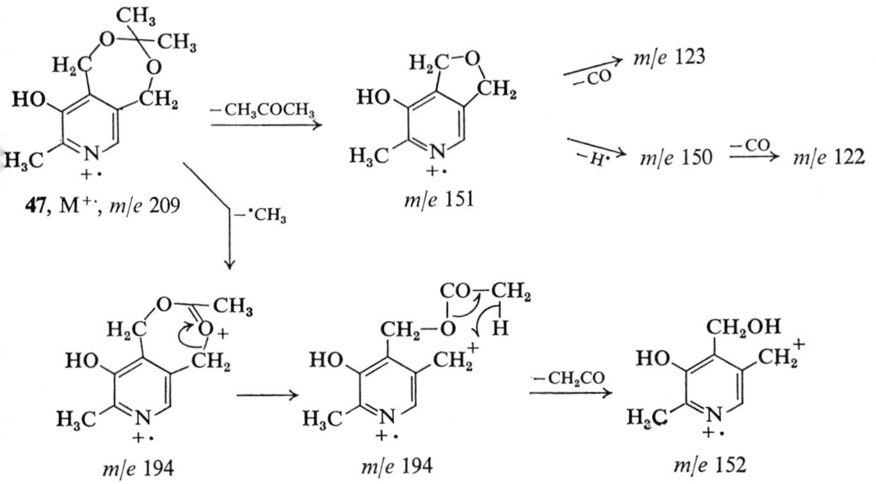

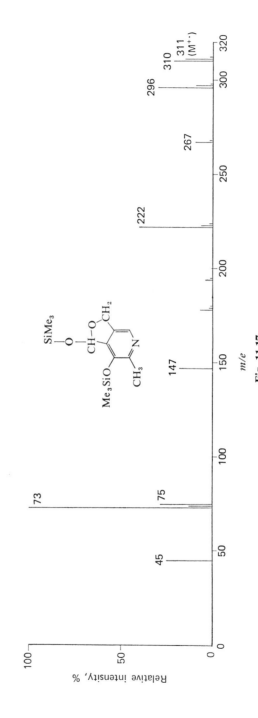

Fig. 11.17

The mass spectrum of the tris(trimethylsilyl) ether (**48**) of pyridoxol has been reported and the main fragmentations are shown (31). It will be noted that no parent ion is observed and that the fragmentation is very different from that of the free triol.

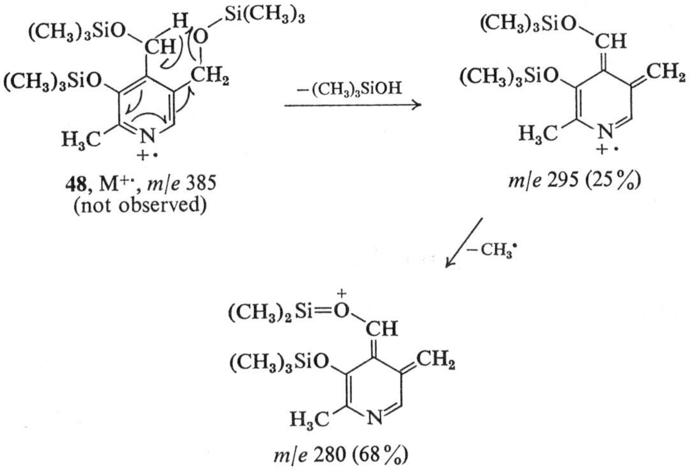

(CH₃)₃SiO
H
Si(CH₃)₃
CH
O
(CH₃)₃SiO
CH₂
H₃C
N
+•
48, M⁺˙, m/e 385
(not observed)
−(CH₃)₃SiOH
(CH₃)₃SiO
CH
(CH₃)₃SiO
CH₂
H₃C
N
+•
m/e 295 (25%)
−CH₃•
(CH₃)₂Si=O
+
CH
(CH₃)₃SiO
CH₂
H₃C
N
m/e 280 (68%)

No further significant ions containing the pyridoxol nucleus occur in the spectrum. The base peak is the ion Me_3Si^+ (*m/e* 75) and another strong silicon-containing ion is $Me_2Si{=}O^+{-}SiMe_3$ (*m/e* 147, 30%).

Trimethylsilylation of pyridoxal (**49**) with *N*-trimethylsilylacetamide produces the bis(trimethylsilyl) derivative (**51**) of the hemiacetal form (**50**) and this has also been subjected to electron impact (31). The spectrum of **51**

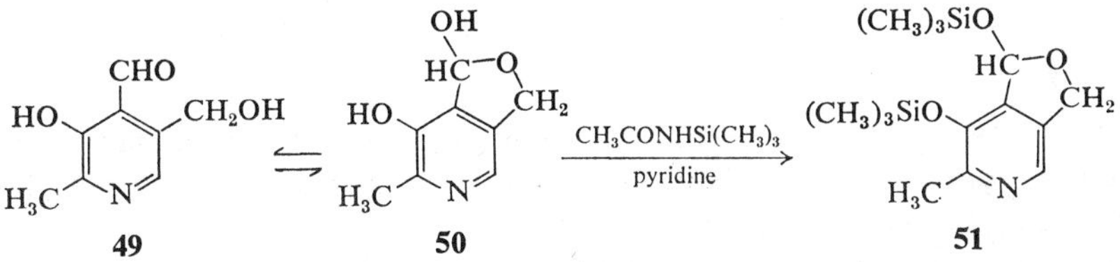

CHO
HO
CH₂OH
H₃C
N
49
OH
HC
O
CH₂
HO
H₃C
N
50
CH₃CONHSi(CH₃)₃
pyridine
(CH₃)₃SiO
HC
O
CH₂
(CH₃)₃SiO
H₃C
N
51

is shown in Fig. 11.17, and it is seen that the compound gives a clear parent ion accompanied by stronger M-1 and M-15 species. A rationale presented to account for the main features of the spectrum is shown. Attention is drawn to the necessity of complex skeletal rearrangements to bring about the elimination of CO_2 in forming *m/e* 267 from the parent ion.

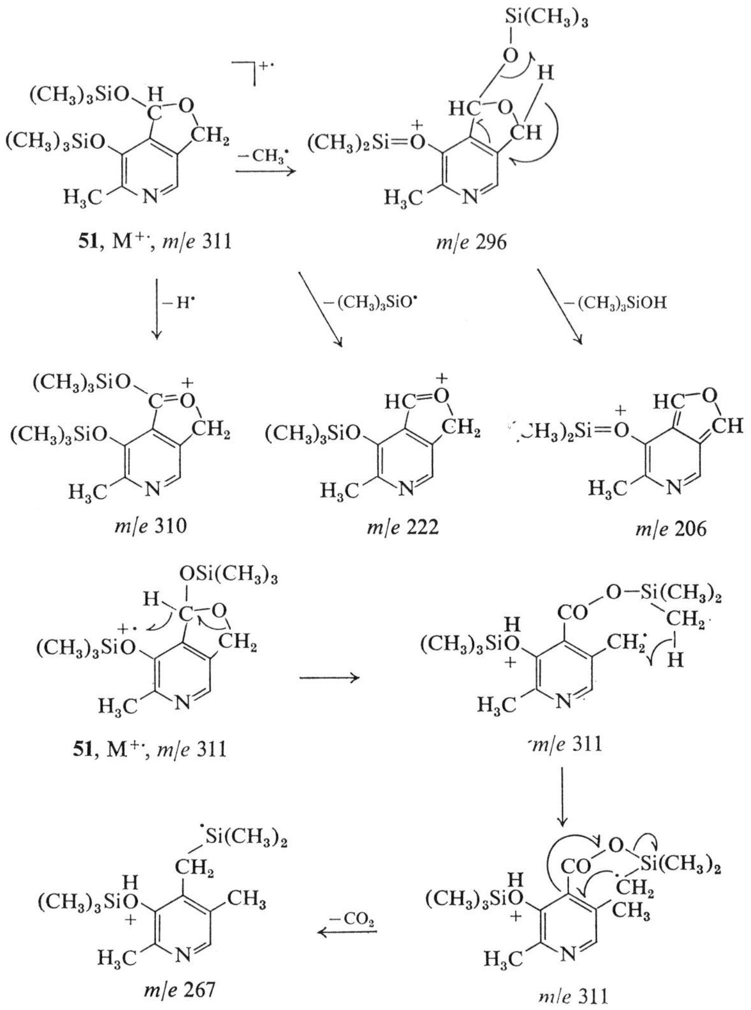
51, M+·, m/e 311
m/e 296
−CH3·
−H·
−(CH3)3SiO·
−(CH3)3SiOH
m/e 310
m/e 222
m/e 206
51, M+·, m/e 311
m/e 311
m/e 311
−CO2
m/e 267

The mass spectrum of pyridoxal itself does not appear to have been reported, although that of the deoxy-derivative (**52**) has been recorded (32) and may be rationalized as shown.

In view of the small samples required for mass-spectrometric investigation, the technique is likely to be of considerable importance in this field.

The mass spectra of the three acetoxy compounds (**53**, **54**, and **55**) obtained from 2-picoline-*N*-oxide and acetic anhydride have been reported, and the rationalization of the major fragmentations is shown (33). Of interest are weak M-CO ions in the spectra of **53** and **54**. The formation of these species necessitates skeletal rearrangements, and it has been suggested that these

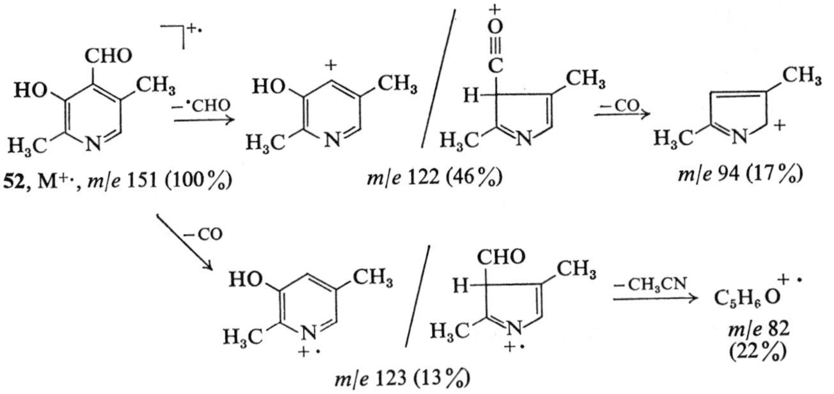

CHO
HO
CH3
H3C
N
−•CHO
−CO
52, M+·, m/e 151 (100%)
m/e 122 (46%)
m/e 94 (17%)
m/e 123 (13%)
−CH3CN
C5H6O
m/e 82
(22%)

O—CO—CH3
CH3—CO—O
C—O—CO—CH3
53, M+·, m/e 151 (34%)
54 (38%)
55 (13%)
−CH2CO
−CH2CO
−CH3CO•
OH
HO
CHO
m/e 109
(100%)
m/e 109 (100%)
m/e 108 (100%)
−CO
−CO
−CO
−H•
m/e 81
(27%, 15%)
m/e 80
(45%, 38%)

−CO
54, M+·, m/e 151
m/e 123 (7%)

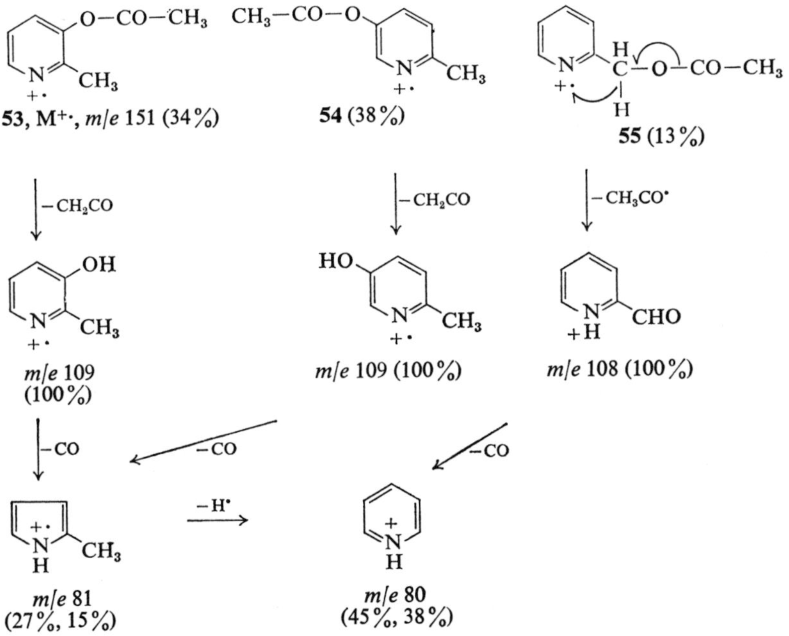

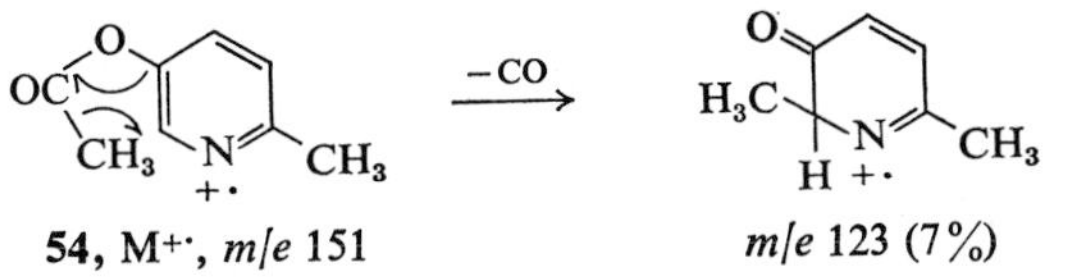

involve methyl migrations. The M-CH$_3$· ion in the spectrum of **55** is probably stabilized by participation of the nitrogen atom, leading to the cyclic ion **55a**.

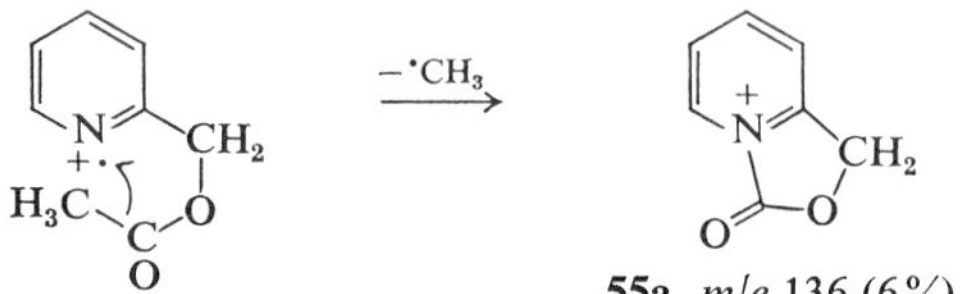

55, M$^+$·, m/e 152

55a, m/e 136 (6%)

D. Pyridines with Nitrogen-Containing Substituents

This area of pyridine chemistry has been subjected to very little mass-spectrometric investigation. No analyses are reported of simple amino-pyridines, nitro-pyridines, or cyano-pyridines. The mass spectrum of nicotine (**56**) has already been described (Chap. 10, Sec. III.A) and it will be recalled that the main fragment ion is **56a**, characteristic of the 2-substituted pyrrolidine ring.

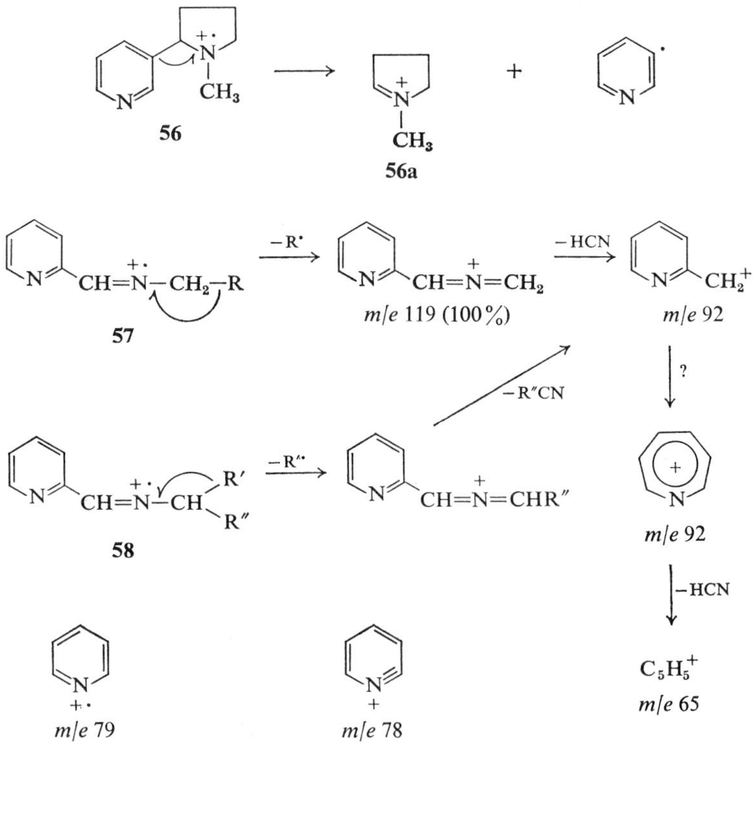

56

56a

The mass spectra of a number of Schiff's bases derived from pyridine-2-aldehyde have been described. Such compounds are of importance as ligands and so their spectra are of some interest.

The main features of the spectra of Schiff's bases of pyridine-2-aldehyde and normal (57) and α-branched (58) aliphatic amines are summarized as shown (34). The pyridine radical cation (m/e 79) and its dehydro-derivative (m/e 78) are prominent in the spectra and presumably result from loss of the side chain from the molecular ion with and without hydrogen transfer, respectively.

The Schiff's base from aniline and pyridine-2-aldehyde (59) shows intense molecular ion and M-1 peaks. The latter may well be 59a and/or 59b.

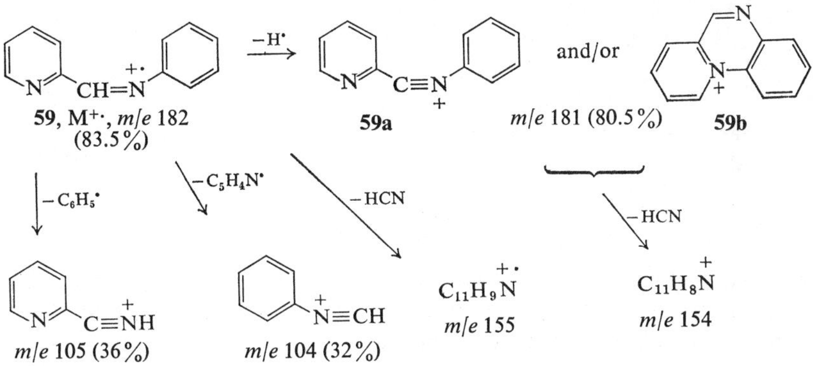

59, M$^{+\cdot}$, m/e 182 (83.5%)

59a

m/e 181 (80.5%)

59b

m/e 105 (36%)

m/e 104 (32%)

$C_{11}H_9N$ m/e 155

$C_{11}H_8N$ m/e 154

E. Halopyridines

The spectrum of pentafluoropyridine (60) is shown in Fig. 11.18 (35). Fragmentation is clearly much modified by the presence of the fluorine atoms, and losses of :CF and ·CF$_3$ are of importance in the fragmentation. The compositions of the various ions are shown on the spectrum.

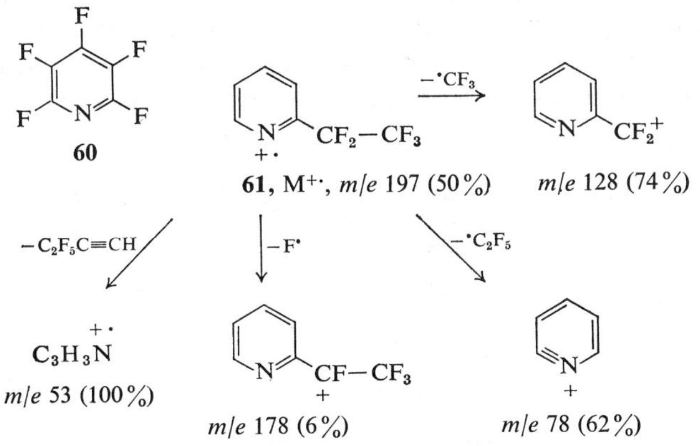

60

61, M$^{+\cdot}$, m/e 197 (50%)

m/e 128 (74%)

C_3H_3N m/e 53 (100%)

m/e 178 (6%)

m/e 78 (62%)

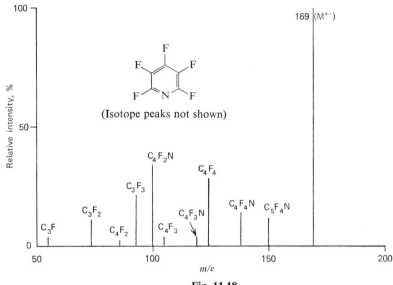

Fig. 11.18

The mass spectrum of 2-(pentafluoroethyl)pyridine (**61**) resembles that of a simple alkylpyridine. Similar fragmentation is observed with 2-(heptafluoropropyl)pyridine, although in this case the β-fission ion (m/e 128) is the base peak of the spectrum (36).

III. QUINOLINES AND ISOQUINOLINES

A. Unsubstituted Quinoline and Isoquinoline

The spectra of the two parent heterocycles (**62** and **63**) are shown in Figs. 11.19 and 11.20 (37, 38). It will be noticed that they are very similar and that the only fragmentation of importance is loss of HCN from the molecular ion. No ^{13}C- or deuterium-labeling experiments have been carried out for either compound, but it seems reasonable that $C_{(2)}$ is involved in the loss

62, M+·, m/e 129 m/e 102   **63**, M+·, m/e 129

$-C_2H_2$

$C_6H_4^+$·
m/e 76

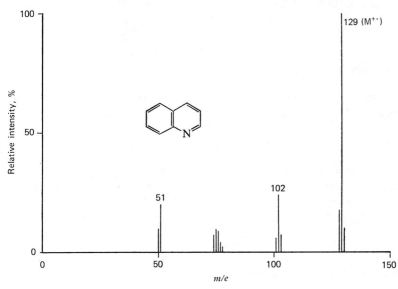

Fig. 11.19

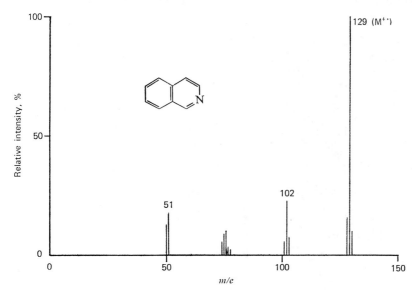

Fig. 11.20

in quinoline. Both $C_{(1)}$ and $C_{(3)}$ may be lost in isoquinoline, and ^{13}C-labeling would be very interesting in this case. The fragmentation may be summarized as shown.

B. Alkylquinolines and -isoquinolines

All the monomethylquinolines examined show the same fragmentation sequence which may be summarized:

$$m/e\ 143\ (\text{M}^{+\cdot}) \xrightarrow{-\text{H}\cdot} m/e\ 142 \xrightarrow{-\text{HCN}} m/e\ 115$$

We shall first consider 2-, 3-, and 4-methylquinolines (20). The M-1 species is most intense for the 3-methyl compound, and this is most easily explained in terms of the stability of the carbonium ion **65a** (compare the three methylpyridines, Sec. II.B). For the subsequent fragmentation the ratio

intensity [M-(H + HCN)]/intensity [M-H]

is practically constant for the three compounds methylated in the heterocyclic ring, suggesting that a common intermediate is being formed and decomposing by loss of HCN. This decomposing species is most economically described as the azabenzotropylium ion **64a**, and can be written as shown.

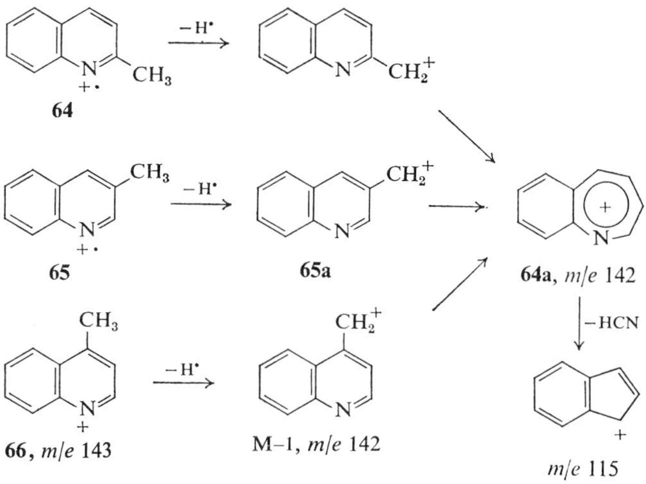

The M-1 peaks for the 5-, 6-, 7-, and 8-methylquinolines (e.g., Fig. 11.21) (39) are more intense than those of the previously discussed isomers and show little variation in intensity with the position of the substituent (20). In all

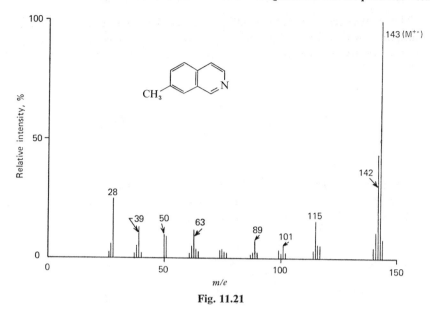

Fig. 11.21

four the ion is probably best represented as the pyridotropylium cation **67a**. It will be noticed that loss of HCN from the m/e 142 species **67a** is a much

less important process than that from its isomer **64a**. This is to be expected since loss of HCN from a pyridine ring is not a highly favored process, and in fact it has been suggested that **67a** isomerizes to **64a** before elision of HCN occurs (20).

The decomposition of ethyl- and higher alkyl-quinolines parallels that of the pyridine analogs. As noted there, when a C_2 or longer side chain is situated adjacent to the heteroatom, γ-cleavage becomes important.

The McLafferty rearrangement dominates the spectra of 2-substituted quinolines with three-carbon or longer side chains. That the γ-hydrogen atom is transferred specifically has been demonstrated by deuterium labeling (20).

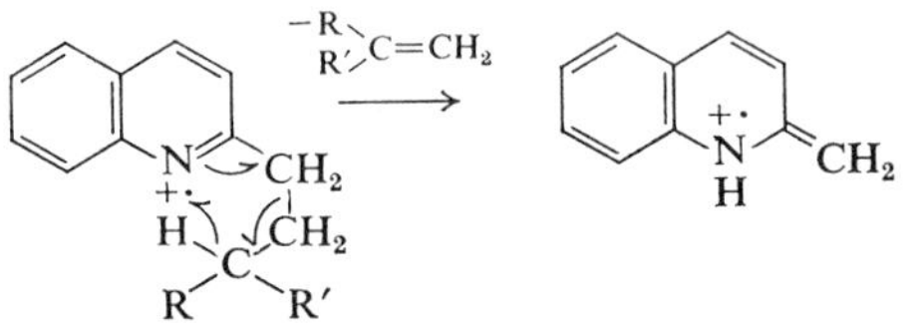

The mass spectra of 6- and 7-n-butylquinolines (68) are dominated by α,β-cleavage and McLafferty rearrangements (20), a situation similar to that seen with the corresponding n-butylbenzene (40).

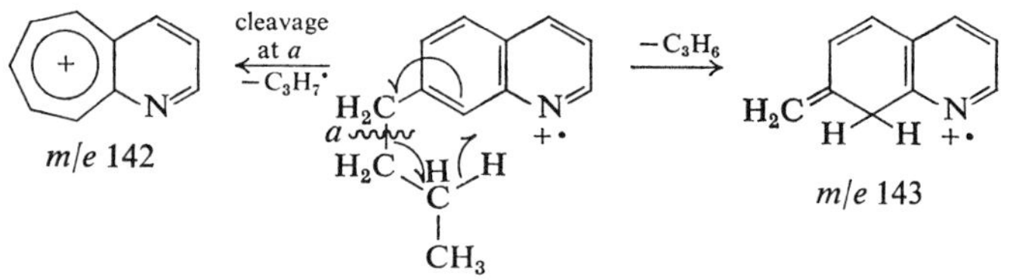

68, M$^{+\cdot}$, m/e 185

The mass spectrum of 8-n-propylquinoline (69) shows intense peaks corresponding to losses of H·, CH$_3$·, and C$_2$H$_4$ from the parent ion and may be rationalized as shown. The McLafferty rearrangement in this case occurs by an unusual but not unprecedented seven-center mechanism (20).

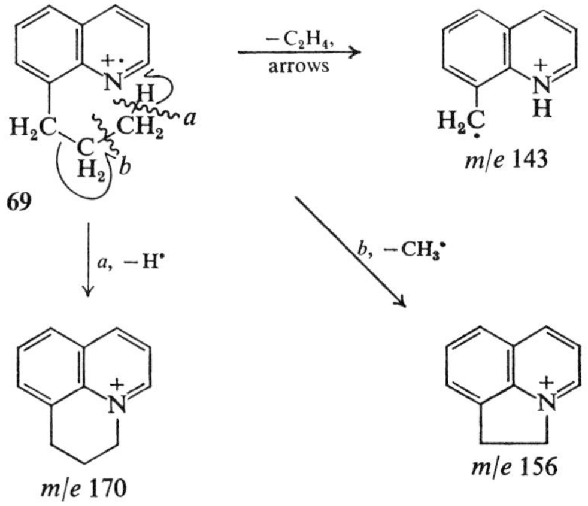

The mechanism of the fragmentation of 1- and 3-methylisoquinolines (70 and 71) is of special interest since the question of ring expansion in the

M-1 species has been examined by ^{13}C-labeling (41). The spectra show that the usual sequence $M \rightarrow M\text{-}1 \rightarrow m/e$ 115 occurs. The ratio

$$\text{intensity } [M\text{-}H]/\text{intensity } [M\text{-}(H + HCN)]$$

is essentially the same for both compounds, suggesting that the common M-1 species **70a** loses HCN. The 1-methyl compound (**70**) was labeled at $C_{(1)}$ (**72**) and at CH_3 (**73**) with ^{13}C.

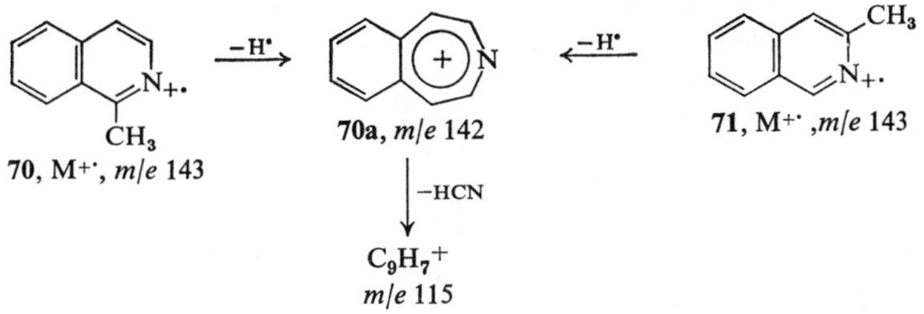

It will be seen that the m/e 142 species **70a** can be formed by either a phenyl-to-carbon migration [pathway (i)] or a nitrogen-to-carbon migration [pathway (ii)]. If it is assumed that the rearranging species is the M-1 ion **70b**, these two possibilities may be represented mechanistically as shown.

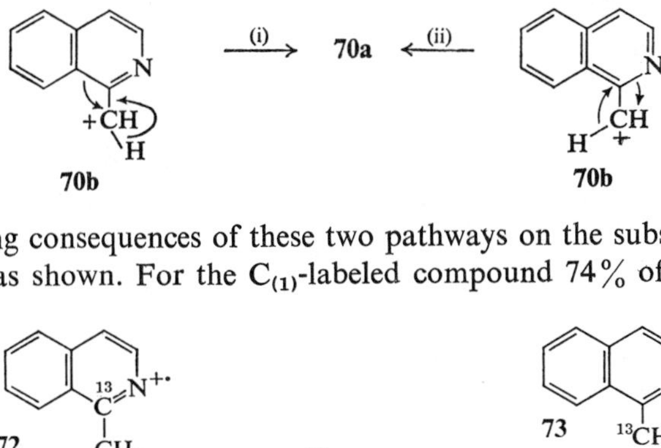

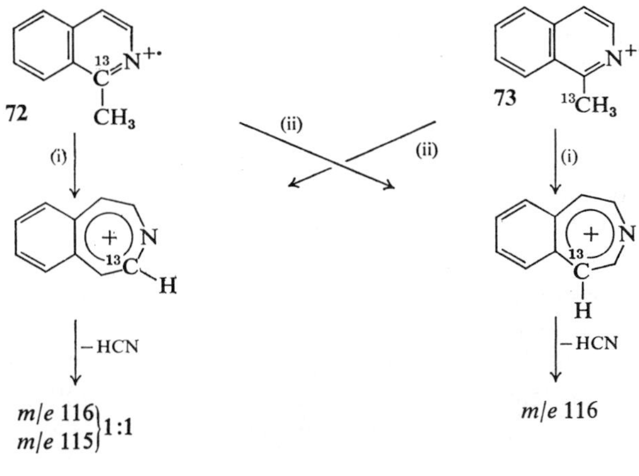

The labeling consequences of these two pathways on the subsequent loss of HCN are as shown. For the $C_{(1)}$-labeled compound 74% of the label was retained (m/e 116) and 26% lost (m/e 115), whereas for the methyl-labeled compound 86% was retained and 14% lost. These results show that migration of the methyl carbon atom, regardless of mechanism, is a very significant process in the formation of the M-[H + HCN] fragment since 26% of it is lost as HCN. It is also clear that the phenyl-migration pathway is preferred to the nitrogen-migration pathway by a factor of nearly 2:1 in the formation of the azabenzotropylium ion.

Other fragmentations of alkylisoquinolines closely parallel those of the analogous quinolines (20). Thus γ-cleavage is a preferred process for 1-alkylisoquinolines with two-carbon or longer chains together with the McLafferty rearrangement (where possible). This is well-demonstrated in the spectrum of 1-n-butylisoquinoline (74).

74, M$^{+\cdot}$, m/e 185

m/e 143 (100%)

m/e 156 (20%)

m/e 156

C. Quinolines with Oxygen Substituents

The mass spectra of the seven isomeric monomethoxyquinolines have been analyzed.

The 3-, 4-, 5-, 6-, and 7-methoxy compounds have very similar spectra and may be exemplified by the fragmentation of the 7-substituted compound (75) shown in Fig. 11.22 (42). It will be seen that the important fragmentations are losses of ·CHO and CH$_2$O from the molecular ion together with the sequential loss of CH$_3$· then CO to give the m/e 116 species. These fragmentations have adequate precedents, and are outlined in the scheme.

The spectra of the 2- and 8-methoxy compounds (76 and 77) (Figs. 11.23 and 11.24) are quite similar to one another and markedly different to those of the other isomers. It will be seen that they are characterized by intense

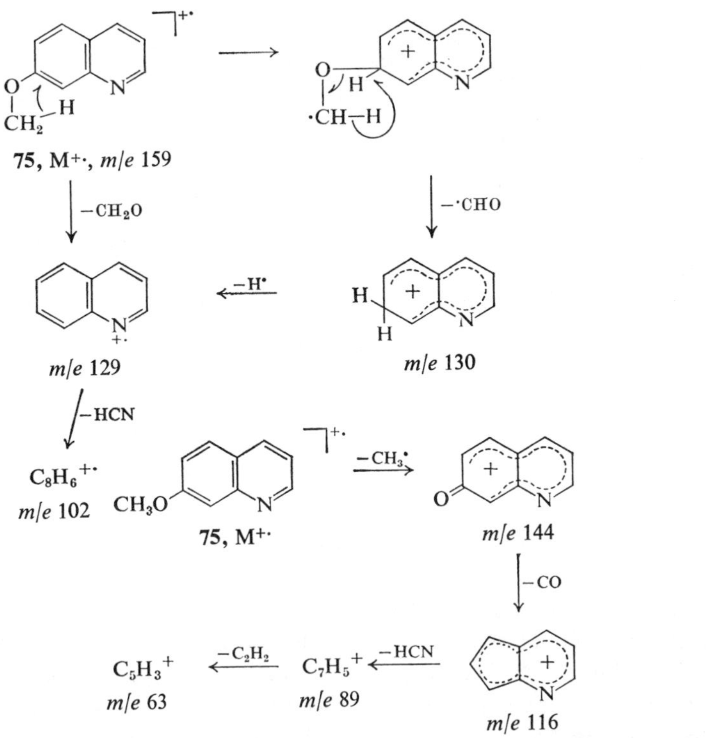

75, M$^{+\cdot}$, m/e 159

$-CH_2O$

m/e 129

$-H^{\cdot}$

$-\cdot CHO$

m/e 130

$-HCN$

$C_8H_6^{+\cdot}$
m/e 102   CH$_3$O

75, M$^{+\cdot}$

$-CH_3^{\cdot}$

m/e 144

$-CO$

$C_5H_3^+$
m/e 63

$-C_2H_2$

$C_7H_5^+$
m/e 89

$-HCN$

m/e 116

M-1 ions whereas the sequence M → m/e 144 → m/e 116 is greatly reduced in importance. It has very reasonably been suggested that participation by the neighboring nitrogen atom with consequent ring formation may be responsible for the intense M-1 ions, and the facility of this reaction probably represses the formation of the M-$\cdot$CH$_3$ species with these compounds.

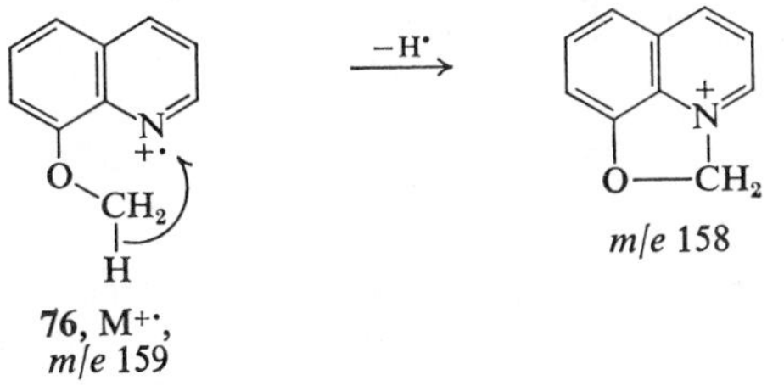

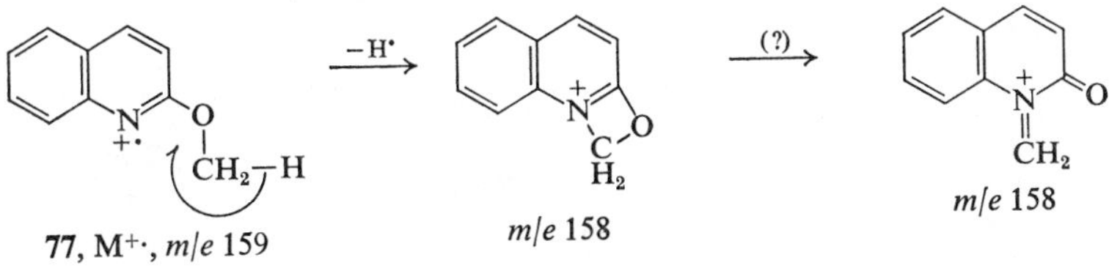

76, M$^{+\cdot}$,
m/e 159

$-H^{\cdot}$

m/e 158

77, M$^{+\cdot}$, m/e 159

$-H^{\cdot}$

m/e 158

$(?)$

m/e 158

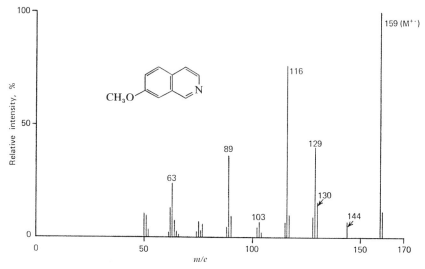

Fig. 11.22

The mass spectra of trideuteromethoxyquinolines support the hydrogen-transfer reactions that accompany the losses of ·CHO and CH_2O whereas ^{13}C-labeling of the methoxy carbon in 8-methoxyquinoline disproves the possibility of methyl migration before fragmentation. The M-29 and M-30 peaks are undisplaced in the spectrum of the labeled compound, showing that only the methoxy carbon atom is lost in these fragmentations.

2-Hydroxyquinoline (78) (carbostyril) decomposes by loss of CO followed by HCN (42, 43). This suggests the decomposition is largely occurring through the tautomer (78a) and may be represented as shown. Supporting this

78, $M^{+·}$, m/e 145
(100%)

78a

m/e 117 (90%)

$C_7H_6^{+·}$
m/e 90
(52%)

rationale is the very similar decomposition of N-methyl-2-quinolone (79), although here the product of loss of CO is N-methylindole (79a) which

79, $M^{+·}$, m/e 159

79a, m/e 131

79b, m/e 130

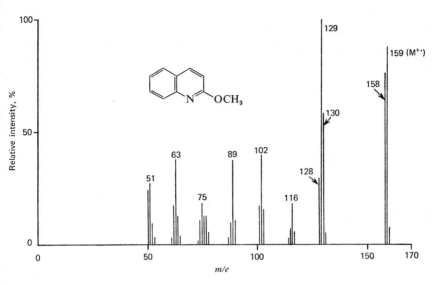

Fig. 11.23

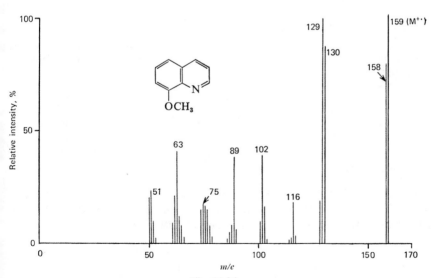

Fig. 11.24

undergoes the expected loss of a further atom of hydrogen to give **79b** (42, 43). A similar decomposition has been observed with 3-methyl-, 4-methyl-, and 3,4-dimethylcarbostyrils (43).

In 3-hydroxyquinoline (**80**), which of course cannot tautomerize to a quinolone, the intensity of the M-28 ion is greatly reduced (42). As expected,

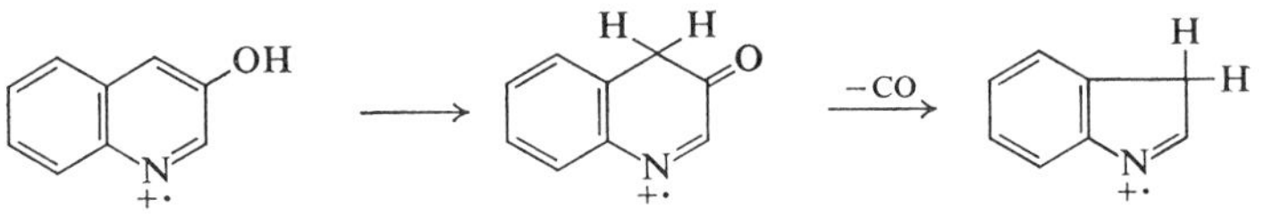

80, M$^{+\cdot}$,
m/e 145 (100%)

m/e 145

m/e 117 (12%)

1-methyl-4-quinolone (**81**) undergoes loss of CO from the molecular ion and like 4-pyridone (Sec. II.C.2) shows no tendency to lose C$_2$H$_2$ in a retro-Diels-Alder reaction (42).

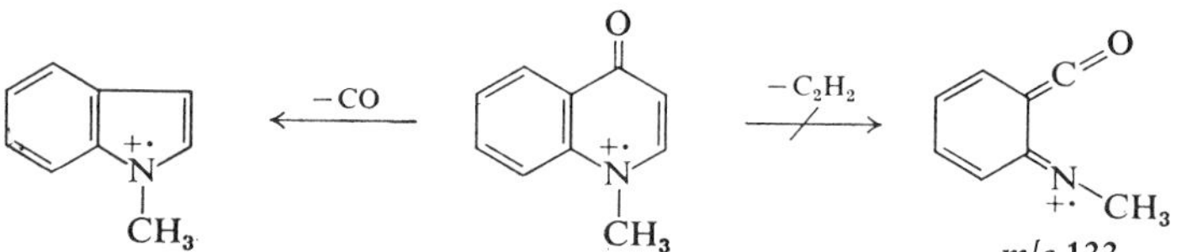

m/e 131 (82%)

81, M$^{+\cdot}$, m/e 159
(100%)

m/e 133

Like the 3-hydroxy compound, the 5-, 6-, and 7-hydroxy isomers exhibit weak M-CO ions, but the process is an important one with the 8-hydroxy compound (**82**), possibly as a result of hydrogen transfer to the nitrogen atom (42).

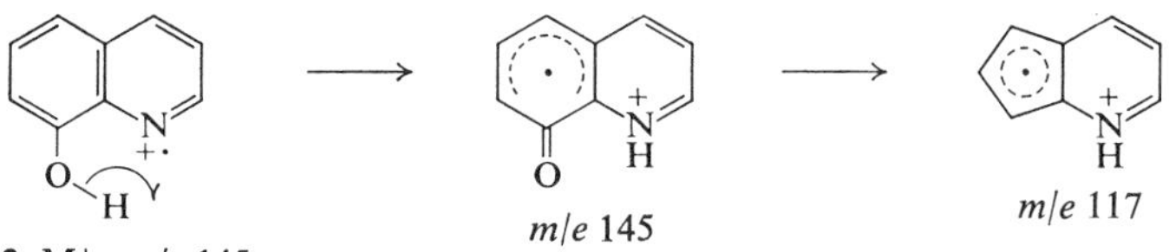

82, M$^{+\cdot}$, m/e 145

m/e 145

m/e 117

In 1-n-propylcarbostyril (**83**) the length of the alkyl chain is sufficient for a McLafferty rearrangement to occur and, in fact, this process yields the base peak in the spectrum of this compound (20).

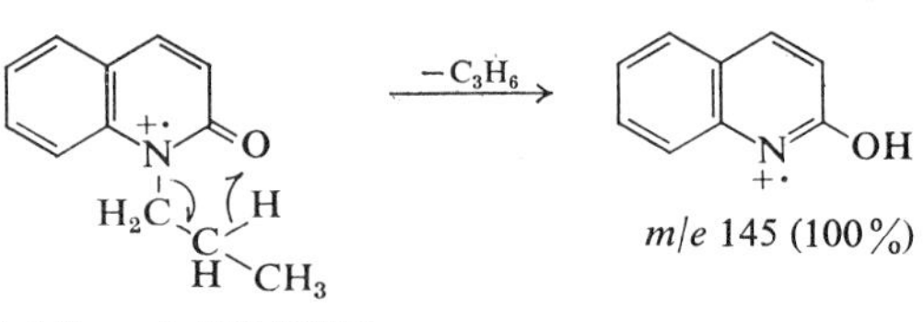

83, M$^{+\cdot}$, m/e 187 (25%)

m/e 145 (100%)

Little work has been carried out on the mass spectra of hydroxyisoquinolines, but it seems that fragmentation is essentially similar to that of the corresponding quinolines. 1-Methyl-3-hydroxyisoquinoline (**84**) shows the expected loss of CO, then H· (44), and this can be expressed by the rationale shown.

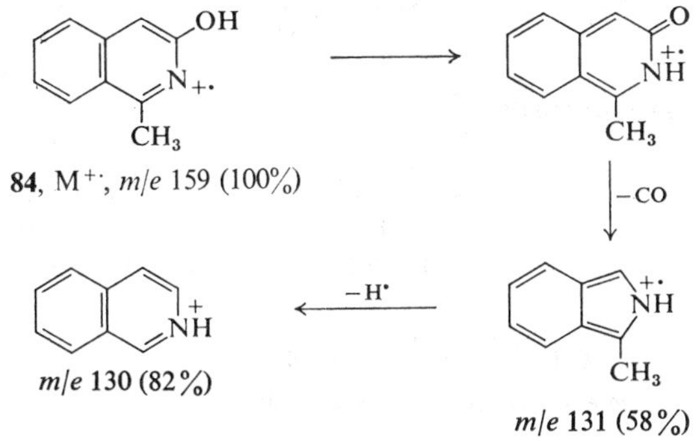

OH
N+·
CH3
84, M+·, m/e 159 (100%)
O
+·
NH
CH3
–CO
–H·
+
NH
m/e 130 (82%)
+·
NH
CH3
m/e 131 (58%)

The fragmentation of the ethyl ether of **84** is of some interest. It shows (44) a strong M-·CH_3 ion (as well as M-C_2H_4) which may well be stabilized by interaction with the nitrogen atom as shown. It will be recalled that a similar interaction has been postulated in connection with the M-1 ions in 2- and 8-methoxyquinolines.

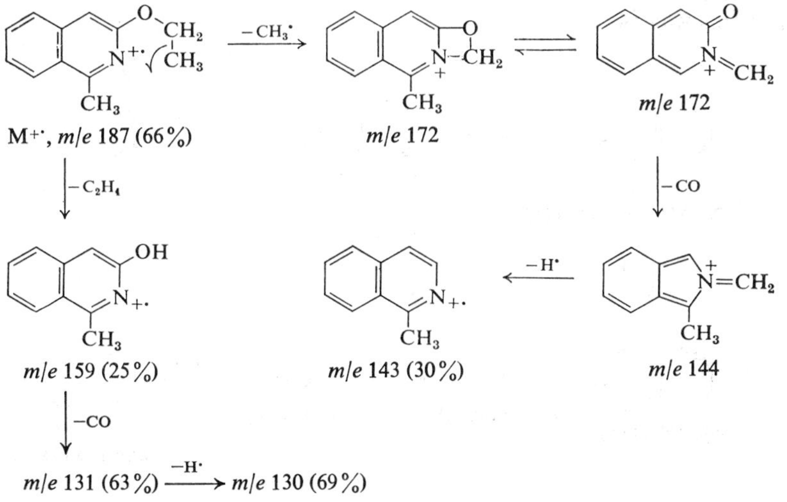

O
CH2
N+·
CH3
CH3
M+·, m/e 187 (66%)
–CH3·
O
N–CH2
+
CH3
m/e 172
O
N
+
CH2
m/e 172
–C2H4
–CO
OH
N+·
CH3
m/e 159 (25%)
N+·
CH3
m/e 143 (30%)
–H·
+
N=CH2
CH3
m/e 144
–CO
m/e 131 (63%)
–H·
m/e 130 (69%)

D. *N*-Oxides

Quinoline *N*-oxide (**85**) shows an intense molecular ion and, apart from loss of oxygen as a minor process, decomposes (45) by the sequence

$$M^{+\cdot} \xrightarrow{-CO} m/e\ 117 \xrightarrow{-HCN} m/e\ 90$$

This sequence necessitates a rearrangement in which oxygen becomes bonded to carbon, and the rationale shown has been suggested.

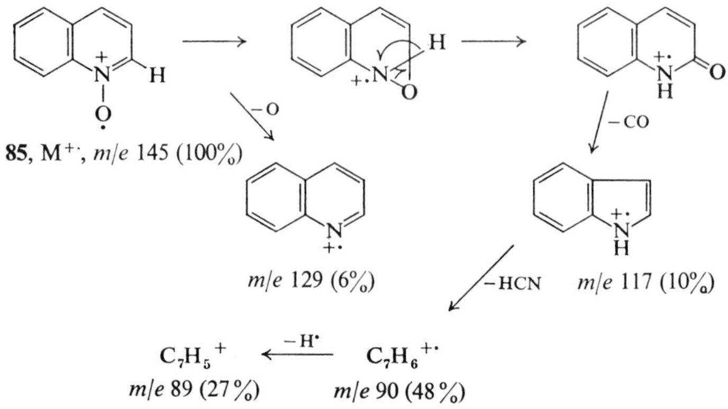

85, M$^{+\cdot}$, m/e 145 (100%)

m/e 129 (6%) m/e 117 (10%)

$C_7H_5^+$ $\xleftarrow{-H^\cdot}$ $C_7H_6^{+\cdot}$
m/e 89 (27%) m/e 90 (48%)

Quinoline N-oxides photoisomerize to carbostyrils (46, 47), and the analogy with the impact-induced rearrangement is striking.

The sequence is reversed with isoquinoline N-oxide (**86**), HCN being the first fragment lost, then CO. This decomposition can be rationalized as rearrangement to the isocarbostyril radical cation, which decomposes in just this way (45).

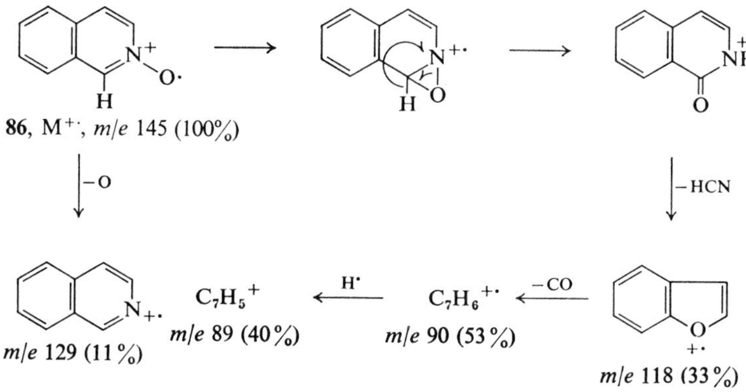

86, M$^{+\cdot}$, m/e 145 (100%)

m/e 129 (11%) $C_7H_5^+$ $\xleftarrow{H^\cdot}$ $C_7H_6^{+\cdot}$ $\xleftarrow{-CO}$
m/e 89 (40%) m/e 90 (53%)

m/e 118 (33%)

As is the case with 2-methylpyridine N-oxide (Sec. II.C.1), introduction of a methyl group adjacent to the nitrogen atom in quinoline or isoquinoline N-oxide results in marked "ortho effects" and production of intense M-·OH ions. The process is typified in the rationale of the spectrum of 2-methyl-quinoline N-oxide (**87**) (45).

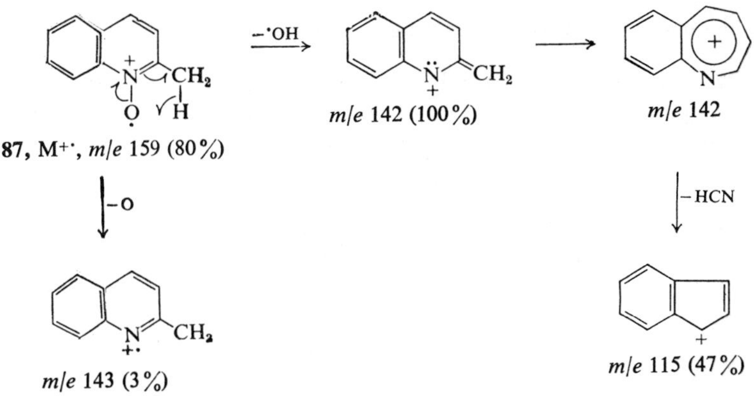

–·OH
CH2
N
O
H
87, M+·, m/e 159 (80%)
–O
m/e 142 (100%)
m/e 142
–HCN
CH2
m/e 143 (3%)
m/e 115 (47%)

It is interesting to note that 4- and 6-methylquinoline *N*-oxide show loss of ·CHO rather than the dominant loss of CO observed with the unsubstituted compound. The reason for this effect of the remote methyl group is not fully clear. The rationale shown has been produced (outlined for the 6-methyl-compound, **88**). It should be noted that the appropriate metastable ion shows that the loss is of a ·CHO unit and not sequential loss of H·, then CO.

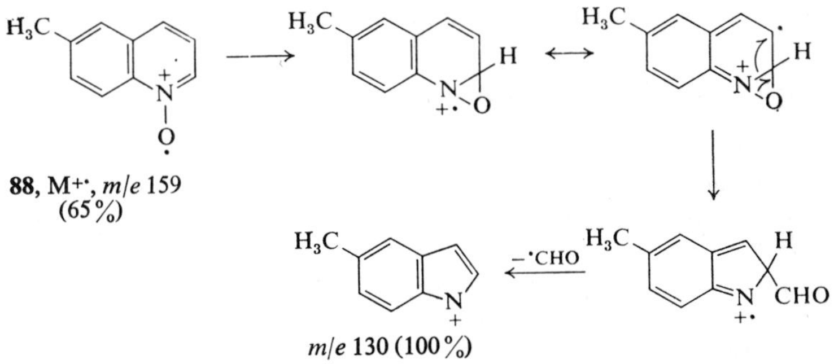

H3C
88, M+·, m/e 159 (65%)
–·CHO
CHO
m/e 130 (100%)

Introduction of an aryl group adjacent to the *N*-oxide function results in a spectrum dominated by the M-1 ion, and the sequence shown involving cyclization has been suggested (exemplified by 2-phenylquinoline *N*-oxide, **89**) (45).

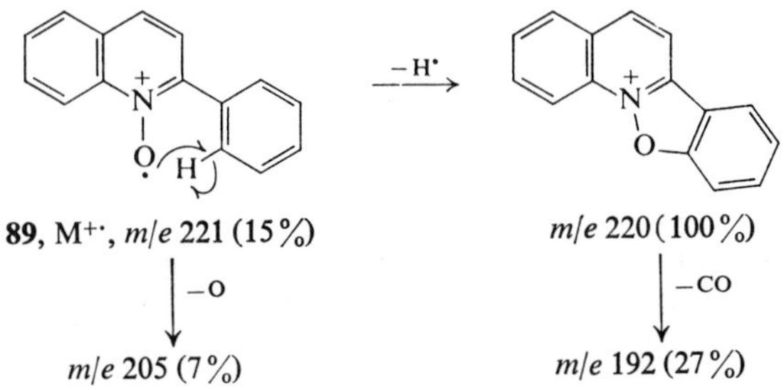

–H·
89, M+·, m/e 221 (15%)
m/e 220 (100%)
–O
–CO
m/e 205 (7%)
m/e 192 (27%)

E.   Quinolines with Nitrogen-Containing Substituents

Relatively little work has been published in this field. Spiteller (48) reports that aminoquinolines fragment in a manner analogous to the hydroxy compounds. For 2-aminoquinoline (90) the process can be represented (16b) as shown.

90, $M^{+\cdot}$, m/e 144 m/e 144

$C_7H_6^{+\cdot}$ ←—

m/e 90 m/e 117

Budzikiewicz, Djerassi, and Williams (49) report that 8-nitroquinolines show loss of NO_2 and also loss of NO followed by CO. This result can be explained by the same isomerization to a nitrite ester that has been postulated to occur in the spectrum of nitrobenzene (50, 51).

F.   Reduced Quinolines and Isoquinolines

The occurrence of the isoquinoline nucleus at varying stages of reduction in many alkaloids makes it essential for there to be a good understanding of the fragmentation of simple derivatives. In fact, remarkedly little fundamental work on the spectra of simple reduced isoquinolines has been carried out, and this is an open field for useful research.

1.   Dihydroquinolines

Kynurenine yellow (91), a compound obtained by cyclizing the amino acid kynurenine with sodium bicarbonate, is formally 2-carboxy-2,3-diyhdro-4-quinolone.

91, $M^{+\cdot}$, m/e 191   91a, m/e 146

The mass spectrum of 91 (52) is dominated by the ion 91a, resulting from loss of $\cdot CO_2H$ from the molecular ion. This result clearly shows that charge localization on the nitrogen atom produces the main directive influence in the fragmentation of this compound.

2. Dihydroisoquinolines

The only report of the mass spectrum of a simple dihydroisoquinoline is in the use of the technique to identify the 1,2-dihydro compound (92) as the oxidation product of 1,2,3,4-tetrahydroisoquinoline. The compound apparently gives an intense M-1 ion (*m/e* 130) (53) which can be represented as shown. In the 1,2-dihydroisoquinoline 93 (54) the fragmentation is initiated

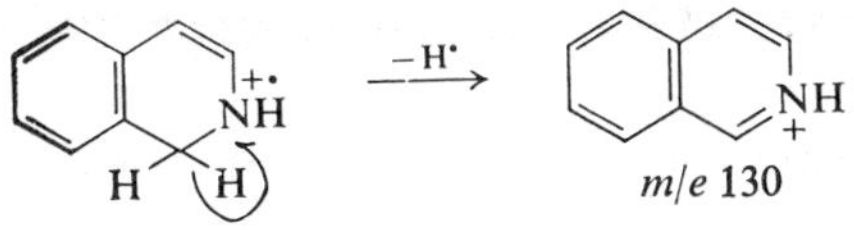

$$-\text{H}^{\cdot}$$

92, M+·, *m/e* 131   *m/e* 130

by the carbonyl group and the base peak (*m/e* 172) corresponds to the ion 93a.

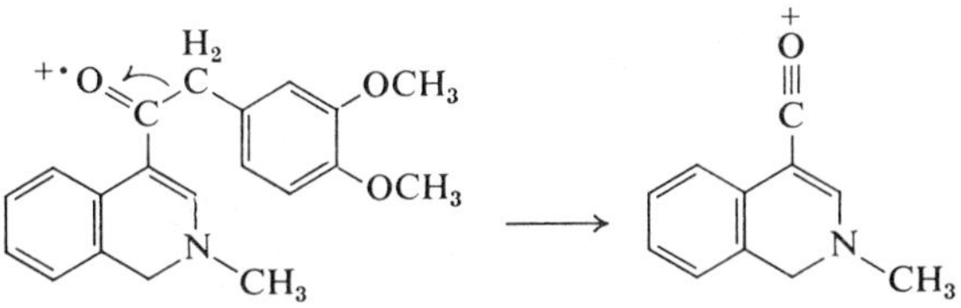

93, M+·, *m/e* 323   93a, *m/e* 172

The 3,4-dihydroisoquinoline ring appears in the alkaloid psychotrine methyl ether (94) (55), but the multiplicity of functional groups present inhibits any simple fragmentation of the dihydroisoquinoline ring. The major fragmentation involves a McLafferty rearrangement with the nitrogen of this ring as the migration terminus.

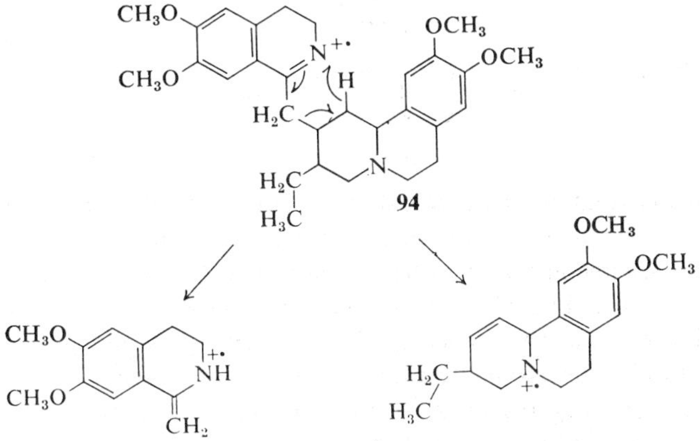

94

m/e 205   *m/e* 273

3. Tetrahydroisoquinolines

Many alkaloids containing the 1,2,3,4-tetrahydroisoquinoline nucleus have been examined mass-spectrometrically, and the technique has found considerable value as a tool in structure determination. The dominant feature of the spectra of such compounds is loss of the substituent in the 1-position to yield an immonium cation which is usually the base peak of the spectrum.

Thus 1-isobutyl-2-methyl-6-methoxy-1,2,3,4-tetrahydroisoquinoline has only a weak molecular ion (95) which by loss of an isobutyl radical leads to the base peak of the spectrum (95a). This ion further decomposes by loss of a methyl group to an ion which can be represented as 95b (56).

95, M⁺·, m/e 233

95a, m/e 176 (100%)

95b, m/e 161

Both the two possible fissions of this kind are observed in the bis-(tetrahydroisoquinolyl)methane 96 (57), a result which would clearly differentiate this compound from an isomer such as 97.

96   m/e 132   97

m/e 146

Simple benzylisoquinoline alkaloids also show this dominant fission, no doubt aided by the high stability of the neutral fragment (a benzyl radical), and the major fragmentations may be exemplified by the spectrum of romneine (98) (58), shown in Fig. 11.25. It will be seen that all ions other than the benzyl-fission product (98a) and m/e 191 are very weak and provide little useful structural information. The ion at m/e 191 must arise from hydrogen transfer accompanying the benzyl fission and may be represented as 98b. It will be noticed that some dehydrogenation to the isoquinolinium cation 98c also occurs.

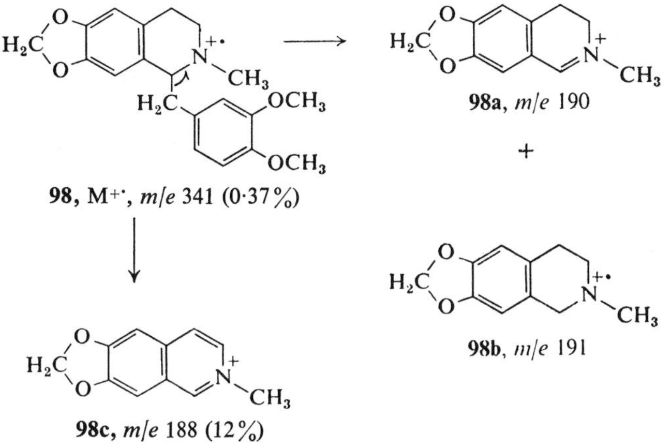

98, M⁺·, m/e 341 (0·37%)

98a, m/e 190

+

98b, m/e 191

98c, m/e 188 (12%)

When the benzyl group is directly attached to the nitrogen atom of the tetrahydroisoquinoline ring, as in the alkaloid sendaverine (**99**), the fragmentation is quite different (59). The published spectrum shows that the molecular ion is of reasonable intensity and the methoxybenzyl cation (**99a**) [or the isomeric methoxytropylium ion (**99b**)] provides the base peak of the spectrum. The rationale for the fragmentation of this compound is presented here.

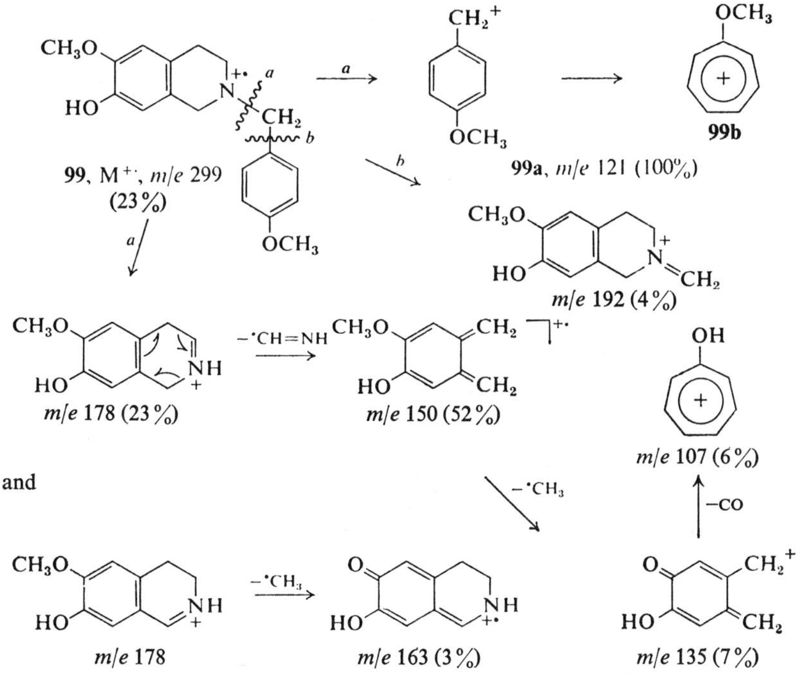

99, M⁺·, m/e 299 (23%)

99a, m/e 121 (100%)

99b

m/e 192 (4%)

m/e 178 (23%)

m/e 150 (52%)

m/e 107 (6%)

and

m/e 178

m/e 163 (3%)

m/e 135 (7%)

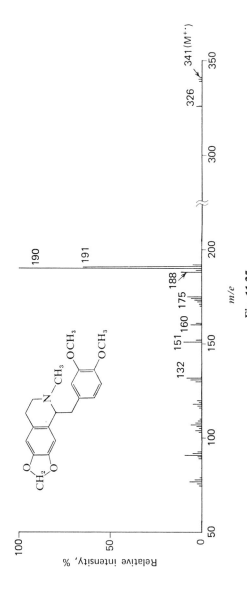

Fig. 11.25

IV. QUINOLIZIDINE DERIVATIVES

The mass spectra of a number of alkaloids containing the quinolizidine ring system have been reported. The spectrum of lupinic acid (100) is shown in Fig. 11.26 (60). Like those of pyrrolizidine derivatives, it can be explained on the basis of initial β-fissions. It will be seen that many hydrogen transfer processes are required in the fragmentation rationale presented. These have not as yet been supported by deuterium labeling, so they must be regarded as tentative.

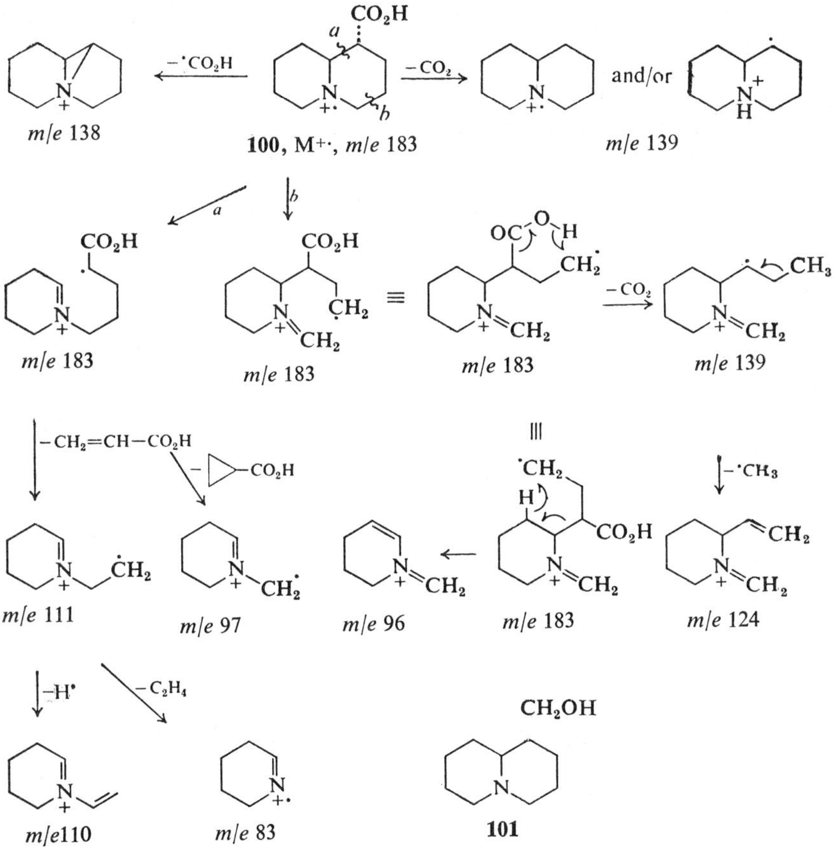

A very similar rationale may be produced for lupinine (101), and the reader is referred to the original publication for further details (60).

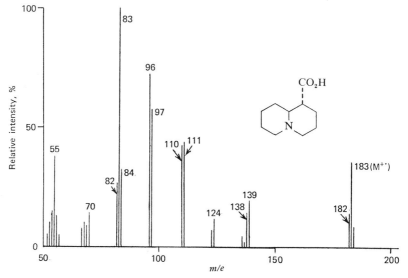

Fig. 11.26

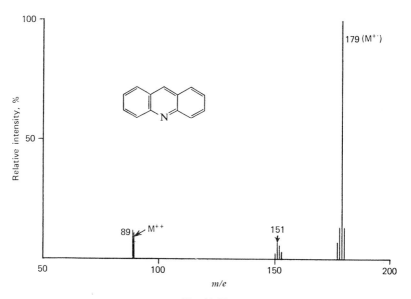

Fig. 11.27

V. TRICYCLIC COMPOUNDS CONTAINING ONE OR MORE PYRIDINE RINGS

A. Acridine and Derivatives

The published spectrum (61) (Fig. 11.27) of acridine (**102**) is simple and, as expected, dominated by the parent ion. Loss of HCN from this species is a less important process than with pyridine or quinoline. Presumably the necessity of fission of three bonds to remove this fragment considerably raises the energy requirements for the process. A possible rationale for the process is shown.

102, M$^{+\cdot}$, m/e 179
(100%)

$\xrightarrow{-\text{HCN}}$

m/e 152 (6.6%)

$\downarrow -\text{H}\cdot$

$C_{12}H_7^+$
m/e 151 (6.5%)

The mass spectrum of 9-ethoxy-N-hydroxyacridan (**103**) is dominated by the loss of the 9-substituent (62), and this would presumably be the main fragmentation route of various 9-substituted acridans. A weaker ion (m/e 212) corresponding to loss of the ethyl group is possibly stabilized by an O—N interaction as shown.

m/e 196 (100%)   **103**, M$^{+\cdot}$, m/e 241 m/e 212

B. Acridones

The mass spectrum of acridone (**104**) is simple and dominated by loss of carbon monoxide from the parent ion (63). The resultant species is best formulated as the carbazole radical cation (**104a**) and its further decomposition is the same as that already observed for that compound.

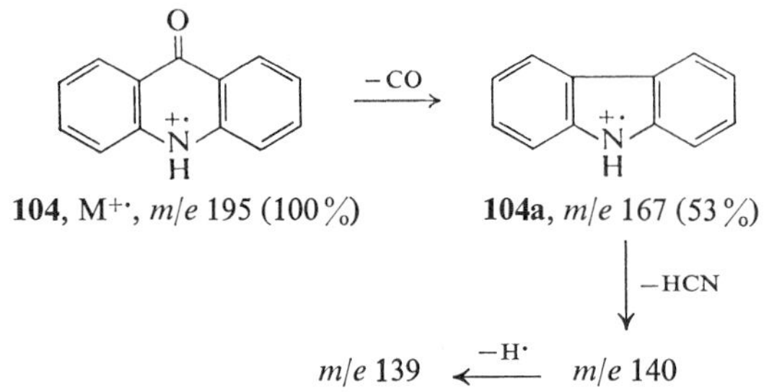

−CO
104, M+·, m/e 195 (100%)
104a, m/e 167 (53%)
−HCN
−H·
m/e 139
m/e 140

The spectrum of *N*-methylacridone (**105**) is rather more complex. The importance of loss of a molecule of CO decreases from that observed with the parent compound whereas decompositions initiated from the M-·CH_3 and M-·CHO ions are significant. A rationale for the major fragmentations is shown.

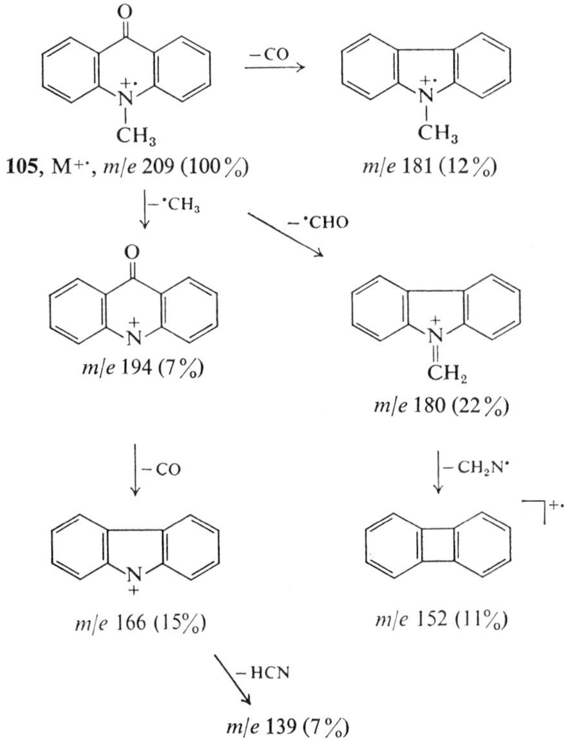

−CO
105, M+·, m/e 209 (100%)
m/e 181 (12%)
−·CH3
−·CHO
m/e 194 (7%)
m/e 180 (22%)
−CO
−CH2N·
m/e 166 (15%)
m/e 152 (11%)
−HCN
m/e 139 (7%)

A considerable amount of work has been carried out on the mass spectra of acridones containing oxygenated substituents (63). It has been shown by

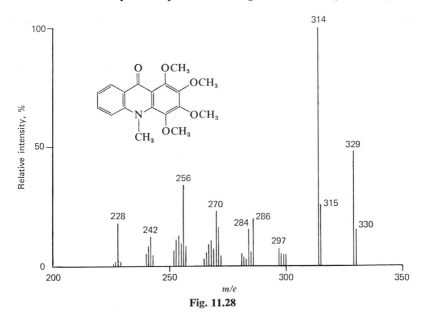

Fig. 11.28

deuterium labeling that in the spectrum of 1,2,3,4-tetramethoxy-10-methyl-acridone (**106**) (Fig. 11.28) 60% of the M-CH$_3$· ion results from loss of the methyl group at the 2-position and 40% from that at the 4-position. The resultant ions (**106a** and **106b**) are well-stabilized, and it is noteworthy that similar structures for losses from the 1- or 3-positions cannot be written.

106, M$^+$·, m/e 329 $-$·CH$_3$ **106a**, m/e 314

$-$·CH$_3$

106b, m/e 314

1-Ethoxyacridones such as **107** show a specific decomposition sequence M → M-OH· → M-(·OH + H$_2$). This sequence is not observed with the ethoxy group in any other position and is thus of diagnostic value for 1-hydroxyacridones (which may be easily ethylated with ethyl iodide in the presence of silver oxide). The process has been rationalized as a specific "ortho effect."

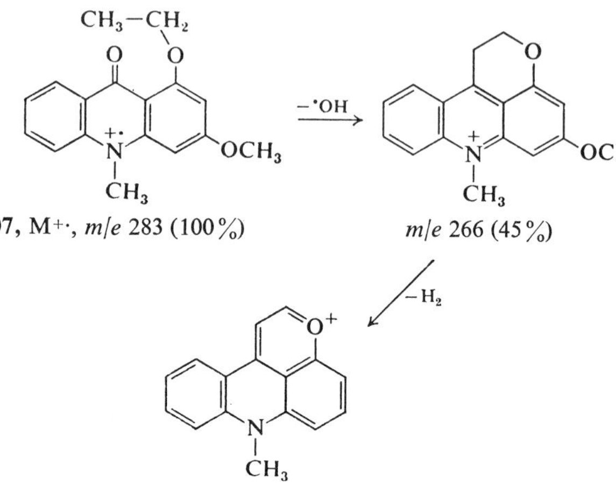

107, M$^{+\cdot}$, *m/e* 283 (100%)   *m/e* 266 (45%)

m/e 264 (15%)

For further details of the fragmentation of a large number of oxygenated acridones the reader is referred to the original publications (63).

C.   Aza- and Diazaphenanthrenes

Little has been published about the mass spectrometry of ring systems of this kind. Benzo[f]quinoline (**108**), like quinoline itself, has an intense molecular ion (the base peak of the spectrum) and shows relatively little fragmentation (64). It is of some interest to note that in this case the M-H$_2$CN ion (*m/e* 151) is more intense than M-HCN (*m/e* 152).

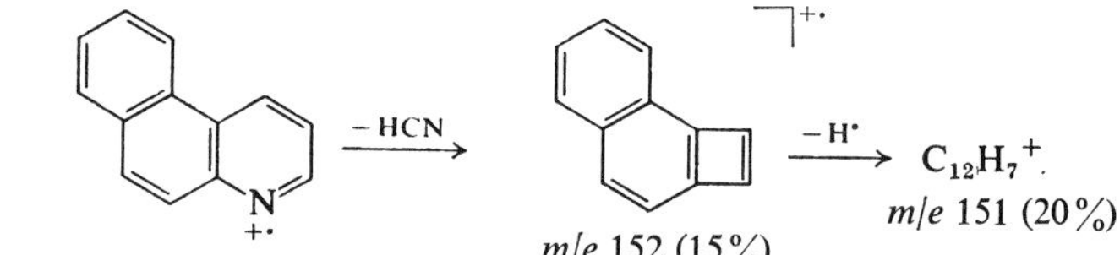

108, M, *m/e* 179 (100%)   *m/e* 152 (15%)   C$_{12}$H$_7^+$ *m/e* 151 (20%)

The diaza compound benzo[f][1,7]naphthyridine (**109**) shows successive losses of two molecules of HCN, but otherwise the spectrum is unexceptional (64). Its 3-methyl derivative (**110**) shows losses of HCN from both the molecular ion and the M-1 species (possibly **110a**) together with loss of CH$_3$CN from the parent ion.

1,5-Dimethyl-4-ethyl-7-methoxy-1,2,3,4,5,10,11,12-octahydrobenzo[g]-quinoline (**111**) has been obtained by methylation of the corresponding

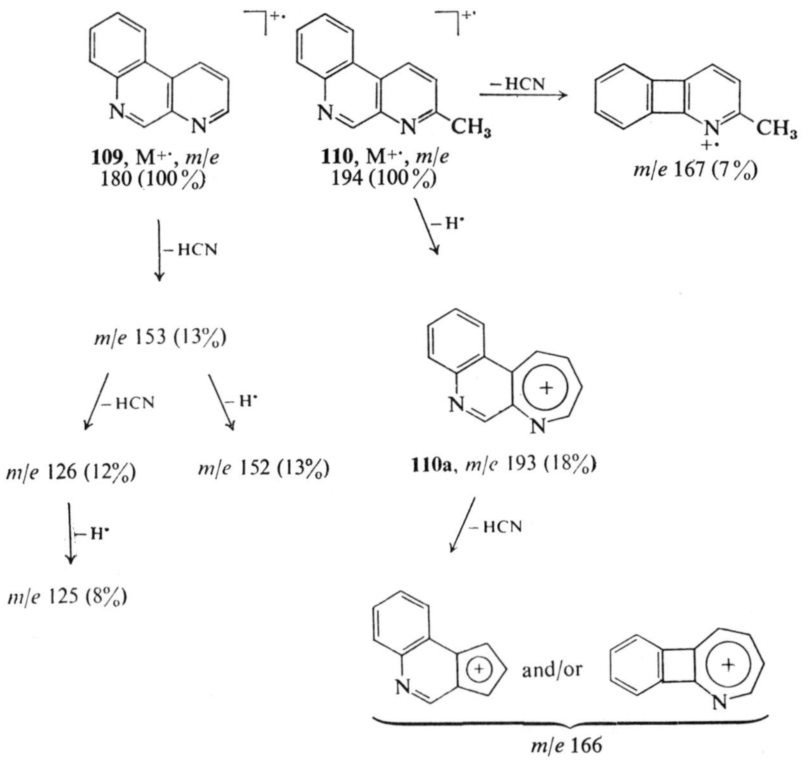
109, M+·, m/e 180 (100%)
110, M+·, m/e 194 (100%)
−HCN
m/e 167 (7%)
−HCN
−H·
m/e 153 (13%)
−HCN
−H·
m/e 126 (12%)
m/e 152 (13%)
110a, m/e 193 (18%)
−H·
m/e 125 (8%)
−HCN
and/or
m/e 166

hydroxy compound, the first octahydrobenzo[*g*]quinoline to exhibit analgetic activity (65). Apart from M-·CH_3 (13%) and M-·C_2H_5 (58%) ions, the main feature of the mass spectrum is a retro-Diels-Alder reaction to give the ions **111a** and **111b**. Loss of an ethyl group from the latter yields the base peak of the spectrum (*m/e* 125).

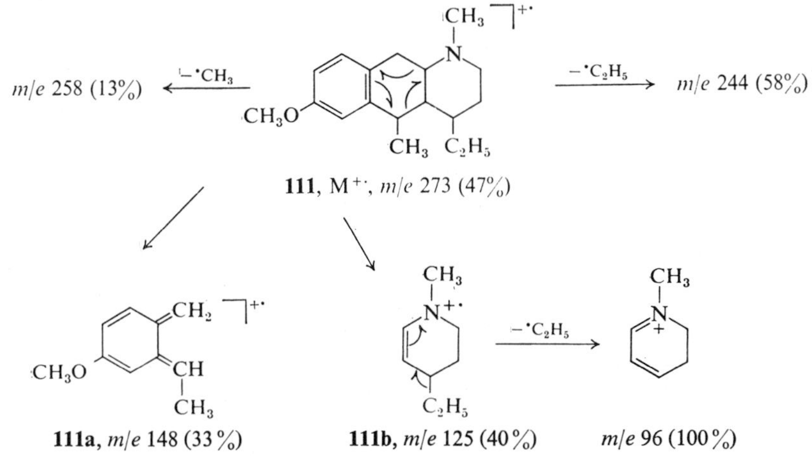
m/e 258 (13%)
−·CH₃
−·C₂H₅
m/e 244 (58%)
111, M+·, m/e 273 (47%)
−·C₂H₅
111a, m/e 148 (33%)
111b, m/e 125 (40%)
m/e 96 (100%)

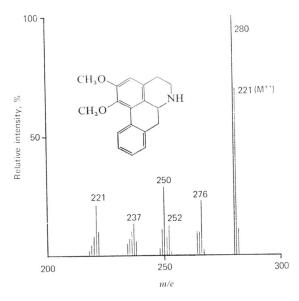

Fig. 11.29

VI. TETRA- AND POLYCYCLIC COMPOUNDS CONTAINING ONE OR MORE PYRIDINE RINGS

A. Aporphine Alkaloids

In view of their simple structure (only one heterocyclic ring) and their biogenetic relationship to the benzyltetrahydroisoquinoline alkaloids, the mass spectrometry of aporphine alkaloids will be briefly reviewed.

The important features of the fragmentation of the group may be seen in the rationalization of the partial spectrum (Fig. 11.29) of N-nornuciferine (**112**) (66). It will be noticed that the base peak of the spectrum is the M-1 ion, presumably the well-stabilized conjugated immonium cation **112a**. It has been suggested that the M-·CH$_3$ (m/e 266) and M-·OCH$_3$ (m/e 250) ions are stabilized by the interaction shown with the benzylic radical center resulting from an internal α-fission (**112b** → **112c** or **112d**). Loss of CH$_2$=NH in a retro-Diels-Alder process gives m/e 252 as shown. In N-methyl aporphines this ion is of course shifted to M-43 and is of obvious diagnostic value. The rationale of the important fragmentations is shown.

Characteristic ions of unknown structure occur at m/e 152 and 165 in all aporphine alkaloid spectra and are of diagnostic value.

CH₃O, CH₃O $\xleftarrow{-H\cdot}$ CH₃O, CH₃O   NH⁺· $\xrightarrow{\alpha\text{-fission}}$ CH₃O, CH₃O   NH⁺   ·CH₂—

112a, m/e 280   **112**, M⁺·, m/e 281   **112b**, m/e 281

$-CH_2NH$ $-\cdot OCH_3$ $\downarrow -\cdot CH_3$

CH₃O, CH₃O   CH₂   CH₃O, H₂C   NH⁺   CH₃O, H₂C   O   NH⁺

m/e 252   **112c**, m/e 250   **112d**, m/e 266

$\downarrow -\cdot OCH_3$

m/e 221

It has been shown that for aporphine alkaloids oxygenated at *both* the 1- and 11-positions the M-1 ion is weaker than the molecular ion (intensity 30–60%), whereas where either the 1- or 11-position is occupied by a hydrogen atom M-1 is stronger than M (67). This observation has been explained as a result of steric hindrance between the 1- and 11-substituents preventing coplanarity of the diphenyl system, and hence also preventing the associated resonance-stabilization of the M-1 ion. The effect of such steric interference on the M-43, M-58, and M-74 ions in the spectra of many aporphines has also been discussed, and the original paper should be consulted for more detail.

B. Spermatheridine Alkaloids

These alkaloids may be regarded as oxidized aporphines and their fragmentation exemplified by that of atherospermidine (**113**) (68). It will be seen that all fragmentation is initiated by formation of the M-CH₃· ion (which has been represented as **113a**) followed by a succession of losses of CO or CH₂O. A rationale for part of the spectrum (Fig. 11.30) is shown. (In the

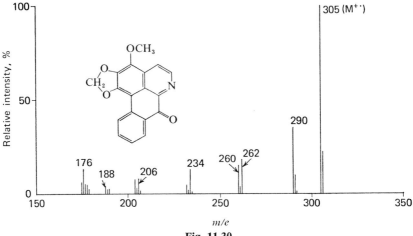

Fig. 11.30

original rationale the authors showed the structure **113b** for the M-·CH₃ species. This is, of course, merely a canonical form of **113a**, but we prefer the latter formulation as a major contributor since it has the positive charge as a stable oxonium center rather than on an oxygen atom carrying an energetically unfavored sextet of electrons.)

C.   Berbine Alkaloids

The fragmentations of this group may be exemplified using the mass spectrum of xylopinine (114) (66). The spectrum is simple and is dominated by fission of ring C in a retro-Diels-Alder process.

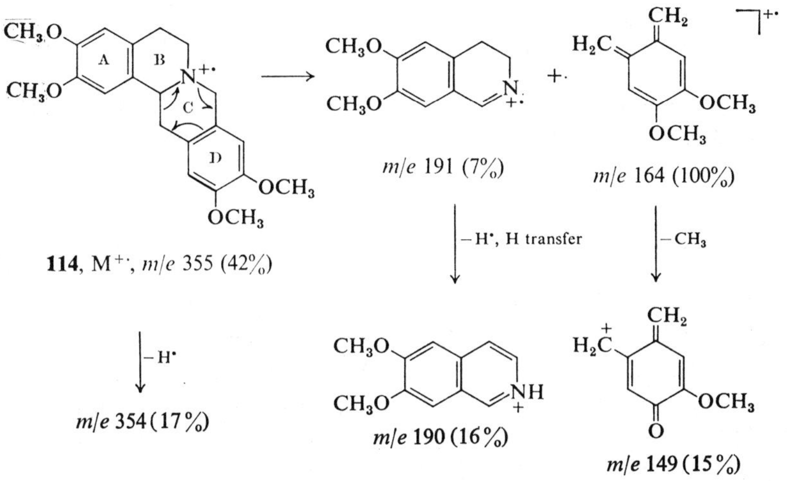

114, M$^+$, m/e 355 (42%)

m/e 191 (7%)

m/e 164 (100%)

m/e 354 (17%)

m/e 190 (16%)

m/e 149 (15%)

D.   Benzo[a]- and Benzo[c]acridines

The mass spectrum of benzo[c]acridine (115) and a number of methyl and ethyl derivatives of this heterocycle and the isomeric benzo[a]acridine have been reported (69). As expected, the molecular ion is the base peak of all the spectra reported, and losses of HCN occur from the molecular ions [although for the alkylated derivatives it was not possible in the absence of high-resolution data to differentiate the isobaric M-HCN and M-(H + C$_2$H$_2$) ions].

With benzo[c]acridine (115) the M-1 ion was of considerable intensity (25% compared with 14% from acridine and 5% from benzo[a]anthracene) and it may be speculated that the ion produced is the cyclic ion 115a.

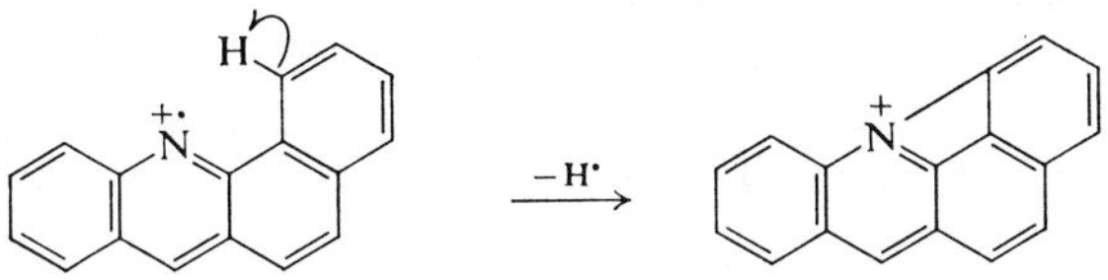

115, M$^+$, m/e 229

115a, m/e 228

The methyl and ethyl compounds show the expected β-cleavages (of H· or ·CH$_3$, respectively) to give species which may well contain tropylium ion rings.

M$^{+\cdot}$, m/e 285 (100%) m/e 270 (17%)

As is common with polycyclic aromatic compounds, doubly and triply charged ions are of significant intensities in the mass spectra of the various benzacridines.

E. Azasteroids

1. 4-Azasteroids

The mass spectrum of 2-carboxy-7aβ-methyl-7-oxo-9aα-H-indano[5,4f]-5aα,10,10aβ,11-tetrahydroquinoline (**116**), a molecule possessing many of the features of aza-estrone, has been reported (70). The compound is remarkably stable to electron impact with the molecular ion providing the base peak of the spectrum and the only important fragmentation being loss of carbon dioxide. Presumably little of the electron deficiency is distributed outside the pyridine ring and thus none of the *a priori* less-stable alicyclic portion of the molecule suffers fragmentation.

116, M$^{+\cdot}$, m/e 299 (100%) m/e 255 (79%)

2. 6-Azasteroids

The mass spectra of 6-azaequilenin (**117**), 6-aza-14(β)-isoequilenin (**118**), and the corresponding alcohols have been reported (71). Those of the two ketones are shown in Figs. 11.31 and 11.32, respectively, and it will be noticed that there are significant differences in the intensities of the ions at m/e 212 and 238 in the two spectra. Assistance in elucidating the fragmentation has been provided by the spectrum of the 14,15-d_2 analogue of 6-azaequilenin.

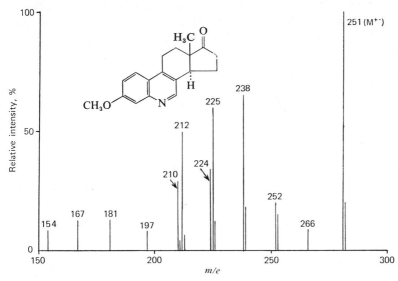

Fig. 11.31

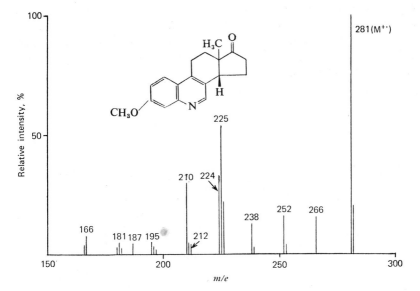

Fig. 11.32

The origin of the M-$^{\cdot}CH_3$ ion (*m/e* 266) has not been clearly defined but probably involves both the *O*-methyl group and the $C_{(18)}$ angular methyl group. It is suggested that the ions at *m/e* 238 and 212 are formed by the pathway shown.

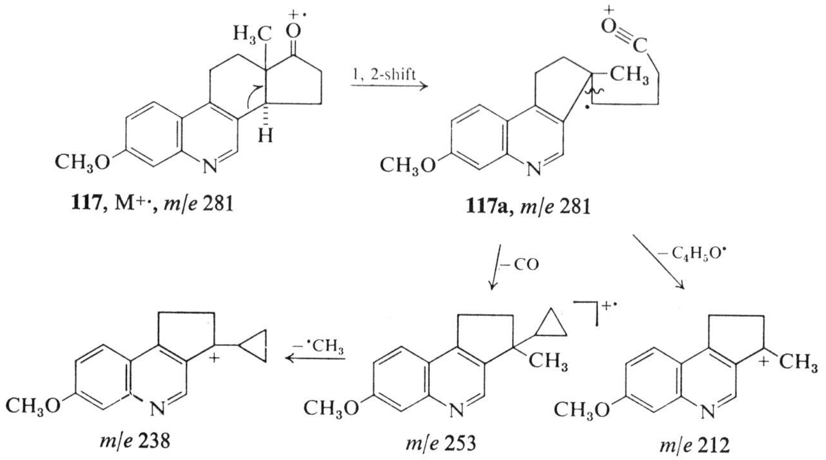

1, 2-shift
117, M+·, m/e 281
117a, m/e 281
−CO
−C4H5O·
−·CH3
m/e 238
m/e 253
m/e 212

It is further suggested that if the 1,2-shift and C—CO fission (shown in converting the molecular ion to its isomeric form **117a**) are concerted, such a process will be aided by *trans* geometry at the ring C-ring D junction, and inhibited by a *cis* arrangement (as in **118**).

The formation of the *m/e* 225 ion in both compounds follows simple α-fission of the $C_{(13)}$—$C_{(17)}$ bond and will not be affected by the stereochemistry at the ring junction.

117 or 118, M+·, m/e 281
m/e 281
−C3H4O
m/e 225

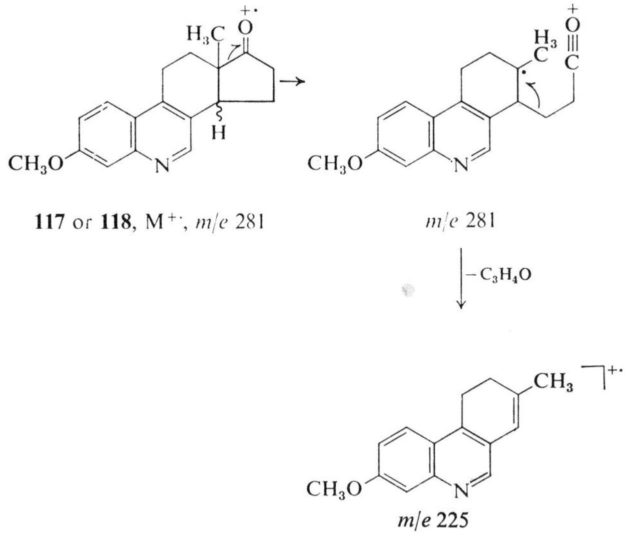

The genesis of the m/e 210 peak is apparently complex. It involves the loss of a C_4H_7O unit (which incorporates the three carbon atoms unique to ring D together with the $C_{(18)}$-methyl group) from the molecular ion.

Rather similar rationales have been proposed for the mass spectra of the alcohols derived from **117** and **118** and the reader is referred to the original paper for full details.

The main fragmentation process in the mass spectrum of the aza-estrone methyl ether **119** is a retro-Diels-Alder reaction leading (initially) to the dihydroquinoline **119a** which gives the stable ion **119b** by loss of a further hydrogen atom (72).

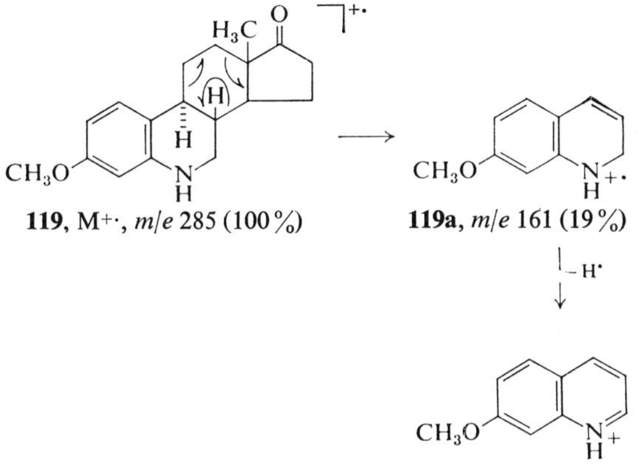

119, M+·, m/e 285 (100%)   **119a,** m/e 161 (19%)

$-$ H·

119b, m/e 160 (55%)

The minor ions present in the spectrum form an overall fragmentation pattern similar to that of estrone methyl ether (73).

F. Tylophorin

This pentacyclic alkaloid, which is a hexahydro-dibenzo[f,h]pyrrolo[1,2-b]-isoquinoline, has a very simple mass spectrum in which the chief fragmentation is the familiar retro-Diels-Alder process leading to the ion **120a** as base peak of the spectrum (74).

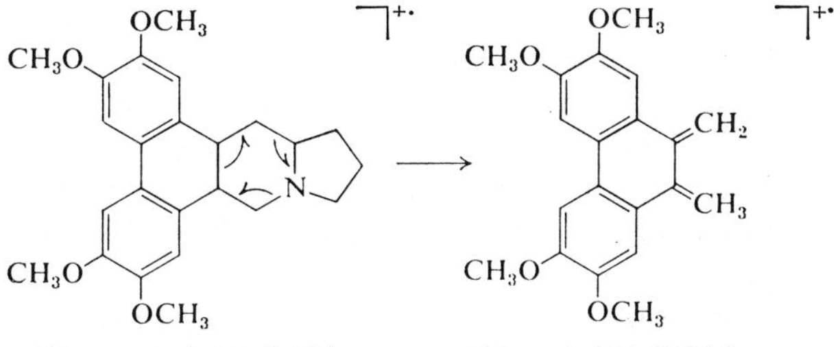

120, M+·, m/e 393 (30%)   **120a,** m/e 324 (100%)

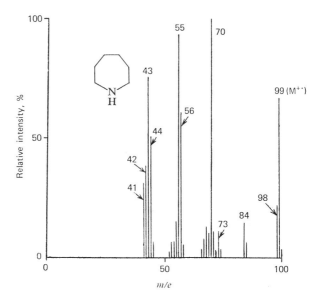

Fig. 11.33

VII.   SEVEN-MEMBERED AND LARGER NITROGEN HETEROCYCLES

A.   Hexamethyleneimine Derivatives

The mass spectrum of hexamethyleneimine (121) is shown in Fig. 11.33 (75). The fragmentation has not as yet been studied by means of deuterium labeling but in general resembles that of piperidine and a possible rationale is suggested in the scheme. A noteworthy difference from the piperidine spectrum is the reduced importance of the M-1 ion which is weaker than the molecular ion. This probably reflects the greater importance here of the internal α-cleavage ion 121a. It will be noticed that the ion at m/e 70, which almost certainly is formed by loss of C_2H_5· from the molecular ion, is much more intense than the ion of the same postulated structure formed by loss of CH_3· from the molecular ion of piperidine (Sec. I).

The spectrum of N-ethylhexamethyleneimine (122) has as its base peak the M-CH₃· ion, and other fragmentations are reduced in importance. The methyl group lost is certainly from the N-alkyl group for, as seen in the spectrum of hexamethyleneimine, loss of the elements of methyl from the ring is not an important process here. This conclusion is supported by the loss of the fragment $^{.}(CH_2)_3NH_2$ from the amine **123** to give the same α-fission ion (76).

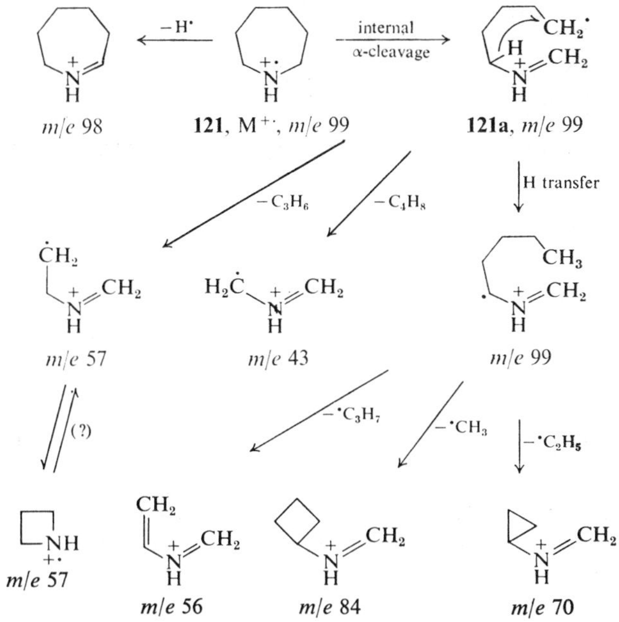

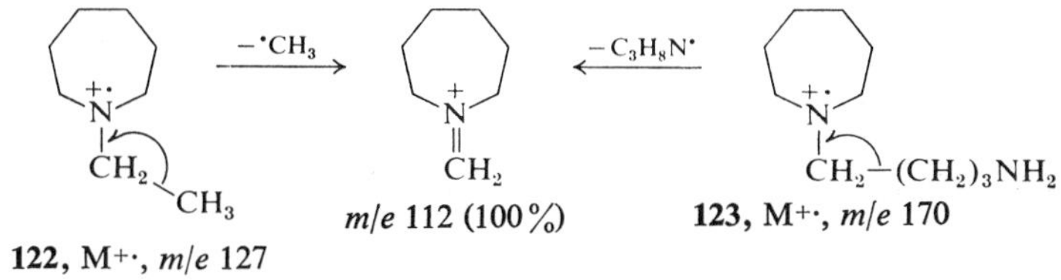

B. Benzazepines and Related Compounds

The mass spectrum of 4-methyl-2,5-dioxo-2,5-dihydro-1*H*-1-benzazepine (**124**) is characteristic of benzazepine diones of this type, and fragmentation occurs by successive losses of two molecules of CO to give the 3-methylindole radical cation (**124a**) (77).

As expected, reduction of the 3,4-double bond of **124** produces a marked change in fragmentation. Thus the base peak of the spectrum of **125** is C_7H_6NO (*m/e* 120) which is formed directly from the molecular ion. A possible structure and genesis is shown, but (in particular) the origin of the

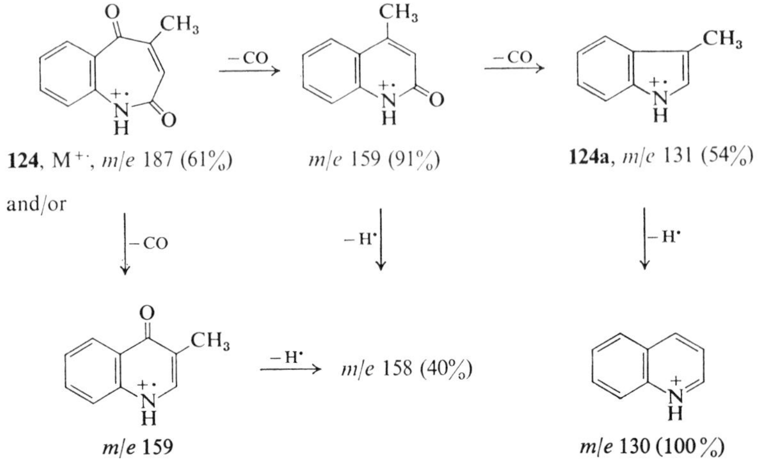

124, M$^{+\cdot}$, m/e 187 (61%) m/e 159 (91%)   **124a**, m/e 131 (54%)

and/or

m/e 159 $\xrightarrow{-H^\cdot}$ m/e 158 (40%) m/e 130 (100%)

transferred hydrogen atom would have to be determined by deuterium labeling. It will be noticed that M-CO ions are much reduced in intensity compared with those in the spectrum of **124**.

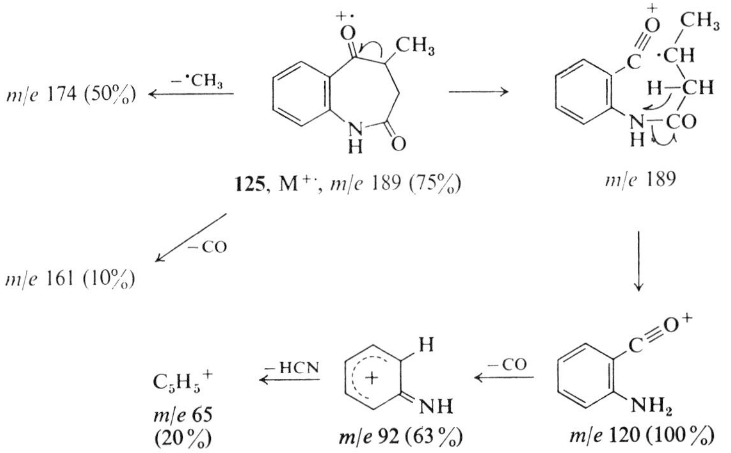

m/e 174 (50%) $\xleftarrow{-\cdot CH_3}$

125, M$^{+\cdot}$, m/e 189 (75%) m/e 189

m/e 161 (10%) $\xleftarrow{-CO}$

$C_5H_5^+$ $\xleftarrow{-HCN}$ $\xleftarrow{-CO}$
m/e 65
(20%) m/e 92 (63%) m/e 120 (100%)

C. Protopine Alkaloids

This group of alkaloids provides a relatively simple example of a ten-membered nitrogen heterocycle, and the mass-spectrometric fragmentation involves an interesting transannular interaction of the nitrogen atom with a carbonyl group. The main features of the fragmentation of the group of compounds are exhibited by protopine (**126**) itself (78). The spectrum is very simple and the base peak (*m/e* 148) is best allocated structure **126a**, resulting from a fragmentation accompanying bond formation as shown.

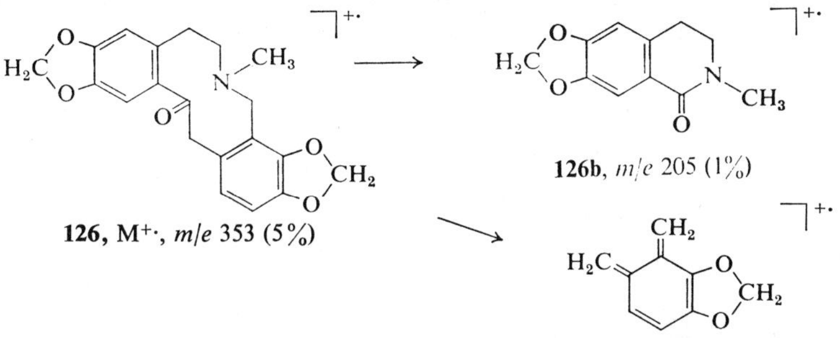

126, M⁺·, *m/e* 353 (5%)

126b, *m/e* 205 (1%)

126a, *m/e* 148 (100%)

The ions at *m/e* 163 and 190 are thought to arise by cleavage of the bond *a* (which is both benzylic and correctly situated for cleavage α to nitrogen), followed by hydrogen transfer in a cyclic six-center process.

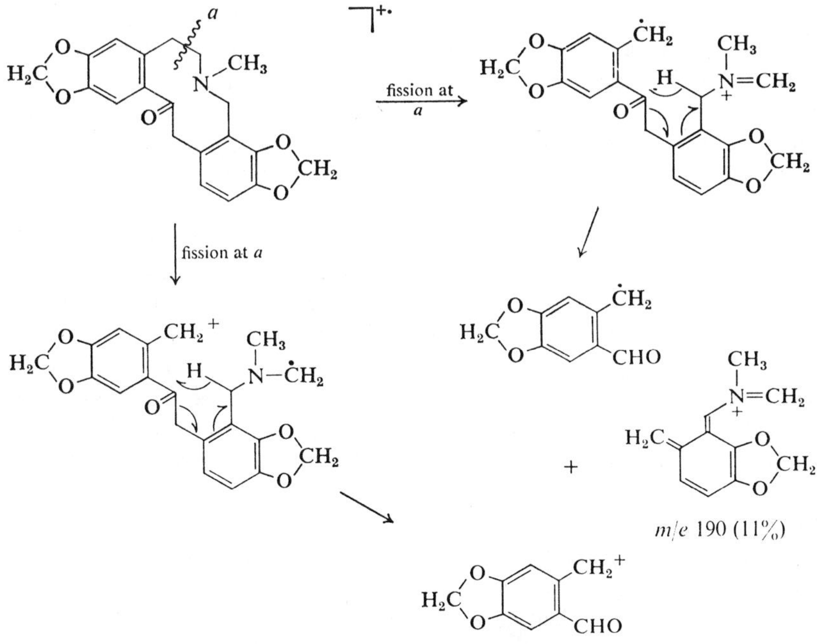

m/e 190 (11%)

m/e 163 (20%)

It is interesting to note that the ion (126b) conjugate to 126a is very weak (~1%) in the spectrum of protopine, but in hunnemannine (127) the cyclic fragmentation with hydrogen transfer results in 127a forming the base peak of the spectrum. The spectrum of hunnemannine O-d_1 shows that some 65% of the transferred hydrogen originates from the phenolic hydroxyl group, and the fragmentation is of obvious diagnostic value (78).

127, $M^{+\cdot}$, m/e 365 (5%)

127a, m/e 206 (100%)

D.   Catenanes

The mass spectrum of the interesting catenane 128 has been reported and analyzed (79) together with the spectra of the noninterlocked macrocycles 129 and 130. The spectrum of the heterocycle 130 is itself of some interest and shows an intense M-·CH$_3$ ion (m/e 406). Deuterium-labeling studies show that this results to a large extent (66%) from loss of the acetyl CH$_3$ group; the importance of this loss may be contrasted with the small losses of CH$_3$· normally observed with N-acetylamines.

The spectrum of the catenane 128 shows a clear parent ion at m/e 1021 which corresponds to the sum of the m/e values of 129 and 130. Fragments in the spectrum correspond to separation of 129 and 130 (and subsequent fragmentation) following ring-opening of either or both rings. Along with these there are fragments in which interlocking is still present, i.e., which correspond to fragmentations without ring-opening. These fragments result from sequential losses of ketene ($M^{+\cdot} \rightarrow m/e$ 979 $\rightarrow m/e$ 937 $\rightarrow m/e$ 895) and correspond to a similar sequence observed with the macrocycle 129. The implication is clear that this sequence does not involve the formation of noncyclic daughter ions.

An interesting fragment ion in the spectrum of 128, which is not present in the spectra of 129 or 130, occurs at m/e 422 and corresponds to the protonated molecular ion of 130. It must result from a hydrogen transfer between the rings of a catenated ion. Metastable-ion data indicates that the ion at

m/e 895 is the precursor of m/e 422 which subsequently decomposes by loss of ketene.

It is clear that the mass spectrometry of catenated compounds can provide fundamental information about the cyclic nature (or otherwise) of fragmentations, and further studies on such compounds will be awaited with interest.

128, $M^{+\cdot}$, m/e 1021

m/e 979 $\xrightarrow[-\text{CH}_2\text{CO}]{}$ m/e 937 $\xrightarrow[-\text{CH}_2\text{CO}]{}$ m/e 895 $\longrightarrow$ m/e 422

129, m/e 600   **130**, m/e 421

Addenda

More work is reported on the mass spectra of pyridoxol and other components of the Vitamin B6 group (Section II-C-2) (80). New mass spectrometric results have appeared for a range of substituted pyridines (Section II) including anilides of nicotinic acid (81), phenylazopyridines (82), thienylpyridines (83), and alkylthiopyridines (84). Analysis of the mass spectra of variously deuterated γ-picolines shows that the M-1 ion does not necessarily have the azatropylium structure (85). Hydrogen scrambling is complete at all ionizing energies before loss of HCN occurs from the molecular ions of 2- and 4-picolines (86).

In the quinoline field (Section III) work has appeared on the anilide of quinoline-2-carboxylic acid (81), 2-substituted-8-hydroxy-quinolines (87), and 1-hydroxy-2-quinolines (88), while interesting comparisons have been made between the photochemistry and mass spectrometry of some 2-substituted quinolines (89). Hydrogen scrambling involving both rings is essentially complete before the quinoline molecular ion loses HCN (86).

Data has appeared on the fragmentations of oxoquinolizidines (Section IV) (90). The spectra of some 1H-azepines have been tabulated (91), and that of 1-methylazepine 2,7-dione discussed (92).

REFERENCES

1. American Petroleum Institute, Research Project 44. Spectrum 618.
2. R. A. Saunders and A. E. Williams in *Advances in Mass Spectrometry*, Vol. 3 (M. L. Mead, Ed.), The Institute of Petroleum, London, 1966, p. 681.
3. A. M. Duffield, H. Budzikiewicz, D. H. Williams, and C. Djerassi, *J. Amer. Chem. Soc.*, **87**, 810 (1965).
4. M. Spiteller-Friedmann and G. Spiteller, *Monatsh. Chem.*, **96**, 104 (1965).
5. A. Rother, J. M. Bobbitt, and A. E. Schwarting, *Chem. Ind.* (London), **1962**, 654.
6. M. Spiteller-Friedmann and G. Spiteller, *Monatsh. Chem.*, **95**, 1234 (1964).
7. C. P. Lewis, *Anal. Chem.*, **36**, 1582 (1964).
8. J. Ø. Madsen, S.-O. Lawesson, A. M. Duffield, and C. Djerassi, *J. Org. Chem.*, **32**, 2054 (1967).
9. A. M. Duffield, H. Budzikiewicz, and C. Djerassi, *J. Amer. Chem. Soc.*, **86**, 5536 (1964).
10. F. Johnson, N. A. Starkovsky, and W. D. Gurowitz, *J. Amer. Chem. Soc.*, **87**, 3492 (1965).
11. E. C. Blossey, H. Budzikiewicz, M. Ohashi, G. Fodor, and C. Djerassi, *Tetrahedron*, **20**, 585 (1964).
12. R. D. Guthrie and J. F. McCarthy, *J. Chem. Soc.*, C, **1966**, 1207.
13. Ref. 1. Spectrum 617.
14. D. H. Williams and J. Ronayne, *Chem. Commun.*, **1967**, 1129.
15. K. Jennings, *Z. Naturforsch.*, **22a**, 454 (1967).
16. (a) Ref. 1, Spectrum 1535; and (b) Q. N. Porter, Unpublished results.
17. K. Biemann, *Mass Spectrometry*, McGraw-Hill, New York, 1962, p. 135.
18. G. Spiteller, "Mass Spectrometry of Heterocyclic Compounds," in *Advances in Heterocyclic Chemistry*, Vol. 7 (A. R. Katritzky, Ed.), Academic Press, New York, p. 318.
19. J. H. Beynon, *Mass Spectrometry and its Applications to Organic Chemistry*, Elsevier, Amsterdam, 1960, p. 403.
20. S. D. Sample, D. A. Lightner, O. Buchardt, and C. Djerassi, *J. Org. Chem.*, **32**, 997 (1967).
21. C. J. Sih, K. C. Wang, D. T. Gibson, and H. W. Whitlock, *J. Amer. Chem. Soc.*, **87**, 1386 (1965).
22. J. D. Anderson, M. M. Baizer, and E. J. Prill, *J. Org. Chem.*, **30**, 1645 (1965).
23. R. Grigg and B. G. Odell, *J. Chem. Soc.*, B, **1966**, 218.
24. N. Bild and M. Hesse, *Helv. Chim. Acta*, **50**, 1885 (1967).
25. G. Spiteller and M. Spiteller-Friedmann, *Monatsh. Chem.*, **93**, 1395 (1962).
26. A. M. Duffield, C. Djerassi, G. Schroll, and S.-O. Lawesson, *Acta Chem. Scand.*, **20**, 361 (1966).
27. R. M. Acheson and P. A. Tasker, *J. Chem. Soc.*, C, **1967**, 1542.
28. G. R. Waller, R. Ryhage, and S. Meyerson, *Anal. Biochem.*, **16**, 277 (1966).
29. E. M. Kaiser, S. D. Wark, J. F. Wolfe and C. R. Hauser, *J. Org. Chem.*, **32**, 1483 (1967).
30. D. C. De Jongh, S. C. Perricone, and W. Korytnyk, *J. Amer. Chem. Soc.*, **88**, 1233 (1966).
31. W. Richter, M. Vecchi, W. Vetter, and W. Walther, *Helv. Chim. Acta*, **50**, 364 (1967).

32. C. Iwata and D. E. Metzler, *J. Heterocycl. Chem.*, **4**, 319 (1967).
33. P. W. Ford and J. M. Swan, *Aust. J. Chem.*, **18**, 867 (1965).
34. E. Schumacher and R. Taubenest, *Helv. Chim. Acta*, **49**, 1455 (1966).
35. J. R. Majer in *Advances in Fluorine Chemistry* (M. Stacey, J. C. Tatlow, and A. G. Sharpe, Eds.), Butterworths, London, 1961, p. 100.
36. G. J. Janz and A. R. Monahan, *J. Org. Chem.*, **29**, 569 (1964).
37. Ref. 1. Spectrum 625.
38. Ref. 1. Spectrum 626.
39. Ref. 1. Spectrum 1419.
40. Ref. 17, p. 122.
41. M. Marx and C. Djerassi, *J. Amer. Chem. Soc.*, **90**, 678 (1968).
42. D. M. Clugston and D. B. MacLean, *Can. J. Chem.*, **44**, 781 (1966).
43. J. Møller and O. Buchardt, *Acta Chem. Scand.*, **21**, 1668 (1967).
44. D. A. Evans, G. F. Smith, and M. A. Wahid, *J. Chem. Soc., B*, **1967**, 590.
45. O. Buchardt, A. M. Duffield, and R. H. Shapiro, *Tetrahedron*, **24**, 3139 (1968).
46. O. Buchardt, *Tetrahedron Lett.*, **1966**, 6221.
47. O. Buchardt, C. Lohse, A. M. Duffield, and C. Djerassi, *Tetrahedron Lett.*, **1967**, 2741.
48. Ref. 18, p. 322.
49. H. Budzikiewicz, C. Djerassi, and D. H. Williams, *Mass Spectrometry of Organic Compounds*, Holden-Day, San Francisco, 1967, p. 580.
50. J. Momigny, *Bull. Soc. Roy. Sci. Liège*, **25**, 93 (1956).
51. J. H. Beynon, R. A. Saunders, and A. E. Williams, *Ind. Chim. Belge*, **1964**, 311.
52. T. Tokuyama, S. Senoh, T. Sakan, K. S. Brown, and B. Witkop, *J. Amer. Chem. Soc.*, **89**, 1017 (1967).
53. W. Bartok and H. Pobiner, *J. Org. Chem.*, **30**, 274 (1965).
54. M. Sainsbury, S. F. Dyke, and A. R. Marshall, *Tetrahedron*, **22**, 2445 (1966).
55. H. Budzikiewicz, S. C. Pakrashi, and H. Vorbrüggen, *Tetrahedron*, **20**, 399 (1964).
56. C. Djerassi, H. W. Brewer, C. Clarke, and L. J. Durham, *J. Amer. Chem. Soc.*, **84**, 3210 (1962).
57. P. Cerutti and H. Schmid, *Helv. Chim. Acta*, **47**, 203 (1964).
58. F. R. Stermitz, L. Chen, and J. I. White, *Tetrahedron*, **22**, 1095 (1966).
59. T. Kametani and K. Ohkubo, *Tetrahedron Lett.*, **1965**, 4317.
60. N. Neuner-Jehle, H. Nesvadba, and G. Spiteller, *Monatsh. Chem.*, **95**, 687 (1964).
61. Ref. 1. Spectrum 639.
62. H. Mantsch and V. Zanker, *Tetrahedron Lett.*, **1966**, 4211.
63. J. H. Bowie, R. G. Cooks, R. H. Prager, and H. M. Thredgold, *Aust. J. Chem.*, **20**, 1179 (1967).
64. W. W. Paudler and T. J. Kress, *J. Org. Chem.*, **32**, 2616 (1967).
65. B. C. Joshi, E. L. May, H. M. Fales, J. W. Daly, and A. E. Jacobson, *J. Med. Chem.*, **8**, 559 (1965).
66. M. Ohashi, J. M. Wilson, H. Budzikiewicz, M. Shamma, W. A. Slusarchyk, and C. Djerassi, *J. Amer. Chem. Soc.*, **85**, 2807 (1963).
67. A. H. Jackson and J. A. Martin, *J. Chem. Soc., C*, **1966**, 2181.
68. I. R. C. Bick, J. H. Bowie, and G. K. Douglas, *Aust. J. Chem.*, **20**, 1403 (1967).
69. N. P. Buu-Hoï, C. Orley, M. Mangane, and P. Jacquignon, *J. Heterocycl. Chem.*, **2**, 236 (1965).
70. R. G. Coombe, Y. Y. Tsong, P. B. Hamilton, and C. J. Sih, *J. Biol. Chem.*, **241**, 1587 (1966).
71. U. K. Pandit, W. N. Speckamp, and H. O. Huisman, *Tetrahedron*, **21**, 1767 (1965).
72. W. N. Speckamp, H. De Koning, U. K. Pandit, and H. O. Huisman, *Tetrahedron*, **21**, 2517 (1965).

73. C. Djerassi, J. M. Wilson, H. Budzikiewicz, and J. W. Chamberlin, *J. Amer. Chem. Soc.*, **84**, 4544 (1962).
74. M. Pailer and W. Streicher, *Monatsh. Chem.*, **96**, 1094 (1965).
75. Ref. 19, p. 391.
76. Ref. 19, p. 393.
77. G. Jones, *J. Chem. Soc.*, *C*, **1967**, 1808.
78. L. Dolejš, V. Hanuš, and J. Slavík, *Collect. Czech. Chem. Commun.*, **29**, 2479 (1964).
79. W. Vetter and G. Schill, *Tetrahedron*, **23**, 3079 (1967).
80. D. C. DeJongh, S. C. Perricone, M. L. Gay and W. Korytnyk, *Org. Mass Spec.*, **1**, 151 (1968).
81. W. Schafer and P. Neubert, *Tetrahedron*, **25**, 315 (1969).
82. E. V. Brown, *J. Het. Chem.*, **6**, 571 (1969).
83. H. Wynberg, T. J. Van Bergen, and R. M. Kellog, *J. Org. Chem.*, **34**, 3175 (1969)
84. K. F. King, F. M. Hershenson, L. Bauer, and C. L. Bell, *J. Het. Chem.*, **6**, 851 (1969).
85. R. Neeter, N. M. M. Nibbering, and Th. J. DeBoer, *Org. Mass Spec.*, **3**, 597 (1970).
86. W. G. Cole, D. H. Williams, and A. N. H. Yeo, *J. Chem. Soc.*, [B], 1284 (1968).
87. W. M. Scott, J. E. Wacks, and R. L. Stevenson, *Org. Mass Spec.*, **2**, 681 (1969).
88. R. T. Coutts and K. W. Hindmarsh, *Org. Mass Spec.*, **2**, 681 (1969).
89. F. R. Stermitz and C. C. Wei, *J. Amer. Chem. Soc.*, **91**, 3103 (1969).
90. M. Hussain, J. S. Robertson, and T. R. Watson, *Austral. J. Chem.*, **23**, 773 (1970).
91. R. Shapiro and S. Nesnow, *J. Org. Chem.*, **34**, 1965 (1969).
92. L. A. Paquette, D. E. Kuhla, J. H. Barrett, and R. J. Haluska, *J. Org. Chem.*, **34**, 2866 (1969).

12 COMPOUNDS WITH TWO OR MORE NITROGEN ATOMS IN THE SAME RING

I. DIAZIRINES

The mass spectrum of diazirine (**1**) is shown in Fig. 12.1 (1). Not surprisingly, the base peak of the spectrum corresponds to ionized methylene; however, the relative importances of impact-induced and thermal decompositions are quite uncertain. The molecular ion is remarkably intense for such an unstable compound and the M-1 ion presumably corresponds to the charge-delocalized cation **1a**.

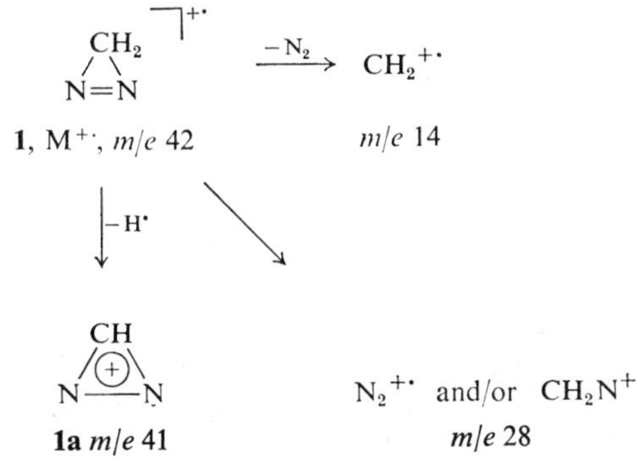

$$
\begin{array}{ccc}
\overset{\displaystyle CH_2}{\underset{\displaystyle N=N}{\diagup\diagdown}} \,\,\rceil^{+\cdot} & \xrightarrow{\;-N_2\;} & CH_2^{+\cdot} \\[2mm]
\textbf{1},\, M^{+\cdot},\, m/e\ 42 & & m/e\ 14
\end{array}
$$

$$
\Big\downarrow{-H^\cdot} \qquad \diagdown
$$

$$
\begin{array}{ccc}
\overset{\displaystyle CH}{\underset{\displaystyle N\!\!-\!\!N}{\diagup\!\!\underset{(+)}{}\!\!\diagdown}} & & N_2^{+\cdot}\ \text{and/or}\ \ CH_2N^+ \\[2mm]
\textbf{1a}\ m/e\ 41 & & m/e\ 28
\end{array}
$$

The mass spectrum of difluorodiazirine (**2**) has also been recorded (2). There the molecular ion is considerably weaker, and again the main decomposition pathway is loss of N_2. The intensity of the M-F ion is noteworthy. It is presumably the cation **2a**, analogous to the ion **1a** already considered in the spectrum of the parent compound.

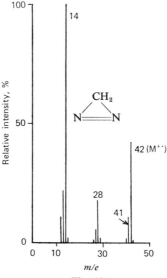

Fig. 12.1

No molecular ion was observed in the mass spectrum of chlorodifluoro-aminodiazirine (**3**), and the M-N_2 species (**3a**) was a surprisingly weak ion. The base peak (*m/e* 61) corresponds to the species $ClCN^{+\cdot}$ (**3**).

$$\text{2, M}^{+\cdot},\ m/e\ 78\ (0\cdot1\%) \xrightarrow{-N_2} CF_2^{+\cdot} \xrightarrow{-F^{\cdot}} CF^+$$

m/e 50 (100%)   *m/e* 31 (63·6%)

$$\xrightarrow{-F^{\cdot}}$$

2a, *m/e* 59 (46·8%)

3, M$^{+\cdot}$ (not observed)

$$\xrightarrow{-N_2} Cl-\overset{+\cdot}{C}-NF_2$$

m/e 99, 101 (4·6%)

$\big|-Cl^{\cdot}$ $\diagdown -F_2N^{\cdot}$ $\diagdown^{a}$ $\big|-F_2\ (?)$

F_2N —   *m/e* 92 (19%)

Cl —   *m/e* 75, 77 (17%)

$Cl-C\equiv N^{+\cdot}$
m/e 61, 63 (100%)

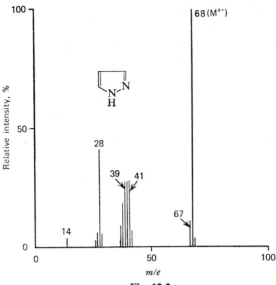

Fig. 12.2

II. PYRAZOLE DERIVATIVES

A. Pyrazoles

The mass spectrum of pyrazole (**4**) is shown in Fig. 12.2 (4). The stability of the molecule is reflected in the intense molecular ion. The species *m/e* 41 most probably results from loss of HCN from the molecular ion whereas

m/e 39 must be $C_3H_3^+$. Three isobaric possibilities exist for m/e 40, and high-resolution mass-spectrometry together with deuterium labeling is required to elucidate the structure of this ion. The ion at m/e 28 may be $N_2^{+\cdot}$ or CH_2N^+; mechanistically the latter seems the most likely possibility and the fragmentation of the compound is rationalized as shown.

Enhanced M-1 ions are present in the spectra of *C*-alkylpyrazoles and much of the fragmentation appears to begin from these species as shown in the tentative rationale for the spectrum of 3-(5-)methylpyrazole (**5**) (4). The importance of ring expansion (**5a** → **5b**) could only be estimated by a

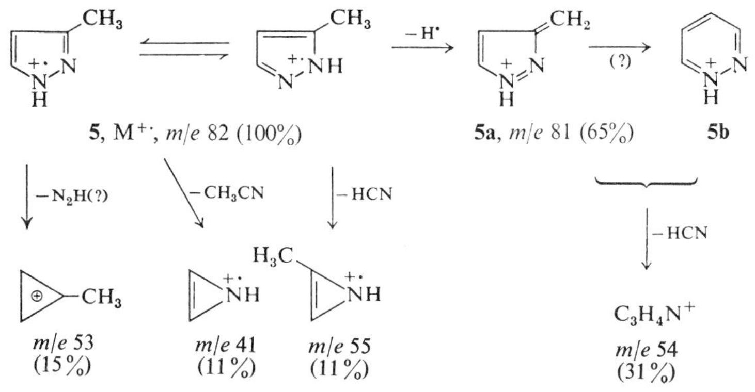

^{13}C-labeling study of suitable examples, and in the absence of information on metastable ions and of high-resolution data, the rationale can only be regarded as tentative.

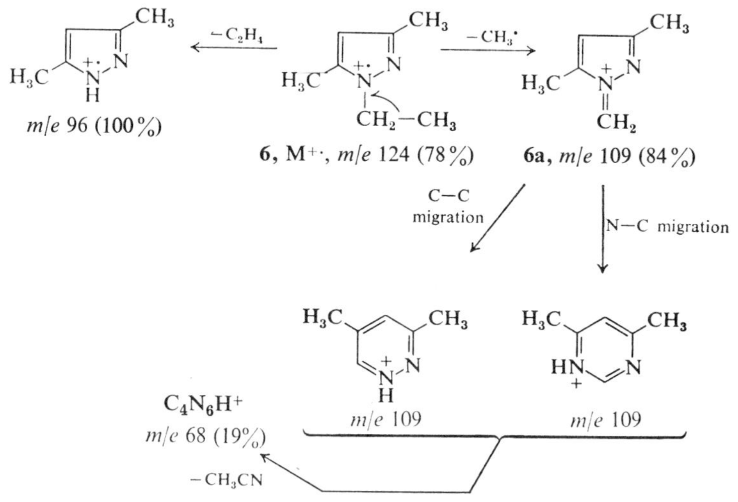

The mass spectrum of 1-ethyl-3,5-dimethylpyrazole (**6**) (4) has an intense M-·CH$_3$ ion, showing the importance of α-cleavage in *N*-alkylpyrazoles. Again, ring expansion of the ion produced (**6a**) is probably important. It would be of considerable interest to see whether C—C or N—C migration occurs in the ring-expansion process. As expected, the M-·CH$_3$ species shows marked loss of CH$_3$CN.

The mass spectrum of the pyrazole alkaloid with asomnine (**7**) (5) has not been analyzed in detail, but the occurrence of prominent ions at M-28 and M-56 suggests that cleavages α and β to one nitrogen atom may be important in this case.

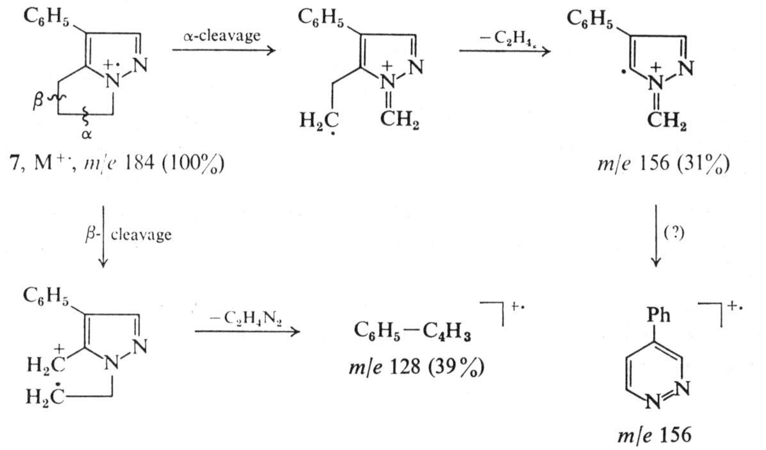

7, M$^+$·, *m/e* 184 (100%)   *m/e* 156 (31%)

m/e 128 (39%)   *m/e* 156

The recorded spectrum of 1-phenyl-3-methyl-5-pyrazolone (**8**) (4) suggests that the main fragmentation is ring fission with formation of the phenyl-diazonium cation (**8a**). Losses of CO and ·CHO from the molecular ion are unimportant processes. A tentative rationale for the spectrum is shown.

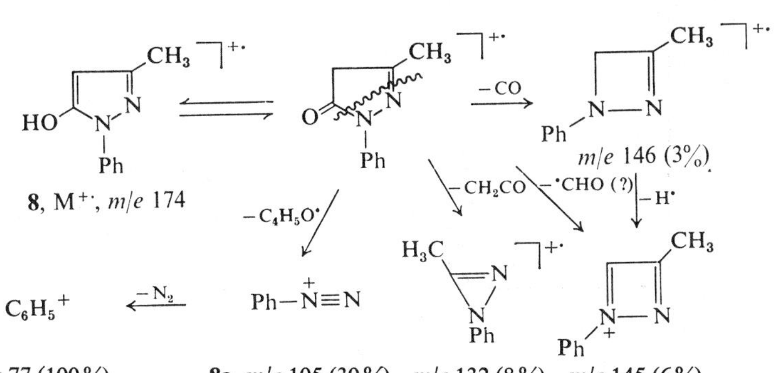

8, M$^+$·, *m/e* 174

m/e 146 (3%)

m/e 77 (100%)   **8a**, *m/e* 105 (30%)   *m/e* 132 (8%)   *m/e* 145 (6%)

B.   Pyrazolines

The mass spectrum of 3,5,5-trimethyl-2-pyrazoline (**9**) is shown in Fig. 12.3. It will be seen that the base peak arises from the expected α-fission, resulting in loss of a methyl group from the 5-position, and that the only other significant fragment ion results from subsequent extrusion of CH_3CN (6).

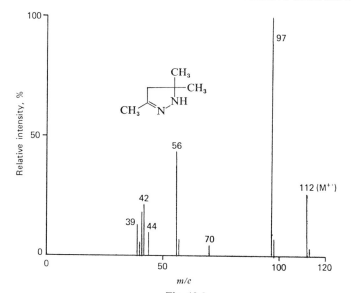

9, M+·, m/e 112 m/e 97 m/e 56

The mass spectra of the carbamoylpyrazolines which result (6) from thermal cyclization of the semicarbazones of α,β-unsaturated aldehydes and ketones have been examined and have also been compared with the spectra of their noncyclized isomers (6). Although the spectra of the cyclized and noncyclized compound are similar, there are significant differences in ion intensities, and it has been suggested that impact-induced rather than thermal cyclization occurs when the semicarbazones are examined mass-spectrometrically.

Typical of the fragmentation of compounds of this kind is that of Scholtz's base (**10**), the pyrazoline obtained by thermal cyclization of mesityl oxide semicarbazone. It will be seen that elimination of HCNO from the molecular

Fig. 12.3.

ion gives the 3,5,5-trimethylpyrazoline which has already been discussed, and fragmentation below m/e 112 follows the pathway described for that compound.

10, M$^+$·, m/e 155 (5%) m/e 112 (20%)

C. Bicyclic Compounds Containing a Pyrazole Ring

1. 3a,6a-Diazapentalene (11)

The intense molecular ion (base peak) of this interesting heterocycle reflects its aromatic character. Intensities of fragment ions in the spectrum of **11** have not been reported, but the presence of an M-C$_2$H$_2$ ion (7) may indicate formation of the pyrimidine or pyrazine radical cation by the routes shown. Peaks at lower masses are small fragments (m/e 53, C$_2$HN$_2$; m/e 52, C$_2$N$_2$; and m/e 39, C$_3$H$_3$).

11, M$^+$·, m/e 106 (100%) m/e 80 m/e 80

2. Indazolones

The mass spectrum of the parent compound (**12**) is shown in Fig. 12.4 (8). It has been shown that in the ground state the compound exists as the enol, and the fragmentation is most easily explained with this formulation for the radical cation. Loss of 29 mass units in the initial fragmentation is due to elisions of ·N$_2$H (90%) and CHO· (10%). The rationale for these decompositions is shown.

1-Methylindazolone (**13**) shows (Fig. 12.5) a fragmentation pattern basically similar to the parent compound, but in addition to the major loss of CH$_3$N$_2$ leading to **12a**, loss of 29 mass units from the parent ion is a significant process (M-29, 6%). This ion is, in fact, a triplet and the fragments lost are ·CHO, ·N$_2$H, and NCH$_3$ in the ratio 2.5:1:1. The loss of ·CHO presumably occurs by a similar mechanism to that outlined for **12**, but the losses of the

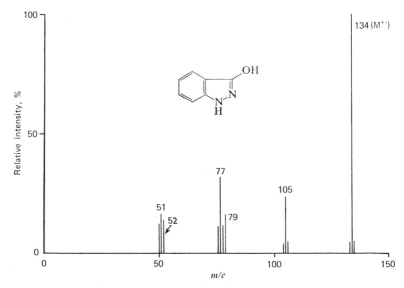

Fig. 12.4

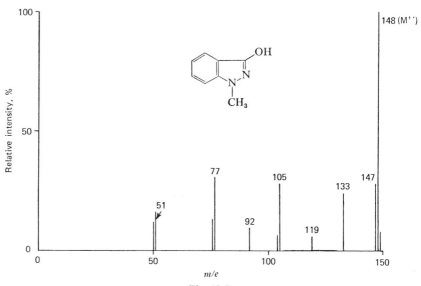

Fig. 12.5

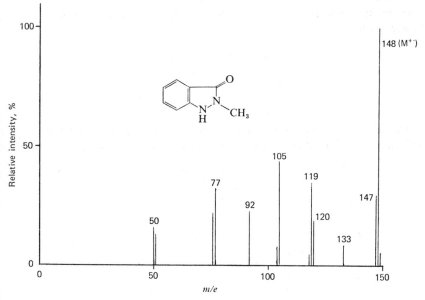

isobaric fragments $\cdot N_2H$ and NCH_3 must require skeletal rearrangements prior to fragmentation.

13   **14**   **15**

The spectra of the isomeric 2-methylindazolone (**14**) (Fig. 12.6) and 3-methoxyindazole (**15**) (Fig. 12.7) are basically similar although there are

Fig. 12.6

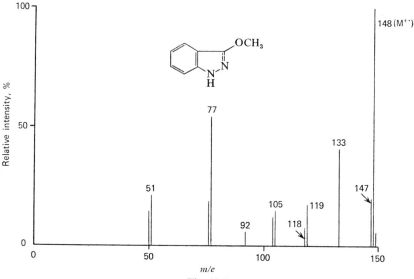

Fig. 12.7

considerable variations in ion intensities. In each the M-29 peak is a triplet representing losses of ·N₂H, ·CHO, and CH₃N, so that skeletal rearrangements are important here too. The M-·CH₃ peak is much more intense in the methoxy compound **15**; presumably the common O—CH₃ fission observed with many aromatic ethers containing electron-donating substituents is important here.

The relatively intense M-1 ion in the *N*-methylindazolones is noteworthy, and the formation of this species may well involve ring expansion of the type frequently invoked. β-Fission (loss of $\cdot CH_3$) leads to the base peak of the spectrum of 1-ethylindazolone (**16**).

III. IMIDAZOLE AND ITS DERIVATIVES

A. Simple Imidazoles

The mass spectrum of imidazole (**17**) (Fig. 12.8) (9) is dominated by elision of HCN to give a species best represented as the aziridine cation (**17a**).

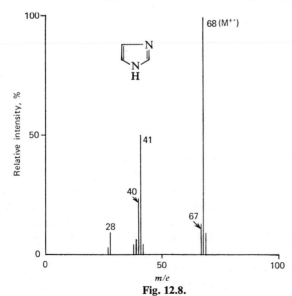

The spectra of the 1-deutero- and 1,2-dideutero-compounds show losses of both HCN and DCN, suggesting the absence of a specific process for this

Fig. 12.8.

elimination or perhaps randomization of hydrogen atoms in the molecular ion.

The elimination of HCN from 1-methylimidazole (18) is specific and involves the 2- and 3-positions only, while with 2-methylimidazole (19) the HCN originates from either the 1,5- or (less likely) the 3,4-positions. 1,2-Dimethylimidazole (20) shows negligible loss of HCN from the molecular ion, supporting the general observation that HCN is not lost from positions 3 and 4. It is noteworthy that loss of HCN from the molecular ion is an

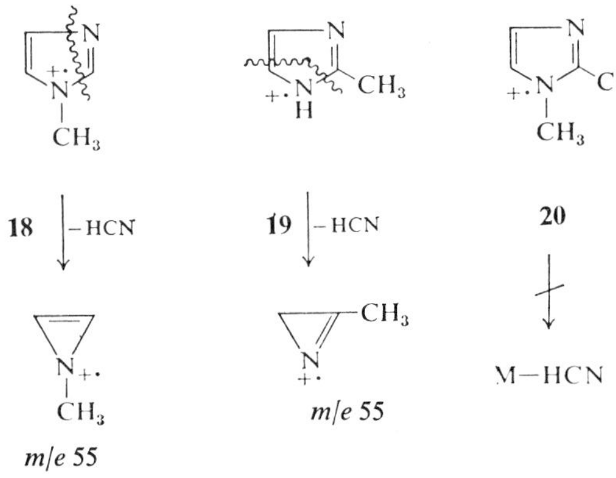

important process even in tribromoimidazole (21) in which bromine migration must precede the elimination.

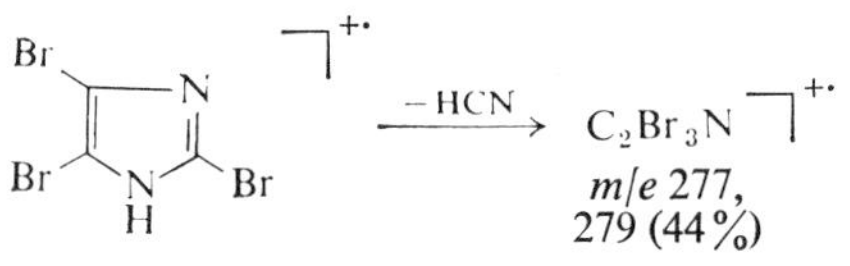

21, M$^{+\cdot}$, m/e 304, 306 (100%)

The spectra of the methylimidazoles contain pronounced M-1 ions. The deuterium-labeling results suggest that randomization of hydrogen atoms between the ring and the methyl group may well be occurring with the 1- and 4(5)-methyl-compounds (22) (9). The M-1 ions may be represented as ring-expanded pyridazinium (23) or pyrimidinium (24) cations as shown. The M-1 species (18a, 19a, 22a, and/or 23/24) show the expected losses of HCN. This loss also occurs from the molecular ions of the monomethylimidazoles, and the 1- and 2-methyl compounds also show the rather unexpected elision of the fragment ·CH$_2$CN from both the molecular ion and the M-1 species (9). These decompositions may be represented (for the 2-methyl compound (19)) by the rationale shown.

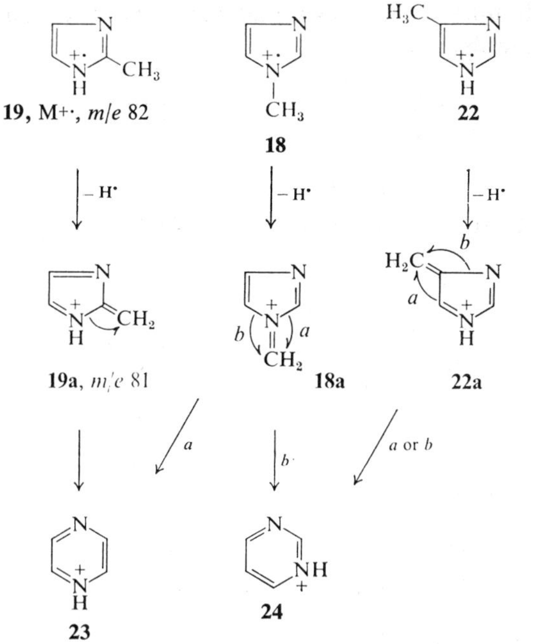

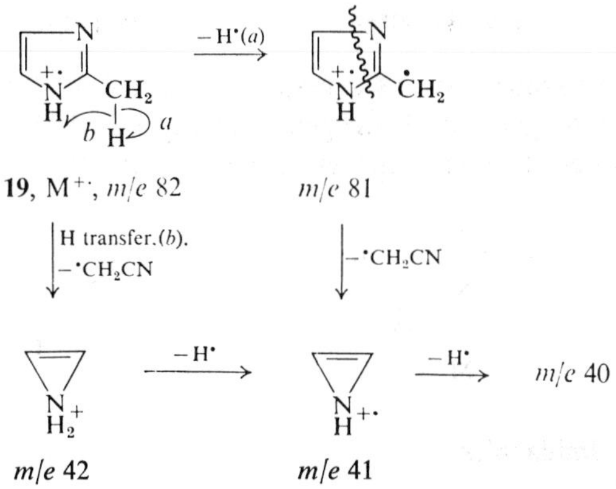

Imidazole-2- and -4(5)-carboxylic acids (25 and 26) show marked losses of H_2O by an "ortho effect" mechanism. Interestingly, however, a similar effect is not seen in the corresponding methyl esters (27 and 28) which also differ one from the other in their fragmentation and thus may be distinguished mass-spectrometrically.

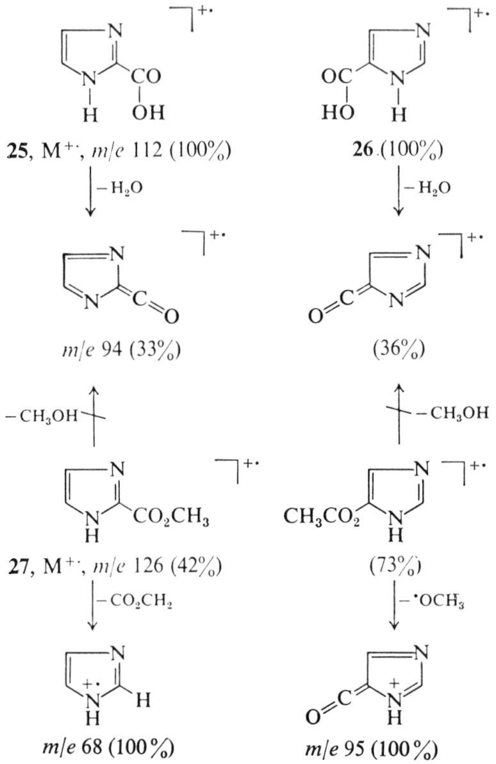

25, M$^{+\cdot}$, m/e 112 (100%)   **26** (100%)

$\downarrow -H_2O$ $\downarrow -H_2O$

m/e 94 (33%)   (36%)

$-CH_3OH \uparrow$ $\uparrow -CH_3OH$

27, M$^{+\cdot}$, m/e 126 (42%)   (73%)

$\downarrow -CO_2CH_2$ $\downarrow -{}^{\cdot}OCH_3$

m/e 68 (100%) m/e 95 (100%)

2-Formylimidazoles have as a major fragmentation loss of CO from the molecular ion to give (by hydrogen rearrangement) the related imidazole radical ion which then decomposes in the appropriate way.

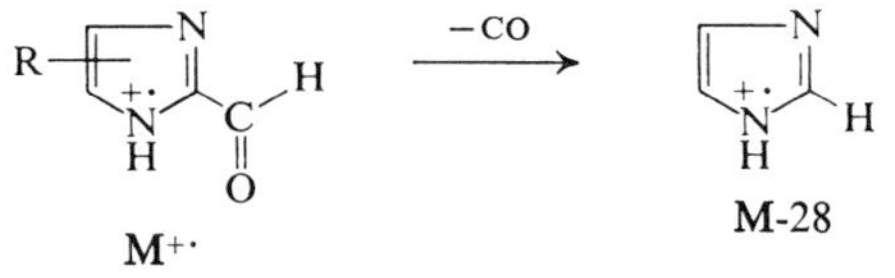

$$\text{M}^{+\cdot} \xrightarrow{-CO} \text{M-28}$$

B. Reduced Imidazoles

The mass spectra of a number of mono- and disubstituted alkyl and aryl imidazolines have been analyzed (10). All principal fragments contain at least one nitrogen atom. The fragmentation of 2-methylimidazoline (**29**) is typical of the monoalkyl derivatives.

The m/e 55 ion has apparently not been subjected to high-resolution investigation so the relative importances of the two possible structures in the above rationale cannot be estimated although the apparent weakness of m/e 56 suggests that pathway *a* may be a minor one.

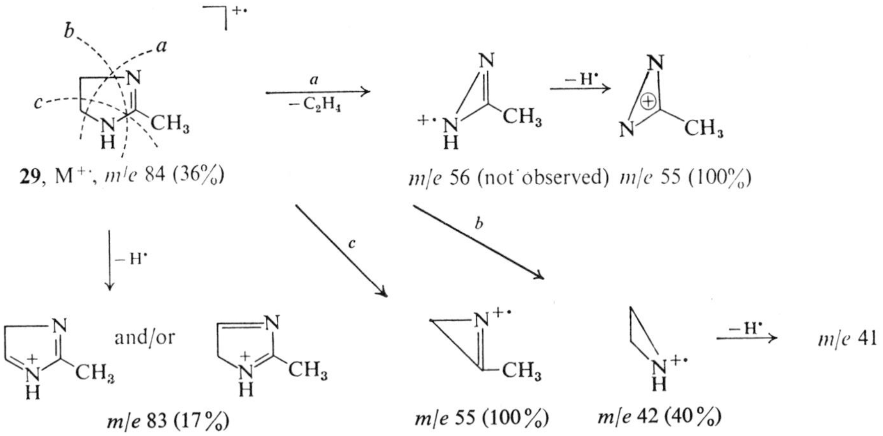

29, M+·, m/e 84 (36%)
a
−C2H4
m/e 56 (not observed)
−H·
m/e 55 (100%)
−H·
m/e 83 (17%)
and/or
c
b
m/e 55 (100%)
m/e 42 (40%)
−H·
m/e 41

The spectrum of 2-phenylimidazoline (**30**) has as its base peak the M-29 ion (m/e 117) which presumably results from elision of CH_2NH (process *c*); while the fragment at m/e 104 shifts to m/e 105 in the monodeuterated compound (10). These results suggest the rationale shown.

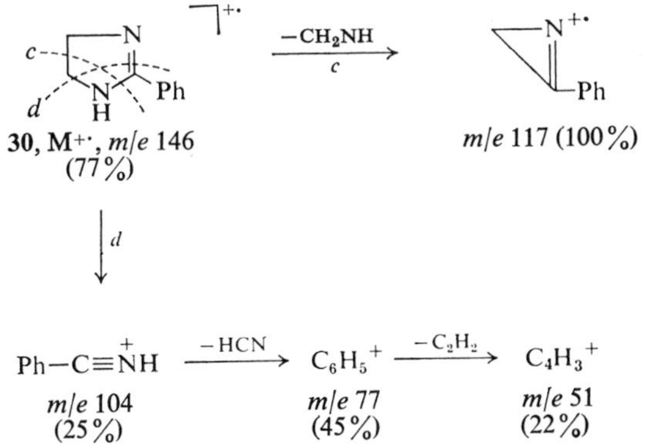

30, M+·, m/e 146 (77%)
−CH2NH
c
m/e 117 (100%)
d
Ph−C≡NH
m/e 104 (25%)
−HCN
C6H5+
m/e 77 (45%)
−C2H2
C4H3+
m/e 51 (22%)

2-Phenyl-4-methylimidazoline has intense ions in its spectrum at M-29 and M-43, and these are best explained as arising from the two tautomeric forms of this molecule (**31** and **32**) as shown (10).

The base peak of the spectrum of 2-undecylimidazoline (**33**) at m/e 84 is best explained as the result of a McLafferty rearrangement whereas the species

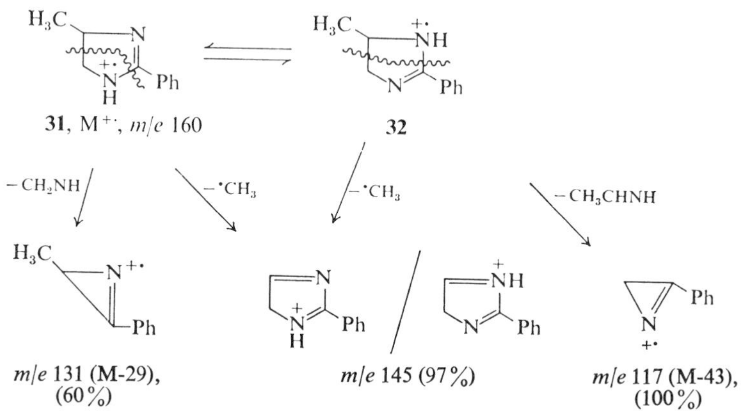

31, $M^{+\cdot}$, m/e 160   **32**

$-CH_2NH$ $-\cdot CH_3$ $-\cdot CH_3$ $-CH_3CHNH$

m/e 131 (M-29), (60%) m/e 145 (97%) m/e 117 (M-43), (100%)

at m/e 97 (γ-cleavage) may be the bicyclic ion **33a** or may perhaps result from allylic cleavage of the tautomer **34** (to give **34a**).

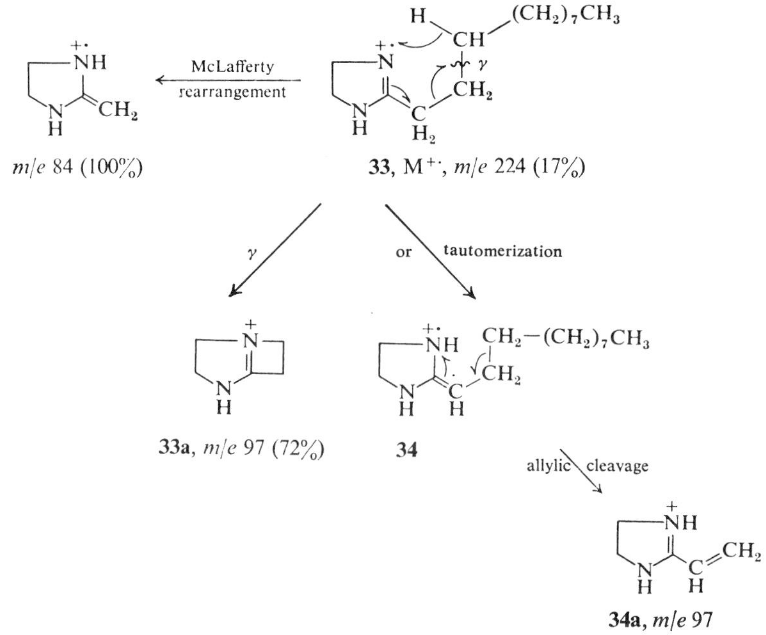

m/e 84 (100%)   **33**, $M^{+\cdot}$, m/e 224 (17%)

McLafferty rearrangement

γ or   tautomerization

33a, m/e 97 (72%)   **34**

allylic cleavage

34a, m/e 97

The spectra of 5,5-diphenylimidazolidone-4 (**35**) and some related compounds have been reported (11). That of **35** shows the predominant fragmentation to be a sequential loss of HNCO and H· to give the well-stabilized fragment **35a**. Substantial M-1 and M-2 species are present and probably have the structures **35b** and **35c** as shown.

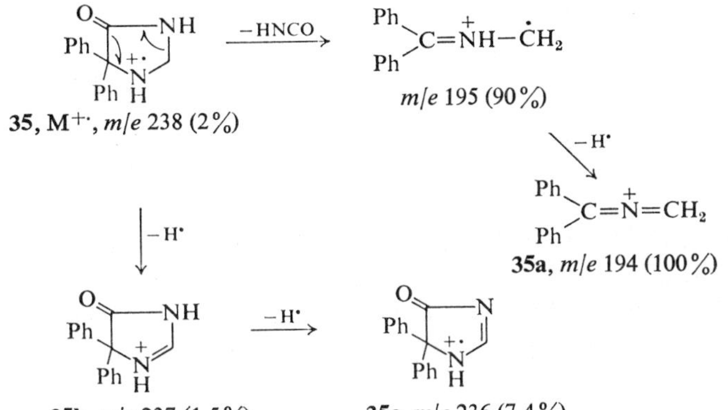

35, M$^{+\cdot}$, m/e 238 (2%) — $-$HNCO $\rightarrow$ m/e 195 (90%)

$-$H· $\rightarrow$ **35a**, m/e 194 (100%)

35b, m/e 237 (1.5%) $-$H· $\rightarrow$ **35c**, m/e 236 (7.4%)

From the very limited data published on the mass spectrum of the *p*-nitrophenylthioimidazolidone **36** (12), it appears that a ring fission with hydrogen transfer, perhaps proceeding as shown, is the major fragmentation.

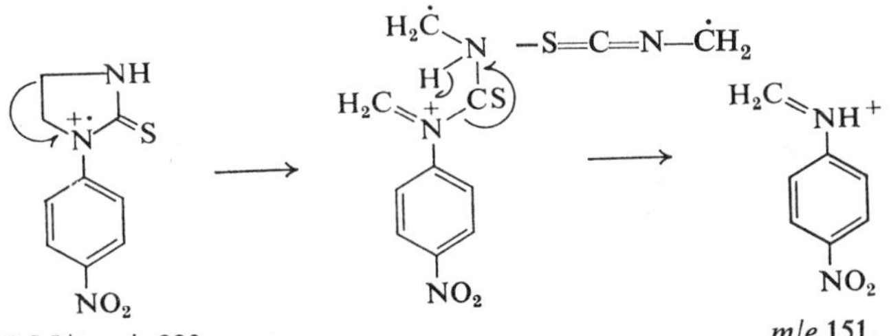

36, M$^{+\cdot}$, m/e 223 m/e 151

Clearly there is scope for a great deal of research on the fragmentations of reduced imidazoles and related compounds.

C. Benzimidazoles

The mass spectrum of the parent compound (**37**) shows successive losses of two molecules of HCN, eventually leading to the species $C_5H_4^{+\cdot}$ (m/e 64) (13).

A possible rationale based on the fragmentation of the simple imidazoles is shown.

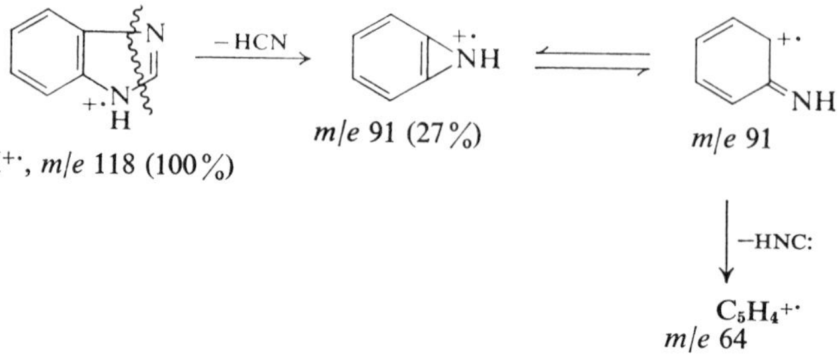

37, M$^{+\cdot}$, m/e 118 (100%)   m/e 91 (27%)   m/e 91

C$_5$H$_4$$^{+\cdot}$
m/e 64

Loss of CH$_3$CN from 2-methylbenzimidazole (**38**) is a less important process (13, 14) and the chief fragmentation is loss of a hydrogen atom to give an intense (60%) M-1 ion. Little labeling has been carried out and the origin of the hydrogen lost is largely conjectural. The possibilities are shown.

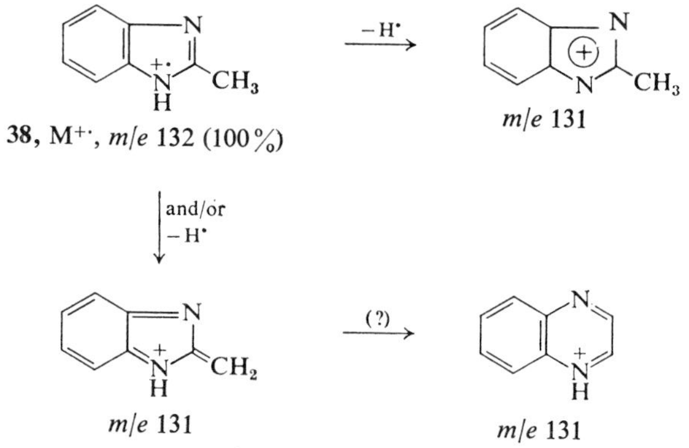

38, M$^{+\cdot}$, m/e 132 (100%)   m/e 131

and/or
−H·

m/e 131   (?)   m/e 131

Only limited data is available on the mass spectrum of 2-ethylbenz-imidazole (**39**) but here the M-1 ion is the base peak of the spectrum (14). This may be due to γ-fission of the type already discussed in connection with 2-ethylpyridines and 2-ethylquinolines (Chap. 11, Sec. II.B)

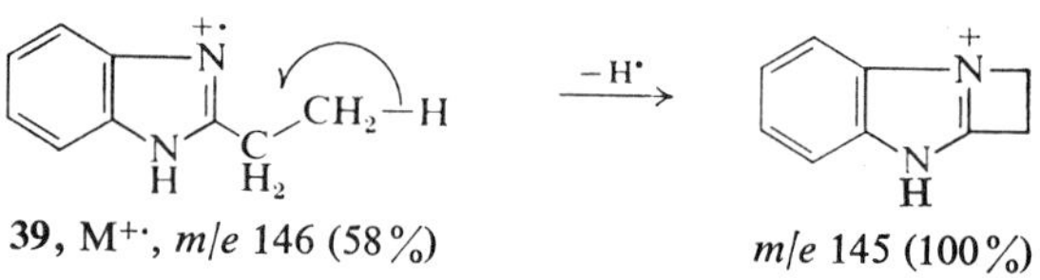

39, M$^{+\cdot}$, m/e 146 (58%)   m/e 145 (100%)

The mass spectra of the *N*-oxides of benzimidazole and several 2-substituted derivatives have been reported, and in all cases the major fragmentation is loss of oxygen from the relatively weak molecular ion to give the parent benzimidazole (14). No "ortho effect" such as that observed with 2-methyl-pyridine *N*-oxide occurs with 2-methylbenzimidazole *N*-oxide (**40**). This is presumably because of less favorable geometry for the hydrogen transfer step in the five-membered ring compound.

38
(M−16)
(100%)

40, M⁺·, *m/e* 148 (10%)

The phenylbenzimidazolylbenzimidazoline (**41**) fragments in a way indicative of charge localization on the reduced ring as shown, and no fragments characteristic of the benzimidazole portion of the molecule are reported (15).

41, M⁺·, *m/e* 459 (low intensity)

m/e 401 (intense)

m/e 400 (intense)

The hydrogen loss from *m/e* 401 to give *m/e* 400 may well be explained in terms of formation of the well-delocalized symmetrical ion **41a**, but a much more detailed examination of the fragmentation would be required before a firm decision could be reached on this point.

The mass spectrum of the interesting betaine (**42**) obtained by condensation of *o*-diethylaminoaniline and alloxan has been investigated mass-spectro-metrically and two major fragmentation pathways (A and B) are observed (16):

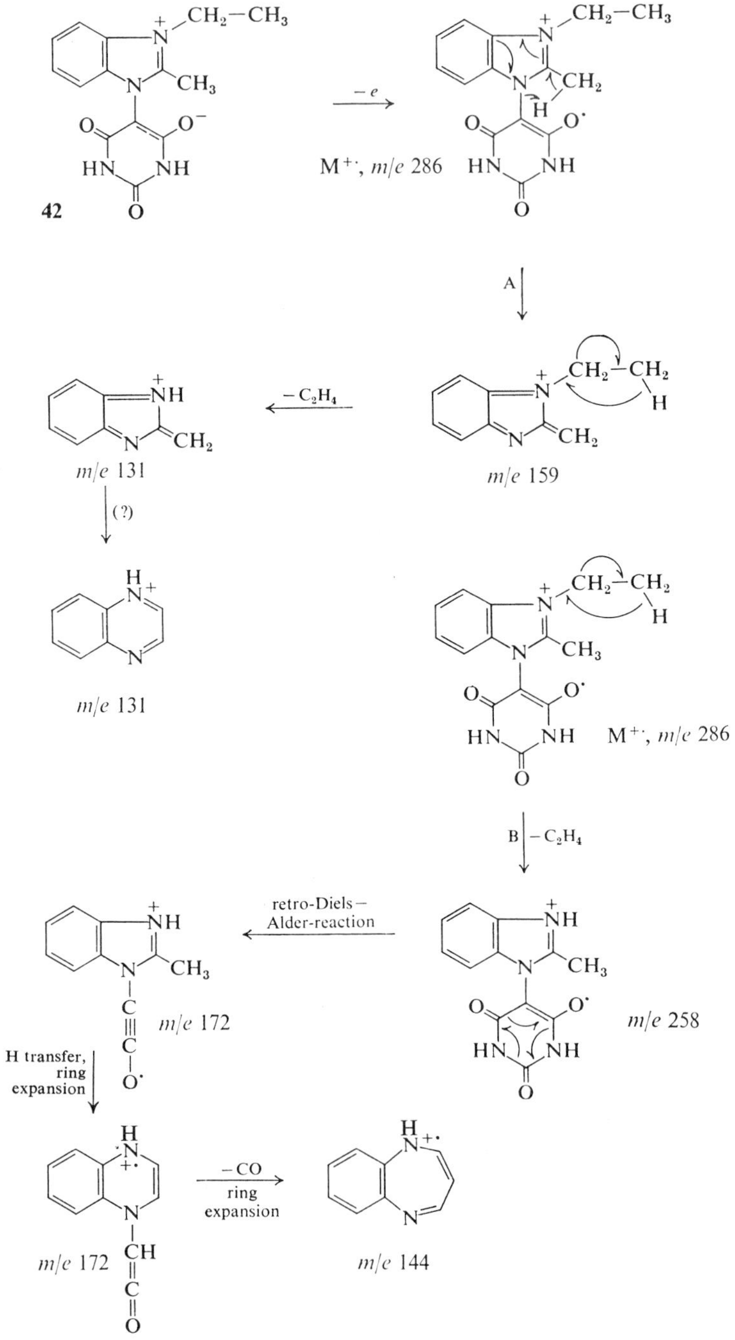

42
−e
M+·, m/e 286
A
− C2H4
m/e 131
m/e 159
(?)
m/e 131
M+·, m/e 286
B − C2H4
retro-Diels–Alder-reaction
m/e 172
m/e 258
H transfer, ring expansion
− CO ring expansion
m/e 172
m/e 144

D. Pyridoimidazoles

Some mass-spectral data have been reported for the alkaloids glochidine (**43**) and glochidicine (**44**) (17). In each case the base peak of the spectrum results from loss of the *n*-hexyl chain (fission α to nitrogen). Little other fragmentation occurs with glochidicine but other fragmentations observed with glochidine are shown in the scheme.

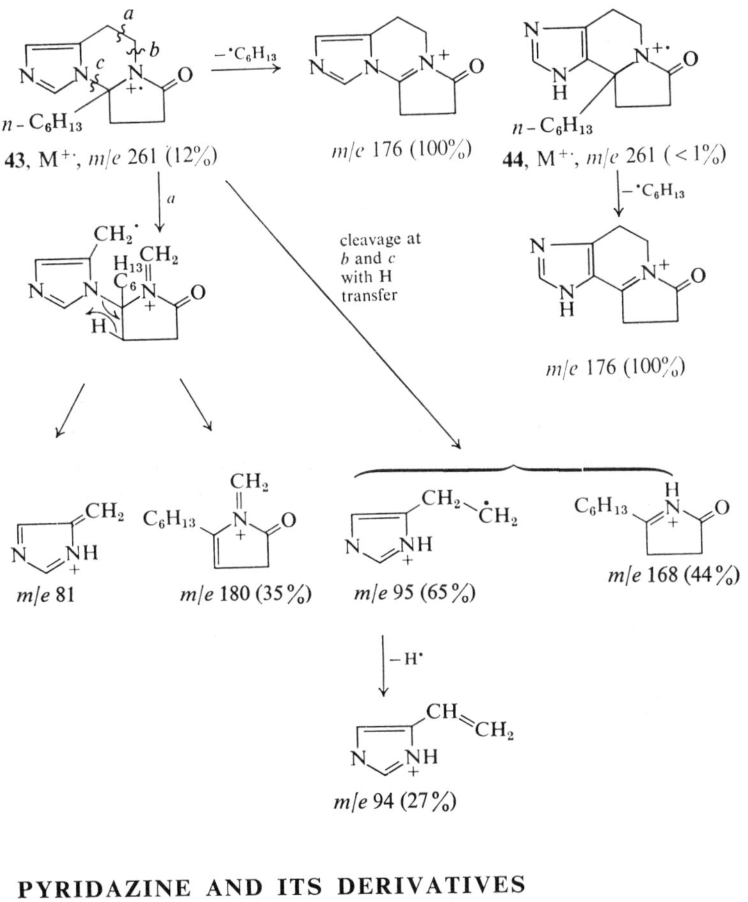

a
b
c
$-\cdot C_6H_{13}$
n-C_6H_{13}
43, $M^{+\cdot}$, m/e 261 (12%)
m/e 176 (100%)
44, $M^{+\cdot}$, m/e 261 (<1%)
$-\cdot C_6H_{13}$
cleavage at
b and c
with H
transfer
m/e 176 (100%)
m/e 81
m/e 180 (35%)
m/e 95 (65%)
m/e 168 (44%)
$-H\cdot$
m/e 94 (27%)

IV. PYRIDAZINE AND ITS DERIVATIVES

The spectrum of pyridazine (**45**) (Fig. 12.9) is simple, and high-resolution measurements have shown that the ion at m/e 52 is composed of C_4H_4 (73.5%) and C_3H_2N (26.5%). Those at m/e 51 and 50 are solely due to the hydrocarbon fragments $C_4H_3^+$ and $C_4H_2^{+\cdot}$. It has been concluded that the

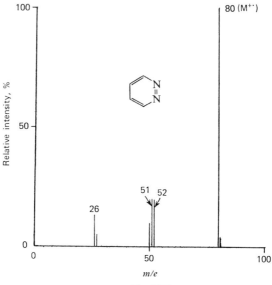

Fig. 12.9

$C_4H_4^{+\cdot}$ ion is probably the cyclobutadiene cation (**45a**) although the tetrahedrane radical cation (**45b**) is also a possibility (18). Recent work on the structure of the M-CO ion in the mass spectra of certain substituted tetracyclones suggests the possibility of a tetrahedrane rather than cyclobutadiene structure (19), and it would be interesting to examine the spectra of the analogous pyridazines.

The simple fragmentation of the parent compound is characteristic also of chloro-, methyl- and aminopyridazines (20), and the spectra of these compounds will not be discussed further.

Pyridazones fragment initially by loss of carbon monoxide followed by loss of N_2 (20). A typical spectrum is that of 3-pyridazone (**46**) for which the rationale shown may be suggested.

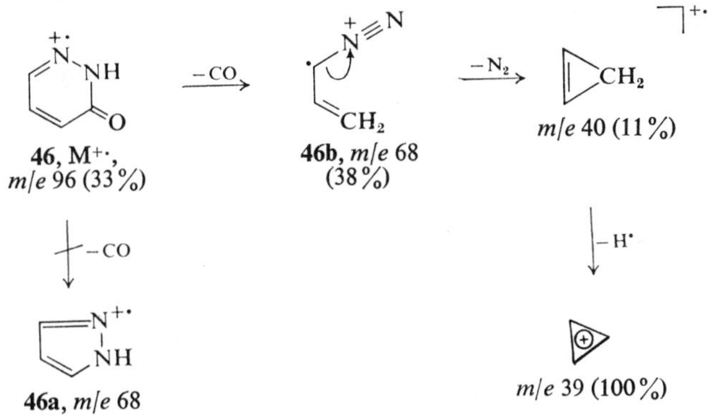

46, M$^{+\cdot}$,
m/e 96 (33%)

46b, *m/e* 68
(38%)

m/e 40 (11%)

46a, *m/e* 68

m/e 39 (100%)

It is clear that the *m/e* 68 species is not the pyrazole radical cation **46a** since this fragments in quite a different manner (Sec. II.A). The aliphatic diazonium radical cation (**46b**) suggested for *m/e* 68 seems a likely candidate since it can easily decompose to the hydrocarbon ion *m/e* 40, thence to *m/e* 39. The effect of *N*-methylation on the spectrum of **46** would be very interesting; in particular, the importance of methyl migration in such a compound would be worthy of study.

For a discussion of the important features of the mass spectra of a number of other pyridazines, the reader is referred to the original publication (20).

The spectrum of phthalazine (**47**) is of interest since it allows examination of the effect of further aromatic fusion on the basic pyridazine fragmentation (20). Quantum-mechanical calculations have shown that the "fixed bond" structure shown in **47a** provides a better representation of the ground state of the molecule than the alternative form **47b** in which the benzene ring is in a quinonoid form.

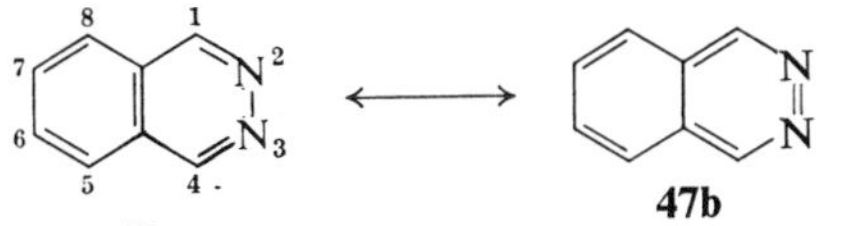

47a

47b

In fact, the tendency for fragmentation by loss of N_2 is much reduced in phthalazine compared with pyridazine, and sequential losses of HCN become important. It will be recalled that loss of HCN is the minor process with the pyridazine molecule, and presumably the greater double-bond character of the 1,2 and 3,4 bonds in **47a** (and hence their greater strength) inhibits their fission. The important fragmentations of **47a** are shown in the rationale.

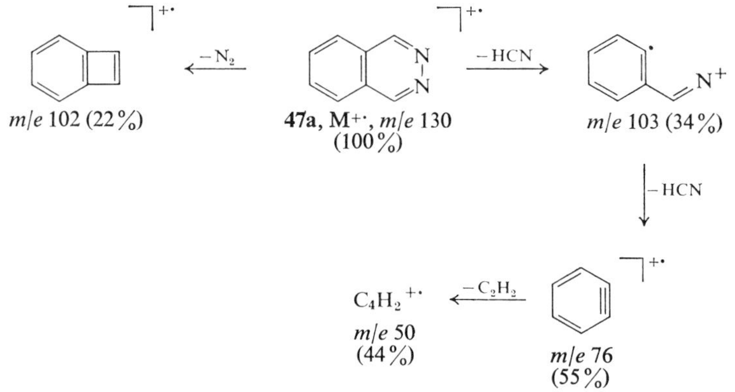

m/e 102 (22%) ← $-N_2$ ← **47a**, M$^{+\cdot}$, m/e 130 (100%) — HCN → m/e 103 (34%)

$\downarrow$ — HCN

$C_4H_2^{+\cdot}$ ← $-C_2H_2$ ← m/e 76 (55%)
m/e 50 (44%)

Similar sequential loss of HCN dominates the spectrum of pyrido[2,3-d]-pyridazine (**48**) and quinoxalino[2,3-d]pyridazine (**49**), and indeed in the

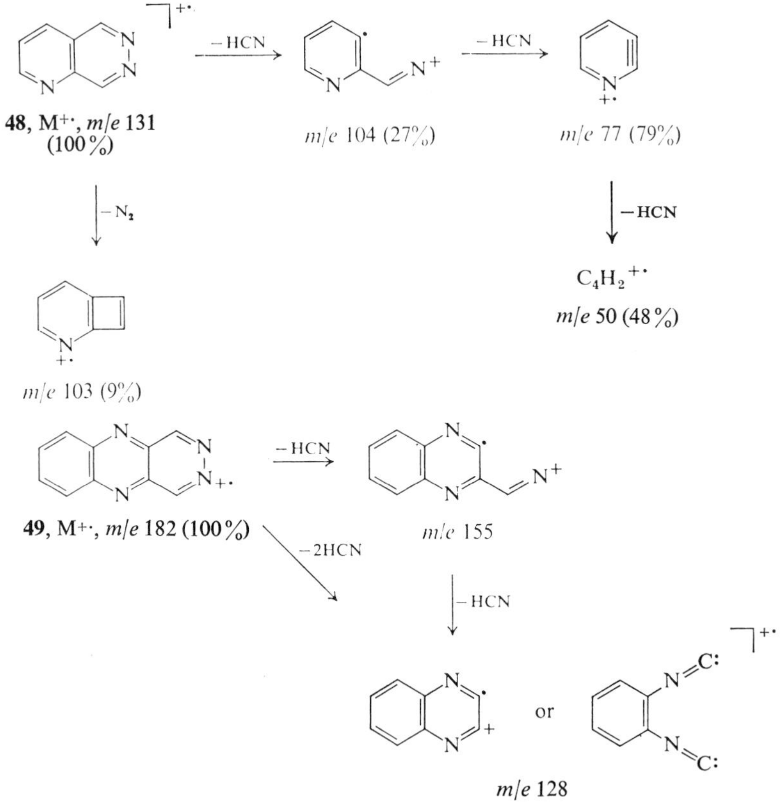

48, M$^{+\cdot}$, m/e 131 (100%) — HCN → m/e 104 (27%) — HCN → m/e 77 (79%)

$\downarrow$ — N$_2$

m/e 103 (9%)

$\downarrow$ — HCN

$C_4H_2^{+\cdot}$
m/e 50 (48%)

49, M$^{+\cdot}$, m/e 182 (100%) — HCN → m/e 155

-2HCN

$\downarrow$ — HCN

m/e 128   or

latter the appearance of the appropriate metastable ion indicates concerted loss of the two HCN molecules.

Phthalazone (50) fragments by losses of CO, N_2, and $\cdot N_2H$ from the molecular ion (20), and the observed subsequent behavior of the M-CO ion (loss of HCN) does not in this case preclude cyclization to the benzpyrazole radical cation (50a). A possible rationale for the major fragments is shown.

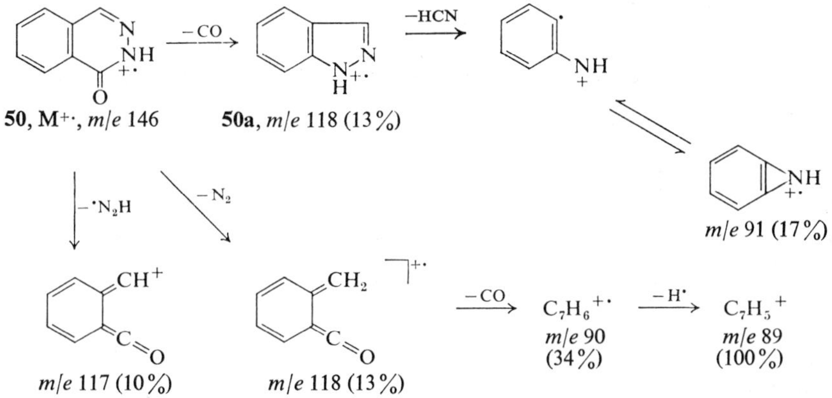

50, $M^{+\cdot}$, m/e 146   50a, m/e 118 (13%)

m/e 91 (17%)

m/e 117 (10%) m/e 118 (13%)

$-CO$ → $C_7H_6^{+\cdot}$ → $-H^\cdot$ → $C_7H_5^+$
m/e 90 m/e 89
(34%)   (100%)

"Proximity effects" are noticed in the mass-spectra of phthalazines containing amino, methyl, phenyl, or methoxyl substituents adjacent to nitrogen (20). These are evidenced by enhanced M-H· ions, and in the methoxy and phenyl compounds (51 and 52) the resultant species may well be stabilized by cyclization.

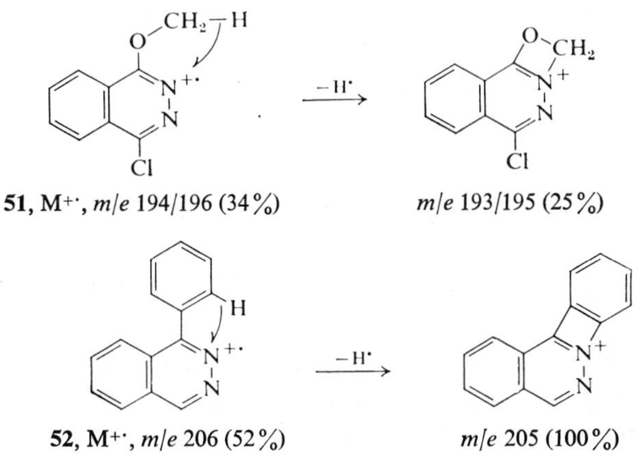

51, $M^{+\cdot}$, m/e 194/196 (34%) m/e 193/195 (25%)

52, $M^{+\cdot}$, m/e 206 (52%) m/e 205 (100%)

Loss of hydrogen from a methyl group in the 1-position of the phthalazine nucleus may well be accompanied by ring expansion of the now familiar type to give a diazatropylium cation (53 → 53a → 53b) whereas the intensity of the M-1 ion in the spectrum of 1,4-diaminophthalazine (54) is due to formation of the well-stabilized cation 54a.

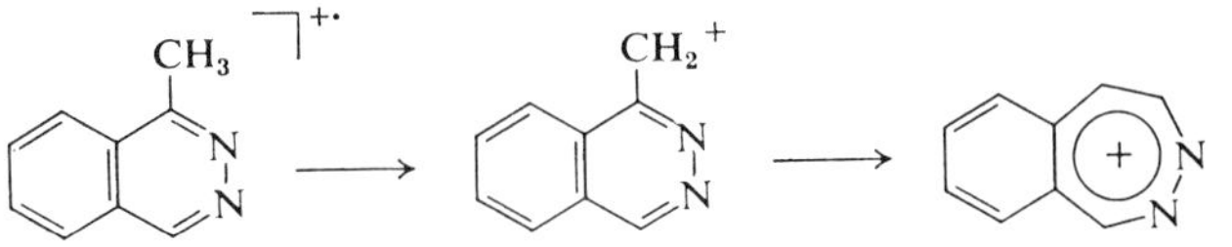

53, M$^{+\cdot}$, m/e 144 (100%)   53a, m/e 144 (25%)   53b

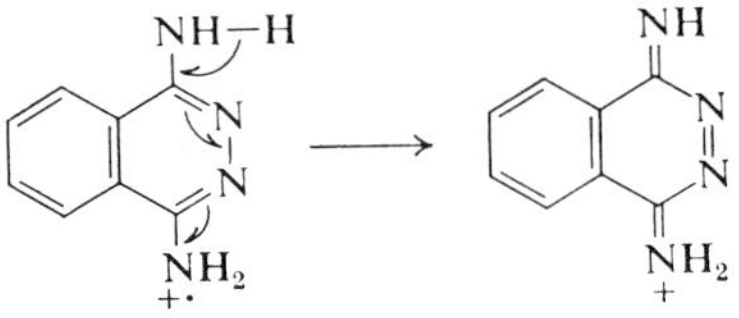

54, m/e 160 (89%)   54a, m/e 159 (100%)

Phthalazine N-oxide (55) decomposes by sequential losses of NO· and HCN. Skeletal rearrangements are not observed (21).

55, M$^{+\cdot}$, m/e 146 (94%) m/e 116 (44%) $C_7H_5^+$ m/e 89 (100%)

The spectrum of dibenzopyridazine (benzo[c]cinnoline, 56), like that of pyridazine itself, is dominated by loss of nitrogen, in this case leading to the species m/e 152 (56a) as the base peak of the spectrum (22, 23).

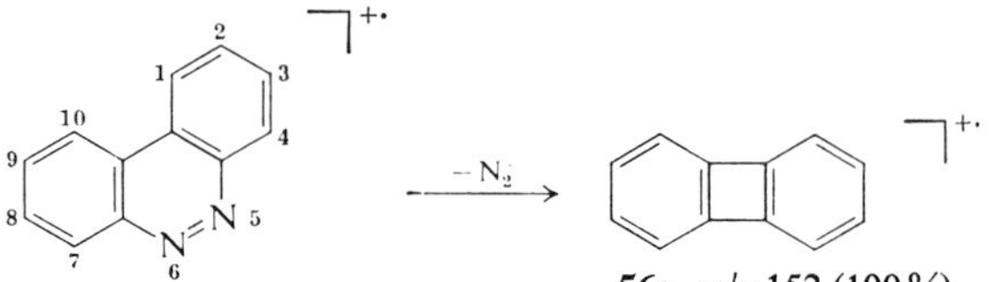

56, M$^{+\cdot}$, m/e 180 (67%)   56a, m/e 152 (100%)

Loss of the elements of nitrogen is characteristic of substituted benzo[c]-cinnolines although fragmentation in a substituent group may take precedence in appropriate cases. Thus the 2-diethylamino derivative (57) (23)

sequentially loses $CH_3\cdot$, C_2H_4, then N_2, and the chain of decomposition may be logically represented as shown.

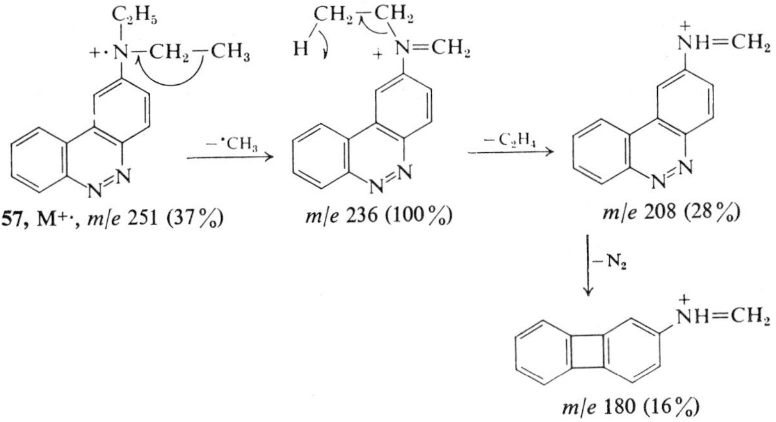

57, M+·, m/e 251 (37%)
m/e 236 (100%)
m/e 208 (28%)
m/e 180 (16%)
−·CH3
−C2H4
−N2

Marked "proximity effects" of the type already encountered in the phthalazines are noticed in 4-substituted benzo[*c*]cinnolines, and the spectrum of the 4-diethylamino-derivative (**58**) is quite different from that of the 2-substituted isomer (**57**) (23). Thus the base peak is the M-$\cdot C_2H_5$ ion (*m/e* 222) which may be represented as **58a**. (An alternative formulation involving ring expansion of the four-membered ring in **58a** to a five-membered one has also been considered.) Similar "proximity effects" have also been noted with 4-alkoxy-, 4-carboxy-, and 4-alkoxycarbonyl-benzo[*c*]cinnolines, and the full paper should be consulted for further details (23).

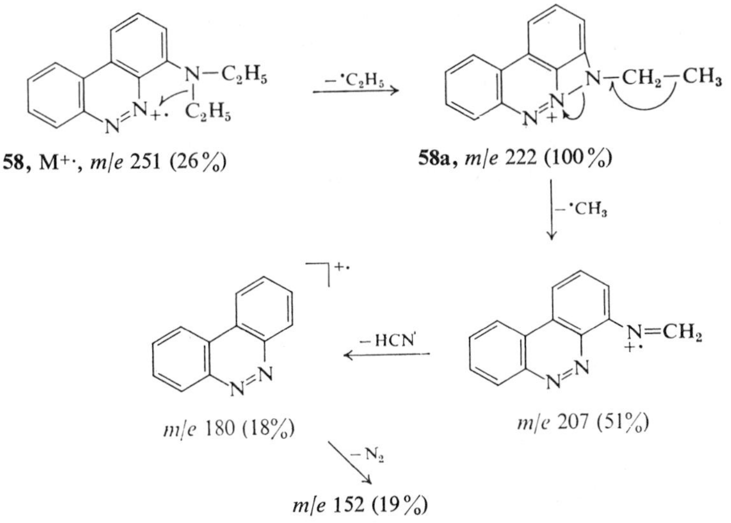

58, M+·, m/e 251 (26%)
58a, m/e 222 (100%)
m/e 207 (51%)
m/e 180 (18%)
m/e 152 (19%)
−·C2H5
−·CH3
−HCN
−N2

The mass spectrum of benzo[*c*]cinnoline *N*-oxide (**59**, R = H) is devoid of the skeletal rearrangements observed with other *N*-oxides (21), and the compound fragments by loss of ·NO, then HCN. The 4,7-dimethyl *N*-oxide (**59**, R = CH₃), however, shows a very complex spectrum in which loss of CO from the molecular ion is observed. This process involves a triple hydrogen rearrangement and full rationalization is difficult, but the structure **59a** has been suggested for the M-CO species (23).

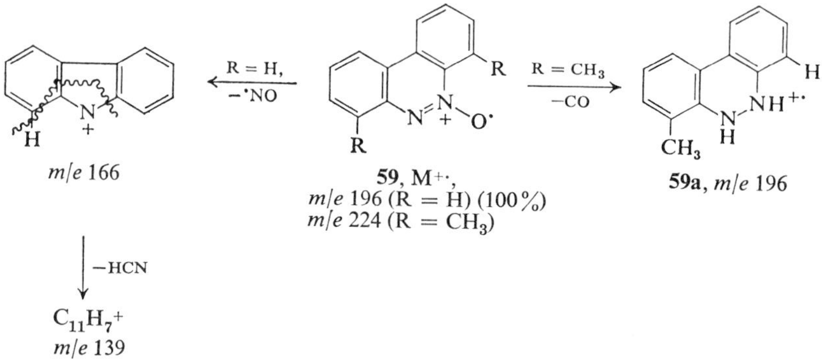

m/e 166

R = H,
−·NO

59, M⁺·,
m/e 196 (R = H) (100%)
m/e 224 (R = CH₃)

R = CH₃
−CO

59a, m/e 196

−HCN

$C_{11}H_7^+$
m/e 139

V. PYRIMIDINES

A. Simple Pyrimidines

The mass spectrum of the parent compound (**60**) is shown in Fig. 12.10 (24). It will be seen that extrusion of HCN from the molecular ion is, as expected, the dominant fragmentation mode. Labeling experiments to determine whether $C_{(2)}$ or $C_{(6)}$ is involved (or possibly both) have not yet been reported. Subsequent loss of a further molecule of HCN leads to the base peak of the spectrum, which is due to ionized acetylene (m/e 26).

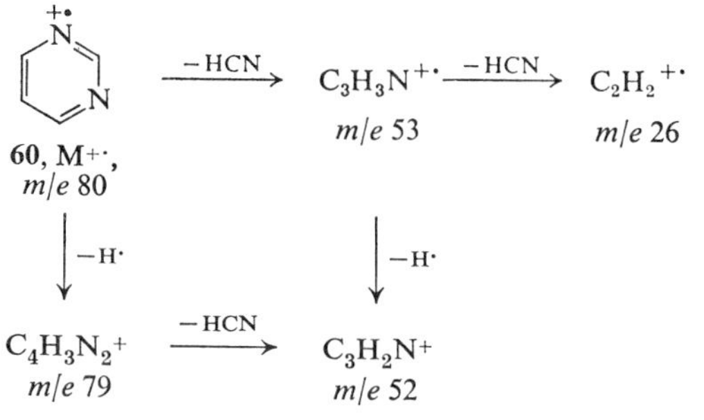

60, M⁺·,
m/e 80

−HCN

$C_3H_3N^{+·}$
m/e 53

−HCN

$C_2H_2^{+·}$
m/e 26

−H·

$C_4H_3N_2^+$
m/e 79

−HCN

$C_3H_2N^+$
m/e 52

−H·

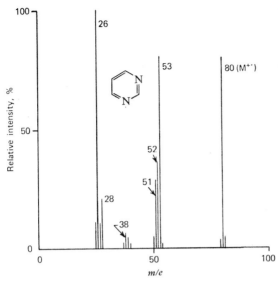

Fig. 12.10

Spectra of simple alkylpyrimidines have not yet been reported, but several polyfunctional alkylpyrimidines have been subjected to electron impact and their spectra will be discussed at the appropriate places.

2-Aminopyrimidine (**61**) also decomposes (Fig. 12.11) by sequential losses of two molecules of HCN (24). The product of the first decomposition may well be the pyrazole or imidazole radical cation, and further decomposition of this species agrees qualitatively with that already described for these compounds (Sec. II.A and III.A). Two pathways, initiated by the imino tautomer of **61**, satisfactorily describe the isotope distribution in the spectrum of dideutero-2-aminopyrimidine, and are outlined in the scheme.

61, M$^{+\cdot}$, m/e 95

$-$HNC:

$-$ HCN

m/e 68

m/e 68

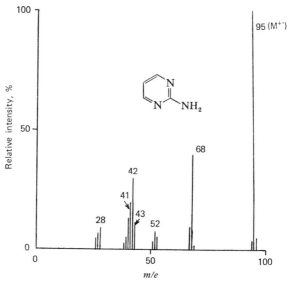

Fig. 12.11

Similar behavior has been observed with other 2-aminopyrimidines (25). Thus the decomposition of the 2-amino-4,6-dimethyl compound (62) can largely be explained in terms of elision of HCN to give the 3,5-dimethyl-pyrazole radical cation (62a) although the m/e 82 ion, not present in the spectrum of the pyrazole, may originate from the open-chain isomer of the molecular ion as shown and may well be the 3-methylimidazole radical cation (62b).

Fragmentation is much reduced in 2-amino-4-phenyl-6-methylpyrimidine (63) and most of the ion current is carried by the $M^{+\cdot}$ and $(M-1)^+$ species. This latter may well be the cyclic ion 63a and, if so, it presents another example of cleavage γ to a heterocyclic nitrogen atom. The intensity of the M-HCN ion in the spectrum of 63 is only 3% (26).

63, $M^{+\cdot}$, m/e 185   63a, m/e 184 (85%)
(100%)

The effect on the mass spectrum of modification of the amino group in 2-aminopyrimidines has been examined (26). 2-Acetamidopyrimidines eliminate ketene in the usual manner and the subsequent fragmentation is that of the free amine (26). It is possible that the elimination proceeds via a McLafferty rearrangement to give the imino form of the pyrimidine directly.

R = CH_3 or C_6H_5

The fragmentation of representative 2-piperidino- and 2-morpholino-pyrimidines has also been reported (26). Fragmentations begin in the non-aromatic ring, and are typical of the particular heterocycle involved. For further details the original publication should be consulted.

An amino group in the 4-position of the pyrimidine ring modifies the fragmentation pathways somewhat, and the scheme shown for the fragmentation (Fig. 12.12) of 2,6-dimethyl-4-aminopyrimidine (64) is typical (25). It is convenient to visualize fragmentation commencing with both the amino and imino forms of the molecule in this case. The ion at m/e 83 may correspond to a species such as 64a, resulting from loss of $\cdot CH_2CN$ from the molecular ion.

The mass spectrum of 4-hydroxy-6-methylpyrimidine (65) shows little loss of HCN from the molecular ion but marked loss of CO (25). This is most conveniently formalized as arising from the keto form of the molecular ion, and the resultant species (m/e 82) can be written as the open-chain ion

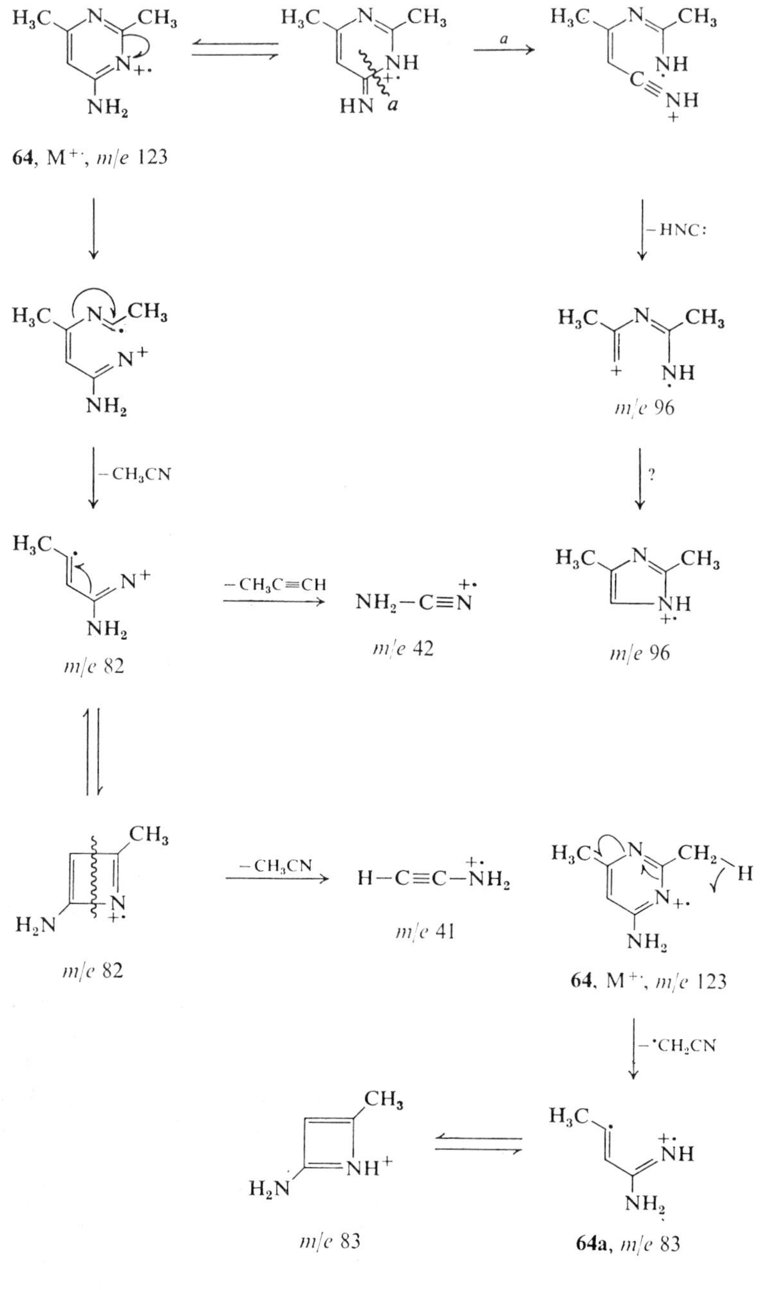
64, M+·, m/e 123
a
−HNC:
m/e 96
?
−CH3CN
−CH3C≡CH
NH2−C≡N
m/e 42
m/e 82
m/e 96
−CH3CN
H−C≡C−NH2
m/e 41
m/e 82
64, M+·, m/e 123
−·CH2CN
m/e 83
64a, m/e 83

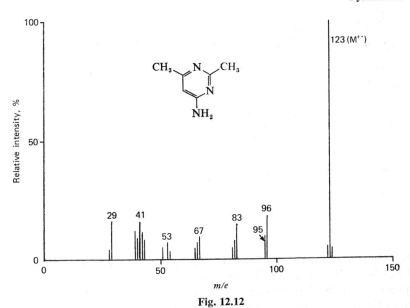

Fig. 12.12

65a or the imidazole **65b**. Loss of 40 mass units from m/e 82 produces the intense ion at m/e 42, and it has been suggested that the neutral fragment is methylacetylene.

65, M+·, m/e 110 (100%)

65a, m/e 82 (25%)

65b

$-CO$

(?)

$|-CH_3C\equiv CH$

$|-H·$

m/e 42 (65%)

m/e 81 (21%) (?)

The mass spectra of the important pyrimidine derivatives uracil (**66**), thymine (**67**), and cytosine (**68**) have been determined and their characteristic fragmentations adequately explained (24). Uracil decomposes by a retro-Diels-Alder reaction (loss of HNCO), and subsequent decompositions of the ion produced (**66a**) are easily understood.

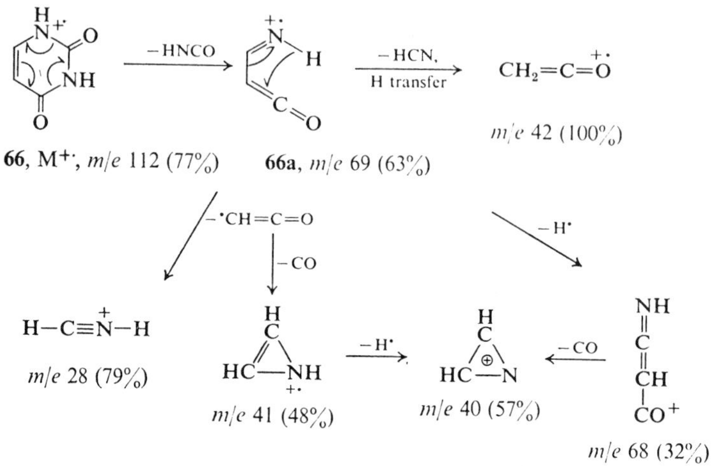

−HNCO
−HCN,
H transfer
$CH_2=C=O^{+\cdot}$
m/e 42 (100%)
66, M+·, m/e 112 (77%)
66a, m/e 69 (63%)
−·CH=C=O
−CO
−H·
H−C≡N−H (+)
m/e 28 (79%)
m/e 41 (48%)
−H·
m/e 40 (57%)
−CO
m/e 68 (32%)

The fragmentation of thymine (**67**) exactly parallels that of uracil and need not be discussed in detail. The isomeric 6-methyluracil (**69**) also decomposes initially by the retro-Diels-Alder reaction but the next fragmentation is loss of the methyl group, facilitated by formation of the stable protonated nitrile (**69a**).

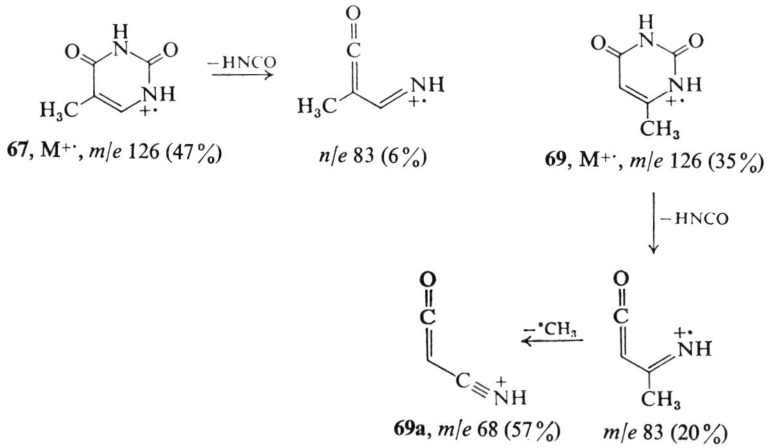

−HNCO
67, M+·, m/e 126 (47%)
n/e 83 (6%)
69, M+·, m/e 126 (35%)
−HNCO
−·CH3
69a, m/e 68 (57%)
m/e 83 (20%)

The mass spectrum of cytosine (**68**) is much more complex than those of the dioxygenated compounds, and three distinct pathways for fragmentation can be distinguished. Of interest in this compound is expulsion of carbon monoxide from the parent ion, which does not occur significantly with uracil and thymine. The tentative rationale shown explains the main decompositions of **68**.

pathway A

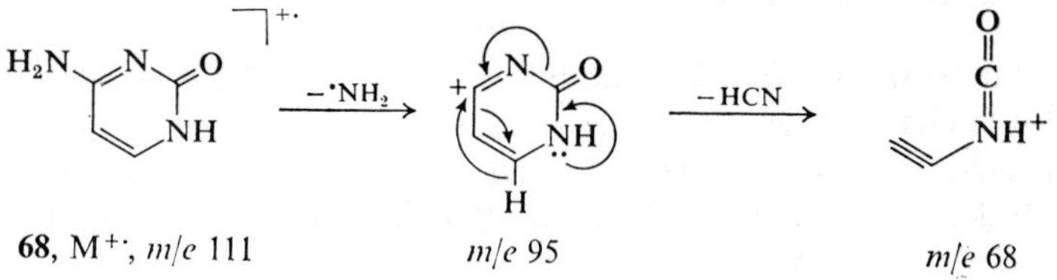

$$68, \text{M}^{+\cdot}, m/e \; 111 \qquad m/e \; 95 \qquad m/e \; 68$$

pathway B

$$68, \text{M}^{+\cdot}, m/e \; 111 \qquad m/e \; 83 \qquad m/e \; 56$$

$$m/e \; 55$$

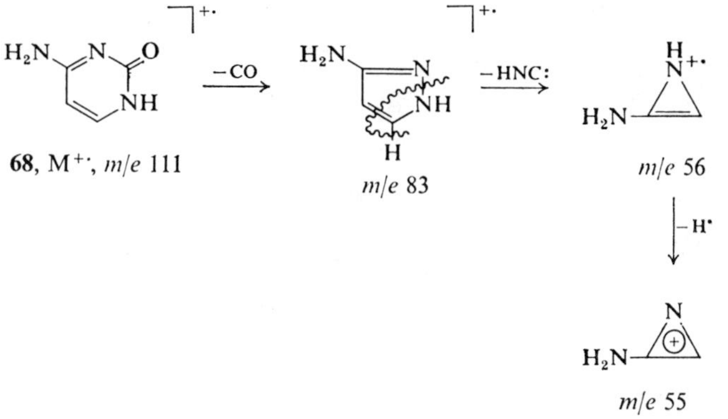

pathway C

$$68, \text{M}^{+\cdot}, m/e \; 111 \qquad m/e \; 111 \qquad m/e \; 68$$

$$m/e \; 110 \qquad m/e \; 110 \qquad m/e \; 69$$

$$m/e \; 67$$

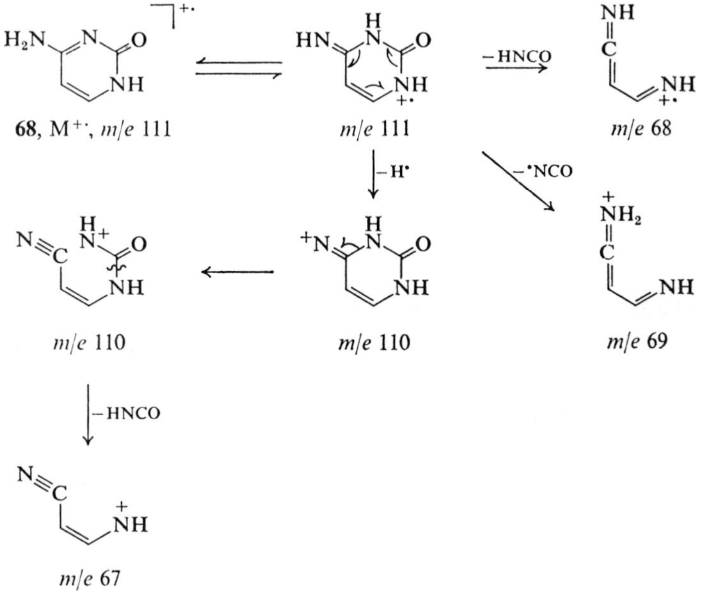

B. Reduced Pyrimidines

The mass spectrum of 5-acetamido-1,3-diacetyl-1,2,3,4-tetrahydropyrimidine (**70**) is characterized by losses of ketene, an acetyl group, and the elements of acetic acid (27). The following rationale has been suggested with the well-delocalized nature of the ions **70a** and **70b** as its driving force.

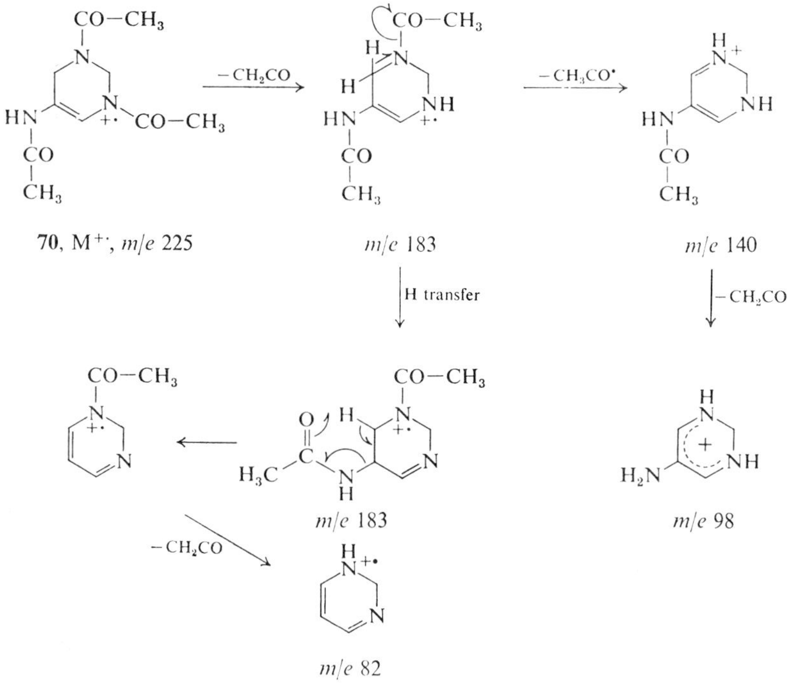
70, M+·, m/e 225
−CH₂CO
m/e 183
−CH₃CO·
m/e 140
H transfer
−CH₂CO
m/e 183
m/e 98
−CH₂CO
m/e 82

The mass spectrum of hexahydropyrimidine (**71**, R = H) is characterized by formation of an intense M-1 ion which may be represented by the cyclic

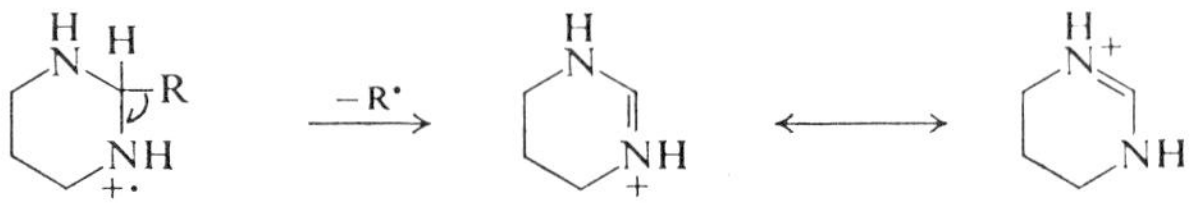
−R·

71, M+·   **71a**, *m/e* 85 (100%)

amidinium structure **71a** (28). With 2-methylhexahydropyrimidine (**71**, R = CH_3) the methyl radical is expelled in preference to the hydrogen atom, in agreement with the stabilities of the neutral fragments lost in the two processes.

C. Barbiturates and Related Compounds

Barbituric acid (72) has a simple spectrum dominated by the parent ion and the species $CH_2=C=O^{+\cdot}$ (m/e 42) which is the base peak. Loss of CO from the molecular ion is only of minor importance and ions at m/e 69, 70, and 85 can be explained in terms of simple ring fissions as shown (29).

$$a, -(HNCO+HNCO) \longrightarrow CH_2=C=O^{+\cdot}$$
$$m/e\ 42\ (100\%)$$

72, $M^{+\cdot}$, m/e 128 (47%)

$b, -HNCO$ $c, - \overset{NH}{\underset{NH}{|}} CO$

m/e 85 (8%)

$$\overset{+}{C}O-CH_2-\dot{C}O \quad \overset{-H^{\cdot}}{\longrightarrow} \quad O=C=CH-C\equiv O^{+}$$
$$m/e\ 70\ (2\%) \qquad\qquad m/e\ 69\ (5\%)$$

A wide range of 5-substituted and 5,5-disubstituted barbituric acids have been examined mass-spectrometrically, and for full details the original publications should be consulted (29, 30). Only a few typical examples will be discussed here.

The n-butyl compound (73) shows strong ions corresponding to M-C_3H_6 (m/e 142), M-$C_3H_7\cdot$ (m/e 141), M-C_4H_8 (m/e 128), and M-$C_4H_7\cdot$ (m/e 129). The presence of a double hydrogen rearrangement (m/e 129) is of interest and merits a more complete examination with use of deuterium labeling. A possible rationale for the main fragmentations is shown.

Suitably-substituted 5,5-dialkylbarbituric acids follow the initial McLafferty rearrangement by an allylic cleavage, showing that the initially formed enol does not ketonize rapidly compared with this cleavage. This type of fragmentation is seen in the mass spectrum of the ethyl-n-butyl-compound 74.

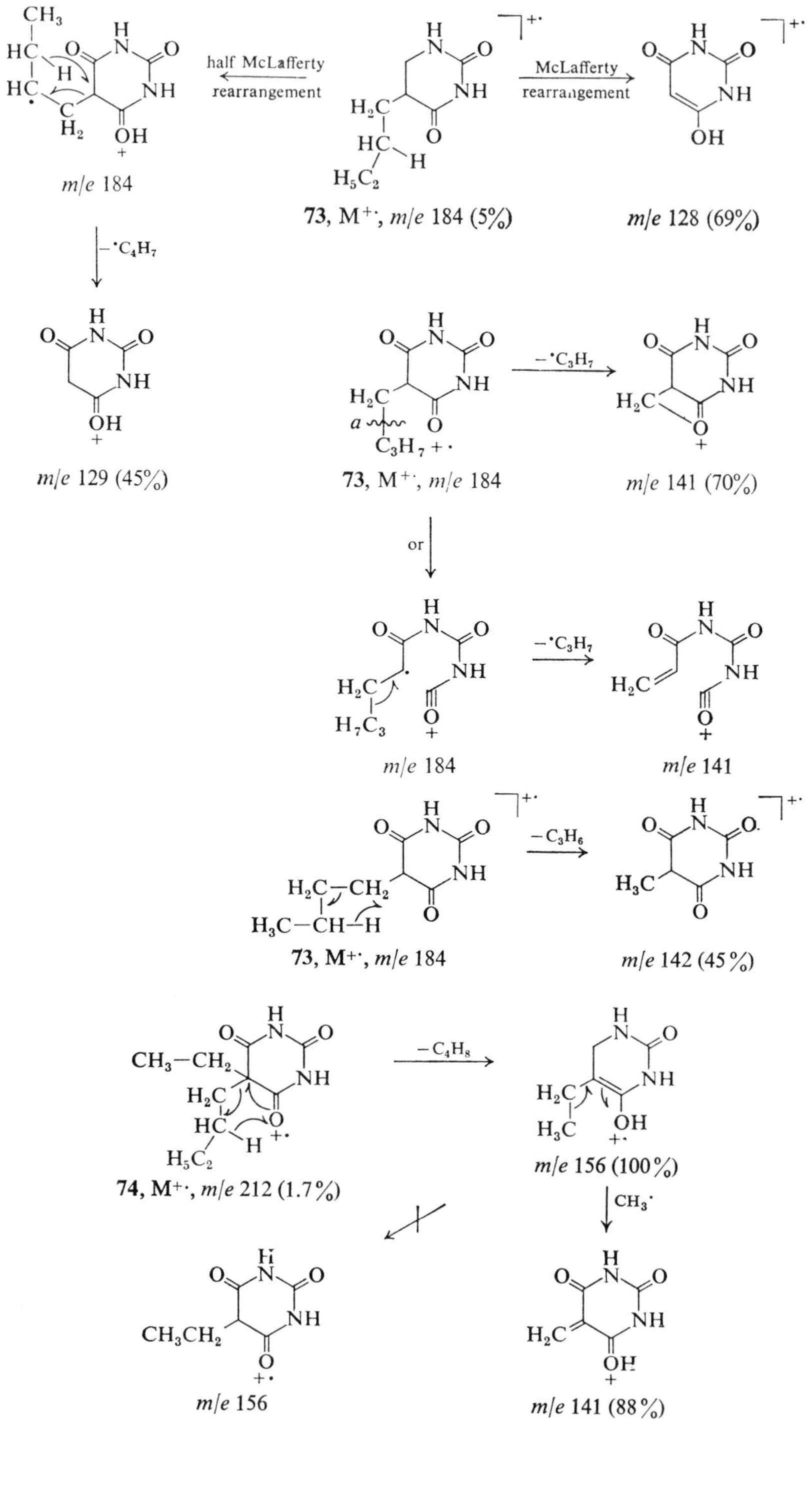

m/e 184

half McLafferty rearrangement

73, M$^{+\cdot}$, m/e 184 (5%)

McLafferty rearrangement

m/e 128 (69%)

$-\,^{\cdot}C_4H_7$

m/e 129 (45%)

73, M$^{+\cdot}$, m/e 184

$-\,^{\cdot}C_3H_7$

m/e 141 (70%)

or

m/e 184

$-\,^{\cdot}C_3H_7$

m/e 141

73, M$^{+\cdot}$, m/e 184

$-C_3H_6$

m/e 142 (45%)

74, M$^{+\cdot}$, m/e 212 (1.7%)

$-C_4H_8$

m/e 156 (100%)

$CH_3^{\cdot}$

m/e 156

m/e 141 (88%)

D. Pyrimidine Nucleosides

Free pyrimidine nucleosides are sufficiently volatile to be examined mass-spectrometrically provided direct insertion into the source is used (31).

The main fragmentation occurs at the bond between the base and the sugar moiety with accompanying transfer of one or two hydrogens to the base. These hydrogen atoms appear to originate from the hydroxyl hydrogens of the sugar, judging from the spectra of the deuterated compounds. Peaks occur also at m/e values corresponding to (base + 30) and M-89; the probable origin of these species is shown in the rationale for the fragmentation of uridine (75) which is quite typical of this class of compound.

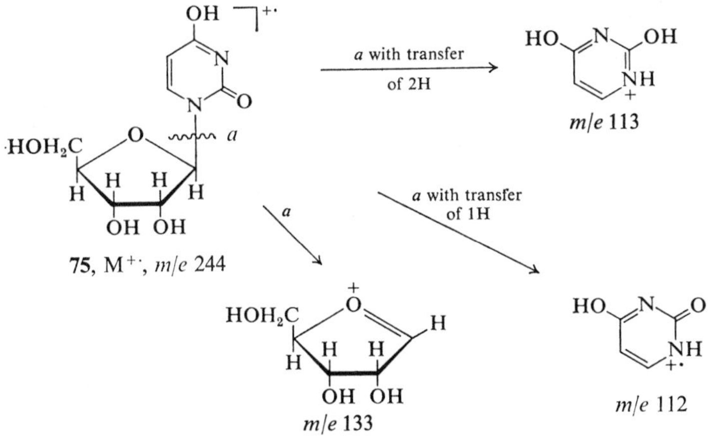

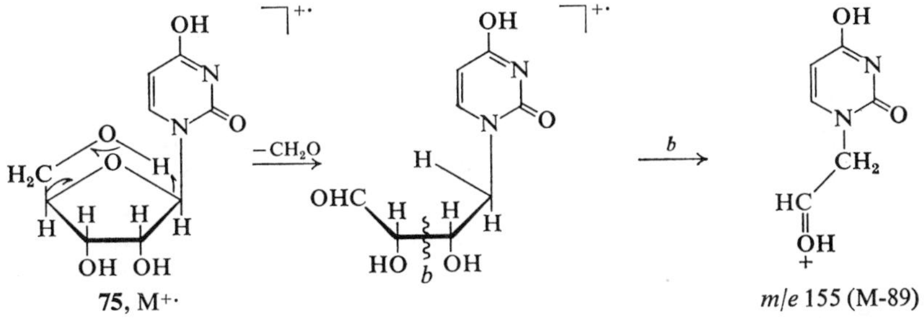

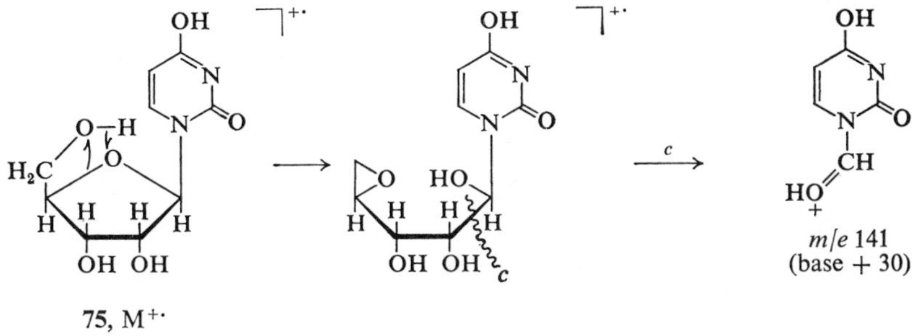

E.   Quinazolines (Benzopyrimidines)

The mass spectrum of quinazoline (76) is very simple and involves the successive loss of two molecules of HCN from the molecular ion to give ultimately an ion which may be written as ionized benzyne (m/e 76) (32).

76, M$^{+\cdot}$, m/e 130 (100%) m/e 103 (58%) m/e 76 (38%)

The spectrum of 4-deuteroquinazoline confirms that more than 90% of the HCN molecules eliminated in the first step originate from N$_{(3)}$ and C$_{(4)}$. Apparently randomization of hydrogen in the molecular ion is not an important process here (at least as far as this fragmentation is concerned).

The four methylquinazolines (77) in which the methyl group is in the carbocyclic ring have superimposable spectra and decompose in exactly the same way as the parent compound to give, ultimately, "methylbenzyne," m/e 90 (32).

77, M$^{+\cdot}$, m/e 144 (100%) m/e 117 (28%) m/e 90 (27%)

Similar sequential losses of HCN and CH$_3$CN are observed with the 2- and 4-methyl compounds (78 and 79), but both also lose a methyl radical from the parent ion to give a species which then decomposes to m/e 102 by loss of HCN. The 4-methyl compound loses CH$_3$CN from the molecular ion, showing that here, too, C$_{(4)}$ and N$_{(3)}$ are eliminated before C$_{(2)}$ and N$_{(1)}$.

78, M$^{+\cdot}$, m/e 144 (100%) m/e 129 (6%) m/e 102 (14%)

m/e 103 (28%)   **79**, M$^{+\cdot}$, m/e 144 (100%) m/e 129 (22%) m/e 102 (14%)

The 5-, 6-, and 7-hydroxyquinazolines decompose by the familiar pathway involving successive losses of two molecules of HCN, but the 2-, 4-, and 8-hydroxy compounds decompose by initial loss of CO, followed by loss of HCN (32). The process may be illustrated for the 4-hydroxy compound (**80**) as shown. The formal product of elision of CO from the 4-hydroxy compound is benzimidazole and the subsequent fragmentation is typical of that compound (Sec. III.C).

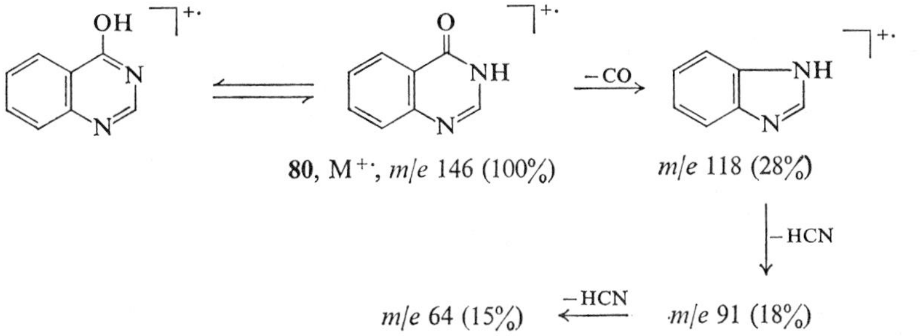

OH
N
N
O
NH
N
80, M+·, m/e 146 (100%)
−CO
NH
N
m/e 118 (28%)
−HCN
m/e 91 (18%)
−HCN
m/e 64 (15%)

It is interesting to note that the quinazolone alkaloid anisotine (**81**), which has the $C_{(4)}$-oxygen function fixed in the keto form, shows no loss of

81, M+·, m/e, 349 (100%)
−H·
−CH3OH
−·CO2CH3
−CH3OH
OCH3
H3C
NH
CH3
m/e 317 (6%)
m/e 290 (6%)
m/e 316 (15%)

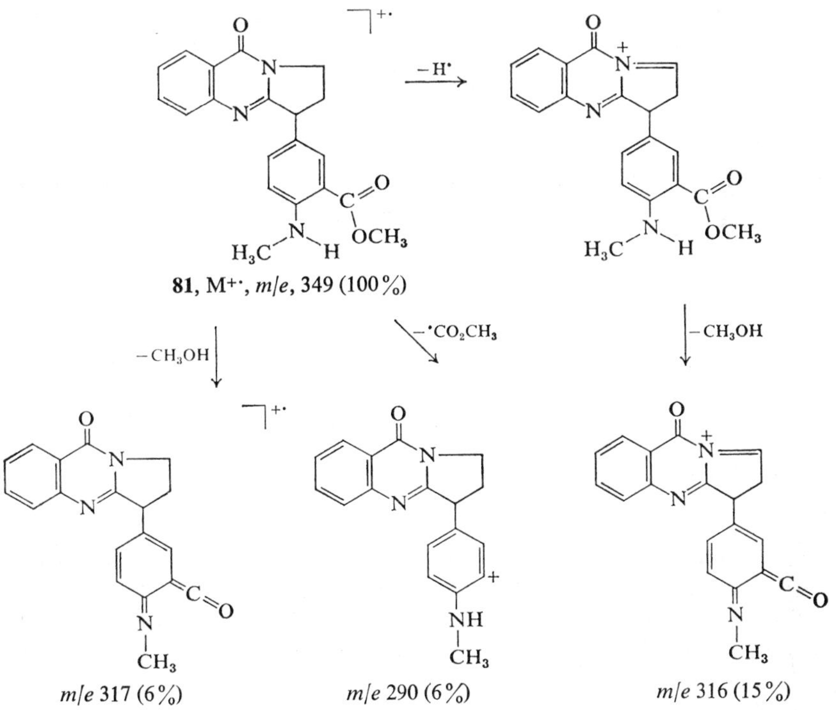

CO from the molecular ion, and fragmentations involve parts of the molecule other than the quinazolone nucleus (33). Different behavior again is seen in the mass spectrum of the alkaloid glycorine (1-methyl-4-quinazolone, **82**) (34). Here the initial result of electron impact is a retro-Diels-Alder reaction with loss of the elements of HCN. The fragmentation subsequent to this loss

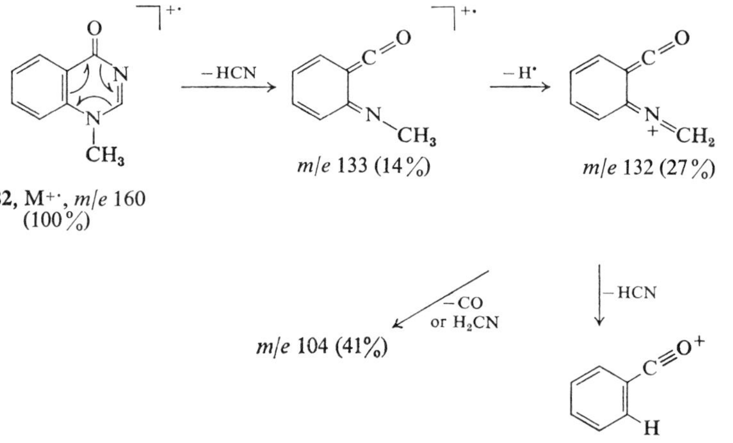

82, M$^{+\cdot}$, m/e 160
(100%)

m/e 133 (14%)

m/e 132 (27%)

m/e 104 (41%)

m/e 105 (34%)

is shown in the rationale presented here. Glycosminine (2-benzyl-4-quinazolone, **83**) (34) also fragments by the retro-Diels-Alder process despite the fact that the molecule is not "fixed" in the keto form. Most of the ion current is carried by the M-1 ion, however, which is probably the substituted tropylium ion **83a**.

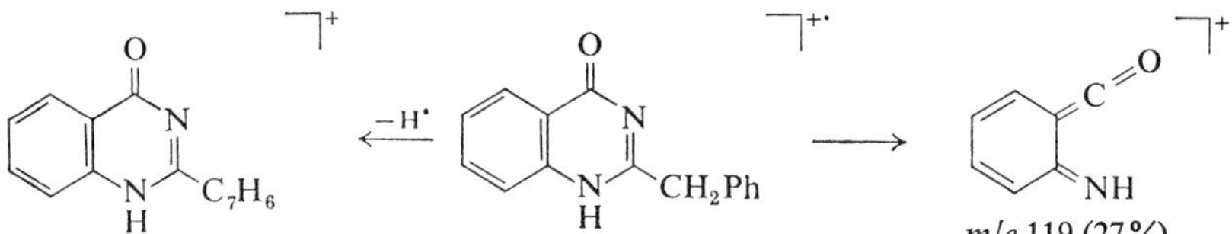

83a, m/e 235 (100%)

83, M$^{+\cdot}$, m/e 236 (65%)

m/e 119 (27%)

The tricyclic quinazolone **84** likewise shows no elision of carbon monoxide from the molecular ion and the chief fragmentations appear to be formation of M-1 and M-15 ions (35). The finer details of these decompositions could only be ascertained by means of deuterium labeling, but the tentative rationale shown is put forward.

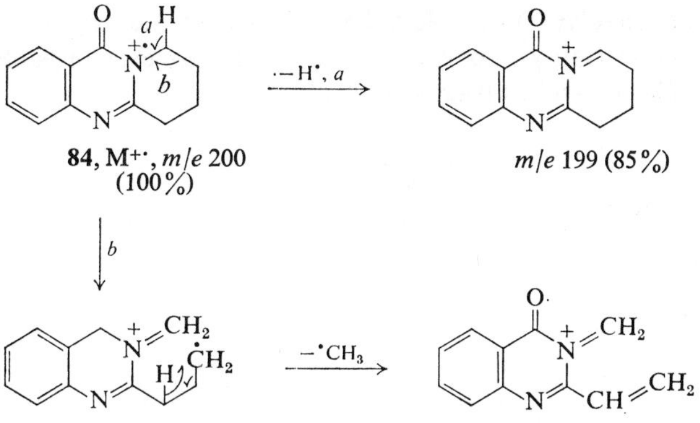

84, M$^{+\cdot}$, m/e 200 (100%) $\xrightarrow{\cdot-H^{\cdot},\, a}$ m/e 199 (85%)

$\downarrow b$

m/e 200 $\xrightarrow{-\cdot CH_3}$ m/e 185 (54%)

The mass spectrum of methyl 2-*t*-butylquinazoline-4-sulfenate (**85**) is dominated by losses of CH$_3$O$^{\cdot}$ and CH$_2$O, illustrating the lability of the S—O linkage to electron impact. No significant fragmentations of the quinazoline ring occur in the partial spectrum reported (36).

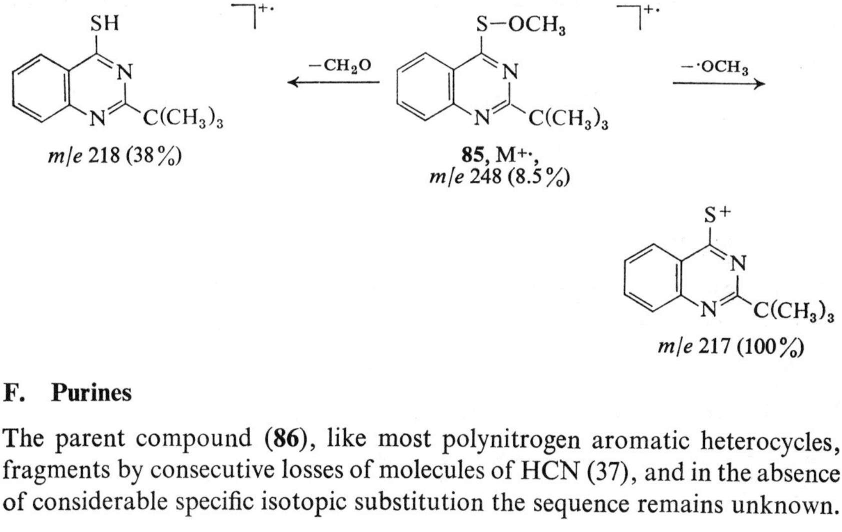

m/e 218 (38%) $\xleftarrow{-CH_2O}$ **85**, M$^{+\cdot}$, m/e 248 (8.5%) $\xrightarrow{-\cdot OCH_3}$ m/e 217 (100%)

F. Purines

The parent compound (**86**), like most polynitrogen aromatic heterocycles, fragments by consecutive losses of molecules of HCN (37), and in the absence of considerable specific isotopic substitution the sequence remains unknown.

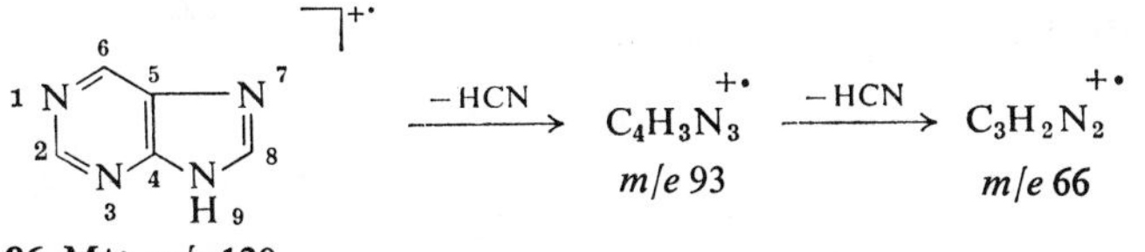

86, M$^{+\cdot}$, m/e 120 $\xrightarrow{-HCN}$ C$_4$H$_3$N$_3^{+\cdot}$ m/e 93 $\xrightarrow{-HCN}$ C$_3$H$_2$N$_2^{+\cdot}$ m/e 66

6-Aminopurine (adenine, **87**) similarly fragments by consecutive losses of HCN (38). Here deuterium labeling suggests that C$_{(6)}$ and the attached amino nitrogen are involved in the first HCN loss, but the origin of the second molecule is uncertain; it does not, however, involve exchangeable hydrogen. The rationale shown has been suggested.

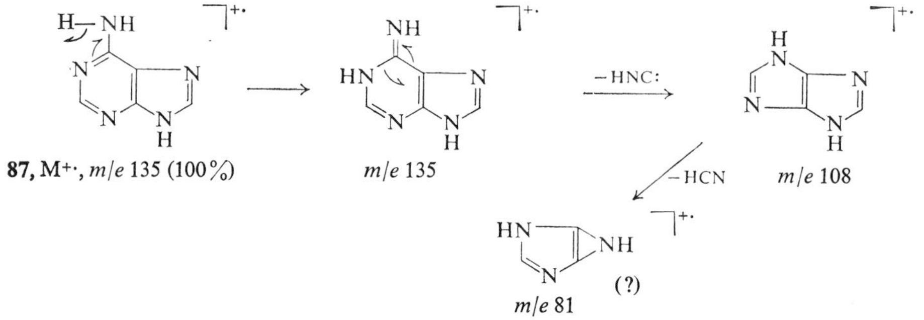
87, M+·, m/e 135 (100%)
m/e 135
−HNC:
m/e 108
−HCN
(?)
m/e 81

The spectra of a number of 6-alkylaminopurines have also been examined. That of the 6-*n*-hexylamino-compound (**88**) contains a number of interesting features, the most surprising of which is loss of $\cdot NH_2$ from the molecular ion. The deuterated analog loses $\cdot NHD$ in the same fragmentation, and the

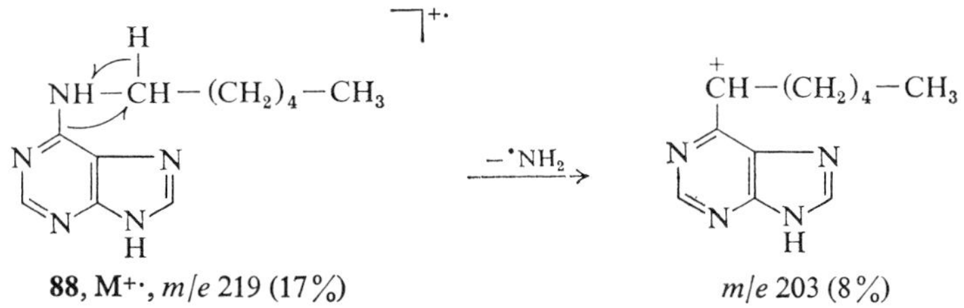
−·NH₂
88, M+·, m/e 219 (17%)
m/e 203 (8%)

rationale shown is suggested for this rearrangement (38). A series of ions (*m/e* 190, 176, 162, and 148) correspond to losses of alkyl radicals from the

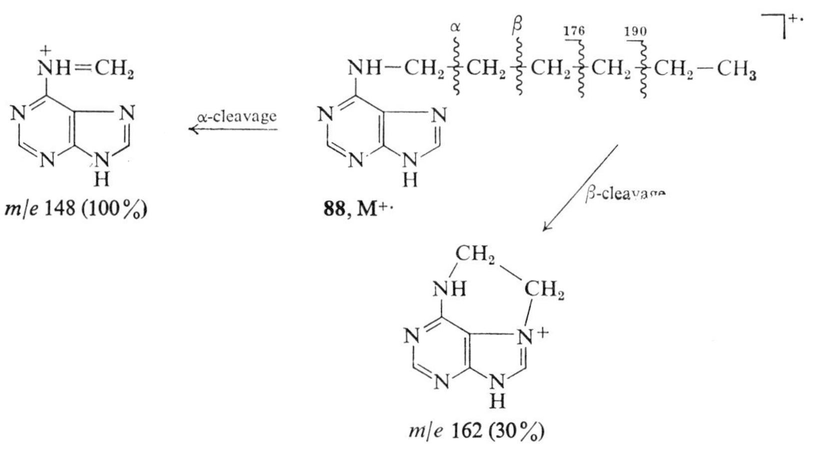
α
β
176
190
α-cleavage
m/e 148 (100%)
88, M+·
β-cleavage
m/e 162 (30%)

side chain. The β-cleavage ion is unusually strong and may well be stabilized by cyclization as shown. Loss of the alkyl chain with hydrogen transfer in a McLafferty rearrangement also occurs, producing an intense ion at m/e 135.

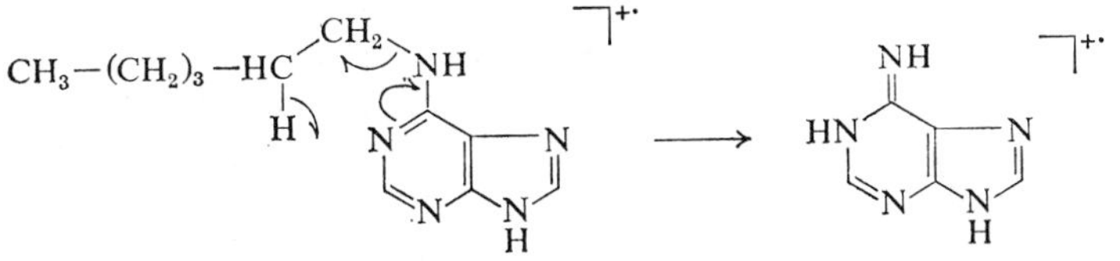

88, M$^{+\cdot}$, m/e 219 m/e 135 (85%)

Mass spectrometry has been used extensively in determining the structure of zeatin (**89**) (39, 40), a substance which regulates cell division in plant tissues. In particular the characteristic intense ion at m/e 135 indicates the presence of a single alkenylamino substituent. The rationale for the major ions in the spectrum is shown.

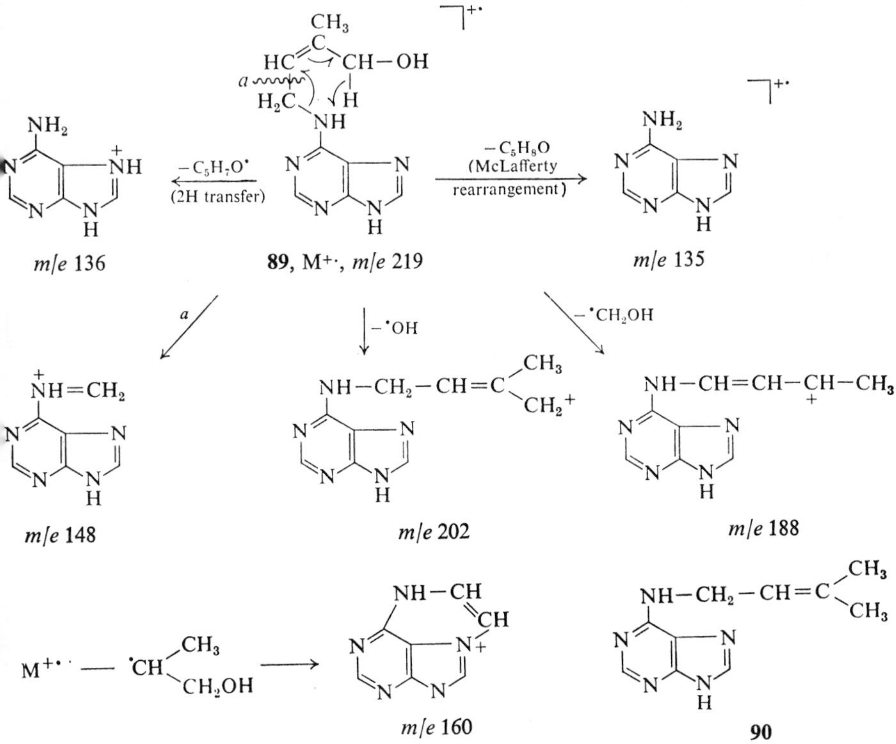

m/e 136   **89**, M$^{+\cdot}$, m/e 219 m/e 135

$-C_5H_7O^\cdot$ (2H transfer) $-C_5H_8O$ (McLafferty rearrangement)

m/e 148 m/e 202 m/e 188

$-{}^\cdot OH$ $-{}^\cdot CH_2OH$

$M^{+\cdot} \longrightarrow {}^\cdot CH \overset{CH_3}{\underset{CH_2OH}{\diagdown}} \longrightarrow$

m/e 160   **90**

The ions at m/e 160 and 188 appear to be formed following a double-bond isomerization in the molecular ion in which a hydrogen atom is transferred in a 4-center process. The same ions occur in the spectrum of 6-(3-methylbut-2-enyl)aminopurine (90) (38).

The mass spectra of caffeine (91), theobromine (92), and theophylline (93) (41) all show intense molecular ions and (like the N-methylquinazolones already discussed) decompose by what is effectively a retro-Diels-Alder reaction although the decomposition occurs in two steps initiated by cleavage of the 1,6-bond. The rationale for the spectrum of caffeine is shown, but similar fragmentation pathways can be constructed for 92 and 93.

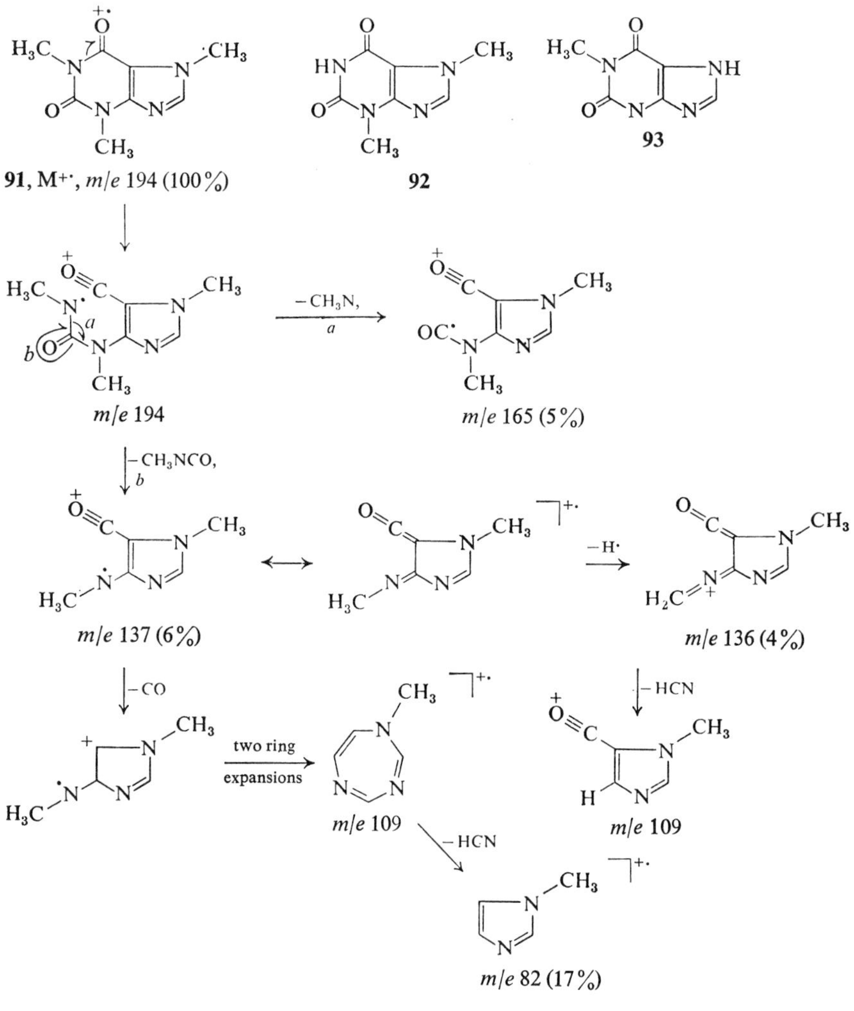

91, M⁺·, m/e 194 (100 %)   92   93

m/e 194 m/e 165 (5 %)

m/e 137 (6 %) m/e 136 (4 %)

m/e 109 m/e 109

m/e 82 (17 %)

The fragmentation of thiotheophylline (**94**) proceeds by a route basically similar to that of its oxygen analog (42). Noteworthy is the loss of CS from the molecular ion to give the species **94a**. It will be recalled that there is little tendency for the diones previously discussed to lose CO from the molecular ion.

94, M+·, *m/e* 196 (100%)   94a, *m/e* 152

The purine nucleoside adenosine (**95**) fragments in a manner similar to the simple pyrimidine nucleosides already discussed except that much less of the ion current is carried by the sugar fragment since the purine, having a

96   95

higher electron density, competes more effectively for the positive charge than can the pyrimidine (31). Likewise the base peak (*m/e* 164) of the spectrum of puromycin (**96**) corresponds to the purine nucleus plus two transferred hydrogen atoms, but the important fragmentation of the purine nucleus appears to originate from the *m/e* 163 species (43).

G.   Pteridines

Because of the generally low concentrations of naturally-occurring pteridines encountered, mass spectrometry is an important analytical method with these compounds. The mass spectra of the parent compound and some alkyl- and hydroxypteridines have been analyzed, but there is clearly scope for more work in this important field.

Pteridine (**97**) undergoes the expected sequential loss of two molecules of HCN, and deuterium labeling shows that 75% of the first molecule eliminated involves $C_{(4)}$ and $N_{(3)}$ whereas 19% comes from $C_{(2)}$ and $N_{(1)}$. The preference for fragmentation to originate in the pyrimidine ring is thus marked (44). A similar sequence is observed with 2-methylpteridine **(98)**

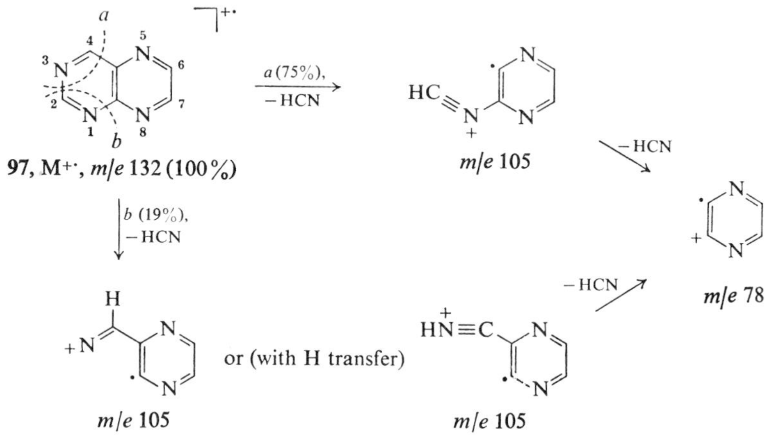
a
b
a (75%), − HCN
b (19%), − HCN
− HCN
− HCN
or (with H transfer)
97, M+·, m/e 132 (100%)
m/e 105
m/e 105
m/e 105
m/e 78

leading to the same ion (*m/e* 78), but an additional pathway involving loss of the radical $\dot{C}H_2CN$ becomes apparent and has been rationalized (44) in

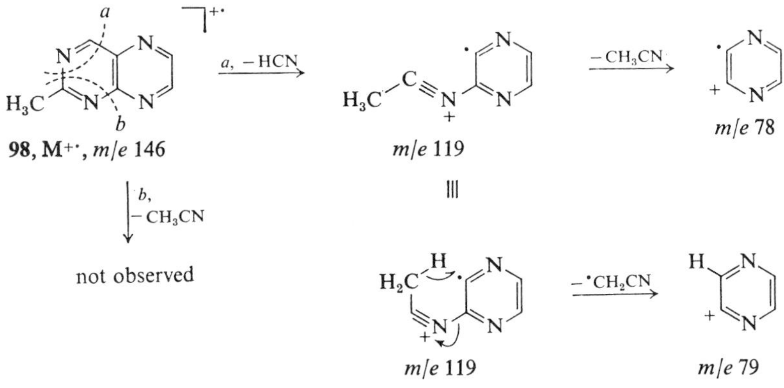
a, − HCN
− CH3CN
b, − CH3CN
not observed
− ·CH2CN
98, M+·, m/e 146
m/e 119
m/e 78
m/e 119
m/e 79

terms of hydrogen transfer as shown. Not surprisingly, in view of the preference for initial elision of $C_{(4)}$ and $N_{(3)}$, 4-methylpteridine loses CH_3CN exclusively as the initial decomposition to give the *m/e* 105 ion already encountered.

7-Methylpteridine **(99)** decomposes by initial loss of HCN, and here deuterium labeling has shown that fragmentation of the pyrazine ring is involved. Apparently the electron-releasing effect of the $C_{(7)}$ methyl group stabilizes the positive charge on the nitrogen of the $C_{(7)}$—$N_{(8)}$ group of ion **99a**.

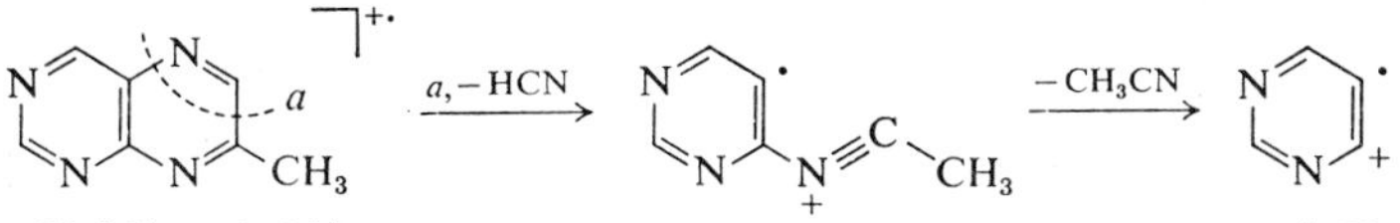

99, M⁺·, *m/e* 146   99a, *m/e* 119   *m/e* 78

In line with this observation, 6,7-dimethylpteridine (**100**) decomposes largely by the sequence M → M-41 → M-(41 + 41) although some loss of HCN in the second stage is also observed. The sequence of loss of the acetonitrile molecules is unknown, but in principle it could readily be determined by deuterium substitution in one of the methyl groups.

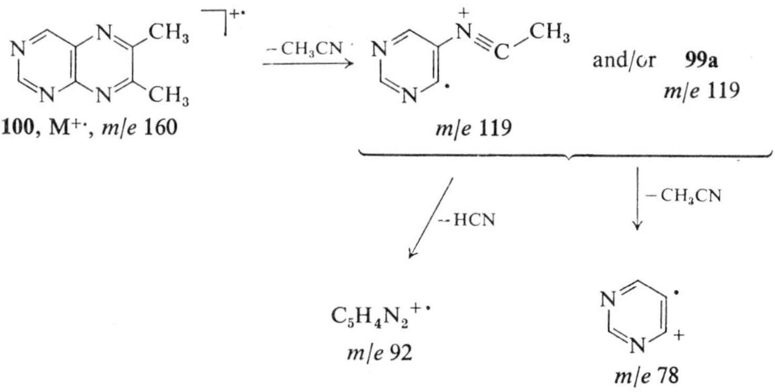

100, M⁺·, *m/e* 160   *m/e* 119   and/or **99a** *m/e* 119

$C_5H_4N_2^{+·}$ *m/e* 92

m/e 78

The mass spectra of the four isomeric hydroxy-pteridines [which in fact exist as shown in the lactam forms (45)] have been analyzed (44). The 6- and 7-hydroxy compounds (**101** and **102**, respectively) decompose by initial loss of CO to give a species which fragments as does purine and is thus represented as **101a/101b**.

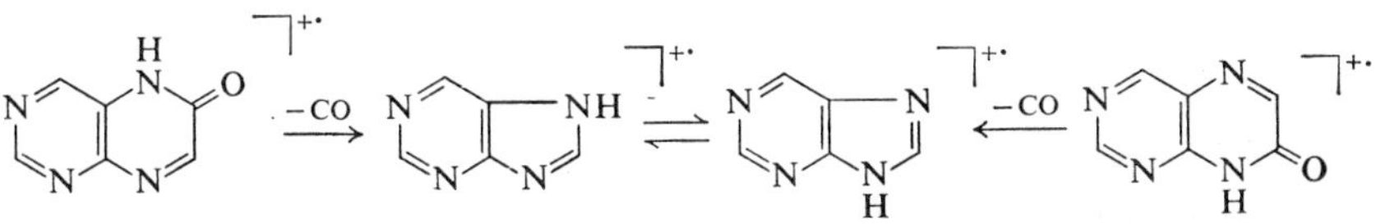

101, M⁺·, *m/e* 148   101a, *m/e* 120   101b   102, M⁺·, *m/e* 148

The 2-hydroxy compound (**103**), on the other hand, initially extrudes HCN from the molecular ion, then loses CO as the second step. The pathway shown seems likely but has not been unequivocally demonstrated by labeling experiments.

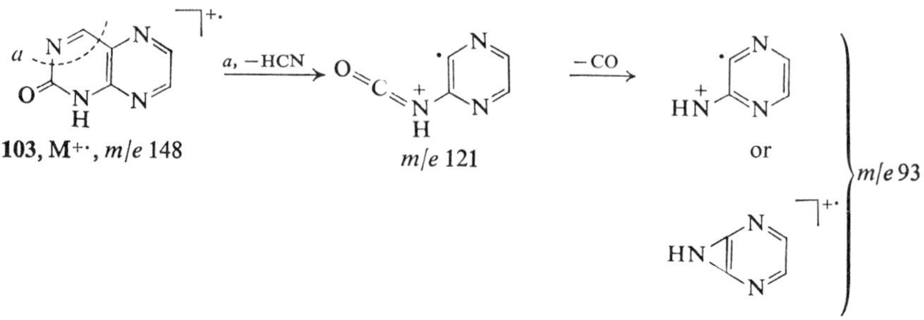

a
a, −HCN
−CO
103, M+·, m/e 148
m/e 121
or
m/e 93

A less important process is sequential extrusion of three molecules of HCN with retention of the oxygen function. This may well involve initial fragmentation of the pyrazine ring.

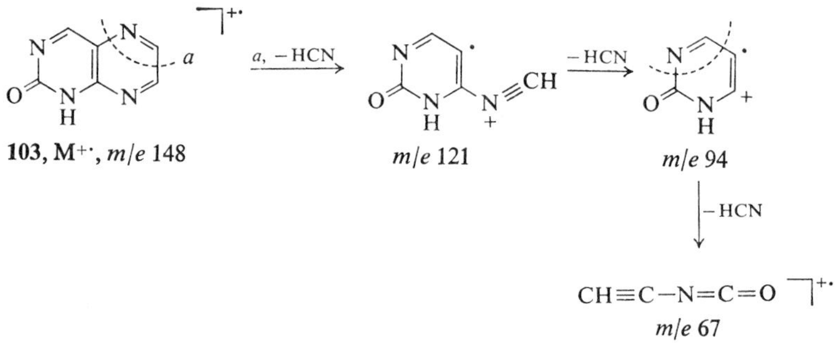

a
a, −HCN
−HCN
103, M+·, m/e 148
m/e 121
m/e 94
−HCN
CH≡C−N=C=O
m/e 67

4-Hydroxypteridine (**104**) shows almost equal tendencies to lose HCN and CO from the molecular ion, and presumably processes such as those shown are responsible.

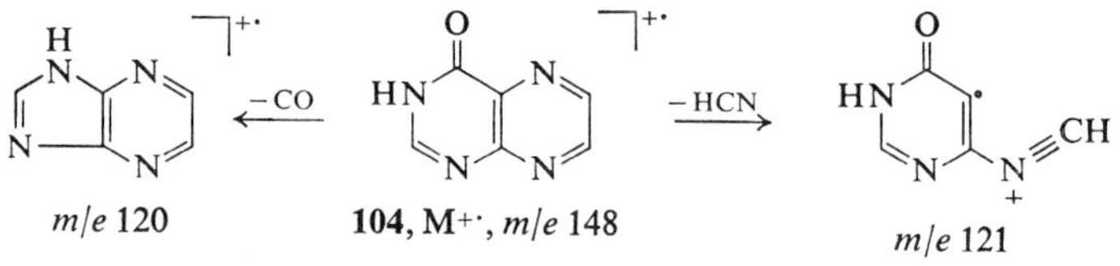

−CO
−HCN
m/e 120
104, M+·, m/e 148
m/e 121

2,4-Dihydroxypteridine (**105**) fragments by initial loss of HNCO. It has been suggested (44) that this involves $N_{(3)}$ and $C_{(4)}$, but the loss of $N_{(3)}$

and $C_{(2)}$ in a process analogous to that seen with the quinazol-4-ones seems equally likely.

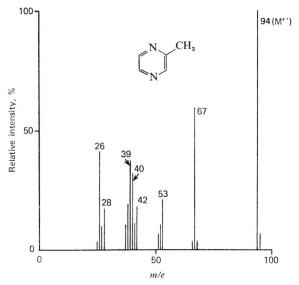

VI. PYRAZINES

A. Simple Pyrazines

The mass spectrum of the parent compound does not appear to have been reported but would no doubt be dominated by the loss of HCN molecules. 2-Methylpyrazine (**106**) (Fig. 12.13) on electron impact loses both HCN and CH$_3$CN from the molecular ion and then undergoes losses of CH$_3$CN and HCN, respectively (46). A possible rationale is shown.

Fig. 12.13

The ion at m/e 42 probably corresponds to $CH_3—C≡N^+H$ formed by fission a with hydrogen transfer from the neutral fragment. It is noteworthy that no significant M-1 ion is present. This is in agreement with earlier observations concerning the low stability of a carbonium ion such as 106a.

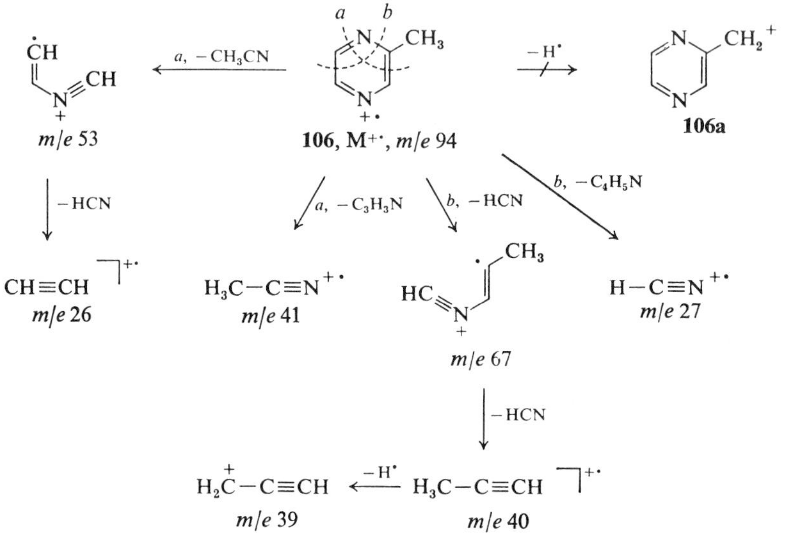

The McLafferty rearrangement becomes an important process in pyrazines with n-propyl or longer side chains (47), as does γ-cleavage of the type already encountered in (for example) 2-alkylpyridines (Chap. 11, Sec. II.B). These processes are well-exemplified in the mass spectrum of 2,5-dimethyl-3-n-propylpyrazine (107).

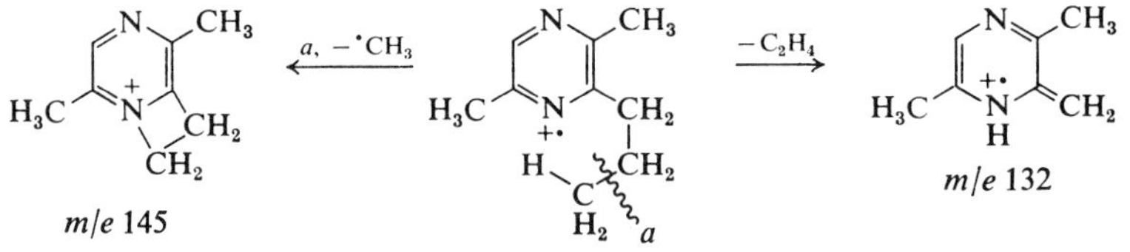

The basic fragmentations described are still prominent in the complex pyrazine etioluciferamine (108), a degradation product of luciferin (48). It is interesting to notice that no significant degradation of the indole nucleus occurs upon electron impact.

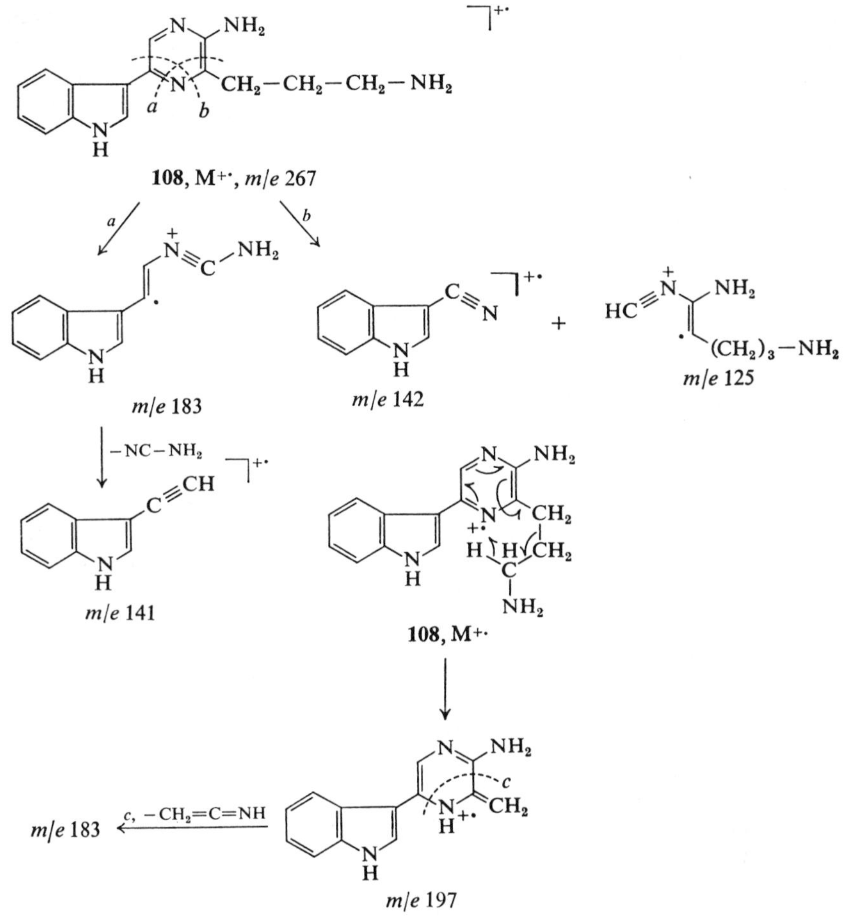

108, M$^{+\cdot}$, m/e 267

m/e 183

m/e 142

m/e 125

m/e 141

108, M$^{+\cdot}$

m/e 183 $\xleftarrow{c, -CH_2=C=NH}$

m/e 197

Pyrazine-2,3-dicarboxylic anhydride (**109**), like phthalic anhydride (Chap. 6, Sec. V.B), decomposes by successive losses of CO_2 and CO to give an ion which may be represented as the heteroaryne **109a** (49).

It is interesting to note that pyrolysis of **109** at 830°/0.05 mm gives a high yield of maleonitrile (**110**) and fumaronitrile (**111**). This result can be explained as thermal production of the heteroaryne which undergoes fission

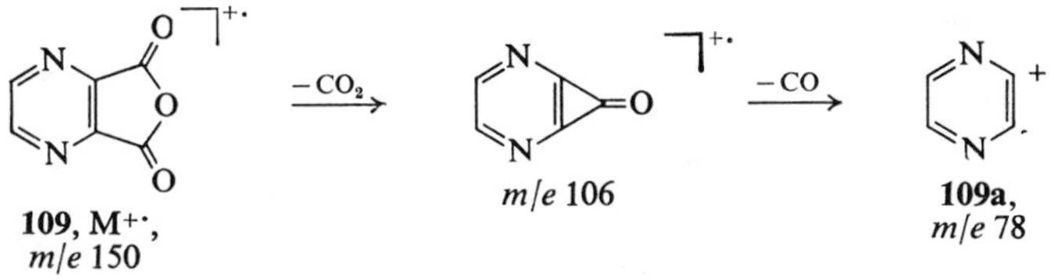

109, M$^{+\cdot}$, m/e 150

m/e 106

109a, m/e 78

to the bisisonitrile, followed by thermal rearrangement to the observed products. The process clearly demonstrates a close parallel between impact-induced and thermal processes.

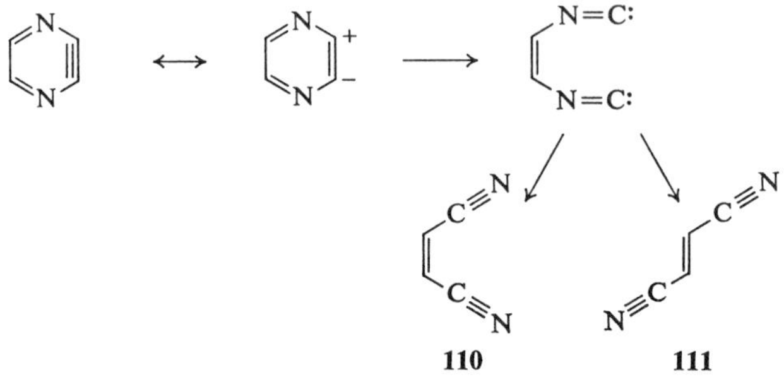

110   **111**

B. Reduced Pyrazines

1. Dihydropyrazines

2,3-Dimethyl-5,6-dihydropyrazine (**112**) is less stable to electron impact than the fully-aromatic pyrazines but decomposes in part by a similar route, namely loss of CH_3CN (50). With the reservation that no high-resolution data or information on metastable ions has been published, the rationale shown explains the spectrum satisfactorily.

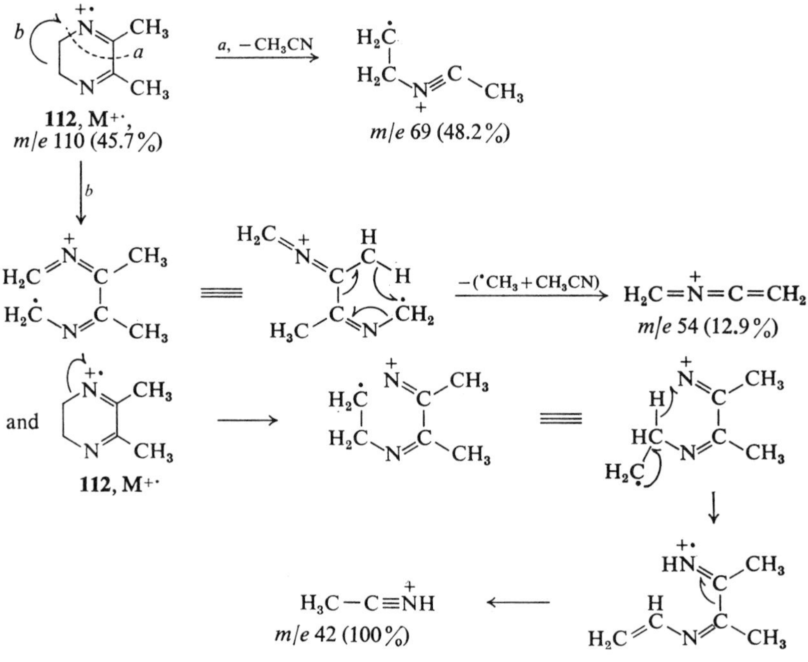

2. Piperazines (Hexahydropyrazines)

The spectrum of piperazine (113) (Fig. 12.14) (51) can be rationalized in a similar manner to those of piperidine and morpholine. Here again internal α-cleavage initiates most fragmentation. The origin of the hydrogen atom transferred in the McLafferty rearrangement (m/e 86 → m/e 44) has been confirmed by deuterium labeling (47).

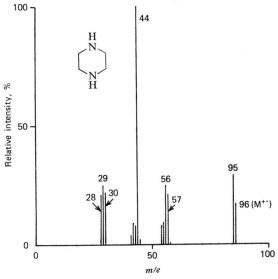

In C-alkylated piperazines loss of a substituent by α-cleavage becomes an important process, but the same fragmentations described for the parent

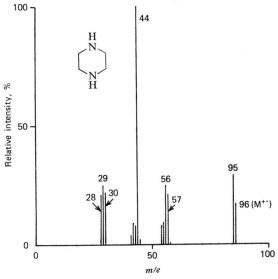

Fig. 12.14

compound still occur and can be very useful in establishing positions of substituents in polysubstituted piperazines (47). *N*-Methylpiperazine fragments in the manner described for the unsubstituted compound, and as is to be expected the M-$\cdot CH_3$ ion is of negligible intensity in the spectrum. The *N*,*N'*-dimethyl compound (**114**) (52) lacks the hydrogen atom which is transferred in the McLafferty rearrangement observed in the spectrum of the parent compound, and the base peak arises from a simple C—N fission in the internal α-cleavage ion **114a**.

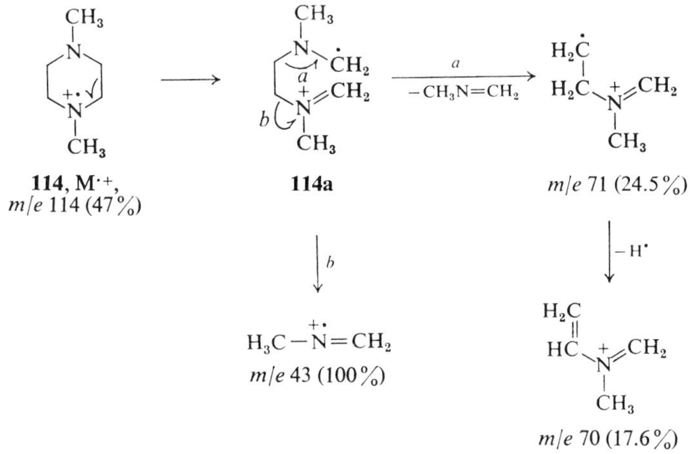

CH3
114, M·+,
m/e 114 (47%)
114a
a
−CH3N=CH2
m/e 71 (24.5%)
b
H3C−N=CH2
m/e 43 (100%)
−H·
m/e 70 (17.6%)

3. 2,5-Diketopiperazines

The basic fragmentations occurring in this group of heterocycles are straightforward (53) and may be exemplified by the fragmentation rationale for the 3,6-dimethyl compound **115**.

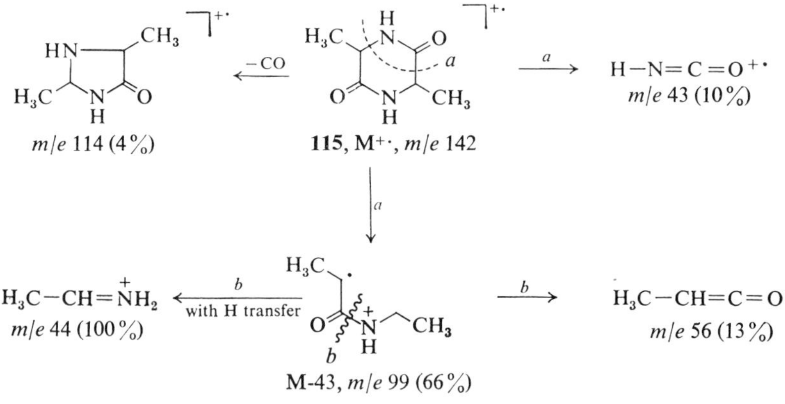

−CO
m/e 114 (4%)
115, M+·, m/e 142
a
H−N=C=O+·
m/e 43 (10%)
b
with H transfer
H3C−CH=NH2
m/e 44 (100%)
M-43, m/e 99 (66%)
H3C−CH=C=O+·
m/e 56 (13%)

McLafferty rearrangements become important when the (alkyl) substituent in the 3- and/or 6-position has two or more carbon atoms. Thus the diisopropyl compound **116** successively loses two molecules of propene to give **116a** and **116b**.

116, M$^{+\cdot}$, m/e 198 (1.2%)

$-C_3H_6 \longrightarrow$

116a, m/e 156 (100%)

$-C_3H_6$

116b, m/e 114 (8%)

C. Quinoxaline Derivatives

Little has appeared concerning the mass spectra of this class of heterocyclic compounds. The major peaks in the spectrum of the tetrahydroquinoxaline dione **117** have been reported (54) and a partial rationale may be constructed, but it will be seen that fragmentations involving the side chains rather than the ring dominate the spectrum at higher masses. High-resolution and metastable-ion data would be required for a more complete analysis.

D. Phenazine Derivatives

As expected, phenazine (**118**) has an intense molecular ion and shows little fragmentation. The ion at m/e 152 corresponds to the loss of HCN + H from the molecular ion (56) and not, as earlier thought, the loss of N_2 (55). The m/e 153 ion is composed of 75% M-HCN and 25% M-C_2H_3. High-resolution data have not been presented for m/e 154 but the ion may well be M-C_2H_2. On the reasonable assumption that one of the "angular" carbon atoms is involved in the elision of HCN, the M-HCN ion may well have an "open chain" structure (**118a**).

1-Methylphenazine displays intense M-1 and M-2 ions, while the 1- and 2-methoxy compounds show surprisingly intense M-31 and M-32 ions (55).

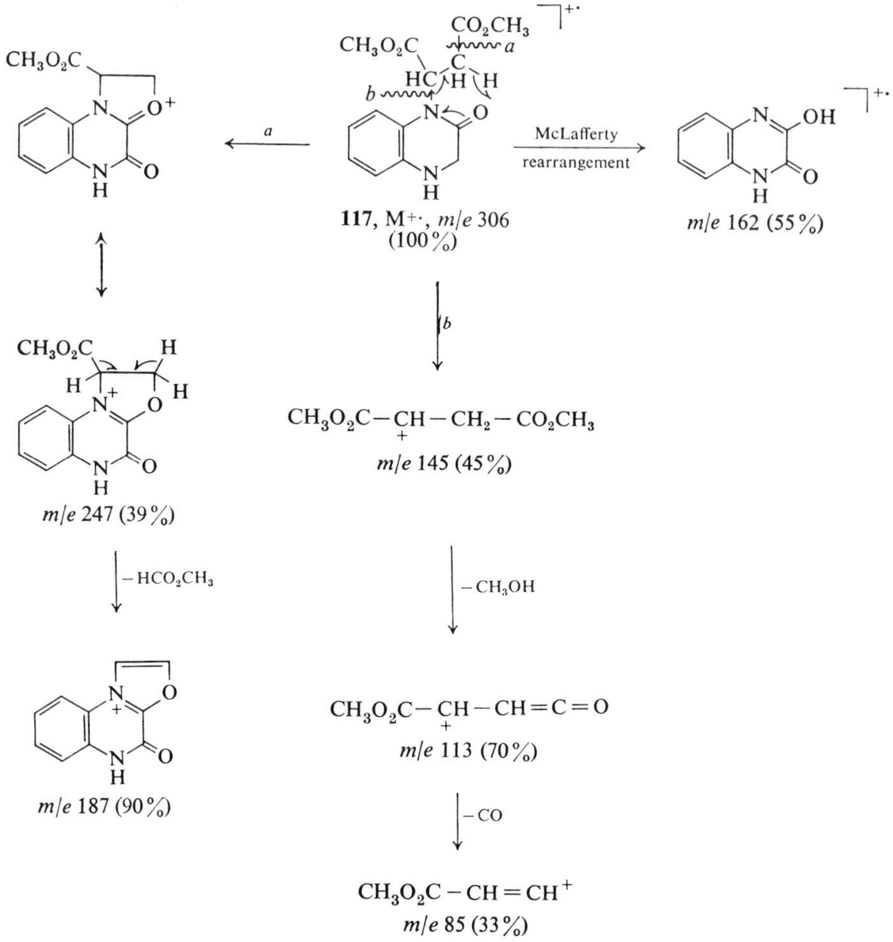

117, M$^{+\cdot}$, m/e 306
(100%)

m/e 162 (55%)

McLafferty rearrangement

m/e 247 (39%)

$CH_3O_2C-CH-CH_2-CO_2CH_3$
m/e 145 (45%)

$-HCO_2CH_3$

$-CH_3OH$

m/e 187 (90%)

$CH_3O_2C-CH-CH=C=O$
m/e 113 (70%)

$-CO$

$CH_3O_2C-CH=CH^+$
m/e 85 (33%)

Satisfactory representations for the hydrogen-deficient species are difficult to produce and it is possible that here, too, open-chain representations for the daughter ions will prove most satisfactory.

Phenazine-N-oxide (**119**) shows the expected M-O ion together with an intense M-C$_2$H$_2$ ion and also an M-CO ion (m/e 168) which must result

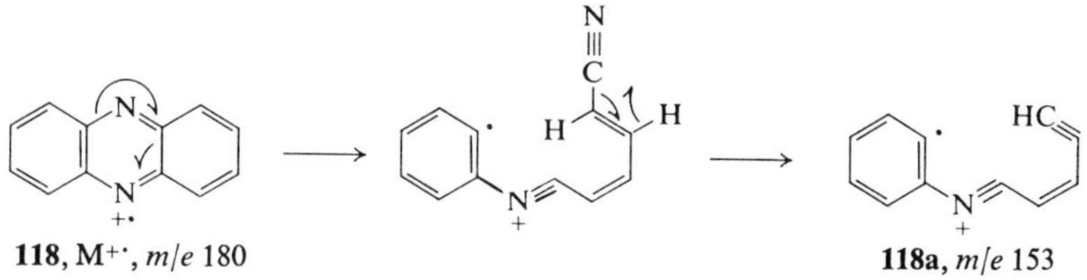

118, M$^{+\cdot}$, m/e 180

118a, m/e 153

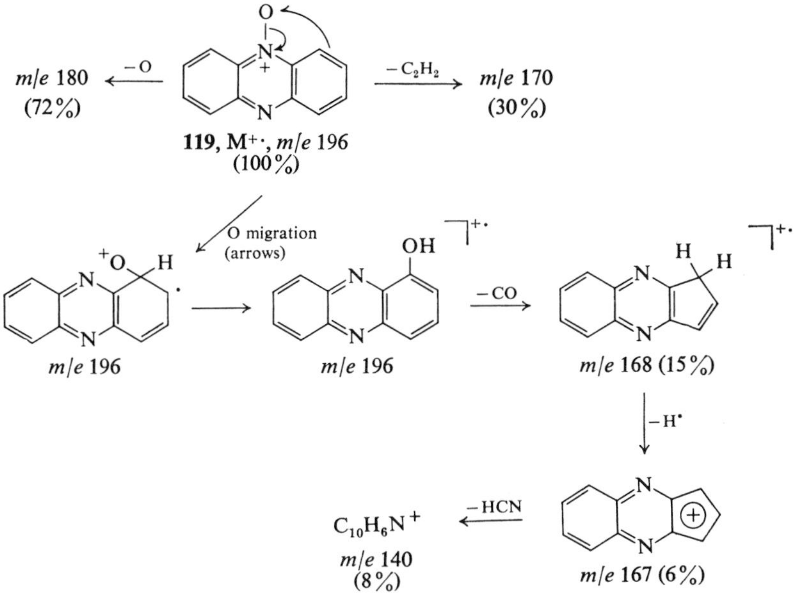

m/e 180 (72%) ← $-O$ — **119, M$^{+\cdot}$, m/e 196 (100%)** — $-C_2H_2$ → m/e 170 (30%)

m/e 196 → O migration (arrows) → m/e 196 → $-CO$ → m/e 168 (15%)

m/e 168 (15%) → $-H^\cdot$ → m/e 167 (6%)

m/e 167 (6%) → $-HCN$ → $C_{10}H_6N^+$ m/e 140 (8%)

from a skeletal rearrangement of the molecular ion. Similar rearrangements occur when benzo[a]phenazine-12-oxide (**120**), phenazine-di-N-oxide (**121**) and benzo[a]phenazine-di-N-oxide (**122**) are subjected to electron impact (21). The di-N-oxides display two pathways involving rearrangement, namely M → M-CO and M → M-O → M-O-CO.

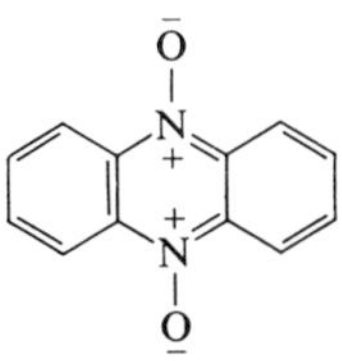

121 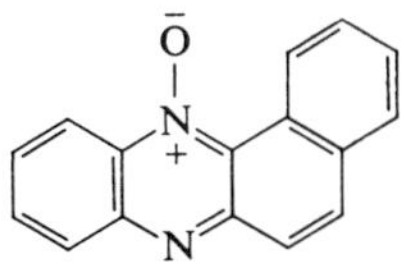   **120** 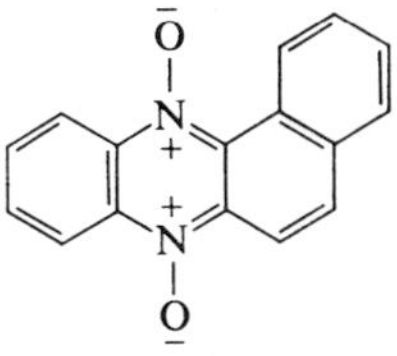   **122**

VII.   FIVE- AND SIX-MEMBERED HETEROCYCLES WITH THREE NITROGEN ATOMS IN THE RING

A.   Triazoles

No data on the fragmentation of simple triazoles appears to have been reported. The triazolino-quinuclidine derivative **123** appears to lose the ethano-bridge as first fragmentation (57) although no high-resolution measurements have been reported to substantiate this assignment.

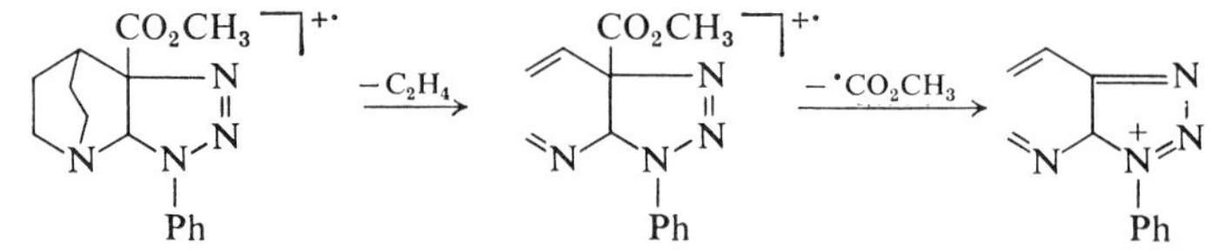

123, M⁺·, *m/e* 286   *m/e* 258   *m/e* 191

The triazolopyrazine **124** is reported to lose the fragment CN_2 from the molecular ion to give the base peak of the spectrum (58). This may well be either an acyclic ion such as **124a** or the pyrazine formed by recyclization (**124b**), but much more work would be required to make definite structural assignments to the ion formed.

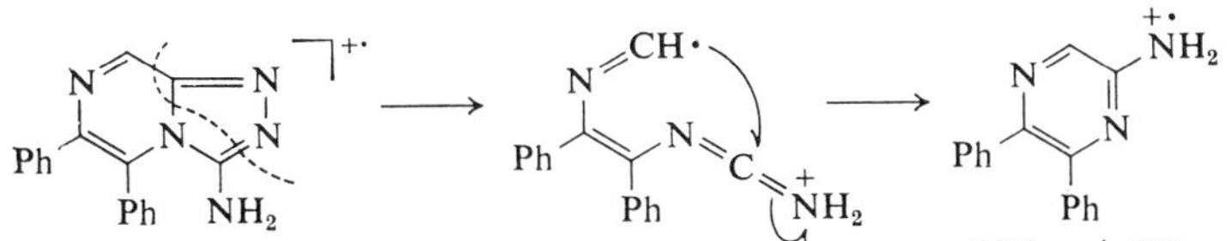

124, M⁺·, *m/e* 287   **124a**, *m/e* 247 (100%)   **124b**, *m/e* 247

B.   1,2,4-Triazines

1,2,4-Triazine (**125**) fragments by initial loss of nitrogen, followed by loss of HCN, and this and minor decomposition pathways are indicated in the rationale shown (59).

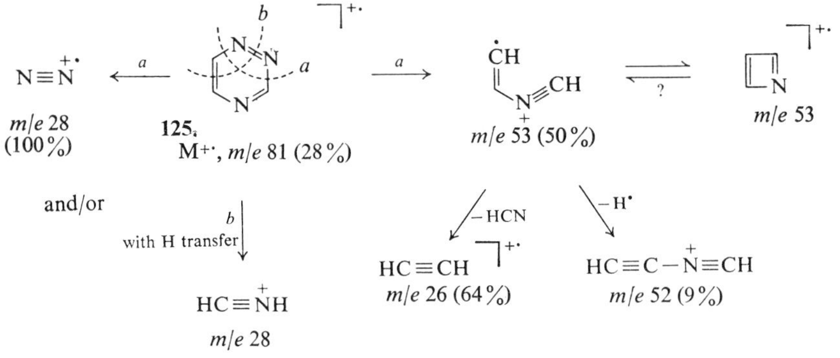

$N\equiv N$ *m/e* 28 (100%)

125, M⁺·, *m/e* 81 (28%)

m/e 53 (50%)

m/e 53

and/or

with H transfer

$HC\equiv \overset{+}{N}H$ *m/e* 28

−HCN

$HC\equiv CH$ *m/e* 26 (64%)

−H·

$HC\equiv C-\overset{+}{N}\equiv CH$ *m/e* 52 (9%)

The appropriate mass shifts occur in the spectrum of 3-deutero-1,2,4-triazine, and thus significant hydrogen randomization between the hydrogen at $C_{(3)}$ and the remaining two hydrogen atoms does not occur.

The mass spectra of a number of esters of 1,2,4-triazine-3-carboxylic acid have also been reported. Their fragmentations are straightforward, and the original paper should be consulted for further details (59).

C. 1,3,5-Triazines

The base peak of the mass spectrum of 1,3,5-triazine (126) is the molecular ion, and as expected the only fragmentation involves successive losses of

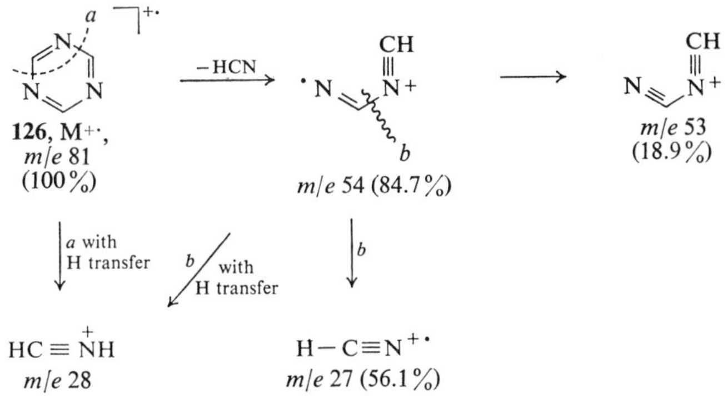

HCN molecules (60). Weak ions at m/e 24 and 25 (C_2 and C_2H) may demonstrate minor skeletal rearrangements to an unsymmetrical triazine, but it is clear from the distinctive spectra of 125 and 126 that no major equilibrium between these species is set up in their respective molecular ions although this might well have been expected to occur via the intermediacy of triaza-benzvalenes and triazaprismanes.

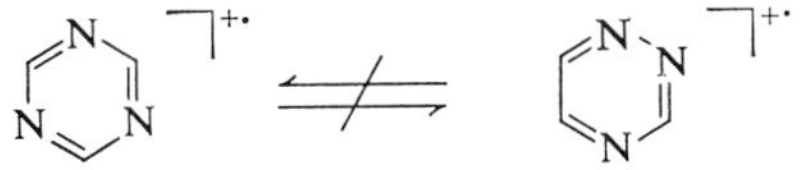

Mass spectra of the triazine herbicides simazine (127) and propazine (128) have been reported (61). Major fragmentation in both compounds involves the substituents rather than the ring and is exemplified by the rationale for the higher-mass peaks in the spectrum of 127.

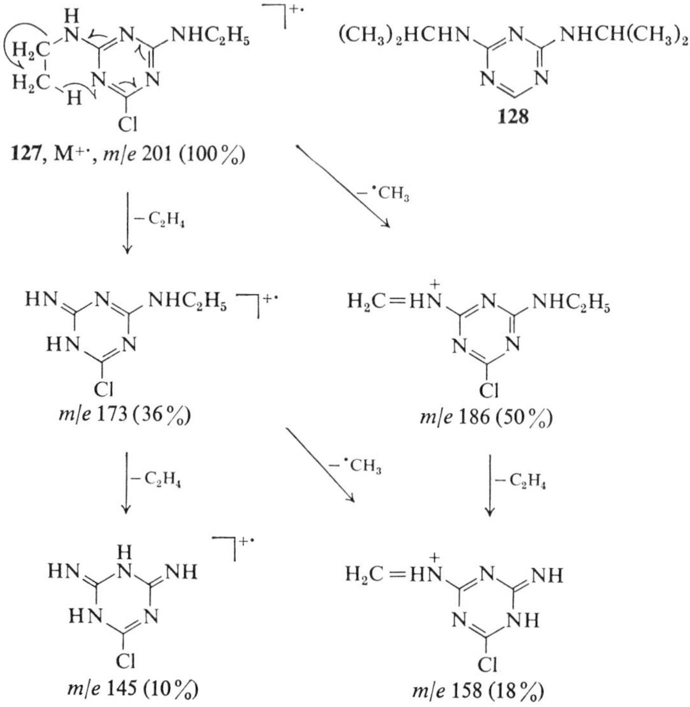

127, $M^{+\cdot}$, m/e 201 (100%)

128

m/e 173 (36%)

m/e 186 (50%)

m/e 145 (10%)

m/e 158 (18%)

The mass spectra of a series of hexahydro-1,3,5-triazines of the general formula **129** have been reported (62). The observed spectra are complicated by depolymerization which occurs under reduced pressure with the establishment of equilibria of the type shown.

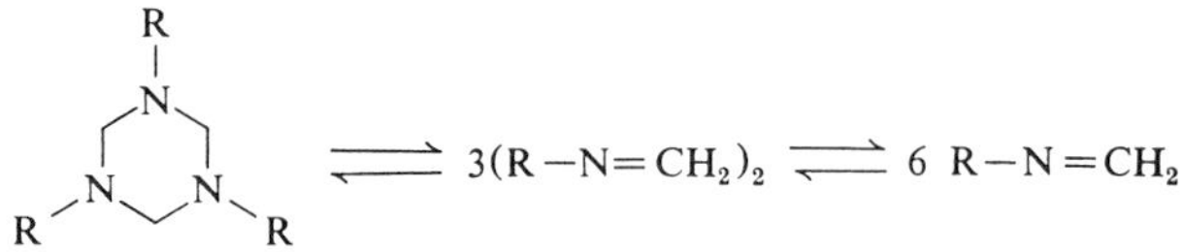

$$\rightleftharpoons \; 3(R-N=CH_2)_2 \; \rightleftharpoons \; 6 \; R-N=CH_2$$

129

All the hexahydrotriazines reported showed M-1 ions that were considerably more intense than the molecular ions on account of formation of a well-delocalized amidinium cation. It will be recalled (Sec. V.B) that a similar situation obtains with the hexahydropyrimidines. A rationale for the formation of the principal ions in the spectrum of the trimethyl compound (**129**, R = CH$_3$) is shown. Noteworthy is the unusual loss of :CH$_2$ from the parent ion.

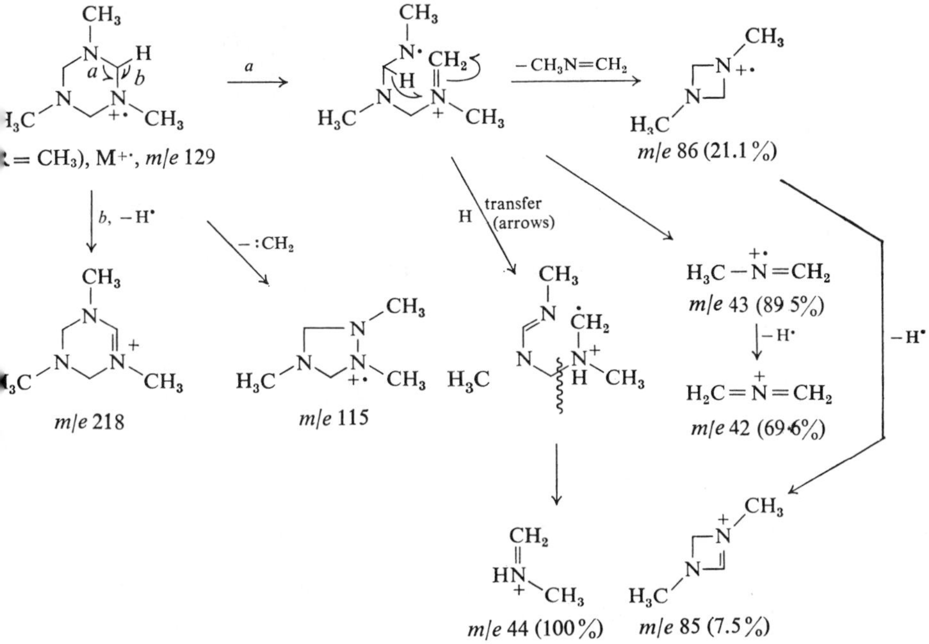

A group of ions at m/e 72, 73, and 74 (9.9, 22.4, and 43.0%) clearly require high-resolution data and metastable-ion information for their origins and structures to be elucidated.

When the substituents R contain two or more carbon atoms, elimination of alkene becomes an important process. Thus the triethyl compound (**129**, $R = C_2H_5$) has a relatively intense ion at M-28 in its mass spectrum.

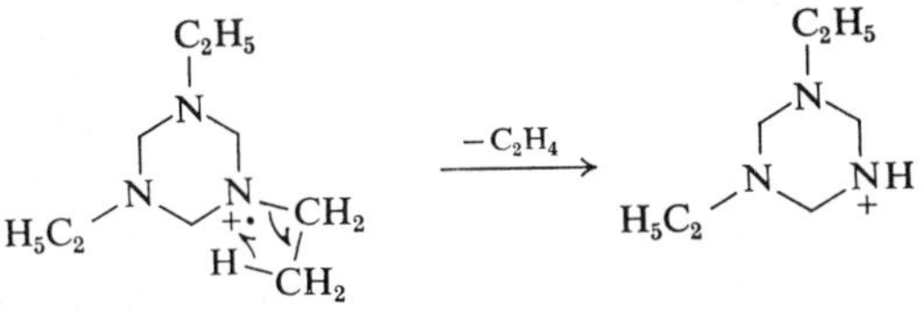

129 (R = C_2H_5), M$^{+\cdot}$, m/e 171 (26.1%) m/e 143 (21%)

VIII. SEVEN-MEMBERED AND LARGER HETEROCYCLES CONTAINING TWO NITROGEN ATOMS IN THE RING

A. Diazepines

The pentaphenyldiazepine **130** fragments by loss of the elements of benzonitrile to give an ion best represented as tetraphenylpyrrole (**130a**). Loss of N_2 is not an important process (63).

130, M+·   —PhCN→   130a

Dibenzo[c,f]-[1,2]-diazepin-11-one (**131**) is an interesting system since it allows a comparison of tendencies to lose the stable fragments N$_2$ and CO. In fact, N$_2$ is the initial unit lost, followed by CO (64).

131, M+·, m/e 208   m/e 180   m/e 152

B. Derivatives of 3,9-Diazabicyclo[4,2,1]nonan-4-one

The mass spectra of several derivatives of this ring system (**132**), which contains two nitrogen atoms in a seven-membered ring, have been examined in order to compare the effect on fragmentation patterns of the amine and amide functional groups in close proximity to one another. Most of the fragmentation can be explained in terms of charge localization on the amino function, as indicated by the rationale shown, based on analysis of the spectra of a number of deuterated derivatives of (**132**, R = CH$_3$) with ion compositions confirmed by high-resolution mass-spectrometry (65).

C. Diazacyclodecane Derivatives

The mass spectra of 6,10-dioxo-1,5-diazacyclodecane (**133**) and 5,10-dioxo-1,6-diazacyclodecane (**134**) are conveniently rationalized in terms of nitrogen-carbonyl transannular interactions (66). Thus loss of water from the weak molecular ion of **133** is apparently a very favored process initiating all other fragmentations. It is most easily explained by the sequence **133** → **133a** → **133b**.

The transannular reaction leading to the cyclol **133a** is the first step in acid-catalyzed decompositions of **133** and provides an interesting analogy between solution and electron-impact chemistry.

The isomer **134** also undergoes impact-induced dehydration, presumably via the bicyclic ion **134a**, but the process is not as important as with **133**. Formation of the base peak (m/e 86) may be initiated by the hydrogen transfer and acyl shift shown.

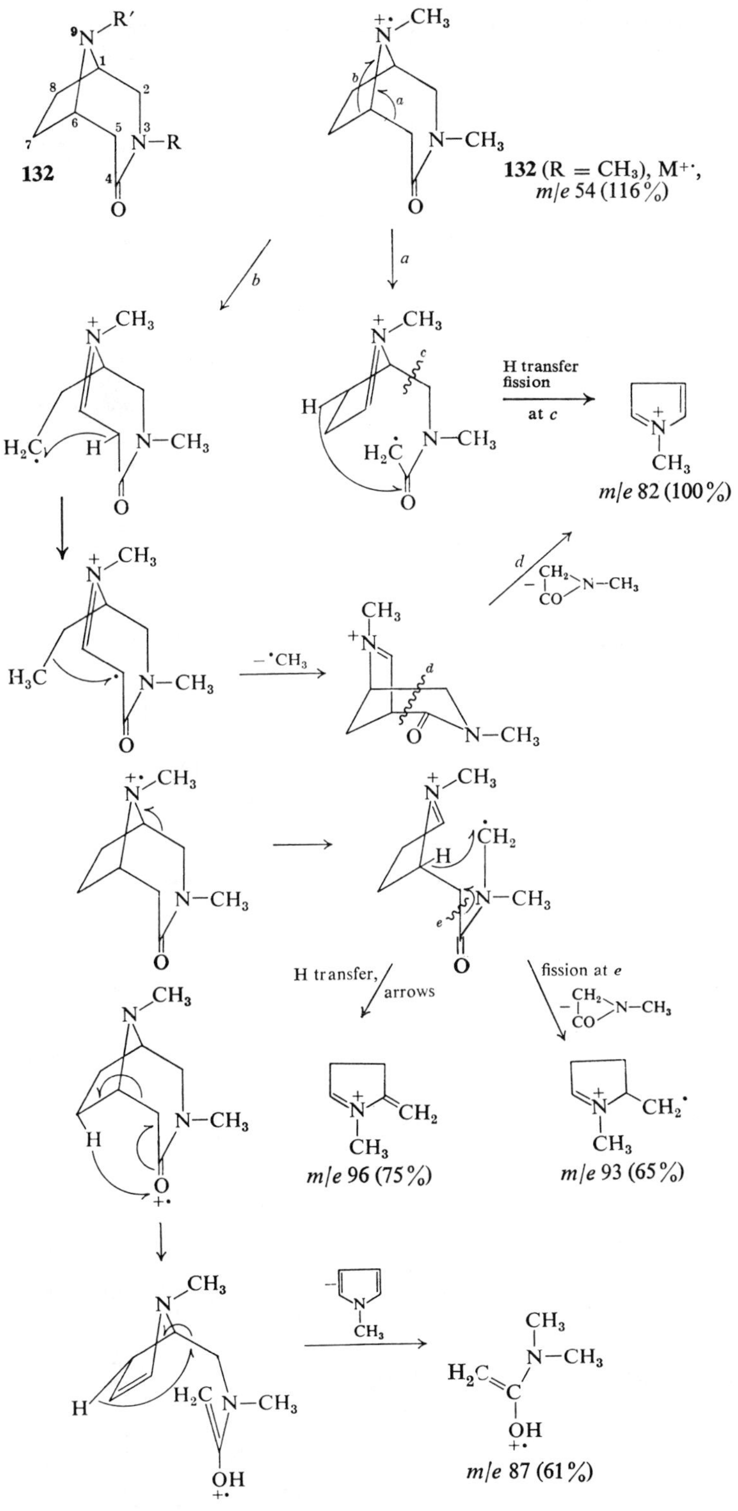

132

132 (R = CH₃), M⁺·, m/e 54 (116 %)

$\mathbf{132}$ (R = CH$_3$), M$^{+\cdot}$, m/e 54 (116 %)

H transfer fission at c

m/e 82 (100 %)

$-\,^{\bullet}$CH$_3$

H transfer, arrows

fission at e

m/e 96 (75 %)

m/e 93 (65 %)

m/e 87 (61 %)

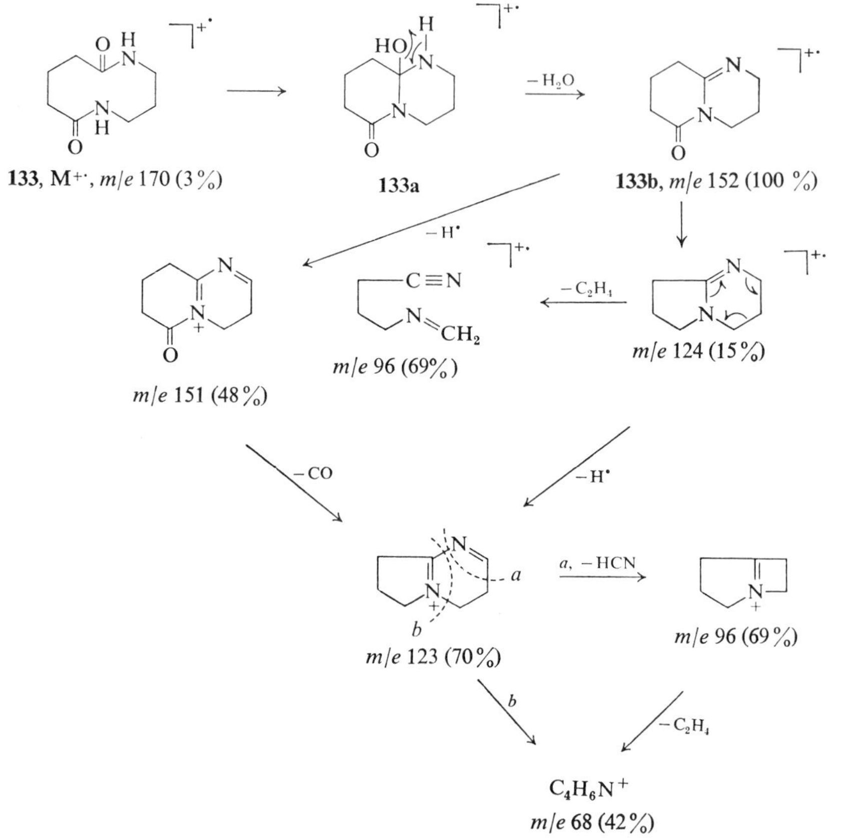

133, M⁺·, m/e 170 (3 %)　　　　**133a**　　　　**133b**, m/e 152 (100 %)

− H·

− C₂H₄

m/e 96 (69%)　　　m/e 124 (15%)

m/e 151 (48%)

− CO　　　　　− H·

a, − HCN

b

m/e 123 (70%)　　　m/e 96 (69%)

b　　　− C₂H₄

C₄H₆N⁺

m/e 68 (42%)

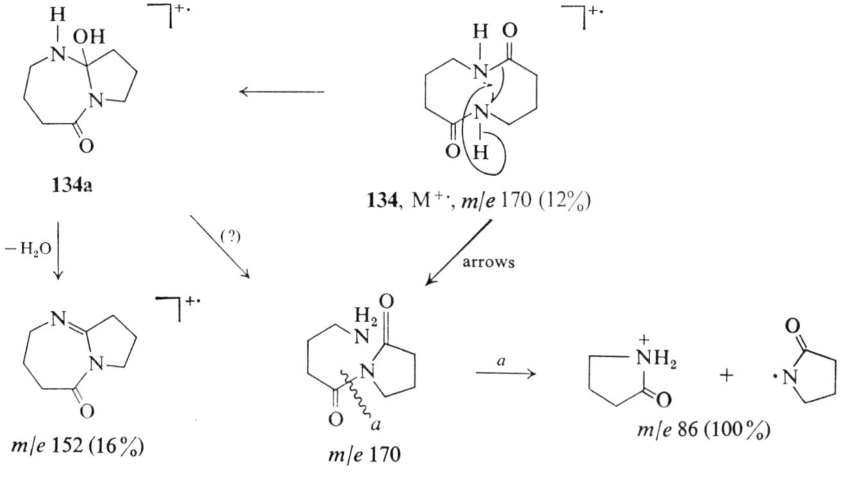

134a　　　　**134**, M⁺·, m/e 170 (12%)

− H₂O　　(?)　　　　arrows

a

m/e 152 (16%)　　　m/e 170　　　m/e 86 (100%)

It should be noted that the rationales presented for the dioxo-diazacyclodecanes **133** and **134** have not been supported by either deuterium-labeling or high-resolution measurements, and there is clearly scope for more work in this interesting field of transannular impact-induced reactions.

Addenda

Further work, including much labelling, helps to elucidate the skeletal rearrangements occurring in diphenylpyrazoles (Section II-A) (67) and more work has been done on pyrazolones (68) and pyrazolidines (69). Further studies on 2-alkylbenimidazoles suggests decomposition *via* quinoxalinium ions (Section III-C) (70), while the fragmentations of some thiazolo[3,2-*a*]-benzimidazoles have been discussed (71). Imidazolopyridines and -pyrimidines decompose by losses of HCN and H_2C_2N (72).

The fragmentations of 2H-cyclopenta[*d*] pyridazines (Section IV) are straightforward (73), and further work has been reported on cinnolines and their N-oxides (Section IV) (74). Further studies on uracil and thymine derivatives (Section V-A) have been published (75, 76), and the fragmentations of the dimeric photoproducts of biologically important pyrimidine derivatives have been analyzed (77). The mass spectra of the nucleoside antibiotics formycin, formycin-B, and showdomycin (which contain the pyrazolo[4,3-*d*] pyrimidine system) have been reported (78), as have those of a number of naturally occurring pteridines (Section V-G) (79). Several publications have appeared on derivatives of quinoxaline (Section VI-C) (80, 81, 82), and the fragmentations of some polyhydroxyphenazines have been discussed (82).

Work has also appeared on triazoles (Section VII-A) (83), tetrazoles (84, 85), 1,2,4-triazines (Section VII-B) (86), benzotriazinones (87), and 1,2,4,5-tetrazines (88).

REFERENCES

1. W. H. Graham, *J. Amer. Chem. Soc.*, **84**, 1063 (1962).
2. R. A. Mitsch, *J. Heterocycl. Chem.*, **1**, 59 (1964).
3. R. A. Mitsch, E. W. Neuvar, R. J. Koshar, and D. H. Dybvig, *J. Heterocycl. Chem.*, **2**, 371 (1965).
4. T. Nishiwaki, *J. Chem. Soc.*, B, **1967**, 885.
5. H.-B. Schröter, D. Neumann, A. R. Katritzky, and F. J. Swinbourne, *Tetrahedron*, **22**, 2895 (1966).
6. D. J. Blythin and E. S. Waight, *J. Chem. Soc.*, B, **1967**, 583.
7. S. Trofimenko, *J. Amer. Chem. Soc.*, **88**, 5588 (1966).
8. J. Desmarchelier and R. B. Johns, *Org. Mass Spectrosc.*, **2**, 37 (1969).
9. J. H. Bowie, R. G. Cooks, S.-O. Lawesson, and G. Schroll, *Aust. J. Chem.*, **20**, 1613 (1967).
10. M. Ohashi, N. Ohno, H. Kakisawa, A. Tatematsu, and H. Yoshizumi, *Org. Mass Spectrosc.*, **1**, 703 (1968).

11. J. T. Edwards and I. Lantos, *Can. J. Chem.*, **45**, 1925 (1967).
12. E. Cherbuliez, Br. Baehler, S. Jaccard, H. Jindra, G. Weber, G. Wyss, and J. Rabinowitz, *Helv. Chim. Acta*, **49**, 807 (1966).
13. S.-O. Lawesson, G. Schroll, J. H. Bowie, and R. G. Cooks, *Tetrahedron*, **24**, 1875 (1968).
14. E. Tatematsu, H. Yoshizumi, E. Hayashi, and H. Nakata, *Tetrahedron Lett.*, **1967**, 2985.
15. C. R. Ganellin, H. F. Ridley, and R. G. W. Spickett, *J. Heterocycl. Chem.*, **3**, 278 (1966).
16. J. W. Clark-Lewis, J. A. Edgar, J. S. Shannon, and M. J. Thompson, *Aust. J. Chem.*, **17**, 877 (1964).
17. S. R. Johns and J. A. Lamberton, *Aust. J. Chem.*, **20**, 555 (1967).
18. M. H. Benn, T. S. Sorensen, and A. M. Hogg, *Chem. Commun.*, **1967**, 574.
19. M. M. Bursey, R. D. Rieke, T. A. Elwood, and L. R. Dusold, *J. Amer. Chem. Soc.*, **90**, 1557 (1968).
20. J. H. Bowie, R. G. Cooks, P. F. Donaghue, J. A. Halleday, and H. J. Rodda, *Aust. J. Chem.*, **20**, 2677 (1967).
21. J. H. Bowie, R. G. Cooks, N. C. Jamieson, and G. E. Lewis, *Aust. J. Chem.*, **20**, 2545 (1967).
22. J. H. Elad and C. J. Danby, *J. Chem. Soc.*, **1965**, 5935.
23. J. H. Bowie, G. E. Lewis, and J. A. Reiss, *Aust. J. Chem.*, **21**, 1233 (1968).
24. J. M. Rice, G. O. Dudek, and M. Barber, *J. Amer. Chem. Soc.*, **87**, 4569 (1965).
25. T. Nishiwaki, *Tetrahedron*, **22**, 3117 (1966).
26. T. Nishiwaki, *Tetrahedron*, **23**, 1153 (1967).
27. R. F. Evans and J. S. Shannon, *J. Chem. Soc.*, **1965**, 1406.
28. R. F. Evans, *Aust. J. Chem.*, **20**, 1643 (1967).
29. A. Costopanagiotis and H. Budzikiewicz, *Monatsh. Chem.*, **96**, 1800 (1965).
30. H.-F. Grützmacher and W. Arnold, *Tetrahedron Lett.*, **1966**, 1365.
31. K. Biemann and J. A. McCloskey, *J. Amer. Chem. Soc.*, **84**, 2005 (1962).
32. T. J. Batterham, A. C. K. Triffett, and J. A. Wunderlich, *J. Chem. Soc., B*, **1967**, 892.
33. R. R. Arndt, S. H. Eggers, and A. Jordaan, *Tetrahedron*, **23**, 3521 (1967).
34. S. C. Pakrashi, J. Bhattacharyya, L. F. Johnson, and H. Budzikiewicz, *Tetrahedron*, **19**, 1011 (1963).
35. J. S. Fitzgerald, S. R. Johns, J. A. Lamberton and A. H. Redcliffe, *Aust. J. Chem.*, **19**, 151 (1966).
36. W. Walter and J. Voss, *Justus Liebigs Ann. Chem.*, **698**, 113 (1966).
37. T. Goto, A. Tatematsu, and S. Matsuura, *J. Org. Chem.*, **30**, 1844 (1965).
38. J. S. Shannon and D. S. Letham, *New Zealand J. Sci.*, **9**, 833 (1966).
39. D. S. Letham, J. S. Shannon, and I. R. McDonald, *Proc. Chem. Soc.*, **1964**, 230.
40. D. S. Letham, J. S. Shannon, and I. R. C. McDonald, *Tetrahedron*, **23**, 479 (1967).
41. G. Spiteller and M. Spiteller-Friedmann, *Monatsh. Chem.*, **93**, 632 (1962).
42. M. Chaigneau, G. Valdener, and J. Seyden-Penne, *Compt. Rend. Acad. Sci. Paris*, **1965**, 3965.
43. S. H. Eggers, S. T. Biedron, and A. O. Hawtrey, *Tetrahedron Lett.*, **1966**, 3271.
44. T. Goto, A. Tatematsu, and S. Matsuura, *J. Org. Chem.*, **30**, 1844 (1965).
45. D. J. Brown and S. F. Mason, *J. Chem. Soc.*, **1956**, 3443.
46. American Petroleum Institute, Research Project 44. Spectrum 1417.
47. K. Biemann, *Mass Spectrometry*, McGraw-Hill, New York, 1962, pp. 184–186.
48. Y. Kishi, T. Goto, S. Eguchi, Y. Hirata, E. Watanabe, and T. Aoyama, *Tetrahedron Lett.*, **1966**, 3437.
49. R. F. C. Brown, W. D. Crow, and R. K. Solly, *Chem. Ind.* (London), **1966**, 343.

50. I. Flament and M. Stoll, *Helv. Chim. Acta*, **50**, 1754 (1967).

51. Ref. 46. Spectrum 618.

52. R. A. Saunders and A. E. Williams in *Advances in Mass Spectrometry*, Vol. 3, (M. L. Mead, Ed.), The Institute of Petroleum, London, 1966, p. 681.

53. N. S. Vul'fson, V. A. Puchkov, Yu. V. Denisov, B. V. Rozynov, V. N. Bochkarev, M. M. Shemyakin, Yu. A. Ovchinnikov, and V. K. Antonov, *Khim. Geterotsikl. Soedin*, **1966**, 614.

54. R. M. Acheson and M. W. Foxton, *J. Chem. Soc.*, *C*, **1966**, 2218.

55. Y. Morita, *Chem. Pharm. Bull.*, **14**, 426 (1966).

56. H. Budzikiewicz, C. Djerassi, and D. H. Williams, *Mass Spectrometry of Organic Compounds*, Holden-Day, San Francisco, 1967, p. 584.

57. W. A. Remers, G. J. Gibs, and M. J. Weiss, *J. Heterocycl. Chem.*, **4**, 344 (1967).

58. S. E. Mallett and F. L. Rose, *J. Chem. Soc.*, *C*, **1966**, 2038.

59. W. A. Paudler and R. E. Herbener, *J. Heterocycl. Chem.*, **4**, 224 (1967).

60. C. M. Judson, R. J. Francel, and J. A. Weicksel, *J. Chem. Phys.*, **22**, 1258 (1954).

61. J. Jörg, R. Houriet, and G. Spiteller, *Monatsh. Chem.*, **97**, 1064 (1966).

62. E. Schumacher and R. Taubenest, *Helv. Chim. Acta*, **49**, 1439 (1966).

63. M. A. Battiste and T. J. Barton, *Tetrahedron Lett.*, **1967**, 1227.

64. R. G. Amiet and R. B. Johns, *Aust. J. Chem.*, **20**, 723 (1967).

65. A. M. Duffield, C. Djerassi, L. Wise, and L. A. Paquette, *J. Org. Chem.*, **31**, 1599 (1966).

66. G. I. Glover, R. B. Smith, and H. Rapoport, *J. Amer. Chem. Soc.*, **87**, 2003 (1965).

67. B. K. Simons, R. K. M. R. Kallury, and J. H. Bowie, *Org. Mass Spec.*, **2**, 739 (1969).

68. J. M. Desmarchelier and R. B. Johns, *Org. Mass Spec.*, **2**, 697 (1969).

69. S. W. Tam, *Org. Mass Spec.*, **2**, 729 (1969).

70. T. Nishiwaki, *J. Chem. Soc. [C]*, 428 (1968).

71. H. Ogura, T. Itoh, and K. Kikuchi, *J. Het. Chem.*, **6**, 797 (1969).

72. W. W. Paudler, J. E. Kuder and L. S. Helmick, *J. Org. Chem.*, **33**, 1397 (1968).

73. D. M. Forkey, *Org. Mass Spec.*, **2**, 309 (1969).

74. M. H. Palmer, E. R. R. Russell, and W. A. Wolstenholme, *Org. Mass Spec.*, **2**, 1265 (1969).

75. J. Ulrich, T. Téoule, R. Massot, and A. Cornu, *Org. Mass Spec.*, **2**, 1183 (1969).

76. R. W. Reiser, *Org. Mass Spec.*, **2**, 467 (1969).

77. C. Fenselau and S. Y. Wang, *Tetrahedron*, **25**, 2853 (1969).

78. L. B. Townsend and R. K. Robins, *J. Het. Chem.*, **6**, 459 (1969).

79. J. A. Blair and C. D. Foxall, *Org. Mass Spec.*, **2**, 923 (1969).

80. C. W. Koch and J. H. Markgraf, *J. Het. Chem.*, **7**, 235 (1970).

81. S. Skujins and G. A. Webb, *Tetrahedron*, **25**, 3955 (1969).

82. S. Skujins, J. Delderfield, and G. A. Webb, *Tetrahedron*, **25**, 3947 (1969).

83. N. E. Alexandrou and E. D. Micromastoras, *Tetrahedron Lett.*, 231 (1968).

84. D. M. Forkey and W. R. Carpenter, *Org. Mass Spec.*, **2**, 433 (1969).

85. K. F. Bach, J. Karliner, and G. E. Van Lear, *Chem. Commun.*, 1110 (1969).

86. T. Sasaki, K. Minamoto, M. Nishikawa, and T. Shima, *Tetrahedron*, **25**, 1021 (1969).

87. J. C. Tou, L. A. Shadoff, and R. H. Rigterink, *Org. Mass Spec.*, **2**, 355 (1969).

88. P. Yates, O. Meresz, and L. S. Weiler, *Tetrahedron Lett.*, 3929 (1968).

13 COMPOUNDS WITH NITROGEN AND OXYGEN ATOMS IN THE SAME RING

I. OXAZOLES

The spectrum of oxazole (**1**) is shown in Fig. 13.1 (1). The group of ions at m/e 40, 41, and 42 corresponds to the usual losses of CHO·, CO (or less likely ·CH$_2$N), and HCN from the molecular ion, which may be represented as shown in the scheme.

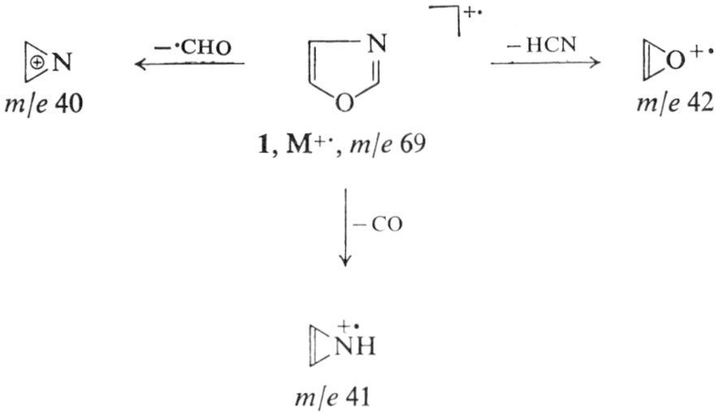

$$\begin{array}{ccc} & \xleftarrow{-\cdot\text{CHO}} & \xrightarrow{-\text{HCN}} \\ m/e\ 40 & \mathbf{1, M^{+\cdot}},\ m/e\ 69 & m/e\ 42 \end{array}$$

$$\Big\downarrow -\text{CO}$$

$m/e\ 41$

The spectra of 2,4- (**2**) and 4,5-dimethyloxazole (**3**) show marked differences (1) and are more akin to the isomeric dialkylpyridines [which show distinct spectra (2)] than to the isomeric xylenes and ethyl-methylbenzenes [which show virtually identical spectra (2–4)]. Of interest is the much more intense M-1 ion (12%) observed with the 4,5-dimethyl isomer. This presumably involves loss of a hydrogen atom from the 5-methyl group and is best represented as **3a/3b**. It will be noticed that there is a close similarity to the formation of M-1 ions from the isomeric methylpyridines and of M-15 ions from their ethyl analogs (Chap. 11, Sec. II.B).

510

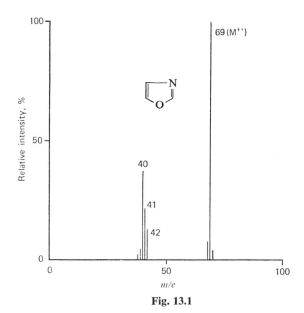

Fig. 13.1

There is apparently a marked tendency for $C_{(5)}$ and $O_{(1)}$ to be lost together as evidenced by the intense CH_3CO^+ (m/e 43) and $C_3H_4N^+$ (m/e 54) ions in the spectrum of **3** and their relative weakness in the spectrum of **2**. In the latter compound loss of CO and ˙CHO is a more marked process than in **3**, further supporting this thesis. The scheme presents a rationale for the major ions in the spectra of **2** and **3**.

Interesting observations have been made concerning the mass spectra of n-hexyloxazoles. Three important fragmentations are observed: (*a*) β-fission, (*b*) β-fission with hydrogen migration (McLafferty rearrangement), and (*c*) γ-fission. These processes are best exemplified by consideration of the appropriate ions in the spectra of 4,5-dimethyl-2-n-hexyloxazole (**4**), 2,5-dimethyl-4-n-hexyloxazole (**5**), and 2,4-dimethyl-5-n-hexyloxazole (**6**).

It will be seen that in **4** and **5**, where there is a nitrogen atom as migration terminus, the McLafferty rearrangement is the preferred fragmentation mode and β-fission is of reduced importance. With **6** the combination of a less-favored McLafferty rearrangement [similar in importance to that observed with n-butylbenzene (5) or 2-n-pentylfuran (6)] and a less-destabilized carbonium ion (m/e 111) renders β-fission the major fragmentation process.

γ-Fission has already been encountered in the spectra of 2-alkylpyridines and -quinolines (Chap. 11, Sec. II.B, III.B), and cyclization (**4a** → **4b**) may also explain the importance of the process here, although if this is so it is surprising that γ-fission is not more important in the fragmentation of **5**,

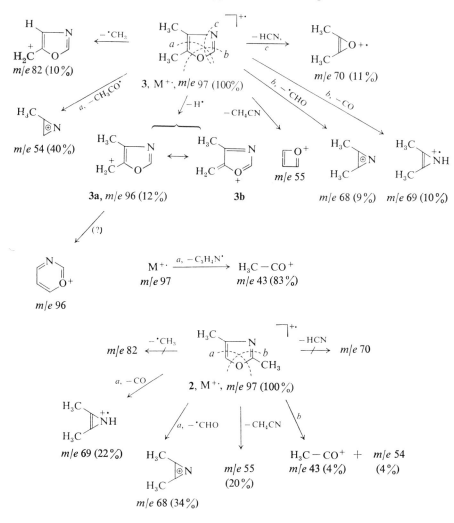

5a cyclizing to 5b. Perhaps the alternative explanation which has been advanced (1) [allylic fission of the tautomeric form (4c) of the molecular ion, 4] provides the best rationale, since 5 cannot form a suitable allylic tautomer.

2- and 4-Phenyloxazole (7 and 8) have similar spectra and in both there is sequential elimination of CO, then HCN (1), as presented in the rationale written for the observed spectra. 4,5-Diphenyloxazole (9) decomposes by a similar sequence to produce the well-stabilized cation of m/e 165, and the rationale shown has been presented (7). 2,5-Diphenyloxazole (10) decomposes

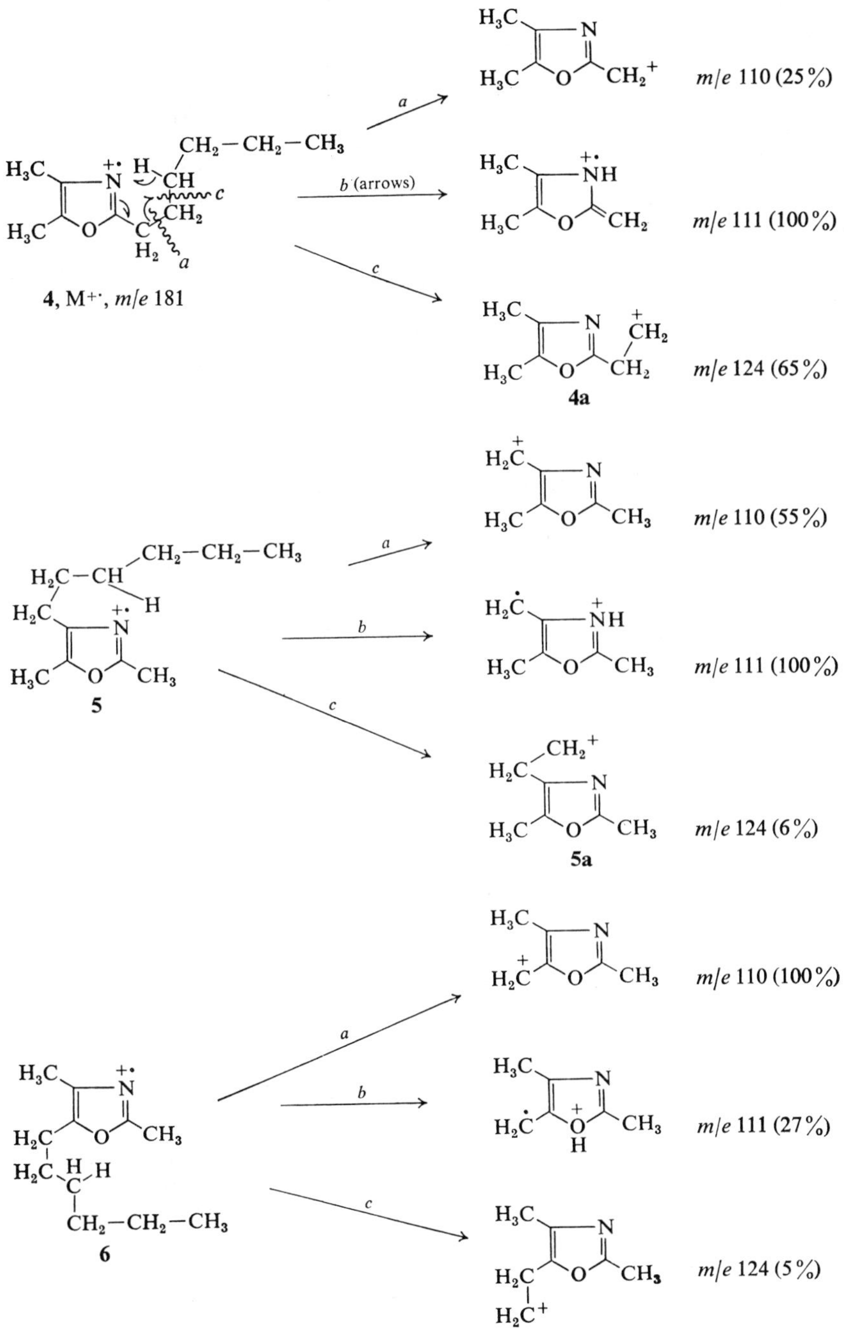
a
b (arrows)
c
4, M+·, m/e 181
m/e 110 (25%)
m/e 111 (100%)
m/e 124 (65%)
4a
5
m/e 110 (55%)
m/e 111 (100%)
m/e 124 (6%)
5a
6
m/e 110 (100%)
m/e 111 (27%)
m/e 124 (5%)

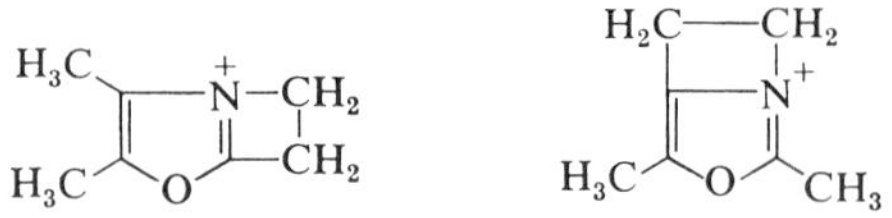

4b

5b

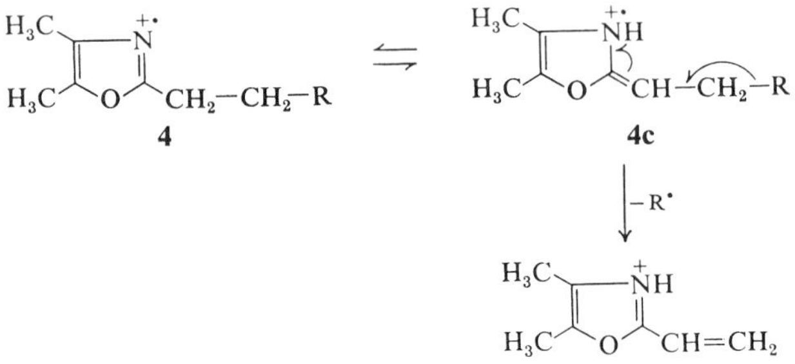

4 ⇌ **4c**

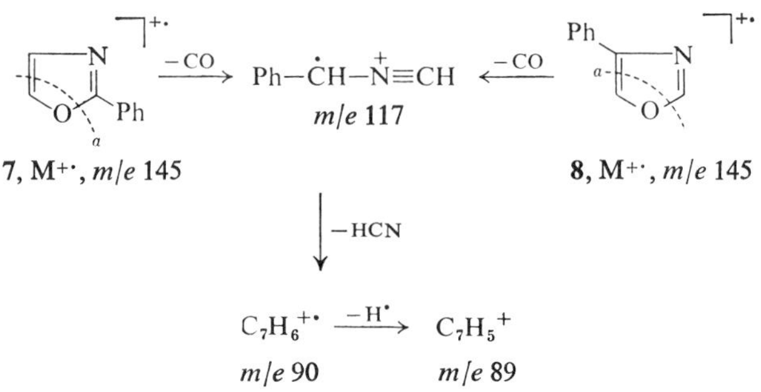

7, M⁺·, *m/e* 145   Ph—ĊH—⁺N≡CH   **8**, M⁺·, *m/e* 145
m/e 117

$$\text{C}_7\text{H}_6^{+\cdot} \xrightarrow{-\text{H}^\cdot} \text{C}_7\text{H}_5^+$$
m/e 90   *m/e* 89

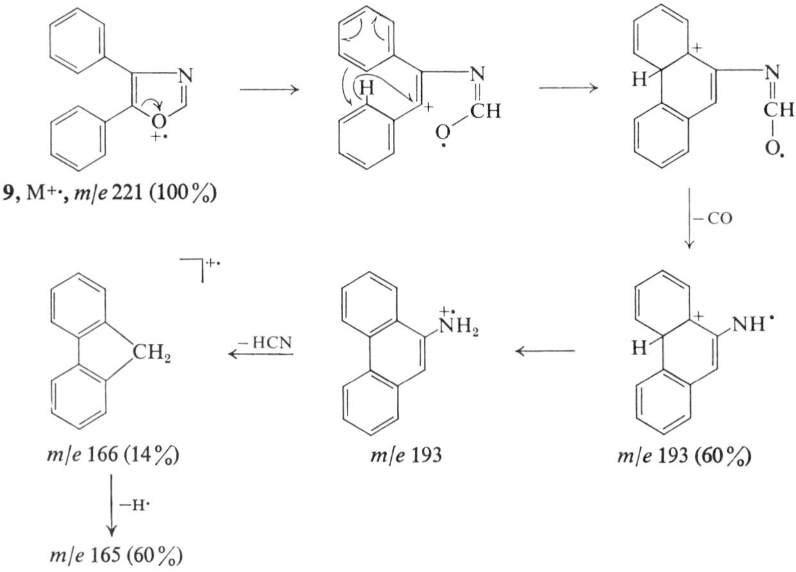

9, M⁺·, *m/e* 221 (100%)

m/e 166 (14%)   *m/e* 193   *m/e* 193 (60%)

m/e 165 (60%)

by concerted elimination of CO and HCN. This process clearly requires much bond-reorganization, and the rationale shown has been suggested (7).

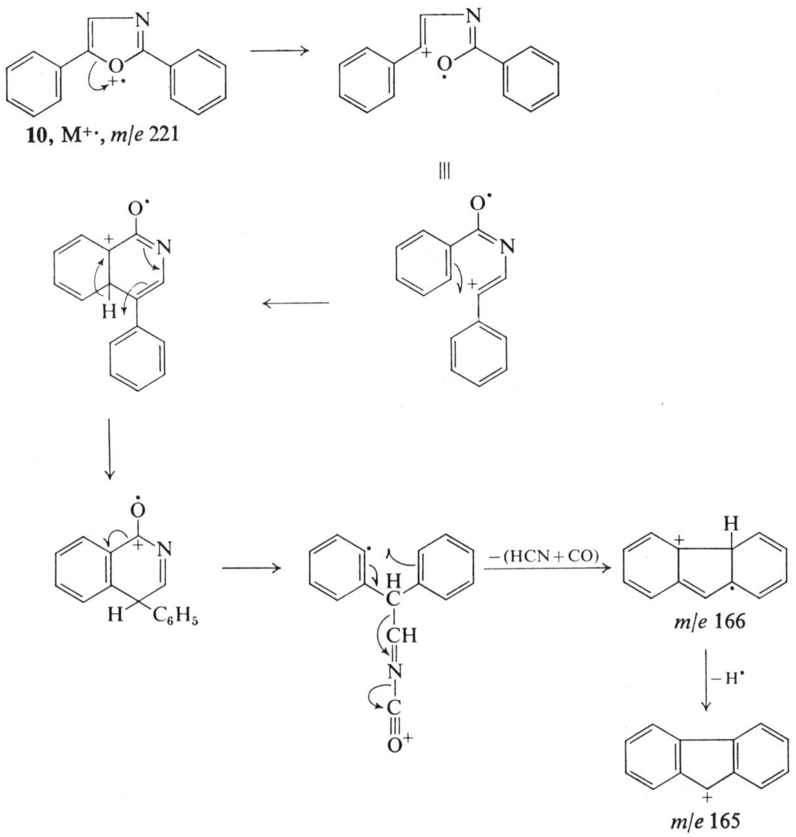

10, M$^{+\cdot}$, m/e 221

m/e 166

$-$ H$^{\cdot}$

m/e 165

Similar pathways have been used to explain the sequential losses of CO and HCN observed in 2,4-diphenyloxazole, and the pathways elucidated for the various diaryloxazoles have been used to confirm the structure of halfordinol (**11**), the dequaternization product of the alkaloid *N*-methyl-halfordinium chloride (**12**) (7).

The only dihydro-oxazoles apparently subjected to electron impact are the isomeric dimethyl-2-oxazolines **13** and **14** (8). Their spectra are shown in Figs. 13.2 and 13.3. The marked differences in fragmentation can be rationalized in terms of the intact structures of the two compounds, and the scheme shown is postulated although deuterium-labeling and high-resolution studies would be required for a full understanding of the fragmentation.

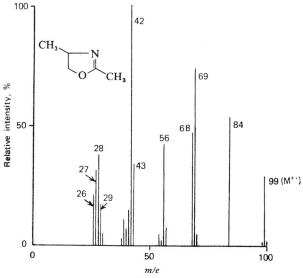

Fig. 13.2

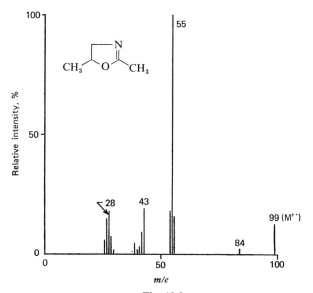

Fig. 13.3

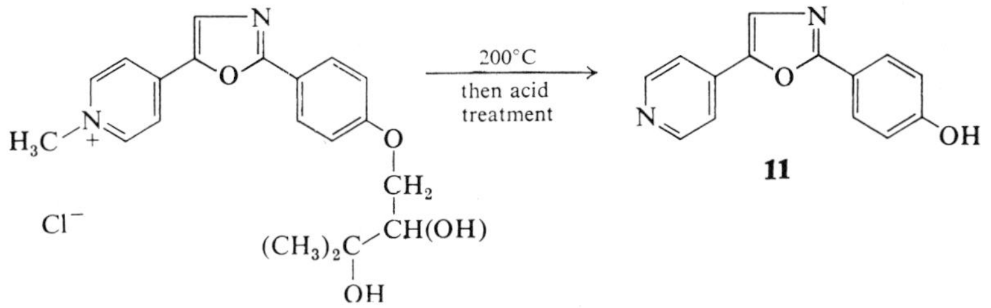

CH_3-CO^+
m/e 43

m/e 56

$CH_3-C\overset{+}{=}NH$
m/e 42

CH_3-CO^+
m/e 43

m/e 56

13, M$^{+\cdot}$, m/e 99

14, M$^{+\cdot}$, m/e 99

$-\cdot CH_3$

$b, -CH_2O$

$b, -CH_3CHO$

m/e 84

m/e 69

m/e 55

14a

$-H\cdot$

$-H\cdot$

m/e 68

m/e 54

14b

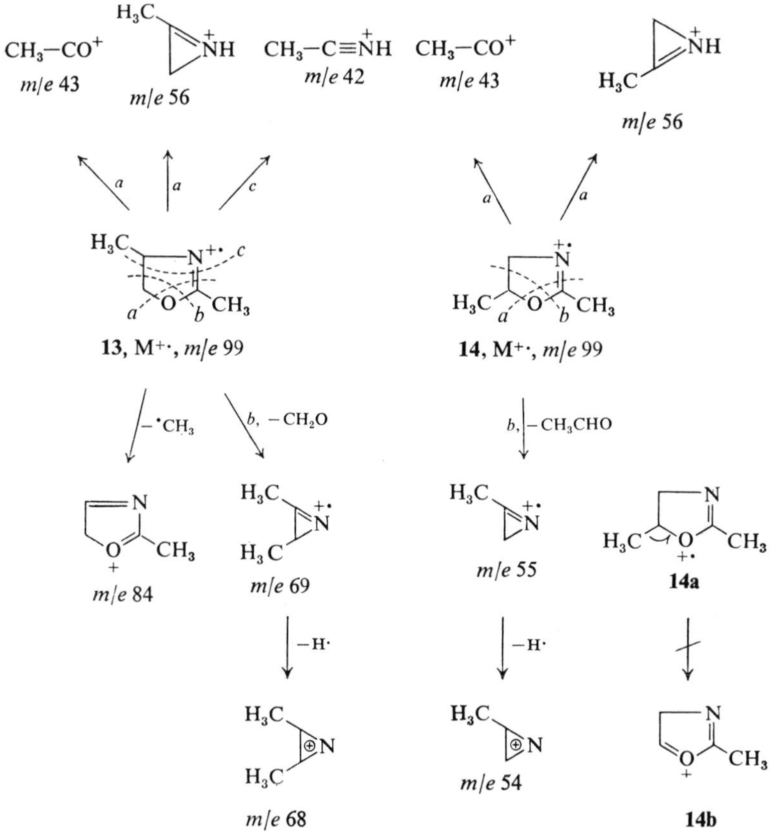

The weakness of the M-CH$_3^{\cdot}$ ion in the spectrum of **14** is interesting and implies that there is little localization of charge on the oxygen atom (**14a** $\nrightarrow$ **14b**).

N-Acetylisomuscazone (**15**) decomposes by elision of CO$_2$ followed by ketene (**9**), and the rationale shown seems likely.

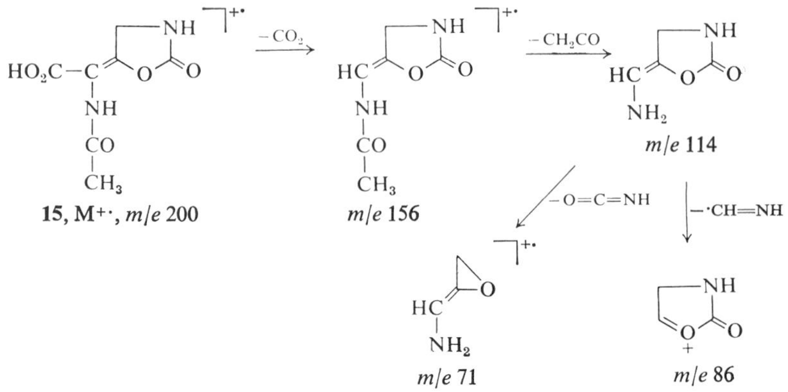

15, M+·, m/e 200 m/e 156 m/e 114

m/e 71 m/e 86

II. ISOXAZOLES

The mass spectrum of 3,5-diphenylisoxazole (**16**) is in some respects similar to that of the azirine **17** (10). The azirine is formed from the isoxazole photochemically· (11), and it has been suggested that the mass-spectrometric fragmentation follows a similar pathway, with the rationale shown.

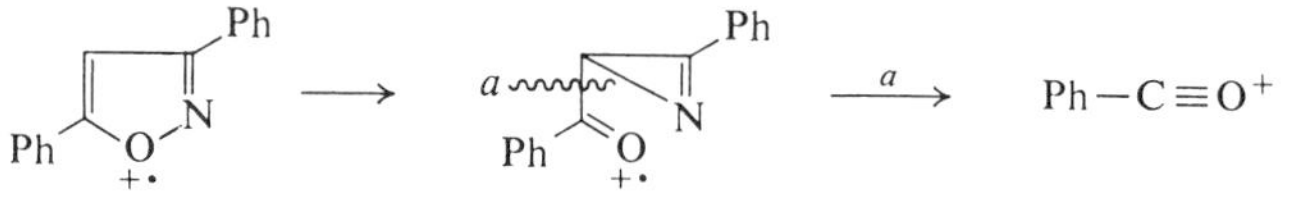

16, M+·, m/e 221   **17**, m/e 221 m/e 105 (100%)

The spectrum of the azirine **17** also contains relatively intense ions at m/e 165 and 166 which are absent from the spectrum of **16**. This has been explained by assuming conversion of the azirine to 2,5-diphenyloxazole (**10**) which, as we have already seen, produces strong hydrocarbon ions in its mass spectrum at m/e 165 and 166.

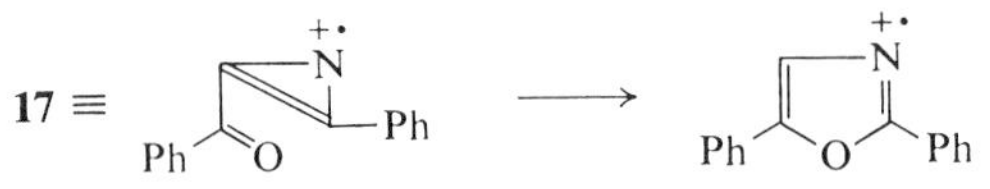

10, m/e 221

The azirine, in fact, rearranges photochemically to **10** if 2537 Å light is used, while with light at >3000 Å photochemical equilibrium with the isoxazole **16** is set up.

The mass spectra of several cycloserine (4-amino-3-isoxazolidone) derivatives have been discussed (12). The *N*-acetyl compound (**18**) decomposes initially by loss of a fragment of 32 mass units (to give m/e 112). This is an unusual fragment and high-resolution measurements show it to be NH_2O. The same fragment is lost from the acetyl-d_3 compound, limiting the origins of the hydrogen atoms involved to positions 2, 4, and 5. 2-Methyl-*N*-acetylcycloserine (**19**) loses the fragment $CH_3 \cdot NHO$, and these observations lead to the rationale shown for the formation of the *m/e* 112 species.

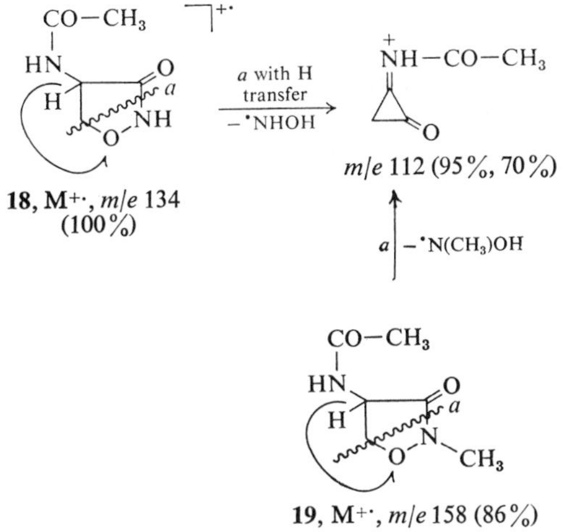

Loss of ketene from the molecular ion of **18** produces only a weak ion at *m/e* 102, but the low abundance of this species is apparently due to its rapid conversion to *m/e* 59. A study of the spectrum of the acetyl-d_3 analog of **18**, together with the metastable ions present in the spectrum, suggest the sequence shown.

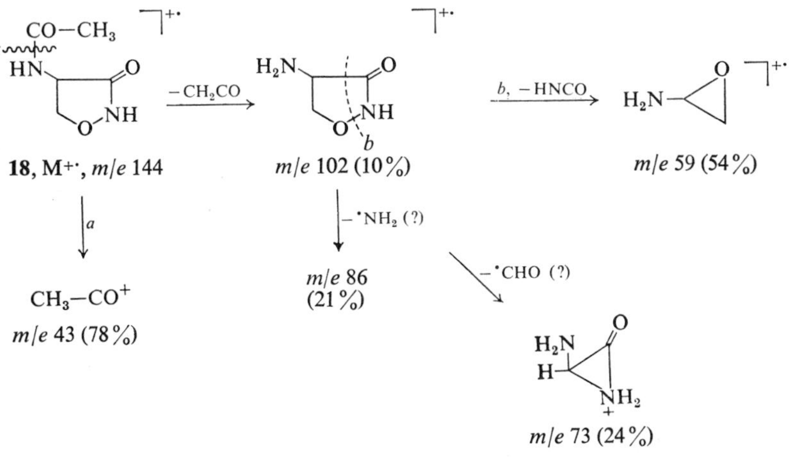

N,2-Diacetylcycloserine does not give a molecular ion but readily eliminates ketene to give **18** whereas 2-methyl-N-acetylcycloserine (**19**) fragments in a generally similar manner to **18**. The original publication should be consulted for further detail.

III. MORPHOLINES

Although only the low-resolution spectrum of morpholine (**20**) is available (13), it is clear that the simple spectrum of this compound can be rationalized by assuming charge localization on nitrogen, with α-cleavage as the dominant fragmentation mode.

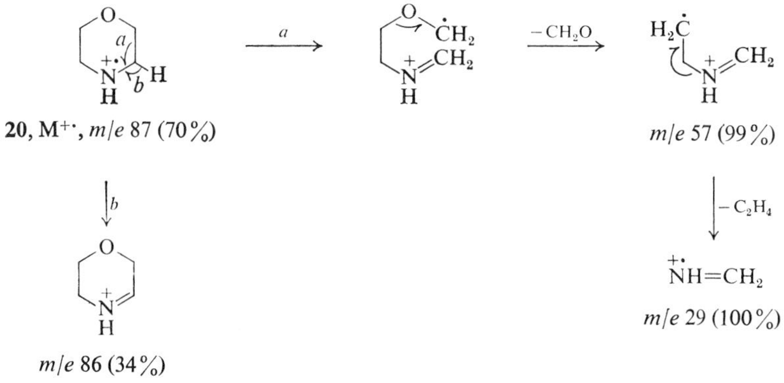

20, M$^{+\cdot}$, m/e 87 (70%)

m/e 57 (99%)

m/e 86 (34%)

$\overset{+\cdot}{N}H=CH_2$

m/e 29 (100%)

The spectrum (Fig. 13.4) (14) of N-methylmorpholine (**21**) shows that similar internal α-cleavage initiates the fragmentation, and here the resultant ion probably decomposes in two ways to give daughter ions at m/e 71 and 43.

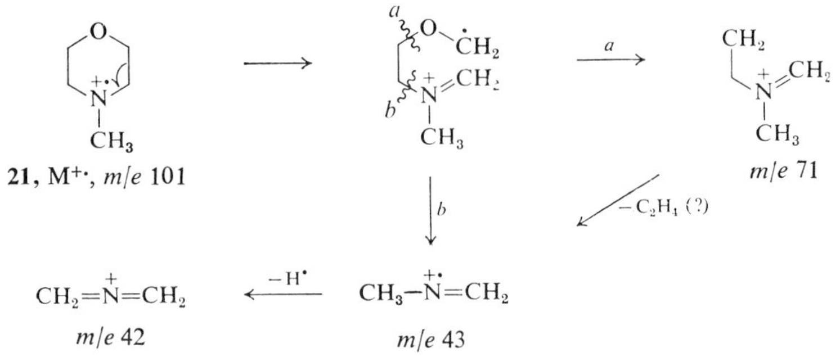

21, M$^{+\cdot}$, m/e 101

m/e 71

$CH_2=\overset{+}{N}=CH_2$

m/e 42

$\xleftarrow{\;-H^\cdot\;}$

$CH_3-\overset{+\cdot}{N}=CH_2$

m/e 43

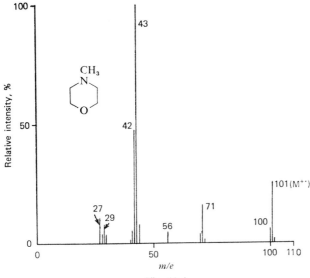

Fig. 13.4

The spectrum of N-ethylmorpholine (15) can be rationalized in a similar way. Here α-cleavage of the ethyl group becomes an important process and leads to a strong M-'CH$_3$ ion (m/e 100, 54%).

The mass spectrum of N-acetylmorpholine (**22**) has been subjected to close scrutiny by means of deuterium-labeling and high-resolution techniques (16). Most interesting is the genesis of the M-CH$_3$' ion which does not have the expected structure **22a**; it results from loss of C$_{(3)}$ and its attached hydrogen atoms together with one hydrogen originally bonded to C$_{(2)}$. The rationale shown has been suggested. The M-43 peak corresponds to elimination of the N-acetyl group in one step without hydrogen transfer between the ionic species produced and the CH$_3$CO radical eliminated, and its formation may well be initiated by the species **22c** as shown. The base peak of the spectrum occurs at m/e 57; the formation of this ion from **22b** (loss of CH$_2$CO then CH$_2$O) and its subsequent conversion to m/e 56 are processes parallel to some already discussed for the parent compound. The m/e 43 ion (73%) is almost entirely CH$_3$CO$^+$ derived from the nitrogen substituent, with only small contributions from C$_2$H$_5$N$^{+\cdot}$ and C$_3$H$_7{}^+$.

The published spectrum of 4-(N-morpholino)-4-hydroxymethyl-n-heptane (**23**) (17) again shows initial fissions of the same basic types discussed for the simpler morpholines although the occurrence of a hydrogen transfer in the decomposition of the internal α-fission ion (process c) should be noted.

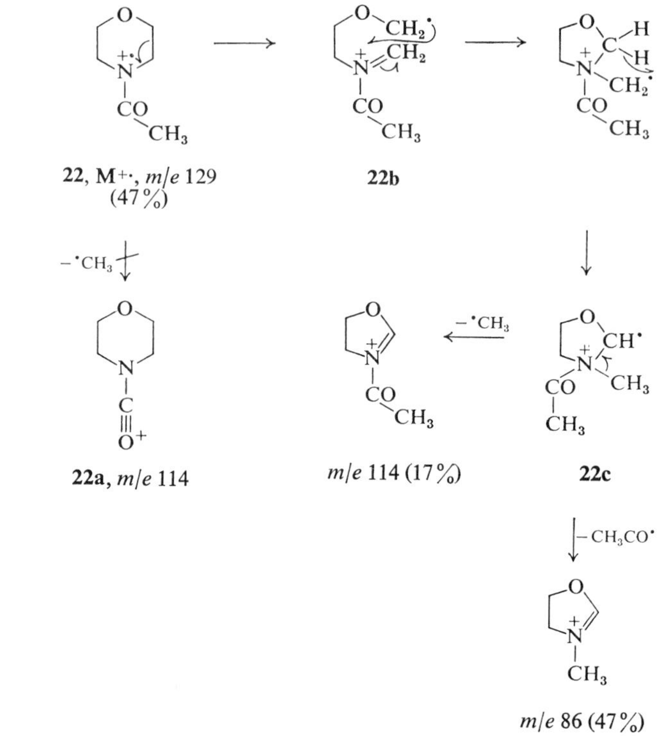

22, M+·, m/e 129 (47%)
22b
− ·CH3
22a, m/e 114
− ·CH3
m/e 114 (17%)
22c
− CH3CO·
m/e 86 (47%)

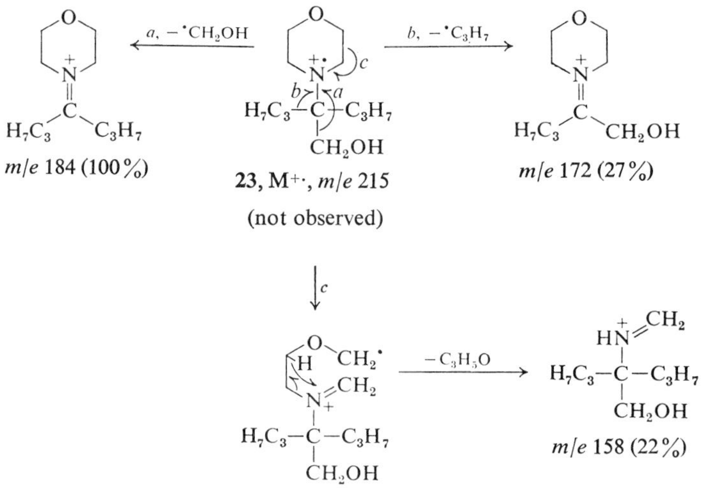

a, − ·CH2OH
b, − ·C3H7
m/e 184 (100%)
23, M+·, m/e 215
(not observed)
m/e 172 (27%)
c
− C3H5O
m/e 158 (22%)

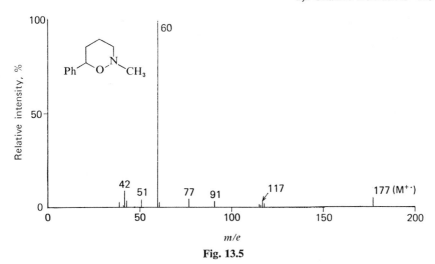

Fig. 13.5

IV. 1,2-OXAZINE DERIVATIVES

A number of 6-aryl-2-methyl-tetrahydro-1,2-oxazines have been prepared by thermal rearrangement of cyclic *N*-oxides (18).

The mass spectrum of the 6-phenyl compound (**24**, Ar = C_6H_5) is shown in Fig. 13.5, and it will be seen that the only intense fragment ion occurs at m/e 60 (C_2H_6NO). This ion is, in fact, the base peak of all the 6-aryl (or heteroaryl) compounds of type **24** examined, and deuterium labeling in the 5-position (of the 6-pyridyl compound) produces a shift to m/e 61. The rationale shown has been produced to explain the genesis of the ion.

Weak ions at *m/e* 115, 116, 117, and 118 in the spectrum of the phenyl compound (**24**, Ar = C_6H_5) result from initial α-cleavage without rearrangement.

24, Ar = C_6H_5, $M^{+\cdot}$ *m/e* 118 *m/e* 117 *m/e* 115

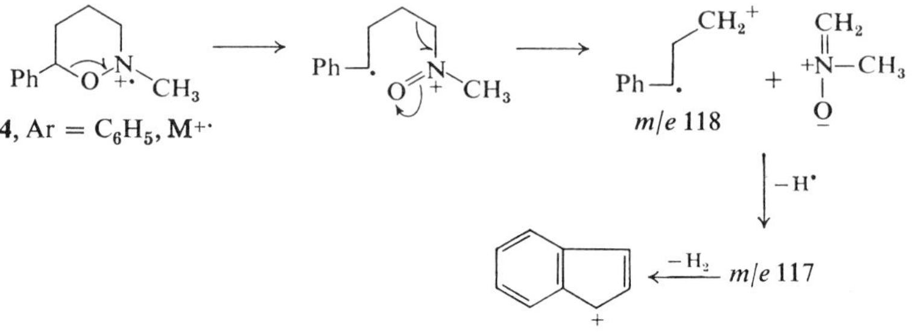

This fragmentation is much more important in the spectrum of the 6-(3-pyridyl) compound (**24**, Ar = C_5H_4N) although the species produced does not eventually cyclize to an aza-indenyl cation.

24, Ar = $3C_5H_4N$, $M^{+\cdot}$, *m/e* 178 (12%) *m/e* 118 (50%) *m/e* 119 (95%)

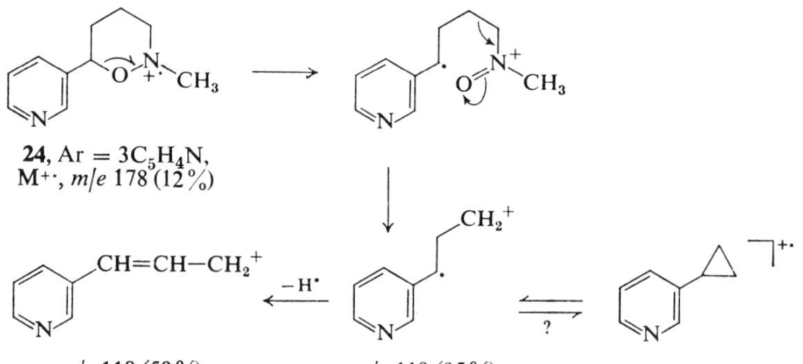

Since it lacks the structural requirements for the hydrogen transfer step the dihydro-oxazine (**25**) does not show an intense ion at *m/e* 60, and its main fragmentation involves loss of the neutral fragment CH_3NO to give *m/e* 131. The intensity of the *m/e* 130 ion would be surprising if it had the structure **25a** (with an unfavored electron sextet on an sp^2-hybridized carbon atom), and it is possible that cyclization and hydrogen transfer have produced the quinolinium cation **25b**.

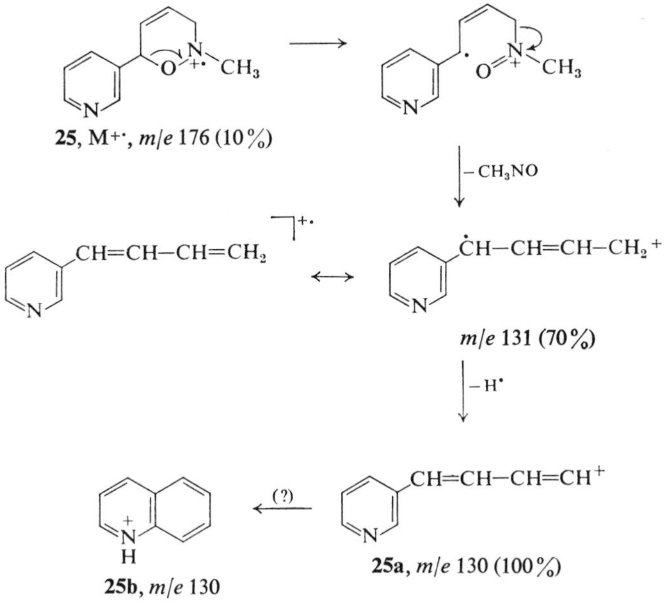

25, M$^{+\cdot}$, m/e 176 (10%)

$-CH_3NO$

m/e 131 (70%)

$-H^\cdot$

25a, m/e 130 (100%)

25b, m/e 130

3,6-Dihydro-2-trifluoromethyl-1,2-oxazine (**26**) has been prepared by the Diels-Alder reaction of butadiene with trifluoronitrosomethane. Apart

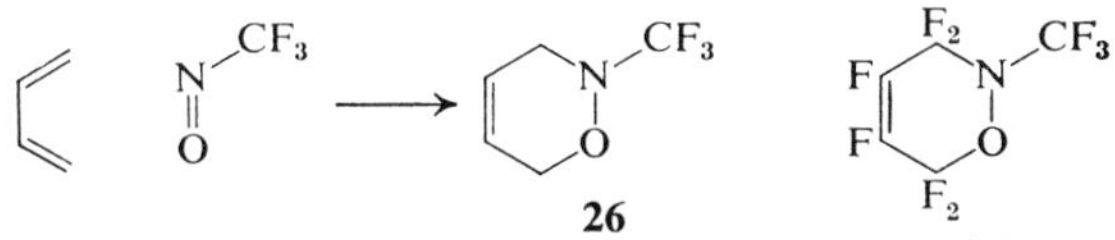

26

27

from the CF$_3^+$ ion (m/e 69, 47.4%), the mass spectrum is dominated by the retro-Diels-Alder reaction and shows no resemblance to that of the oxazines previously discussed (19). Similar behavior is observed with the fully fluorinated analog **27** (20).

A. Seven-Membered Rings Containing Heterocyclic Oxygen and Nitrogen

The seven-membered ring analogs of the 1,2-oxazines **24** have been prepared in a similar manner, and their mass-spectrometric behavior parallels that of their six-membered ring analogs (18). For a full discussion the reader is referred to the original paper, but the rationale for the major fragments is shown for the 7-(3-pyridyl)-derivative **28**.

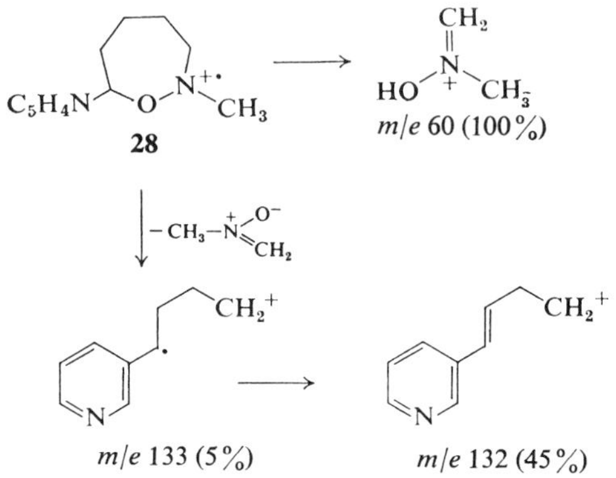

$$m/e\ 133\ (5\%) \qquad\qquad m/e\ 132\ (45\%)$$

V. COMPOUNDS WITH TWO NITROGEN ATOMS AND ONE OXYGEN ATOM PER RING

A.   Furazans (1,2,5-Oxadiazoles)

The parent compound (**29**) has only recently been prepared and its mass spectrum and that of its dideutero-derivative reported. They are described in the rationale shown (21).

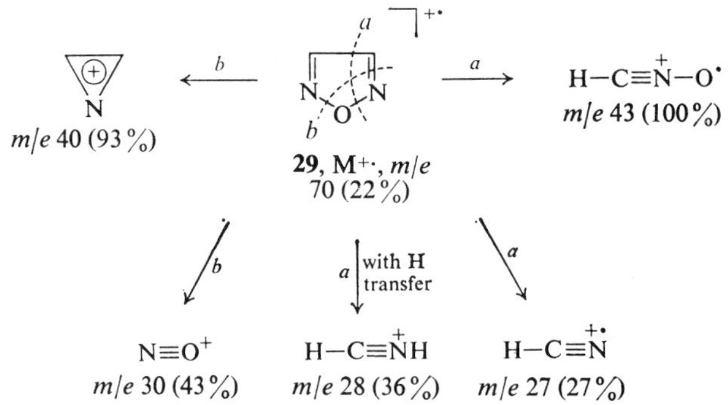

The mass spectrum of the dimethyl derivative **30** (22) can be rationalized on similar lines although the relative intensities of the ions differ considerably. The base peak of the spectrum occurs at m/e 27 and is presumably due to the ion $C_2H_3^+$.

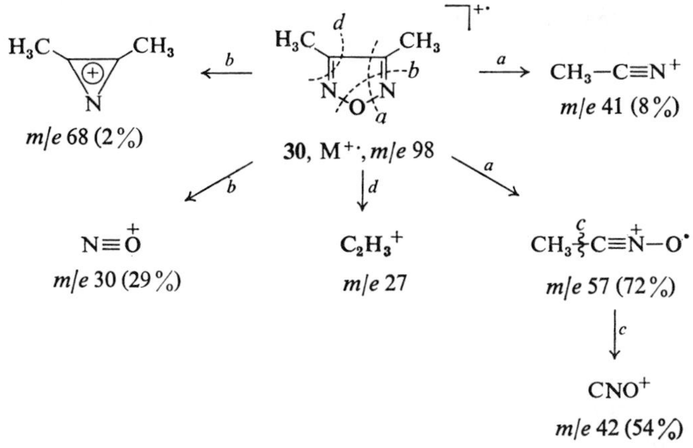

Unlike most N-oxides **13** shows no M-16 ion, and in fact most of its fragmentation may be explained by assuming impact-induced isomerization

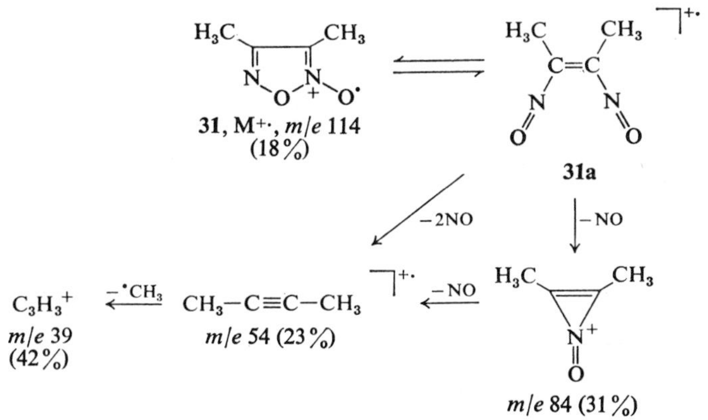

to the dinitroso compound **31a**. The presence of two NO groups in the molecular ion is perhaps responsible for the base peak of the spectrum being the NO$^+$ ion (m/e 30).

The mass spectra of several halo-, halo-alkyl-, and bis-perfluoroalkyl-furazans have also been described, and the original paper should be consulted for more detail (22).

Although the full spectrum of benzofuroxan oxide (**32**) has not been reported, like other quinonoid compounds it shows a marked tendency to produce an (M + 2) ion when introduced into the mass spectrometer by means of a heated inlet. This is characteristic of quinonoid compounds of high redox potential, and the "extra" hydrogen atoms probably originate from adsorbed water (23).

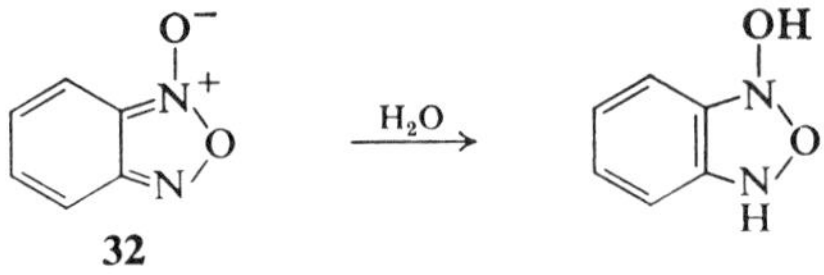

B. 1,3,4-Oxadiazole

This compound (33) is much more stable to electron impact than furazan and the molecular ion is the base peak (24). If, as seems probable, the major fragment ion (m/e 42) has the composition CH_2N_2, the rationale shown seems likely.

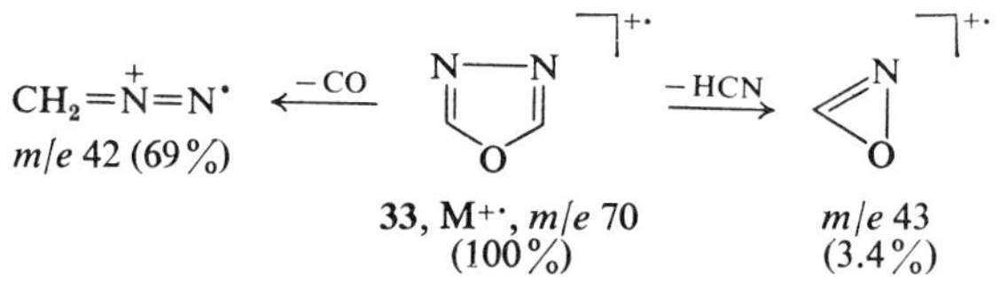

Only major peaks in the spectrum of 2,5-diphenyl-1,3,4-oxadiazole (34) have been reported (25). Ions at m/e 77 and 105 clearly correspond to $C_6H_5^+$ and $C_6H_5CO^+$ and are of obvious origins, but those at m/e 165 and 166 presumably have the composition $C_{13}H_9$ and $C_{13}H_{10}$ and must involve an interesting phenyl migration. The rationale shown seems possible.

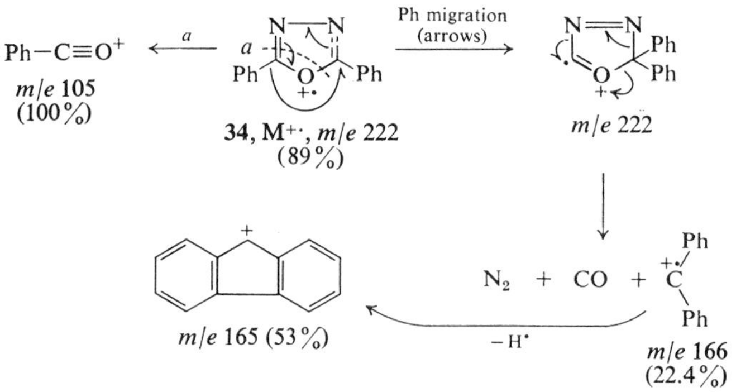

The mass spectra of three 2-phenyl-5-(perfluoroalkyl)-1,3,4-oxadiazoles (35, 36, and 37) have also been discussed. In each case the benzoyl cation (m/e 105) is a prominent ion together with the species $C_8H_5N_2O$ (m/e 145) resulting from loss of the perfluoroalkyl substituent. This ion may well have

have a non-cyclic structure as shown. Fission of the bond situated β to the oxadiazole ring is also a significant process (26).

$$\text{35 } (n = 1), \text{ 36 } (n = 2) \qquad \text{37}$$

Ph–C≡O⁺
(m/e 105)

m/e 195

m/e 145
(89%, 100%, 28·7%)

$$\begin{array}{cc} \text{N==N} \\ \text{Ph} & \text{O} & \text{CF}_2 \\ & + \end{array}$$
m/e 195

C₆H₅⁺ (–CO)
m/e 77

$$\text{Ph–C≡}\overset{+}{\text{N}}\text{–N=C=O} \xrightarrow{?} $$
m/e 145

35, 36, 37

$$\xrightarrow{-\,^{\bullet}\text{NCO}} \text{Ph–C≡N}^{+\bullet}$$
m/e 103
(22.5%, 16.6%, 38.8%)

C. 3,5-Diphenyl-1,2,4-oxadiazole

The fragmentation of this compound (38) may be rationalized as simple ring fission to give benzonitrile oxide (38a) and benzonitrile radical cations. If instead it is assumed that the m/e 119 species represents the phenylisocyanate radical cation (38b), a phenyl migration must be postulated (26), and it is difficult to accommodate the subsequent loss of 30 mass units to give m/e 89.

$$\text{Ph–C≡N}^{+\bullet} \xleftarrow{a} \quad \text{38, M}^{+\bullet} \text{ m/e 222} \quad \xrightarrow{a} \text{Ph–C≡}\overset{+}{\text{N}}\text{–O}^{\bullet}$$
m/e 103 (48.4%)   38a, m/e 119 (100%)

$$\downarrow \substack{-\,\text{HCN} \\ -\,\text{CN}} \qquad \qquad \downarrow -\,\text{NO}$$

C₆H₅⁺   C₆H₄⁺•   C₇H₅⁺
m/e 77   m/e 76   m/e 89 (12.9%)
(18.1%)   (22.5%)

Ph–N=C=O
38b, m/e 119

D. 1,2,3-Oxadiazoles: Sydnones

The mass spectra of several sydnones have been reported and are character-ized by losses of NO$^\cdot$ and CO. The rationale for the fragmentation of 3-phenylsydnone (39) is typical for this class of compound. In the mass spectrum

of 39 there is no evidence to show whether the losses of CO and NO$^\cdot$ are concerted or stepwise, but appropriate metastable ions have been observed for the stepwise loss in the 3-(p-anisyl)- and 3-(p-carboxymethylphenyl)-sydnones (27).

Addenda

Much work has been directed to elucidation of the skeletal rearrangements occurring in diaryl oxazoles (Section I) (20). The spectra of isoxazoles are very different from those of oxazoles, due to predominant O—N cleavage (Section II) (29, 30, 31). Skeletal rearrangements have been demonstrated in substituted isoxazoles (31), and some work has been reported on 5-alkoxy-isoxazoles and isoxazol-5-ones (32). A further study of the fragmentation of sydnones (Section V–D) has been reported (33).

REFERENCES

1. J. H. Bowie, P. F. Donaghue, H. J. Rodda, R. G. Cooks, and D. H. Williams, *Org. Mass Spectrosc.*, **1**, 13 (1968).
2. K. Biemann, *Mass Spectrometry*, McGraw-Hill, New York, 1962, p. 152.
3. P. N. Rylander, S. Meyerson, and H. M. Grubb, *J. Amer. Chem. Soc.*, **79**, 842 (1957).
4. S. Meyerson, *Appl. Spectrosc.*, **9**, 120 (1955).

5. H. Budzikiewicz, C. Djerassi, and D. H. Williams, *Mass Spectrometry of Organic Compounds*, Holden-Day, San Francisco, 1967, p. 82.

6. K. Heyns, R. Stute, and H. Scharmann, *Tetrahedron*, **22**, 2223 (1966).

7. W. D. Crow, J. H. Hodgkin, and J. S. Shannon, *Aust. J. Chem.*, **18**, 1441 (1965).

8. R. T. Lundquist and A. Ruby, *Appl. Spectrosc.*, **20**, 258 (1966).

9. R. Reiner and C. H. Eugster, *Helv. Chim. Acta*, **50**, 128 (1967).

10. H. Nakata, H. Sakurai, H. Yoshizumi and A. Tatematsu, *Org. Mass Spectrosc.*, **1**, 199 (1968).

11. E. F. Ullman and B. Singh, *J. Amer. Chem. Soc.*, **88**, 1844 (1966).

12. G. W. A. Milne and L. A. Cohen, *Tetrahedron*, **23**, 65 (1967).

13. J. H. Beynon, *Mass Spectrometry and its Applications to Organic Chemistry*, Elsevier, Amsterdam, 1960, p. 394.

14. American Petroleum Institute, Research Project 44. Spectrum 1462.

15. Ref. 14. Spectrum 1463.

16. J. M. Tesarek, W. J. Richter, and A. L. Burlingame, *Org. Mass. Spectrosc.*, **1**, 139 (1968).

17. W. Dörscheln, H. Tiefenthaler, H. Göth, P. Cerutti, and H. Schmid, *Helv. Chim. Acta*, **50**, 1759 (1967).

18. R. A. W. Johnstone, B. J. Millard, E. J. Wise, and W. Carruthers, *J. Chem. Soc.*, C, **1967**, 307.

19. R. E. Banks, M. G. Barlow, and R. N. Hazeldine, *J. Chem. Soc.*, **1965**, 4714.

20. R. E. Banks, M. G. Barlow, and R. N. Hazeldine, *J. Chem. Soc.*, **1965**, 6149.

21. R. A. Olofson and J. S. Michelman, *J. Org. Chem.*, **30**, 1854 (1965).

22. H. E. Ungnade and E. D. Loughran, *J. Heterocycl. Chem.*, **1**, 61 (1964).

23. R. T. Aplin and W. T. Pike, *Chem. Ind.* (London), **1966**, 2009.

24. C. Ainsworth, *J. Amer. Chem. Soc.*, **87**, 5800 (1965).

25. J. L. Cotter, *J. Chem. Soc.*, **1964**, 5491.

26. J. L. Cotter, *J. Chem. Soc.*, **1965**, 6842.

27. J. H. Bowie, R. A. Eade, and J. C. Earl, *Aust. J. Chem.*, **21**, 1665 (1968).

28. J. H. Bowie, P. F. Donaghue, H. J. Rodda, and B. K. Simons, *Tetrahedron*, **24**, 3965 (1968).

29. M. Ohashi, H. Kamachi, H. Kakisawa, A. Tatematsu, H. Yoshizumi, H. Kano, and H. Nakata, *Org. Mass Spec.*, **2**, 195 (1969).

30. M. Ohashi, H. Kamachi, H. Kakisawa, A. Tatematsu, H. Yoshizumi, and H. Nakata, *Tetrahedron Lett.*, 379 (1968).

31. J. H. Bowie, R. K. M. R. Kallury, and R. G. Cooks, *Austral. J. Chem.*, **22**, 563 (1969).

32. T. Nishiwaki, *Tetrahedron*, **25**, 747 (1969).

33. R. S. Goudie, P. N. Preston, and M. H. Palmer, *Org. Mass Spec.*, **2**, 953 (1969).

14 COMPOUNDS WITH NITROGEN AND SULFUR ATOMS OR WITH NITROGEN, OXYGEN, AND SULFUR ATOMS IN THE SAME RING

I. THIAZOLES

The molecular ion of thiazole (1) (Fig. 14.1) is the base peak of the spectrum and the only important fragmentation is loss of HCN to give the thiiren radical cation (m/e 58) (1). In the mass spectrum of thiazole-2-d_1 (2) the M-HCN peak is replaced almost entirely by an M-DCN peak (at both 20 and 70 eV), which confirms the involvement of $C_{(2)}$ in this process (2). The result also shows that hydrogen randomization, if it occurs, is a slow process compared with elision of HCN.

A less important fragmentation of 1 leads to the thioformyl cation (m/e 45).

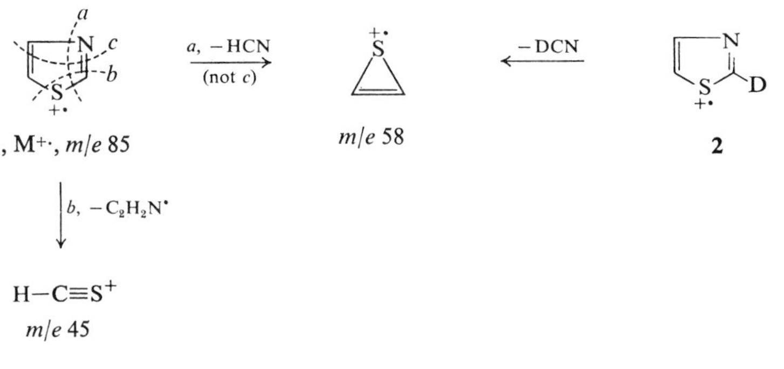

1, M$^{+\cdot}$, m/e 85 m/e 58   2

$$H-C{\equiv}S^+$$
m/e 45

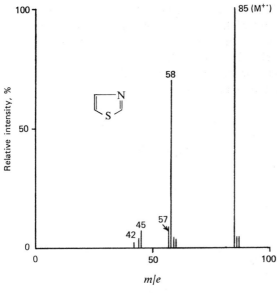

Fig. 14.1

As is expected from the fragmentation of **1**, 4-methylthiazole (**3**) does not lose CH₃CN but only HCN, whereas 2,4-dimethylthiazole (**4**) shows an intense M-CH₃CN ion. These basic fragmentations are displayed by a range of simple thiazoles, and the full paper should be consulted for further details (1).

3, M⁺·, m/e 99
(100%)

$-HCN \longrightarrow$

m/e 72 (40, 97%)

$\xleftarrow{-CH_3CN}$

4, M⁺·, m/e 113
(100%)

$-H·\downarrow$

m/e 71

$\longrightarrow$

m/e 71

$\downarrow -C_2H_2$

$CH{\equiv}S^+$

m/e 45

Ethyl 4-methylthiazole-2-carboxylate (5) shows an interesting fragmentation involving loss of the elements of acetaldehyde from the molecular ion (1). The process can be formulated as a six-center one involving hydrogen transfer, closely related to the McLafferty rearrangement, and the driving force is, no doubt, the presence of the electron-deficient nitrogen atom in the molecular ion 5 to act as hydrogen acceptor.

The mass spectrum of 2-benzoylthiazole (6) contains an intense M-CO ion (m/e 161) which must result from a 1,2-phenyl migration with elision of CO. The intensity of the corresponding rearrangement ion in the mass spectrum of benzophenone is 5%. Molecular rearrangement is also noticed in the spectrum of 2-(α-hydroxybenzyl)thiazole (7), and the reader is referred to the original publication for fuller details (1).

II. BENZOTHIAZOLES

The parent compound (8) gives a mass spectrum displaying an intense molecular ion (base peak) and the most important fragmentation is loss of HCN (3). The spectrum of the 2-d_1 derivative (9) shows predominant (92%) loss of DCN but loss of a significant amount (8%) of HCN. Determination of the metastable-peak intensities for the two losses shows that their ratio (2.5:1.0 at 70 eV) is very different to the ratio of daughter-ion intensities; thus there must be at least two processes with different rate constants for loss of HCN from the molecular ion of benzothiazole. It seems likely that

DCN is directly eliminated from the 2-d_1 derivative in a fast reaction giving an abundant daughter ion and weak metastable peak whereas HCN and DCN are eliminated in a ratio of 4:1, *after* hydrogen-deuterium scrambling, in a slow process giving rise to abundant metastable peaks (2). An actual mechanism for the scrambling reaction cannot yet be presented, but from structural considerations it cannot involve benzvalene- or prismane-type intermediates which have been considered as possible intermediates to explain scrambling in simple monocyclic aromatic compounds.

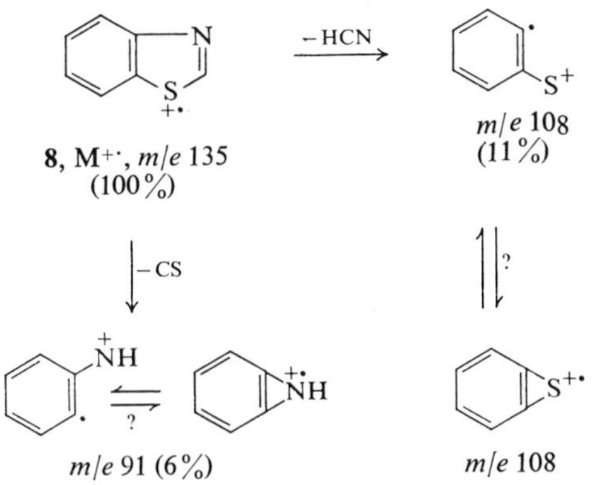

2-Methylbenzothiazole (**10**) shows the expected elision of CH_3CN (*m/e* 108) together with a considerable loss of $\cdot CH_2CN$ (*m/e* 109). Loss of HCN also occurs and its formation must necessarily involve much skeletal rearrangement (3).

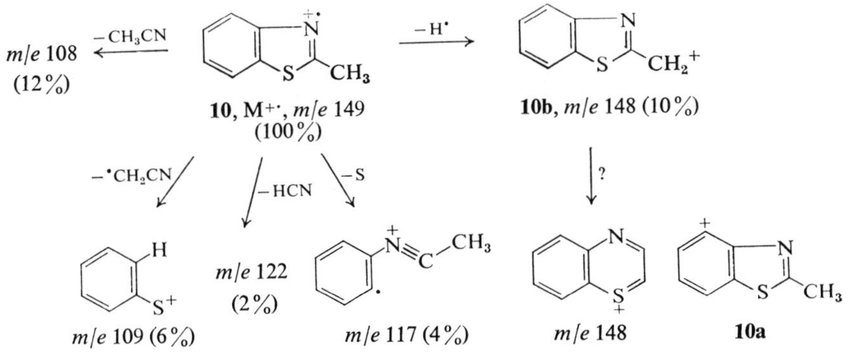

The origin of the hydrogen atom lost in the formation of the M-1 species has not been determined by deuterium labeling, and in fact the resultant ion was formulated as **10a**. But since the M-1 species is considerably weaker in benzothiazole itself, we prefer the formulation **10b** shown for the ion or the aza-thiochromenyl cation **10c**.

An important process occurring with 2-hydroxybenzothiazole **(11)** is elimination of CO, possibly from the lactam form of the compound, while loss of HOCN also occurs (3).

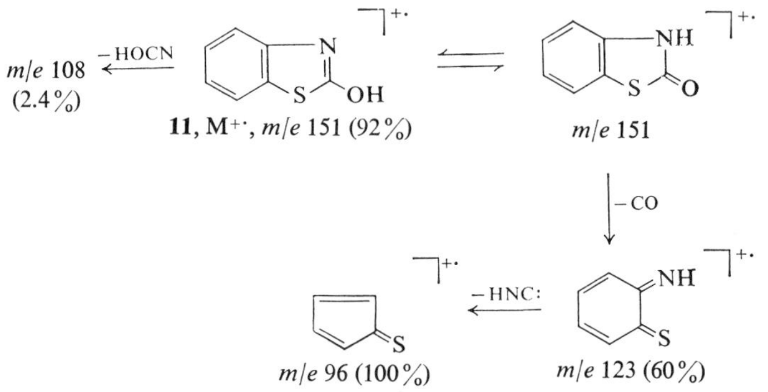

m/e 108 (2.4%)

11, M$^{+\cdot}$, m/e 151 (92%)

m/e 151

m/e 96 (100%)

m/e 123 (60%)

Benzothiazole-2-thiol **(12)** fragments in a rather similar manner but as well as a primary loss of CS, losses of CS$_2$ and S are important.

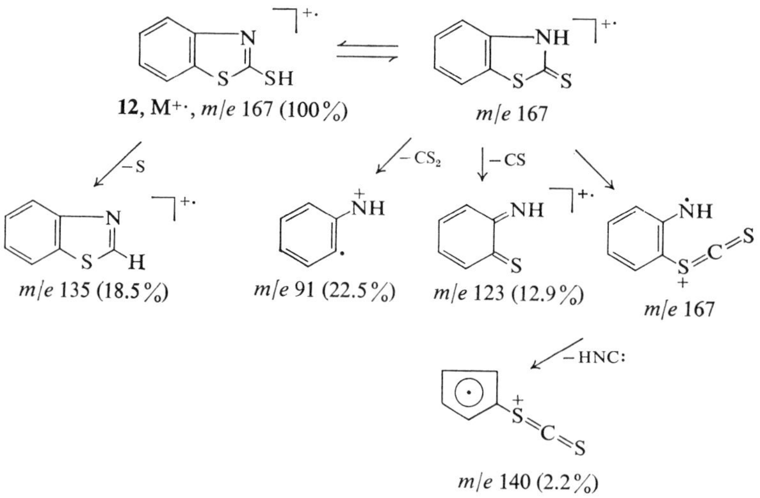

12, M$^{+\cdot}$, m/e 167 (100%)

m/e 167

m/e 135 (18.5%)

m/e 91 (22.5%)

m/e 123 (12.9%)

m/e 167

m/e 140 (2.2%)

Di-(2-benzothiazolyl) ketone **(13)**, like 2-benzoylthiazole **(6)**, is remarkable in showing an extremely intense rearrangement ion involving migration of one benzothiazole ring and loss of carbon monoxide (3). Similarly the M–CO ion is the base peak in the spectrum of 2-formylbenzothiazole.

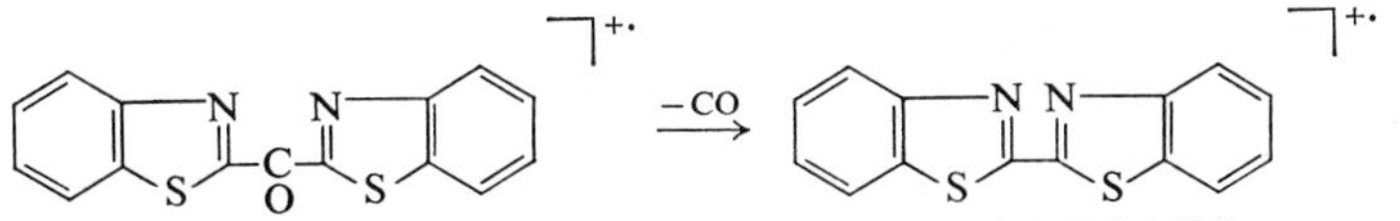

13, M$^{+\cdot}$, m/e 296 (75%)   m/e 268 (100%)

The mass spectra of a number of benzothiazoles with substituents of varying complexity (mostly in the 2-position) have also been described, and the reader is referred to the original paper for full details (3).

III. DIHYDROTHIAZOLES (THIAZOLINES)

The mass spectrum of 2-amino-4,5-dihydrothiazole (**14**) has been analyzed with the aid of the spectrum of its derivative labeled with ^{15}N at the amino group (4). Three fissions across the ring as shown explain the major fragmentations of the molecule. The 2-ureido derivative of **14** loses the elements

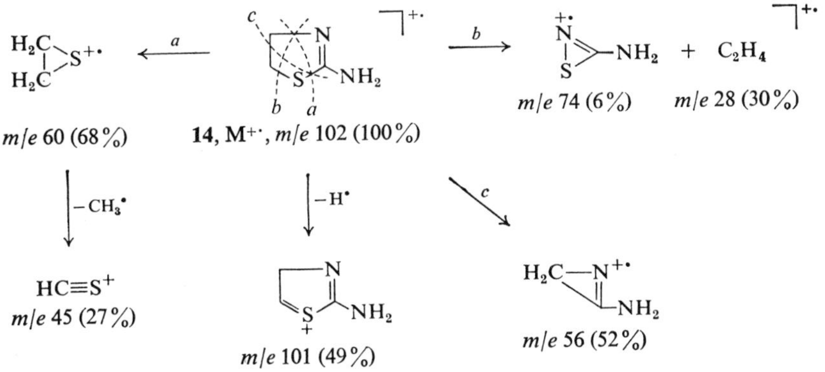

m/e 60 (68%)   **14**, M$^{+\cdot}$, m/e 102 (100%)   m/e 74 (6%)   m/e 28 (30%)

HC≡S$^+$ m/e 45 (27%)

m/e 101 (49%)

H$_2$C—N$^{+\cdot}$ m/e 56 (52%)

of isocyanic acid in a McLafferty rearrangement upon electron impact to give **14a** which of course fragments in the way just described (4).

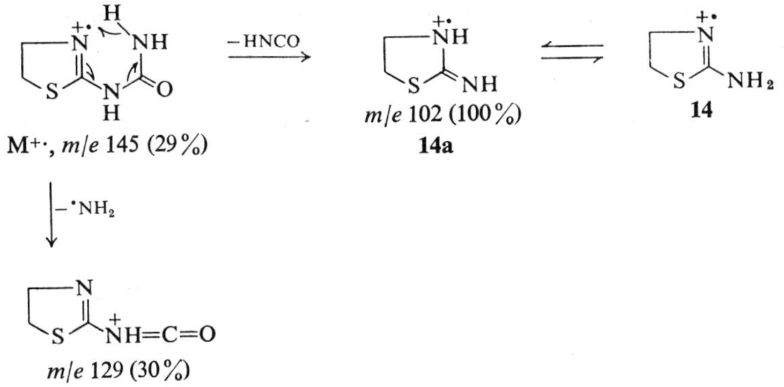

M$^{+\cdot}$, m/e 145 (29%)   m/e 102 (100%)   **14**
          14a

m/e 129 (30%)

It is interesting to note that the 3-(β-hydroxy-β-phenylethyl) derivative (**15**) of the imino tautomer of **14** shows an analog of the McLafferty rearrangement in which a phenyl group migrates (5)

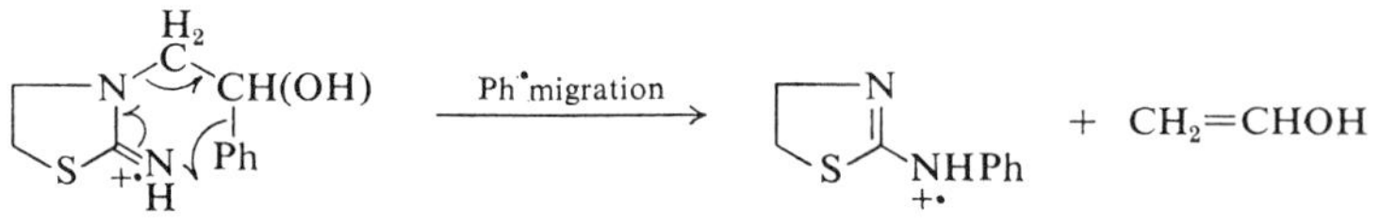

A partial mass spectrum of the bis(methylthio)methylenethiazolone (**16**) has been published (6). The base peak of the spectrum is the ion Ph-CS$^+$, and sequential losses of CH$_3^{.}$ and CH$_2$S can be explained in terms of an ultimate McLafferty rearrangement as shown.

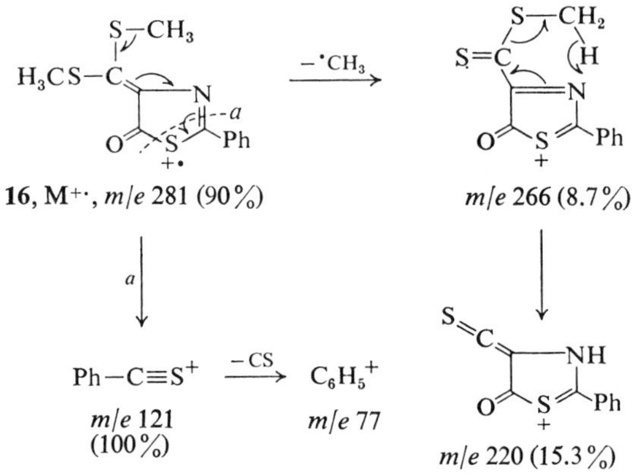

IV. PENICILLINS

Although the spectra of penicillins (examined as their methyl esters) are complex, they show weak but definite molecular ions, and convincing rationales have been advanced for the fragmentations observed (7). The fragmentation of penicillin-G methyl ester (**17**) is used to exemplify the ion reactions which occur. Similar fragmentation occurs with penicillin-V methyl ester.

The base peak of the spectrum arises from a simple fission (with hydrogen transfer) of the β-lactam ring whereas another important pathway involves fission of the thiazolidine ring. Fragmentations initiated by these fissions are rationalized as shown. A second series of fragmentations follow initial loss of the methoxycarbonyl group, and their rationale is given. Third, one

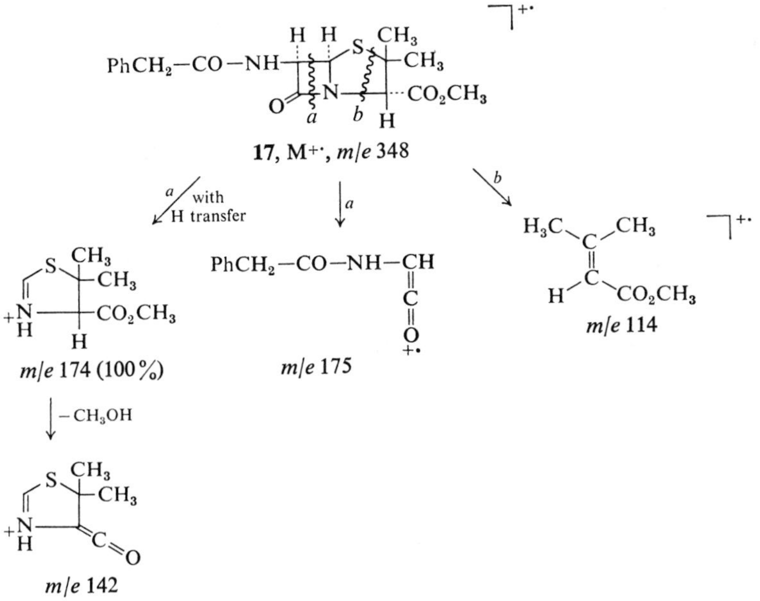

$$PhCH_2-CO-NH-\overset{H\ H}{\underset{O}{\rule{0pt}{0pt}}}\overset{S}{\underset{N}{\rule{0pt}{0pt}}}\overset{CH_3}{\underset{CO_2CH_3}{\rule{0pt}{0pt}}}$$

17, M$^{+\cdot}$, m/e 348

a with H transfer

a

b

$$\overset{S}{\underset{+N}{\rule{0pt}{0pt}}}\overset{CH_3}{\underset{CO_2CH_3}{\rule{0pt}{0pt}}}$$

m/e 174 (100%)

$PhCH_2-CO-NH-\overset{|}{CH}$
  $\overset{\|}{C}$
  $\overset{\|}{O}_{+\cdot}$

m/e 175

$$\overset{H_3C}{\underset{H}{\rule{0pt}{0pt}}}C=C\overset{CH_3}{\underset{CO_2CH_3}{\rule{0pt}{0pt}}}$$

m/e 114

$- CH_3OH$

$$\overset{S}{\underset{+N}{\rule{0pt}{0pt}}}\overset{CH_3}{\underset{C=O}{\rule{0pt}{0pt}}}$$

m/e 142

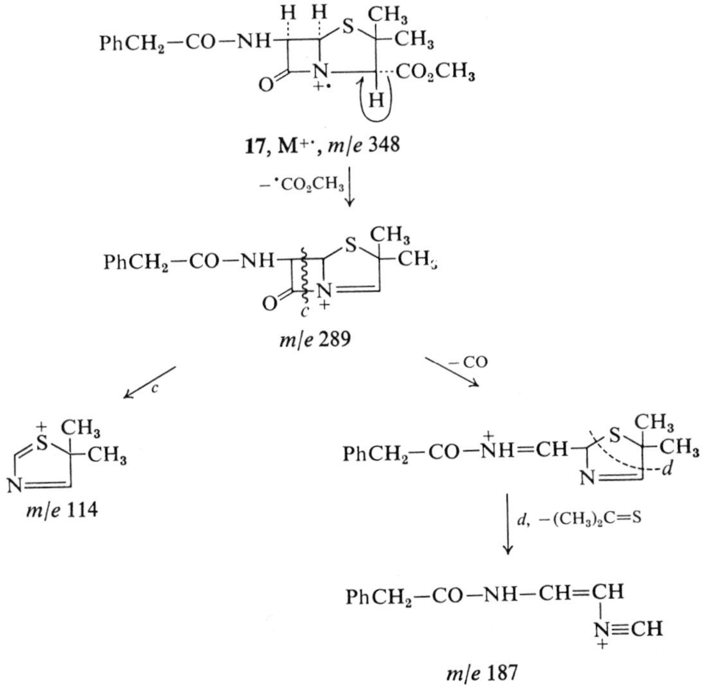

$$PhCH_2-CO-NH-\overset{H\ H}{\underset{O}{\rule{0pt}{0pt}}}\overset{S}{\underset{N}{\rule{0pt}{0pt}}}\overset{CH_3}{\underset{CO_2CH_3}{\rule{0pt}{0pt}}}$$

17, M$^{+\cdot}$, m/e 348

$-\,^{\cdot}CO_2CH_3$

$$PhCH_2-CO-NH-\overset{S}{\underset{N}{\rule{0pt}{0pt}}}\overset{CH_3}{\underset{CH_3}{\rule{0pt}{0pt}}}$$

m/e 289

c

$-CO$

$$\overset{S}{\underset{N}{\rule{0pt}{0pt}}}\overset{CH_3}{\underset{CH_3}{\rule{0pt}{0pt}}}$$

m/e 114

$$PhCH_2-CO-\overset{+}{N}H=CH\overset{S}{\underset{N}{\rule{0pt}{0pt}}}\overset{CH_3}{\underset{CH_3}{\rule{0pt}{0pt}}}d$$

$d,\ -(CH_3)_2C=S$

$$PhCH_2-CO-NH-CH=\overset{|}{CH}$$
  $\overset{+}{N}\equiv CH$

m/e 187

group of fragments is best explained by envisaging C—C fission in the thiazolidine ring and is also shown.

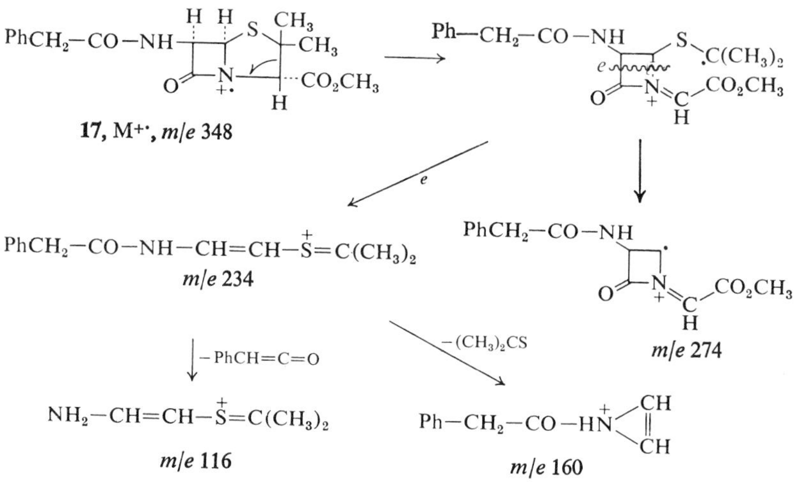

17, M+·, m/e 348

$PhCH_2$—CO—NH—CH=CH—$\overset{+}{S}$=C(CH$_3$)$_2$

m/e 234

NH_2—CH=CH—$\overset{+}{S}$=C(CH$_3$)$_2$

m/e 116

$-PhCH=C=O$

$-(CH_3)_2CS$

PhCH$_2$—CO—NH

m/e 274

Ph—CH$_2$—CO—H$\overset{+}{N}$

m/e 160

Loss of the elements of phenylketene which plays a part in the fragmentation just outlined also occurs from the molecular ion, and the resultant species shows a β-lactam fission in the opposite sense to that observed with the intact molecule.

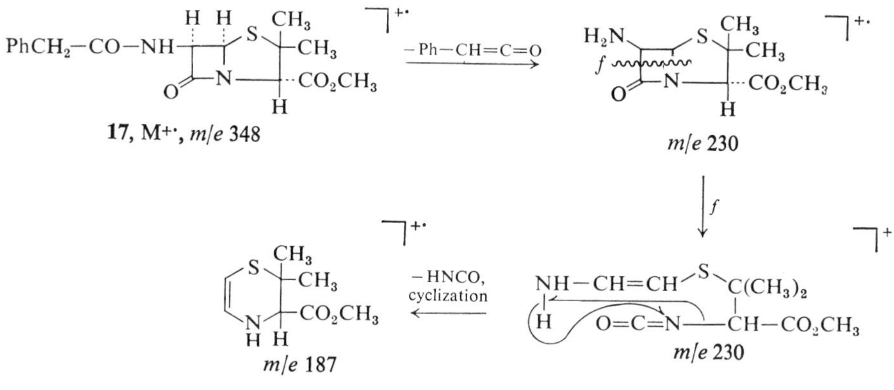

17, M+·, m/e 348

$-Ph-CH=C=O$

m/e 230

m/e 230

$-HNCO$, cyclization

m/e 187

Finally, certain fragments requiring hydrogen transfers or double hydrogen transfers are considered. The actual origins of the transferred hydrogen must remain uncertain in the absence of deuterium labeling, which would obviously be very difficult to carry out in this series of compounds.

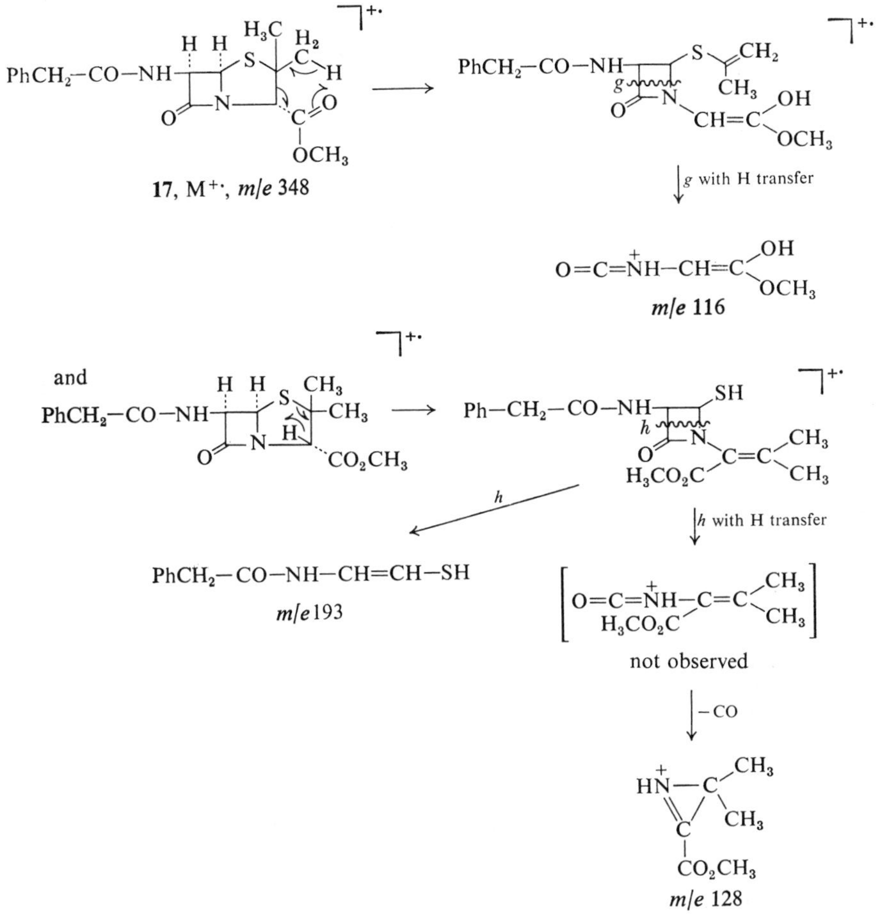

17, M$^{+\cdot}$, m/e 348

g with H transfer

$$O=C=\overset{+}{N}H-CH=C\overset{OH}{\underset{OCH_3}{\diagdown}}$$

m/e 116

and

$$PhCH_2-CO-NH-CH=CH-SH$$

m/e 193

not observed

$-CO$

m/e 128

V. 1,3-THIAZINES

The only members of this class of heterocycle to have been examined mass-spectrometrically are the 2,3-dihydro-1,3-thiazin-4-ones obtained from thioureas and dimethyl acetylenedicarboxylate (8, 9). The fragmentation of **18** can be explained on the basis of three simple ring fissions as shown.

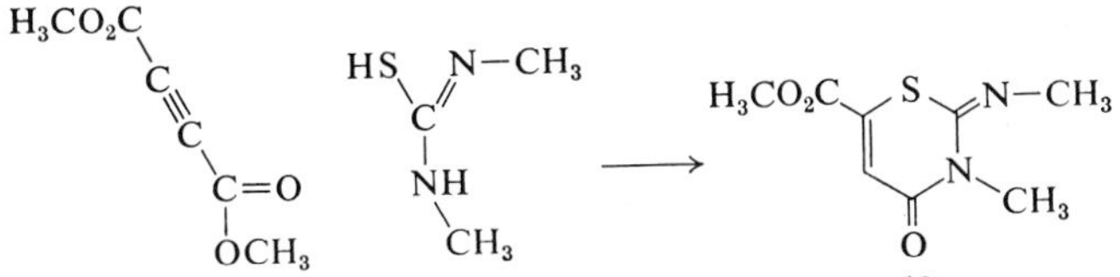

18

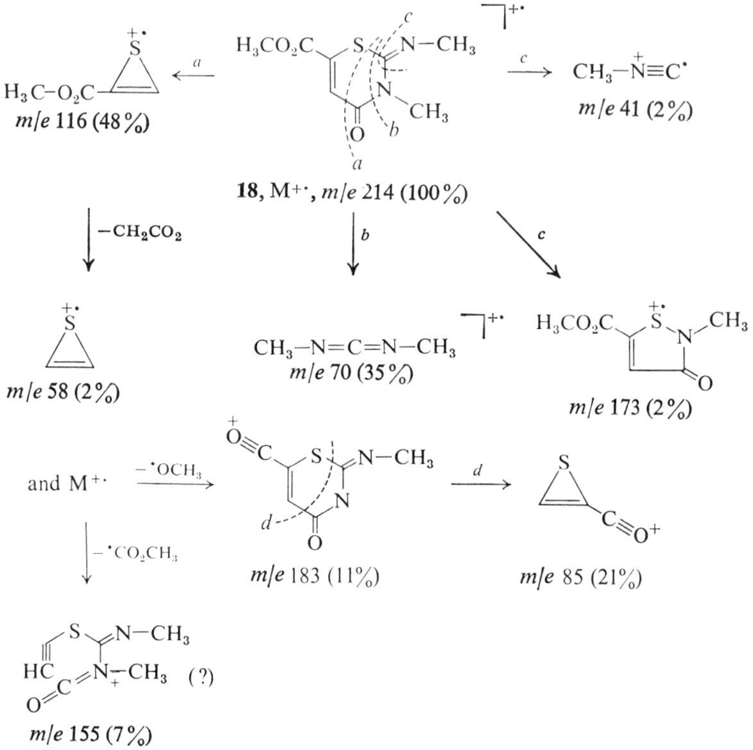

18, M$^{+\cdot}$, m/e 214 (100%)

A basically similar rationale has been presented for the perhydrothiazine analog of **18** and the original publications should be consulted for further details.

VI. DERIVATIVES OF CEPHALOSPORIN-C

The cephalosporin antibiotics are closely related to the penicillins, possessing a dihydrothiazine ring in place of the thiazolidine ring of the latter compounds.

The mass spectra of 7-phenylacetamido- (**19**) and 7-phenoxyacetamido-cephalosporanoic acids (as their methyl esters) have been analyzed with use of high-resolution techniques (10). The fragmentations of the two compounds are very similar and that of **19** will be discussed. The fragmentation is complex and only the genesis of the major ions in the spectrum will be considered.

The first important pathway is initiated by allylic fission (loss of CH$_3$CO$_2$·) and may be represented as shown. The second pathway is initiated by loss of CO from the molecular ion and is best represented as a double ring-opening. The third pathway involves fission of the β-lactam ring with and without hydrogen transfer from the neutral fragment.

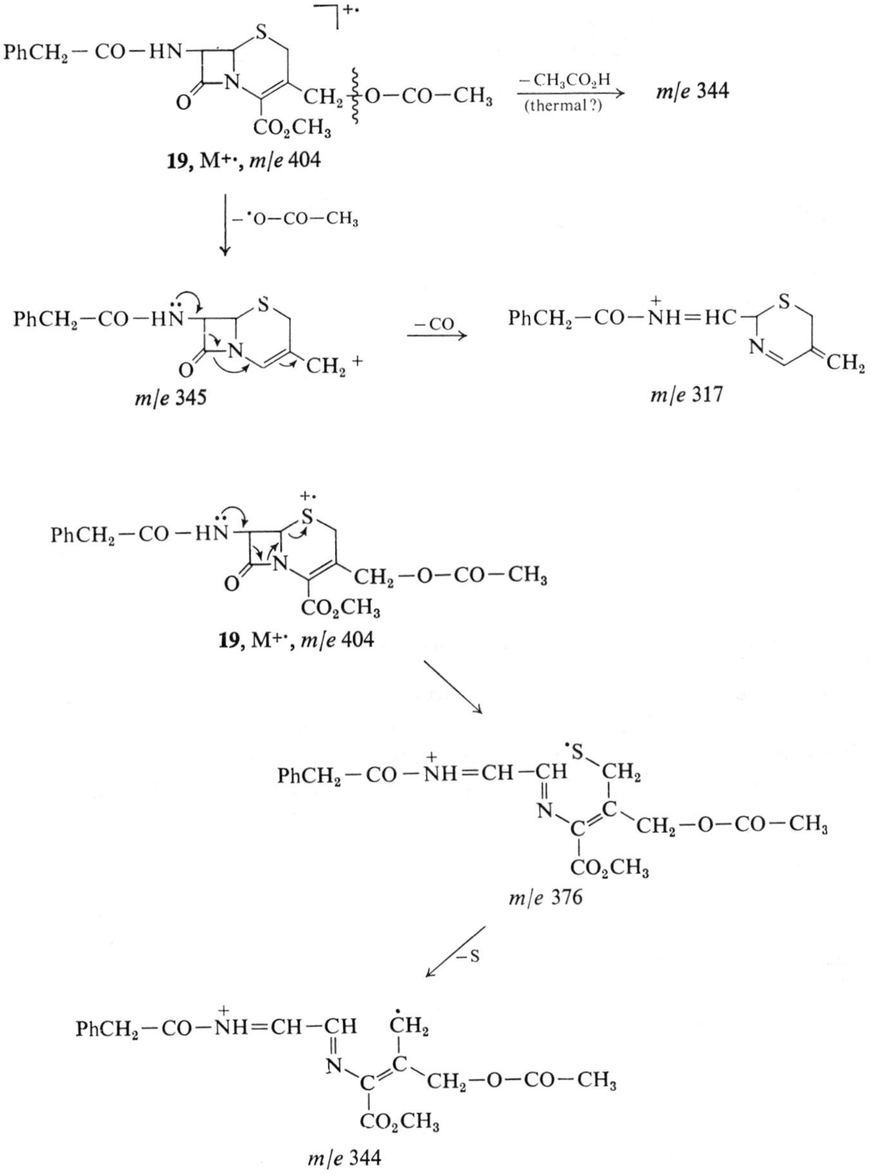

PhCH$_2$—CO—HN ... S ... N ... O ... CH$_2$—O—CO—CH$_3$ $\xrightarrow[\text{(thermal?)}]{-\text{CH}_3\text{CO}_2\text{H}}$ *m/e* 344

CO$_2$CH$_3$

19, M$^{+\cdot}$, *m/e* 404

$\Big|$ $-\cdot$O—CO—CH$_3$

PhCH$_2$—CO—HN ... S ... N ... O ... CH$_2$ $^+$

m/e 345

$\xrightarrow{-\text{CO}}$

PhCH$_2$—CO—$\overset{+}{\text{NH}}$=HC—S ... N ... CH$_2$

m/e 317

PhCH$_2$—CO—HN ... S ... N ... O ... CH$_2$—O—CO—CH$_3$

CO$_2$CH$_3$

19, M$^{+\cdot}$, *m/e* 404

PhCH$_2$—CO—$\overset{+}{\text{NH}}$=CH—CH ... $\cdot$S ... CH$_2$ ... N ... C ... C ... CH$_2$—O—CO—CH$_3$

CO$_2$CH$_3$

m/e 376

$-$S

PhCH$_2$—CO—$\overset{+}{\text{NH}}$=CH—CH ... $\dot{\text{C}}$H$_2$ ... N ... C ... C ... CH$_2$—O—CO—CH$_3$

CO$_2$CH$_3$

m/e 344

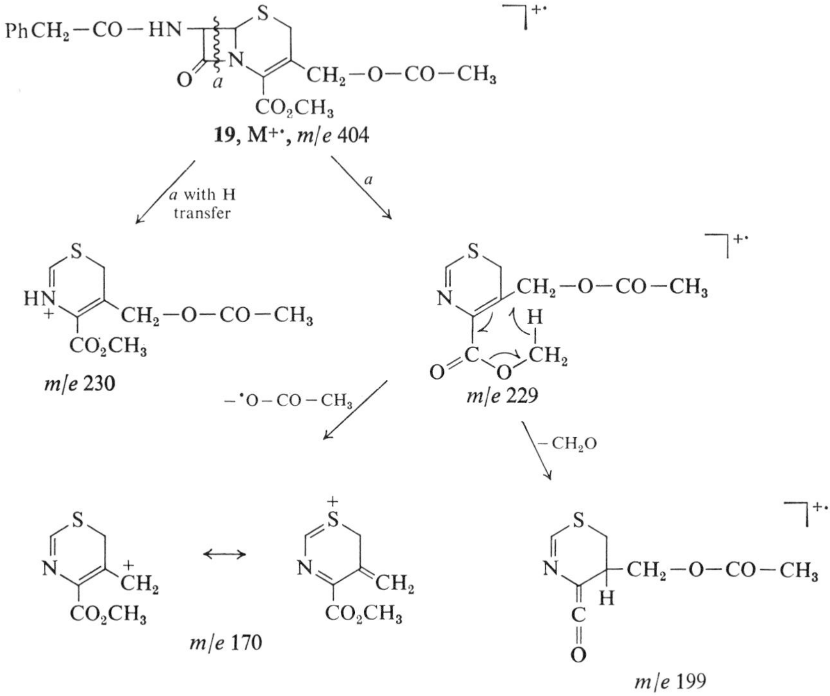

19, M$^{+\cdot}$, m/e 404

a with H transfer

a

m/e 230

$- \cdot O - CO - CH_3$

m/e 229

$- CH_2O$

m/e 170

m/e 199

VII. THIAZOLO[5,4-*d*]PYRIMIDINES

The mass spectrum of the parent compound (**20**) shows consecutive losses of two molecules of HCN. When this is taken in conjunction with the results reported for the monomethyl compounds, it is apparent that the first loss involves positions 4 and 5, the second positions 1 and 2 (11). The sequences

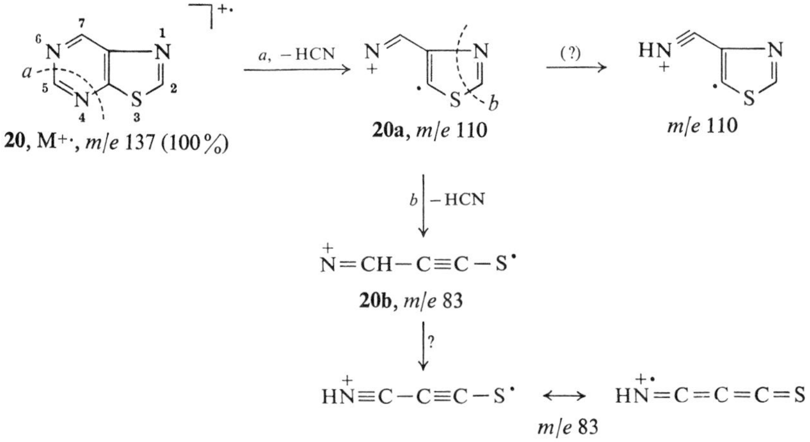

20, M$^{+\cdot}$, m/e 137 (100%)

a, $-$HCN

20a, m/e 110

(?)

m/e 110

b $|$ $-$HCN

$\overset{+}{N} = CH - C \equiv C - S \cdot$

20b, m/e 83

$|$?

$H\overset{+}{N} \equiv C - C \equiv C - S \cdot \longleftrightarrow H\overset{+\cdot}{N} = C = C = C = S$

m/e 83

of elimination of CH_3CN and HCN in the methylthiazolopyrimidines **21**, **22**, and **23** follow the same pattern exactly, as summarized in the scheme.

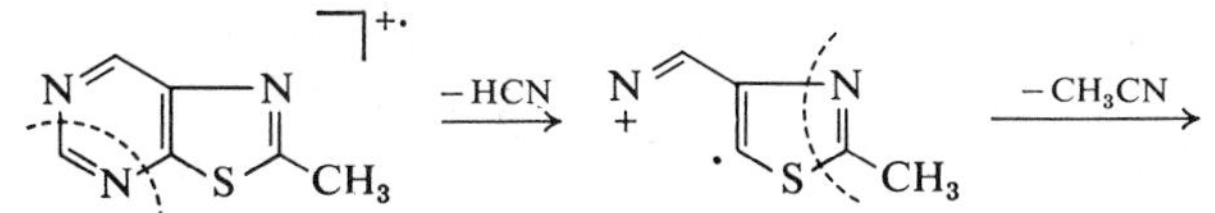

21, M+·, *m/e* 151 (100%)   *m/e* 124 $-CH_3CN$ → **20b** *m/e* 83

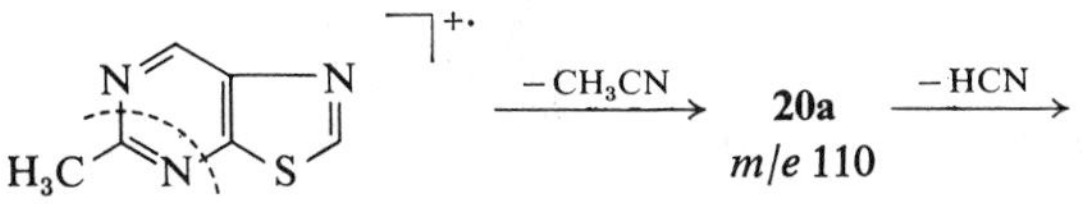

22, M+·, *m/e* 151 (100%) $-CH_3CN$ → **20a** *m/e* 110 $-HCN$ → **20b** *m/e* 83

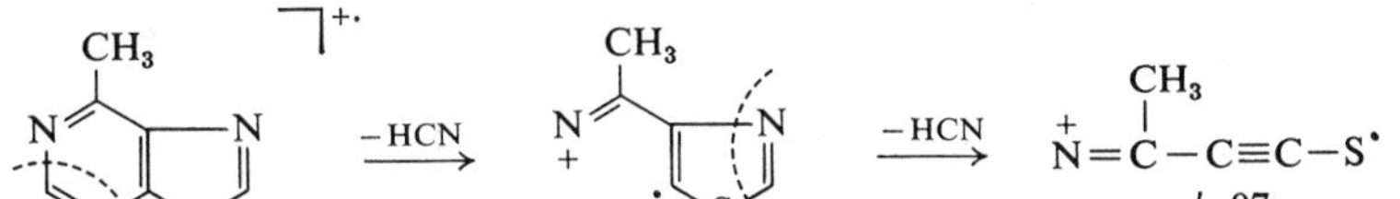

23, M+·, *m/e* 151 (100%)

The fact that fragmentation is significantly less in the 7-methyl isomer **(23)** may perhaps be due to the relative inability of the methyl group to migrate and produce the well-delocalized radical cation **23b**.

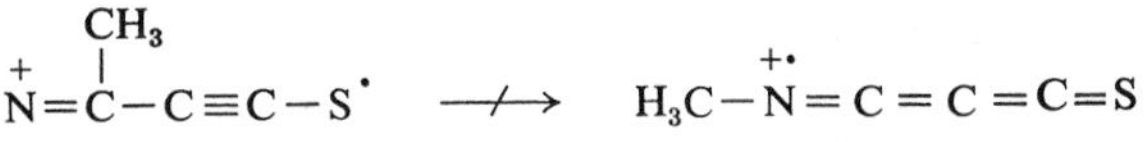

The mass spectra of a number of 2-alkylthio-5-aminothiazolo[5,4-*d*]-pyrimidines (**24**, R = alkyl) have also been reported (12). All show M-˙SH

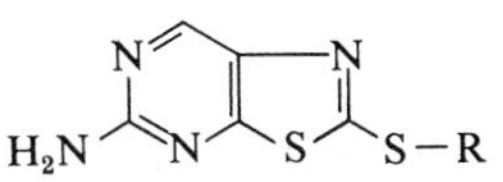

R = CH_3, C_2H_5, n-C_3H_7, n-C_4H_9, n-C_5H_{11}, n-C_6H_{13}, n-C_8H_{17}, n-$C_{10}H_{21}$

24

rearrangement ions although these are intense only for R = CH_3 and C_2H_5 whereas when R ⩾ C_2H_5, M-CH_3S˙ rearrangement ions also occur. A strong ion (usually the base peak) at *m/e* 184 (for R ⩾ C_2H_5) is explicable as the result of a McLafferty rearrangement. Specific deuterium labeling for R = n-butyl has shown that β-hydrogen transfer (presumably to nitrogen) is a

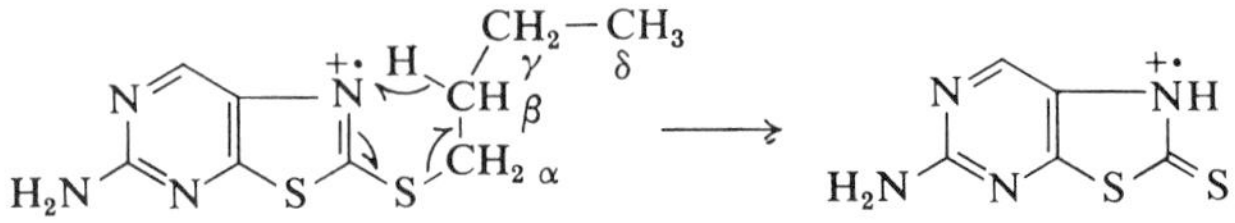

24, R = n-C_4H_9 m/e 184 (100%)

highly specific process. Labeling experiments with the n-butyl compound also show that the hydrogen transferred to sulfur in the formation of the M-$^-$SH ion comes mainly from the α-, β-, and δ-positions, whereas formation of the M-CH_3S^- ion is initiated exclusively by γ-hydrogen migration.

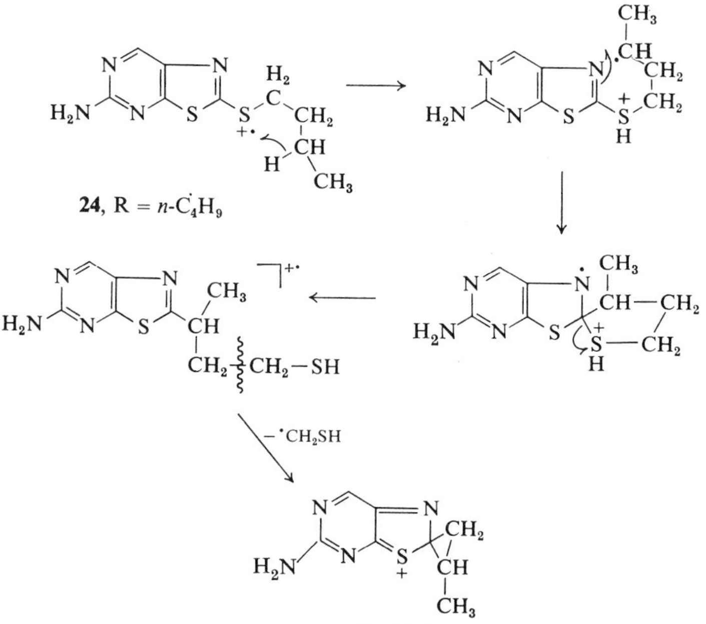

24, R = n-C_4H_9

M-CH_3S

These facile fragmentations largely involving the 2-substituent are presumably responsible for reducing ring fragmentation in these compounds. Those in which the m/e 184 ion occurs (McLafferty rearrangement) also show a relatively strong ion of m/e 142, and this probably corresponds to loss of cyanamide from the aminopyrimidine ring.

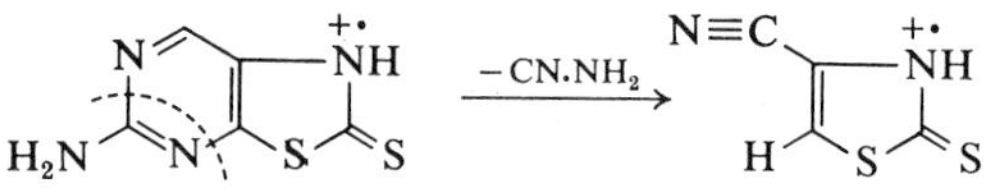

m/e 184   m/e 142

VIII. COMPOUNDS WITH NITROGEN, OXYGEN, AND SULFUR ATOMS IN THE SAME RING

The mass spectra of three 5-substituted 1,3,4-oxathiazol-2-ones (**25**) (obtained by the interaction of amides with trichloromethylsulfenyl chloride) have been reported (13).

$$2\,R{-}CO{-}NH_2 \;+\; Cl_3C{-}S{-}Cl \;\xrightarrow{\;-4HCl\;}\; \underset{O}{\overset{R}{\bigominus}} \;+\; R{-}C{\equiv}N$$

25

$$R = CH_3,\; C_6H_5,\; C_6H_5{-}O{-}CH_2$$

The fragmentations of the three oxathiazolones are very simple, involving extrusions of the stable fragments CO and CO_2 as primary processes, and are exemplified by those of the phenyl compound (**26**).

$$Ph{-}C{\equiv}O^+ \xleftarrow{\;-(CO+\,^{\cdot}NS)\;} \left[\underset{O}{\overset{Ph}{\bigominus}}\right]^{+\cdot} \xrightarrow{\;-CO_2\;} \left[\overset{Ph}{\underset{S}{\diagup}}{N}\right]^{+\cdot}$$

m/e 105 (100%)   **26**, $M^{+\cdot}$, m/e 179 (34%) m/e 135 (4.8%)

| $\downarrow{-CO}$ | $\diagdown^{-\,^{\cdot}NS}$ | $\downarrow{-CO}$ | $\downarrow{-S}$ |

$C_6H_5^+$ $\left[Ph{-}CO{-}N{=}S\right]^{+\cdot}$ $Ph{-}C{\equiv}N^{+\cdot}$

m/e 77 (51%) m/e 151 m/e 103 (33%)

Addenda

Fragmentations of a number of 3-phenylisothiazoles have been reported (14). The rearrangement of a sulphone group to a sulphonate ester is apparent in the mass spectra of benzisothiazole dioxides (15). The mass spectra of phenothiazine itself (16, 17) and a range of substituted phenothiazines of medicinal interest have also been described (17, 18, 19).

REFERENCES

1. G. M. Clarke, R. Grigg and D. H. Williams, *J. Chem. Soc.*, B, **1966**, 339.
2. R. G. Cooks, I. Howe, S. W. Tam, and D. H. Williams, *J. Amer. Chem. Soc.*, **90**, 4064 (1968).
3. B. J. Millard and A. F. Temple, *Org. Mass Spectrosc.*, **1**, 285 (1968).

4. D. L. Klayman, A. Senning, and G. W. A. Milne, *Acta Chem. Scand.*, **21**, 217 (1967).

5. B. R. Webster, *Chem. Commun.*, **1966**, 124.

6. R. F. C. Brown, I. D. Rae, J. S. Shannon, S. Sternhell, and J. M. Swan, *Aust. J. Chem.*, **19**, 503 (1966).

7. W. Richter and K. Biemann, *Monatsh. Chem.*, **95**, 766 (1964).

8. J. W. Lown and J. C. N. Ma, *Can. J. Chem.*, **45**, 939 (1967).

9. J. W. Lown and J. C. N. Ma, *Can. J. Chem.*, **45**, 953 (1967).

10. W. Richter and K. Biemann, *Monatsh. Chem.*, **96**, 484 (1965).

11. A. Tatematsu, S. Inoue, S. Sugiura, and T. Goto, *Nippon Kagaku Zasshi*, **87**, 66 (1966).

12. A. Tatematsu, S. Sugiura, S. Inoue, and T. Goto, *Org. Mass Spectrosc.*, **1**, 205 (1968).

13. A. Senning and P. Kelly, *Acta Chem. Scand.*, **21**, 1871 (1967).

14. T. Naito, *Tetrahedron*, **24**, 6237 (1968).

15. H. Hettler, H. M. Schiebel, and H. Budzıkiewicz, *Org. Mass Spec.*, **2**, 1117 (1969).

16. J. Heiss and K.-P. Zeller, *Org. Mass Spec.*, **2**, 819 (1969).

17. L. Audier, M. Azzaro, A. Cambon, and R. Guedj, *Bull. soc. chim. France*, 1013 (1968).

18. J. N. T. Gilbert and B. J. Millard, *Org. Mass Spec.*, **2**, 17 (1969).

19. R. Guedj, A. Cambon, L. Audier, and M. Azzaro, *Bull. soc. chim. France*, 1021 (1968).

INDEX